AF388661

Computational Methods for Polymers

Computational Methods for Polymers

Special Issue Editor

Masoud Soroush

MDPI • Basel • Beijing • Wuhan • Barcelona • Belgrade • Manchester • Tokyo • Cluj • Tianjin

Special Issue Editor
Masoud Soroush
Drexel University
USA

Editorial Office
MDPI
St. Alban-Anlage 66
4052 Basel, Switzerland

This is a reprint of articles from the Special Issue published online in the open access journal *Processes* (ISSN 2227-9717) (available at: https://www.mdpi.com/journal/processes/special_issues/polymer_method).

For citation purposes, cite each article independently as indicated on the article page online and as indicated below:

LastName, A.A.; LastName, B.B.; LastName, C.C. Article Title. *Journal Name* **Year**, *Article Number*, Page Range.

ISBN 978-3-03928-813-7 (Hbk)
ISBN 978-3-03928-814-4 (PDF)

Cover image courtesy of shutterstock.com user muph. License purchased by the guest editor.

Contents

About the Special Issue Editor

Masoud Soroush received his B.S. (Chemical Engineering) from Abadan Institute of Technology, Iran, and M.S.E. (Chemical Engineering), M.S.E. (Electrical Engineering: Systems), and Ph.D. (Chemical Engineering) all from the University of Michigan, Ann Arbor, USA. After graduating from the University of Michigan, he joined Drexel University, where he is now a Professor of Chemical and Biological Engineering. His research interests include polymer membranes, polymer reaction engineering, model-predictive safety, fault detection and identification, probabilistic modeling and inference, and process systems engineering. He has authored/co-authored more than 340 publications including over 200 refereed papers. He has edited/co-edited 6 books. His awards include the U.S. National Science Foundation Faculty Early CAREER Award and the O. Hugo Schuck Best Paper Award of American Automatic Control Council. He is an elected fellow of AIChE and a senior member of IEEE.

Editorial

Special Issue on "Computational Methods for Polymers"

Masoud Soroush

Department of Chemical and Biological Engineering, Drexel University, Philadelphia, PA 19104, USA; soroushm@drexel.edu

Received: 17 March 2020; Accepted: 17 March 2020; Published: 26 March 2020

Polymers play a key role in our daily lives. Natural polymers include proteins, cellulose, nucleic acids, lignin, natural rubber, and wood resins. Examples of manmade polymers are synthetic rubbers and plastics. To describe polymer properties, processes and reactions mathematically, phenomena at different length and time scales should be accounted for. The length scales are from angstrom to meter, and the time scales are from femtosecond to hour. The wide ranges of the scales require the use of an appropriate modeling method or a combination of appropriate methods. Examples of the methods are: quantum chemical/mechanical methods such as density functional theory (electronic scale); molecular dynamics and Monte Carlo (molecular scale); dissipative particle dynamics, Brownian dynamics, and lattice Boltzmann method (microscopic scale); dissipative particle dynamics and field theoretic polymer simulation (meso scale); and control volume methods and finite element (macroscopic scale) [1–3].

This Special Issue on "Computational Methods for Polymers" includes five articles on the mathematical modeling of polymers, polymer processes and polymerization reactions. The methods used include Monte Carlo simulations, macroscopic-scale modeling, and electronic-scale modeling. These articles are then followed by three on state and parameter estimation, one on model-based control of a molding process, and one on the operability analysis of a polymer membrane reactor. Next, there are three articles on macromolecular structural properties. The Special Issue then ends with two articles, one on the impact, spreading and rebound of a droplet, and the other on the impacts of branching and backbiting reactions on pulsed laser polymerization (PLP).

The first article is on how to produce hyperbranched (HB) polymer architectures in reactors. Tobita [4] studied the fundamental structural characteristics of polymer chains formed in representative types of reactors. In particular, he investigated irreversible step growth polymerization of an AB2 type monomer in a batch and a continuous stirred-tank reactor (CSTR) via Monte Carlo simulations. The simulation results revealed that in a CSTR, a highly branched core region consisting of units with large residence times is formed to give much more compact architecture compared to batch polymerization.

Atan et al. [5] review mathematical models developed for olefin polymerization processes. Their review covers coordination and free-radical mechanisms in different types of reactors such as fluidized bed, horizontal-stirred-bed, vertical-stirred-bed, and tubular. They present guidelines for the mathematical modeling of gas-phase olefin polymerization processes.

Riazi et al. [6] consider more than 40 reactions that are most likely to occur in high-temperature free-radical homopolymerization, and derive moment rate equations for the reactions methodically. Using a step-by-step approach based on the method of moments, their article guides the reader to determine the contributions of each reaction to the production or consumption of each species as well as to the zeroth, first and second moments of chain-length distributions of live and dead polymer chains, in order to derive the overall rate equation for each species, and to derive the rate equations for the leading moments of different chain-length distributions.

Zaccaria et al. [7] used electronic-scale modeling (density functional theory [DFT]) to study four Cl/Me substituted [ONNO] Zr-catalysts for ethene/α-olefin polymerization. They found that replacing electron-donating methyl with isosteric but electron-withdrawing chlorine substituents results in a significant increase in comonomer incorporation. Using DFT, they explored the steric and electronic properties of the ancillary ligand and found that the relative reactivity ratios are mainly determined by the electrophilicity of the metal center. They report that electronic effects observed in these catalysts affect the relative barriers for insertion and a capture-like transition state.

Darvishi et al. [8] use electronic-scale modeling to study undesired polymerization of styrene during distillation, storage, and transportation. They apply DFT and carried out laboratory experiments to study the antipolymer and antioxidant activity of stable nitroxide radicals and phenolics in styrene polymerization. They report that 2,6-di-tert-butyl-4-methoxyphenol, 2,6-di-tert-butyl-4-methylphenol, 4-hydroxy-2,2,6,6-tetramethyl piperidine 1-Oxyl, and 4-oxo-2,2,6,6-tetramethylpiperidine 1-Oxyl are the most effective inhibitors for styrene.

Lathrop et al. [9] report an experimental and theoretical/computational study on modeling and monitoring reversible addition–fragmentation chain–transfer (RAFT) polymerization of methyl methacrylate. They carry out parameter estimation to determine RAFT reaction rate coefficients and an initiator efficiency from in-situ ^{1}H NMR and SEC experimental data. Based on their model, they implement a multi-rate multi-delay observer that include an inter-sample predictor and a deadtime compensator. Their results show that the multi-rate multi-delay observer has satisfactory convergence after a few sampling periods.

Salas et al. [10] present a data-driven strategy for the online estimation of important kinetic parameters in copolymerization of ethylene. The kinetic parameters are chosen based on a global sensitivity analysis. The retrospective cost model refinement algorithm is adapted and implemented to estimate the kinetic parameters in real time. The results demonstrate that the estimates of the kinetic parameters converge to theoretical values of the parameters without requiring prior knowledge of the polymerization model or the theoretical parameter values.

Scott et al. [11] revisit five terpolymerization studies and estimate the 'ternary' reactivity ratios. They compare previously determined binary reactivity ratios and newly estimated 'ternary' reactivity ratios for several systems. They discuss the advantages and challenges associated with 'ternary' reactivity ratio estimation.

Garg et al. [12] present a method of uniquely specifying and robustly achieving user-specified product quality in a complex industrial batch process. They demonstrate this method using a lab-scale uni-axial rotational molding process. The method involves data-driven modeling (subspace state-space model) and model predictive control, and is able to reject raw material variation and achieve product quality which is specified through constraints on quality variables. Results from these experimental studies demonstrate the capability of the proposed method in meeting process specifications and rejecting raw material variability.

Bagalkot et al. [13] present case studies on performing polymer rapid tooling inserts and observing different failures over the life of the tool. They identify critical parameters affecting tool life, and the effect of those parameters on different areas of the tool. They categorize modes of the different failures and the underlying mechanisms including the root causes.

Intensified process units such as polymer membrane reactors pose unique challenges pertaining to design and operation that have not been fully addressed. Bishop and Lima [14] present an approach for modeling membrane reactors. Their model allows for the simulation of polymer membrane reactors under nonisothermal and countercurrent operation. This article demonstrates how operability analyses can be used to identify areas of improvement in membrane reactor design.

Topological indices have been computed for various molecular structures. They are numerical invariants associated with molecular structures and are helpful in that they feature many properties. Among these molecular descriptors, the eccentricity connectivity index is of great importance due to its relationship with pharmaceutical properties. Zheng et al. [15] calculate eccentric connectivity, total eccentricity connectivity, augmented eccentric connectivity, first Zagreb eccentricity, modified eccentric connectivity, second Zagreb eccentricity, and the edge version of eccentric connectivity indices for the molecular graph of a polyethyleneamidoamine dendrimer. They also calculate explicit representations of the polynomials associated with some of these indices.

Dendrimers are branched organic macromolecules with successive layers of branch units surrounding a central core. The molecular topology and the irregularity of their structure plays a central role in determining structural properties like enthalpy and entropy. Hussain et al. [16] determine irregularity indices that are based on the imbalance of edges, for the molecular graphs associated with some general classes of dendrimers. They also provide graphical analyses of these indices for the dendrimers.

In many applications, knowledge on the irregularity of a molecular structure is of importance. The irregularity provides quantitative information on structure–property relationships and structure–activity relationships as well as various physical and chemical properties. Zhao et al. [17] present a study on the computation and comparison of the irregularity measures of different classes of dendrimers. Their investigation focuses on four irregularity indices; the σ irregularity index, the irregularity index by Albertson, the variance of vertex degrees, and the total irregularity index.

Tembely et al. [18] present a mathematical model based on physical principles to simulate droplet impact, spreading, and eventually rebound of a viscoelastic droplet. Their simulations are based on the volume of fluid method in conjunction with a dynamic contact model accounting for the hysteresis between the droplet and substrate. They report that while the kinematic phase of droplet spreading seems to be independent of both the substrate and fluid rheology, the recoiling phase are highly influenced by those operating parameters. Their model can be used to optimize 2D/3D printing of complex fluids.

In PLP, the behavior of the degree of branching and backbiting reactions has not been well-understood yet at high laser frequencies and at relatively high reaction temperatures due to the inherent difficulties in the determination of the degree of branching of polymers. Hamzehlou et al. [19] evaluate the validity of different explanations of the recovery of PLP-molar mass distribution at high laser frequencies using a simulation study. They show that the reduction in the backbiting reaction rate at a high laser frequency, and the consequent decrease in the degree of branching, are not a necessary condition for recovering the PLP-molar mass distribution.

I would like to thank all the contributors for their contributions to this Special Issue, and Processes for supporting this issue.

Conflicts of Interest: The author declares no conflicts of interest.

References

1. Gooneie, A.; Schuschnigg, S.; Holzer, C. A review of multiscale computational methods in polymeric materials. *Polymers* **2017**, *9*, 16. [CrossRef] [PubMed]
2. Askadskiĭ, A.A. *Computational Materials Science of Polymers*; Cambridge Int Science Publishing: Cambridge, UK, 2003.
3. Zeng, Q.; Yu, A.; Lu, G. Multiscale modeling and simulation of polymer nanocomposites. *Prog. Polym. Sci.* **2008**, *33*, 191–269. [CrossRef]
4. Tobita, H. Universal Relationships in Hyperbranched Polymer Architecture for Batch and Continuous Step Growth Polymerization of AB2-Type Monomers. *Processes* **2019**, *7*, 220. [CrossRef]
5. Atan, M.F.; Hussain, M.A.; Abbasi, M.R.; Khan, M.J.H.; Fazly Abdul Patah, M. Advances in Mathematical Modeling of Gas-Phase Olefin Polymerization. *Processes* **2019**, *7*, 67. [CrossRef]
6. Riazi, H.; Arabi Shamsabadi, A.; Grady, M.C.; Rappe, A.M.; Soroush, M. Method of Moments Applied to Most-Likely High-Temperature Free-Radical Polymerization Reactions. *Processes* **2019**, *7*, 656. [CrossRef]
7. Zaccaria, F.; Cipullo, R.; Correa, A.; Budzelaar, P.H.; Busico, V.; Ehm, C. Separating Electronic from Steric Effects in Ethene/α-Olefin Copolymerization: A Case Study on Octahedral [ONNO] Zr-Catalysts. *Processes* **2019**, *7*, 384. [CrossRef]
8. Darvishi, A.; Rahimpour, M.R.; Raeissi, S. A Theoretical and Experimental Study for Screening Inhibitors for Styrene Polymerization. *Processes* **2019**, *7*, 677. [CrossRef]

9. Lathrop, P.M.; Duan, Z.; Ling, C.; Elabd, Y.A.; Kravaris, C. Modeling and Observer-Based Monitoring of RAFT Homopolymerization Reactions. *Processes* **2019**, *7*, 768. [CrossRef]
10. Salas, S.D.; Brandão, A.L.; Soares, J.B.; Romagnoli, J.A. Data-Driven Estimation of Significant Kinetic Parameters Applied to the Synthesis of Polyolefins. *Processes* **2019**, *7*, 309. [CrossRef]
11. Scott, A.J.; Gabriel, V.A.; Dubé, M.A.; Penlidis, A. Making the Most of Parameter Estimation: Terpolymerization Troubleshooting Tips. *Processes* **2019**, *7*, 444. [CrossRef]
12. Garg, A.; Abdulhussain, H.A.; Mhaskar, P.; Thompson, M.R. Handling Constraints and Raw Material Variability in Rotomolding through Data-Driven Model Predictive Control. *Processes* **2019**, *7*, 610. [CrossRef]
13. Bagalkot, A.; Pons, D.; Symons, D.; Clucas, D. Categorization of Failures in Polymer Rapid Tools Used for Injection Molding. *Processes* **2019**, *7*, 17. [CrossRef]
14. Bishop, B.A.; Lima, F.V. Modeling, Simulation, and Operability Analysis of a Nonisothermal, Countercurrent, Polymer Membrane Reactor. *Processes* **2020**, *8*, 78. [CrossRef]
15. Zheng, J.; Iqbal, Z.; Fahad, A.; Zafar, A.; Aslam, A.; Qureshi, M.I.; Irfan, R. Some eccentricity-based topological indices and polynomials of poly (EThyleneAmidoAmine)(PETAA) dendrimers. *Processes* **2019**, *7*, 433. [CrossRef]
16. Hussain, Z.; Munir, M.; Rafique, S.; Hussnain, T.; Ahmad, H.; Chel Kwun, Y.; Min Kang, S. Imbalance-based irregularity molecular descriptors of nanostar dendrimers. *Processes* **2019**, *7*, 517. [CrossRef]
17. Zhao, D.; Iqbal, Z.; Irfan, R.; Chaudhry, M.A.; Ishaq, M.; Jamil, M.K.; Fahad, A. Comparison of Irregularity Indices of Several Dendrimers Structures. *Processes* **2019**, *7*, 662. [CrossRef]
18. Tembely, M.; Vadillo, D.; Soucemarianadin, A.; Dolatabadi, A. Numerical Simulations of Polymer Solution Droplet Impact on Surfaces of Different Wettabilities. *Processes* **2019**, *7*, 798. [CrossRef]
19. Hamzehlou, S.; Aboudzadeh, M.A.; Reyes, Y. On the Recovery of PLP-Molar Mass Distribution at High Laser Frequencies: A Simulation Study. *Processes* **2019**, *7*, 501. [CrossRef]

Article

Universal Relationships in Hyperbranched Polymer Architecture for Batch and Continuous Step Growth Polymerization of AB$_2$-Type Monomers

Hidetaka Tobita

Department of Materials Science and Engineering, University of Fukui, 3-9-1 Bunkyo, Fukui 910-8507, Japan; tobita@matse.u-fukui.ac.jp

Received: 21 March 2019; Accepted: 16 April 2019; Published: 17 April 2019

Abstract: Design and control of hyperbranched (HB) polymer architecture by way of reactor operation is key to a successful production of higher-valued HB polymers, and it is essential in order to clarify the fundamental structural characteristics formed in representative types of reactors. In this article, the irreversible step growth polymerization of AB$_2$ type monomer is investigated by a Monte Carlo simulation method, and the calculation was conducted for a batch and a continuous stirred-tank reactor (CSTR). In a CSTR, a highly branched core region consisting of units with large residence times is formed to give much more compact architecture, compared to batch polymerization. The universal relationships, unchanged by the conversion levels and/or the reactivity ratio, are found for the mean-square radius of gyration Rg^2, and the maximum span length L_{MS}. For batch polymerization, the g-ratio of Rg^2 of the HB molecule to that for a linear molecule conforms to that for the random branched polymers represented by the Zimm-Stockmayer equation. A single linear equation represents the relationship between Rg^2 and L_{MS}, both for batch and CSTR. Appropriate process control in combination with the chemical control of the reactivity of the second B-group promises to produce tailor-made HB polymer architecture.

Keywords: hyperbranched; Monte Carlo simulation; radius of gyration; span length; continuous stirred-tank reactor

1. Introduction

Hyperbranched (HB) polymers are specialty polymeric materials, possessing compact architecture, a vast number of end groups that can be functionalized, and specific space inside the molecule. A wide variety of potential applications have been and are being developed [1]. The HB polymers are macromolecules in between deterministic linear chains and dendrimer structures [2], and their properties are influenced significantly by their detailed branched architecture. The prediction and control of HB architecture is essential to produce higher quality polymers, which opens up a challenging field for the chemical engineers to develop novel production processes.

Basic chemical reaction engineering textbooks emphasize the importance in clarifying the fundamental chemical behavior in three representative reactor types; batch reactor, plug flow reactor (PFR), and continuous stirred-tank reactor (CSTR) [3]. Ideally, a PFR is equivalent to a batch reactor by changing the reaction time to the residence time. In this article, the differences in branched architecture formed in a batch reactor and a CSTR are considered.

For the synthesis of HB polymers consisting of tri-branched monomeric units, the two major chemical methods used are step growth polymerization of AB$_2$-type monomer and self-condensing vinyl polymerization (SCVP). The former is a classical synthetic route originally considered by Flory [4], but recent development in polymer chemistry has made it possible to change the chemical reactivity of the second B group freely [5], and has established the chemical control method for the branching frequency.

Figure 1 shows the reaction scheme of the step growth polymerization of AB_2 type monomer, considered in this article. In the figure, T is the terminal unit with both B's being unreacted, L is the linearly incorporated unit with one of two B's being reacted, and D is the dendritic unit in which both B's have reacted. The reaction rate constant, k_T is for the reaction between an A group and a B group in T, while k_L is for the reaction between A and B in L. The reactivity of the second B group is represented by the reactivity ratio, r defined by:

$$r = k_L/k_T \tag{1}$$

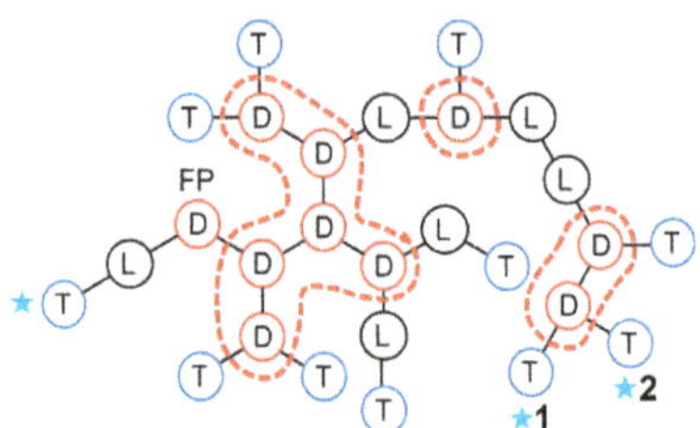

Figure 1. Reaction scheme for the step growth polymerization of AB_2 type monomer.

The magnitude of r can be changed from 0 to infinity at will [5] by using appropriate chemical systems. For instance, it is possible to produce HB polymers with 100% degree of branching (DB, the exact definition will be given shortly), but the quasi-linear polymer can also be produced even with $DB = 1$, as illustrated earlier [5]. Process control in combination with chemical control is needed to synthesize well-designed HB polymers.

In this theoretical study, the branched molecular architecture is investigated by using a Monte Carlo (MC) simulation method proposed earlier for a batch reactor [6,7] and for a CSTR [8,9]. In the MC simulation, the structure of each HB polymer can be investigated, and any desired structural information could be extracted. Figure 2 shows an example of HB polymer generated in the present MC simulation for a CSTR. In Figure 2, FP means the focal point unit that possesses an unreacted A group. Note that in the ring-free model employed in this article, there is only one unit in a molecule that bears unreacted A group.

Figure 2. Schematic representation of a hyperbranched (HB) polymer molecule generated in the present Monte Carlo (MC) simulation for a continuous stirred-tank reactor (CSTR). The tri-branched clusters are shown by the red closed curves, and the T units with a star show the end units for the maximum span length.

This article is aimed at establishing the most fundamental characteristics of HB polymers formed based on the ideal chemical kinetics, and the non-idealities, such as cyclization and shielding [10] are not considered. Both cyclization and shielding depend heavily on the 3D architecture, and it is of prime importance to establish the ideal architecture first. As for the cyclization, only one ring per molecule is allowed. Because smaller rings have a better chance of being formed [11], the effect on the global architecture of large polymers, which is of major interest in this article, may not be significant.

On the other hand, the shielding effect depends on the crowding of the 3D architecture, and inferences on crowding could be obtained from the present type of structural investigation.

One of the most simple and fundamental information of the HB architecture is the degree of branching (DB). The DB of an HB polymer was originally defined by [12]:

$$DB = \frac{T+D}{P} = \frac{2D+1}{P} \tag{2}$$

where P is the degree of polymerization (total number of units in a polymer molecule), i.e., $P = T + D + L$. Note that every time the L-type unit is converted to D, the number of T units increases by one, and therefore, the relationship, $T = D + 1$ always holds true for any HB architecture, as long as the ring formation through the intramolecular reaction between the focal point A group and the unreacted B group in the same molecule is neglected.

When there are no L units, such as for the perfect dendron, $DB = 1$. With the definition given by Equation (2), however, DB cannot go down to 0 because linear polymer structure always possesses one T unit at its tail. To avoid this problem, at the same time, to make balanced comparison with the HB polymers synthesized via SCVP in which the focal point is always the L-type, the following definition for the DB was proposed [6].

$$DB = \begin{cases} 2D/(P-2) & \text{when FP is L, with } P > 3 \\ 2(D-1)/(P-3) & \text{when FP is D, with } P > 3 \\ 0 & \text{for } P \leq 3 \end{cases} \tag{3}$$

where FP means the focal point. In this article, DB defined by Equation (3) is used. For the case of the HB polymer shown in Figure 2, $P = 27$, $D = 10$, and the FP is the D-type, the DB is calculated to be $DB = (2)(9)/(24) = 0.75$. Obviously, the DB of large polymers, i.e., with $P \gg 1$, is given by $DB \cong 2D/P$, in either type of definition.

Figure 3 shows the relationship between the values of DB and P for batch polymerization with $r = 1$ when the conversion of A group is $x_A = 0.9$. The figure was prepared using the unpublished data obtained in the investigation reported earlier [6]. In the figure, each red dot shows a pair of values, DB and P, for each polymer molecule generated in the MC simulation. The circular symbols with a blue line show the average DB within each intervals of P, which shows the expected DB for the given P-value, $\overline{DB}(P)$. The DB-value converges to DB_{inf}, as the degree of polymerization P increases, i.e., for large polymers. On the other hand, the black broken line shows the magnitude of average DB of the whole reaction system. It is clearly shown that the values of DB are distributed around DB_{inf}, rather than the average DB of the whole system, and DB_{inf} is larger than the average DB. These characteristics hold true irrespective of the magnitude of reactivity ratio r, not only for a batch reactor [7] but also for a CSTR [8,9]. In this article, the HB architecture of large polymers, for which $\overline{DB}(P)$ has reached a constant value DB_{inf}, is investigated in detail, both for a batch reactor and a CSTR.

An interesting characteristic of DB_{inf} is that the magnitude of DB_{inf} is essentially kept constant, irrespective of the conversion level for a given reactivity ratio r, both for a batch reactor [6,7] and for a CSTR [8,9]. Note that the average DB of the whole reaction system increases with conversion, but DB_{inf} does not change. The value of DB_{inf} can be estimated by the reactivity ratio r, as shown in Figure 4. Obviously, $DB_{inf} = 1$ for the cases with $r = \infty$, both for batch and CSTR, but the value of DB_{inf} for a CSTR is always larger than that for a batch reactor, as long as r is finite. Incidentally, analytic relationship between DB_{inf} and r was established previously [6,7], and a smooth curve was drawn for batch polymerization in Figure 4. On the other hand, a general equation for a CSTR has not been reported, and five data points reported in the earlier publication [9] were plotted and connected.

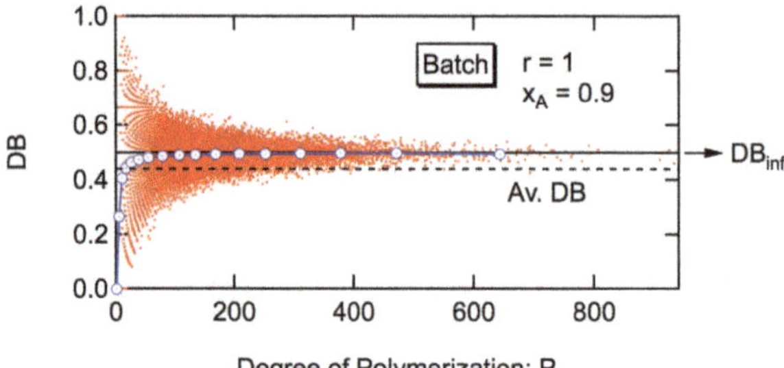

Figure 3. Relationship between degree of branching (DB) and *P* for batch polymerization with *r* = 1 at the conversion level of A, x_A = 0.9. Similar figures can be found in the earlier publications for a batch reactor [6,7] and for a CSTR [8,9].

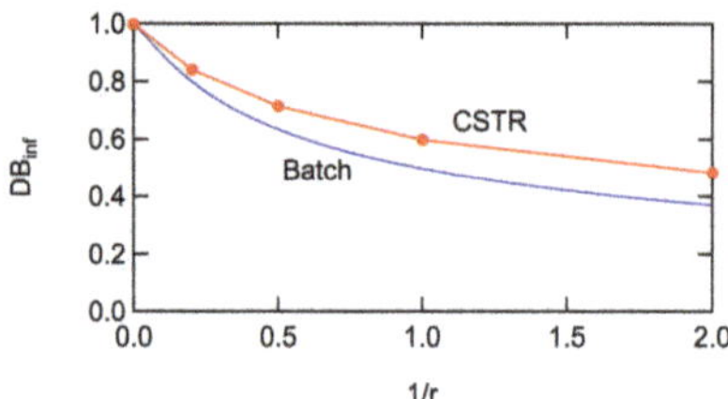

Figure 4. Relationship between DB_{inf} and *r* for batch and CSTR. Data points were taken from the earlier publication [9].

Based on the structural information shown in Figure 2, it is possible to determine the 3D size, represented by the mean-square radius of gyration under the unperturbed condition, $\langle s^2 \rangle_0$. One of the methods used to determine the value of $\langle s^2 \rangle_0$ is the Wiener index (WI) [13]. The WI is related with $\langle s^2 \rangle_0$ through the following relationship [14],

$$\frac{\langle s^2 \rangle_0}{l^2} = \frac{WI}{N^2},$$
(4)

where *l* is the random walk segment length, and *N* is the number of such segments in the polymer molecule. *N* is related with *P* through *N* = *P*/*u*, where *u* is the number of monomeric units in a segment. In the previous investigation [6–9], as well as the present article, *u* = 1 is used. Define Rg^2 by the following equation,

$$Rg^2 = \frac{WI_{u=1}}{N^2},$$
(5)

where $WI_{u=1}$ is the value of WI when *u* = 1. Rg^2 is the value of $\langle s^2 \rangle_0$ normalized by the squared monomer-unit length, and is proportional to $\langle s^2 \rangle_0$, at least for large polymers, *P* >> 1. Note that Rg^2 is equal to the value of $u\langle s^2 \rangle_0/l^2$, which is unchanged irrespective of the magnitude of *u*, as long as the number *N* of steps is large enough.

Figure 5 shows the expected Rg^2 of the polymer molecule whose degree of polymerization is *P* for batch polymerization with the reactivity ratio, *r* = 1. As shown in the figure, the relationship between Rg^2 and *P* does not change with the progress of conversion, x_A [6,7]. The curve moves to smaller Rg^2, as the reactivity ratio *r* is increased [7]. However, even with *r* = ∞ for which *DB* = 1, Rg^2 is much larger than that for the perfect dendron [7], which is shown by the black curve in Figure 4. For large polymers, the power law $Rg^2 \sim P^{0.5}$ is valid, irrespective of the magnitude of *r* [6,7]. The power exponent, 0.5 is the same as for the random branched polymers, represented by the Zimm-Stockmayer equation [15].

In this article, the universal relationship concerning Rg^2, that is independent of r, will be reported, and the relationship with the Zimm-Stockmayer equation will be discussed quantitatively.

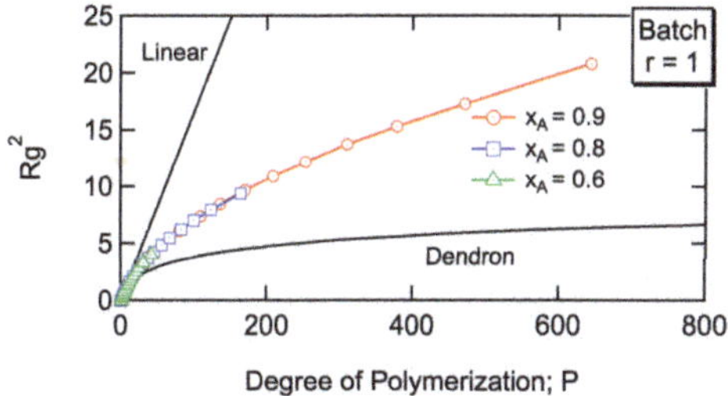

Figure 5. Expected Rg^2-values of the polymers having P for batch polymerization with $r = 1$. Redrawn by using the data reported earlier [6].

For a CSTR, the variance of Rg^2 for large polymers is quite large. It is not perfectly clear, but the relationship between Rg^2 and P does not change significantly, even when the steady state conversion level is changed [8,9]. It was clearly demonstrated that a CSTR produces polymers with much smaller Rg^2 compared with batch polymerization, even when the value of DB_{inf} is deliberately set to be the same for both types of reactors [8]. In this article, the reason for obtaining much more compact architecture in a CSTR is explored by considering the properties of the largest tri-branched cluster in a polymer molecule. The tri-branched clusters are shown by the group of D units surrounded by the red closed frames in Figure 2, and the largest cluster for this example consists of six units. Note that although the focal point unit is the D type, it is connected to only two other units and it is not considered as a tri-branched unit. A universal relationship concerning Rg^2, independent of the steady state conversion level, will also be reported for a CSTR.

Another structural information investigated in this article is the maximum span length, L_{MS}. Here, the span length refers to the distance in the monomeric units [16], and L_{MS} is equivalent to the longest end-to-end distance [17]. In the case of HB polymer shown in Figure 2, $L_{MS} = 12$, which is the distance between the units with a star. There are two routes having $L_{MS} = 12$, starting from the unit with a star to the unit with star 1 or 2. Interesting universal relationships will be reported for the magnitude of L_{MS}.

In this article, the HB architecture formed in a batch and a CSTR is compared and discussed, by investigating the properties of the largest tri-branched cluster, and the magnitudes of Rg^2 and L_{MS}. For Rg^2 and L_{MS}, the universal relationships are sought. Note that the universal relationships reported so far for Rg^2 are with respect to the conversion level, and they change with the reactivity ratio r. In this article, further unification is explored.

2. Methods

The MC simulation method proposed earlier for batch polymerization [6,7], and that for a CSTR [8,9] were used to determine the branched architecture, as shown in Figure 2. This MC simulation method is based on the random sampling technique [18,19], and the polymer molecules were selected from the final product on a weight basis. The whole molecular architecture of each selected molecule was reconstructed by strictly following the history of branched structure formation. By generating a large number of polymer molecules in the simulation, the statistical properties were determined effectively. In the present investigation, the structural properties of large sized polymer molecules are highlighted, and 10^4 polymer molecules with $P > 50$ were collected to determine statistically valid estimates. The reactivity ratios investigated were $r = 0.5, 1, 2, 5$, and ∞.

For a CSTR, the polymerization behavior at steady state is fully described by the reactivity ratio r and the dimensionless number ξ, sometimes referred to as the Damköhler number for the second order reaction, defined by:

$$\xi = k_{\mathrm{T}}[\mathrm{A}]_0 \bar{t}, \tag{6}$$

where $[\mathrm{A}]_0$ is the initial concentration of A group, or equivalently, the initial monomer concentration, and $\bar{t}$ is the mean residence time [3]. The HB polymer shown in Figure 2 was generated for a CSTR with the condition, $r = 2$ and $\xi = 0.35$. The conversion of A group, x_{A} increases with ξ. To set the value of ξ corresponds to fixing the steady state conversion level, x_{A}.

For a CSTR, it was found that the weight-average molecular weight cannot reach the steady state for large $\bar{t}$ cases [8,9], and similar behavior was also reported for the SCVP [20,21] which is another route to synthesize HB polymers. The upper limit ξ-values above which the weight-average molecular weight cannot reach the steady state for a given reactivity ratio r was shown graphically in the earlier publication [9]. For example, the upper limit value of ξ is $\xi_{\mathrm{UL}} = 0.5$ for $r = 1$, and $\xi_{\mathrm{UL}} = 0.25$ for $r = \infty$. In the present investigation, the MC simulations were conducted for the cases with $\xi \leq \xi_{\mathrm{UL}}$.

The WI was calculated for each polymer molecule generated in the MC simulation, by setting up the distance matrix, $\{d_{ij}\}$, where d_{ij} is the distance in the number of monomeric units between the ith and jth unit. The WI when $u = 1$ is given by [13,14]:

$$WI_{u=1} = \frac{1}{2}\sum_{i=1}^{P}\sum_{j=1}^{P} d_{ij}. \tag{7}$$

The maximum span length, L_{MS} was determined by finding the largest value of d_{ij} in the distance matrix.

The statistical properties of various types of clusters in HB polymers were determined from the structural information, shown in Figure 2.

3. Results and Discussion

3.1. Largest Cluster of Tri-Branched Units

A CSTR produces much more compact HB polymers, compared with batch polymerization, as reported earlier [8,9]. First, the reason for this is explored by considering the size of the largest tri-branched cluster.

Figure 6 shows the relationship between the number of units P_{LC} belonging to the largest cluster and the number P of units in the polymer molecule (degree of polymerization). Each dot shows a set of values for each polymer molecule generated in the MC simulation. For a CSTR, the cases with $\xi = \xi_{\mathrm{UL}}$ are shown in the figure, while $x_{\mathrm{A}} = 0.95$ for batch polymerization. General characteristics were the same for the other reaction conditions. For a CSTR, the largest cluster size P_{LC} increases with P, and a very large tri-branched cluster exists in a large polymer molecule. On the other hand, the size of the largest cluster does not increase significantly for batch polymerization, except for $r = \infty$. With $r = \infty$, all units other than the peripheral T units and a focal point are tri-branched units, and the largest cluster size is essentially proportional to P with $P_{\mathrm{LC}} \approx 0.5P$. Except for $r = \infty$, the existence of a large cluster of tri-branched units is an important characteristic of polymers formed in a CSTR, while a large number of small-sized clusters are formed in batch polymerization. It is reasonable to consider that the dimension is smaller for the HB polymers formed in a CSTR. The cases with $r = \infty$ will be discussed later, and consider the properties of various types of clusters for the cases with the reactivity ratio, $r \leq 5$ first.

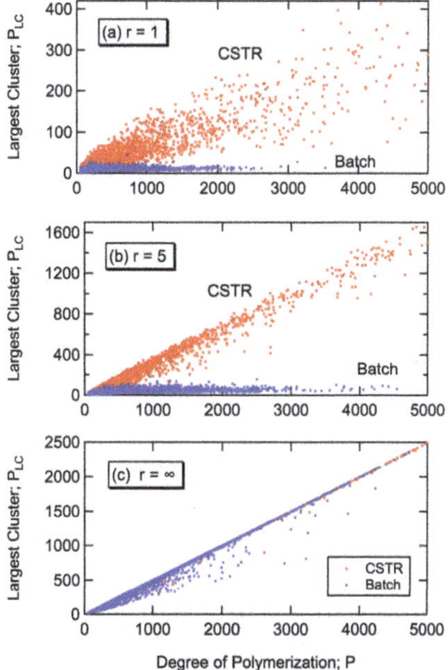

Figure 6. Relationship between the number P_{LC} of units belonging to the largest cluster and the degree of polymerization P. (**a**) $r = 1$; $\xi = 0.5$ for CSTR, and $x_A = 0.95$ for batch. (**b**) $r = 5$; $\xi = 0.306$ for CSTR, and $x_A = 0.95$ for batch. (**c**) $r = \infty$; $\xi = 0.25$ for CSTR, and $x_A = 0.95$ for batch.

The largest tri-branched cluster consists of units whose residence times are different. In the case of the HB polymer shown in Figure 2, the largest tri-branched cluster consists of six units, and in general, the residence time of each unit is different. This kind of detailed information cannot be obtained in experiments, but can be determined in a straightforward manner in the present MC simulation method.

The average residence time of the units belonging to the largest tri-branched cluster is calculated, and is plotted as a red dot in Figure 7. In the same figure, the average residence time of all units in each polymer molecule, as well as that of the peripheral T units, is also shown. Here, the residence time is represented by the dimension residence time defined by [3]:

$$\theta = t/\bar{t}. \tag{8}$$

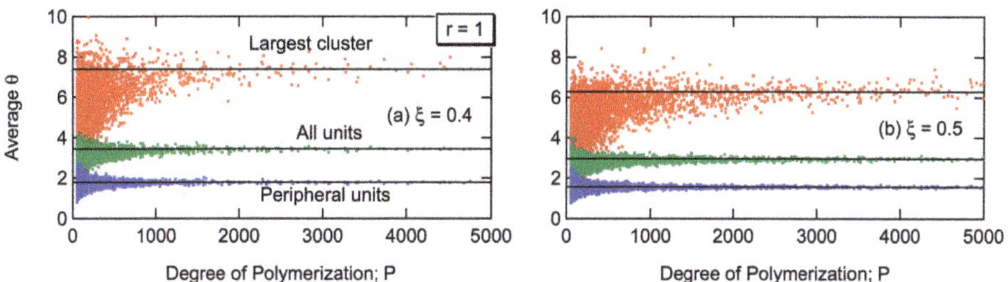

Figure 7. Average residence time of the units in the largest tri-branched cluster (red), of all units (green), and of the peripheral T units (blue) in each polymer molecule for the cases with $r = 1$. (**a**) $\xi = 0.4$, and (**b**) $\xi = 0.5$.

The average residence time approaches a constant value for all three types of units, as the molecular weight (MW) of the polymer increases. The constant value for each type of units decreases as the ξ-value, or equivalently, as the conversion level increases. At steady state for a given value of ξ, the convergent residence time shown by the black line is large for the largest cluster (red), which means that the largest cluster tends to be formed by connecting the units with larger residence times. On the other hand, the peripheral T units consist of units with smaller residence times. It would be reasonable to consider that there exists a gradient in the residence time distribution of units within an HB polymer. The core cluster of tri-branched region consists of units with large residence times, and the residence time of the units decreases toward the peripheral T type units. This tendency in the residence time distribution would be the reason for forming a compact architecture in a CSTR, compared with batch polymerization.

For the cases with $r = \infty$, a higher order tri-branched cluster is considered to differentiate the structure formed in batch and CSTR. Define the second order tri-branched cluster as a group of tri-branched units with all three bonds being connected to the tri-branched units, as shown by the regions enclosed by the blue broken curves in Figure 8. The largest group is named as the largest tri-branched cluster of the second order, and the number of units in such a group is represented by $P_{LC,2}$. In the present example shown in Figure 8, $P_{LC,2} = 2$.

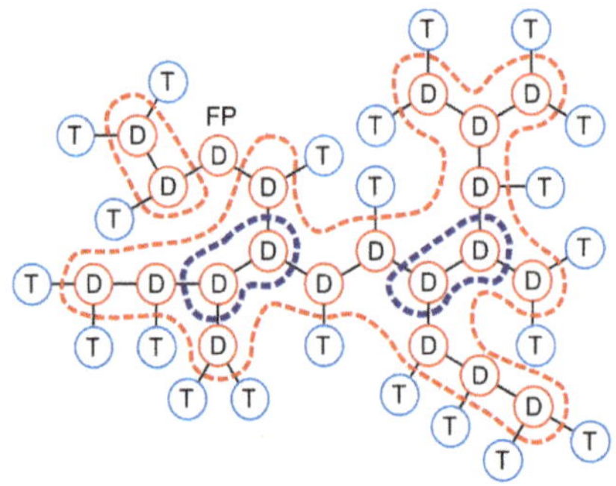

Figure 8. Example of the hyperbranched polymer architecture generated in the present MC simulation for a CSTR with $r = \infty$ and $\xi = 0.2$. The tri-branched clusters of the first order are represented by the groups enclosed by the red broken curves, while the tri-branched clusters of the second order by the blue broken curves. For this polymer, $P = 43$, $D = 21$, and $DB = 1$.

Figure 9 shows the relationship between $P_{LC,2}$ and P for batch and CSTR with $r = \infty$. For a CSTR, $\xi = \xi_{UL} = 0.25$, and $x_A = 0.95$ for batch polymerization. In the case of a CSTR, $P_{LC,2}$ increases with P, and therefore, a very large tri-branched cluster of the second order exists in a large polymer molecule. On the other hand, it does not increase significantly for batch polymerization, which means a large number of smaller-sized tri-branched clusters of the second order are formed. The smaller Rg^2 obtained for a CSTR could be understood from the significant differences in the magnitude of $P_{LC,2}$.

Figure 10 shows the average residence times for various types of units, as a function of P. Again, the average times of each type of units reach constant values for large polymers. The largest tri-branched cluster of the second order consists of units having very large residence times, while the units with smaller residence times tend to be the peripheral T units. There seems to exist a gradient in residence time distribution from the core region to the peripheral units. In a CSTR, a large core tri-branched cluster region is formed, which makes the architecture much more compact, compared with batch polymerization, also for the case with $r = \infty$.

In this section, it was shown that the branched architecture can be controlled by the residence time distribution. To form a core region is key to produce compact architecture. Obviously, the slow monomer addition method [22–24] is another way to form a core region to obtain compact architecture.

On the other hand, if one needs looser structure, the core formation should be avoided. The structural control by using the tanks-in-series process was also discussed previously [25].

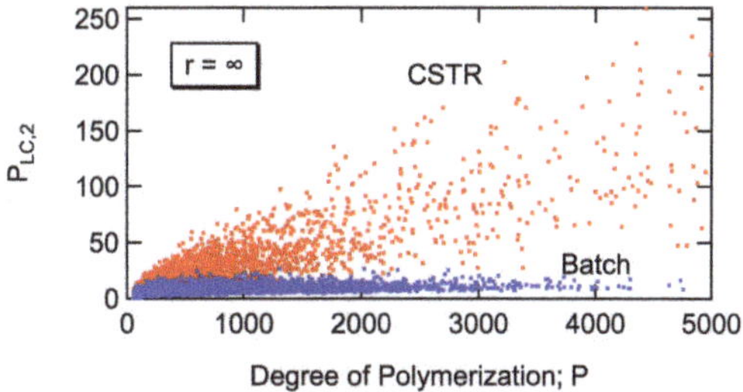

Figure 9. Relationship between the number $P_{LC,2}$ of units belonging to the largest cluster of the second order and the degree of polymerization P, for a batch and a CSTR with $r = \infty$. $\xi = 0.25$ for CSTR, and $x_A = 0.95$ for batch.

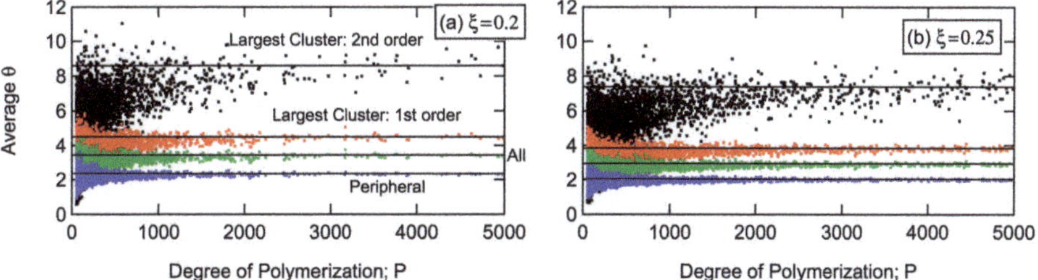

Figure 10. Average residence time of the units in the largest tri-branched cluster of the second order (black dots), of the units in the largest tri-branched cluster of the first order (red), of all units (green), and of the peripheral T units (blue) for the cases with $r = \infty$. (**a**) $\xi = 0.2$, and (**b**) $\xi = 0.25$.

3.2. Radius of Gyration and Maximum Span Length

Both Rg^2 defined by Equation (5) and the maximum span length L_{MS} exemplified in Figure 2 are the characteristic factors describing the spatial size of an HB polymer. In this section, the universal relationships concerning Rg^2 and L_{MS} are explored both for a batch (Section 3.2.1) and a CSTR (Section 3.2.2).

3.2.1. Batch Polymerization

Figure 11 shows the MC simulation results for the relationship between Rg^2 and P with $r = 1$ at $x_A = 0.95$. Each red dot represents a set of Rg^2 and P, generated in the MC simulation. Note that the data were collected for $P > 50$ to clarify the statistical properties of large polymers, where $\overline{DB}(P)$ has reached a constant value, DB_{inf}. Blue circular symbols show the averages within ΔP fractions, and therefore, the blue line connecting these points represents the expected Rg^2-value for a given P.

Figure 12 shows the expected Rg^2-values that correspond to the blue curve in Figure 11 for various combinations of the reactivity ratio r and conversion x_A. The curve for the expected Rg^2-value does not change with the conversion level x_A, and becomes smaller as the reactivity ratio r increases, as already reported earlier [6,7].

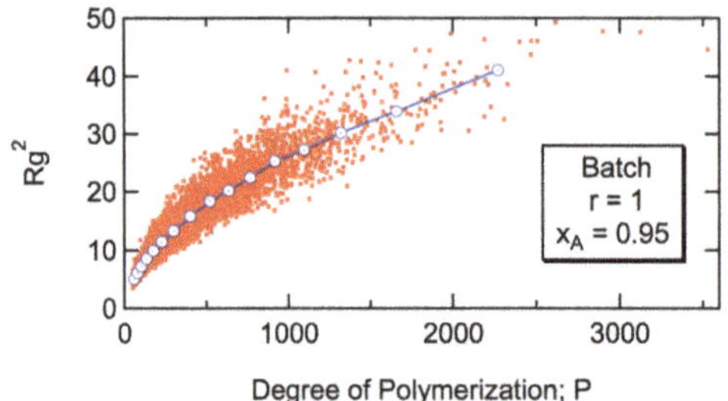

Figure 11. Relationship between Rg^2 and P with $r = 1$ and $x_A = 0.95$ for batch polymerization.

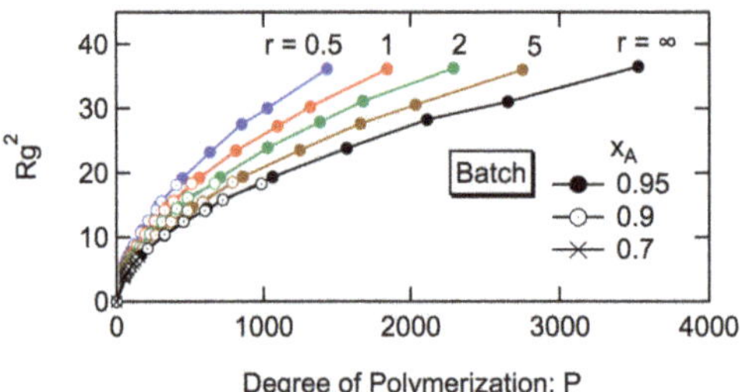

Figure 12. Relationship between Rg^2 and P with various r-values at $x_A = 0.95$ (filled circle), 0.9 (open circle), and 0.7 (cross) for batch polymerization.

The contraction parameter, the ratio g of mean-square radius of gyration of the branched molecule to that for a linear molecule is given by [15]:

$$g = \frac{\langle s^2 \rangle_{0,\text{br}}}{\langle s^2 \rangle_{0,\text{lin}}}\bigg|_{\text{same}P} = \frac{Rg^2_{\text{br}}}{Rg^2_{\text{lin}}}\bigg|_{\text{same}P} = \frac{6Rg^2_{\text{br}}}{P}, \tag{9}$$

where the subscript "br" is for the branched polymer, and "lin" is for linear polymer. Note that Rg^2 is the defined by Equation (5), and therefore:

$$Rg^2_{\text{lin}} = \frac{P}{6}. \tag{10}$$

Figure 13 shows the relationship between g and $DB_{\text{inf}}P/2$. The value of DB_{inf} is a constant for a given reactivity ratio, and the value of $DB_{\text{if}}P/2$ is equal to the average number $\bar{n}_b$ of branch points per molecule for large polymers. Note that $DB = 2D/P$ for large polymers, and $\bar{n}_b = DB_{\text{inf}}P/2$. Because the Rg^2-value for a given P, as well as the magnitude of DB_{inf}, is the same at any conversion level x_A, the calculated results for $x_A = 0.95$ with various r's are shown in Figure 13. All points fall on a single curve, showing a universal relationship, independent of x_A and r.

For the random branched polymers, the g-ratio is represented by the following Zimm-Stockmayer equation [15]:

$$g = \left[(1 + \bar{n}_b/7)^{\frac{1}{2}} + 4\bar{n}_b/9\pi\right]^{-\frac{1}{2}}. \tag{11}$$

where $\bar{n}_b$ is the average number of branch points per molecule.

Figure 14 shows the comparison with the Zimm-Stockmayer equation by using $\bar{n}_b = DB_{\text{inf}}P/2$, which shows an excellent fit. It is suggested that the HB architecture formed in a batch polymerization is random branch, irrespective of the magnitude of reactivity ratio, r. In batch polymerization, the probability that a randomly selected unit from the final product is the D type unit is the same for all units [6,7], and therefore, it is reasonable to obtain HB polymers with random branched architecture. On the other hand, in the case of a CSTR, the probability for a randomly selected unit from the final product being the D type unit is larger for the units with longer residence time, leading to a nonrandom branched architecture, as discussed in the previous section.

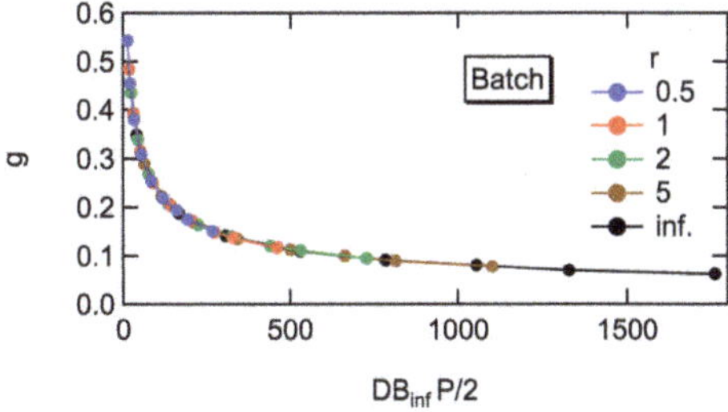

Figure 13. Universal relationship between g and the average number of branch points in a polymer, $DB_{inf}P/2$ for batch polymerization.

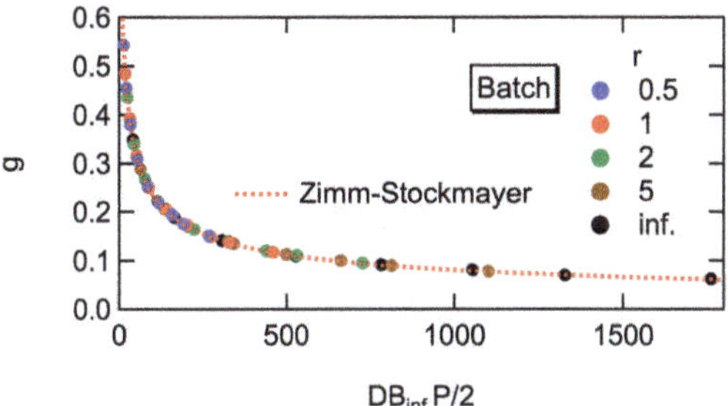

Figure 14. Comparison with the Zimm-Stockmayer equation.

The value of DB_{inf} for a given r can be calculated analytically [6,7], and Figure 4 shows the calculated results graphically. Therefore, the value of Rg^2 can be determined by using Equation (11) in a straightforward manner, without conducting the MC simulation, for any combination of r and x_A.

Next, consider the maximum span length, L_{MS}. Figure 15 shows the MC simulation results for the relationship between L_{MS} and P with $r = 1$ and $x_A = 0.95$ for batch polymerization. Each red dot represents a set of L_{MS} and P, generated in the MC simulation. Blue circular symbols show the averages within ΔP fractions, and the blue line connecting these points represents the expected L_{MS}-value for a given P.

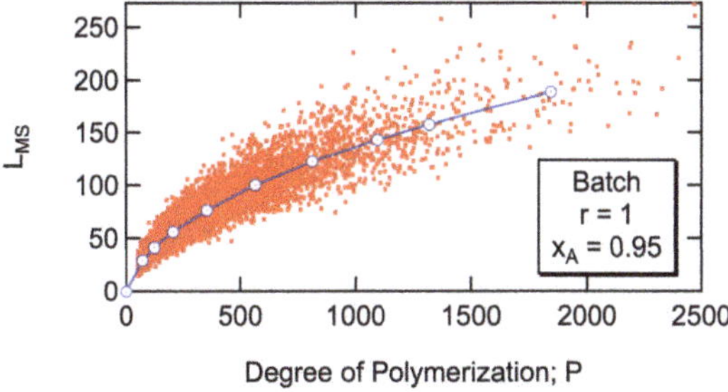

Figure 15. Relationship between L_{MS} and P with $r = 1$ and $x_A = 0.95$ for batch polymerization.

Figure 16 shows the expected L_{MS}-value for a given P, corresponding to the blue curve in Figure 15, for various combinations of the reactivity ratio r and conversion x_A. The curve for the expected L_{MS}-value does not change with the conversion level x_A, and is a function of the reactivity ratio r. The qualitative tendency is quite similar to Rg^2.

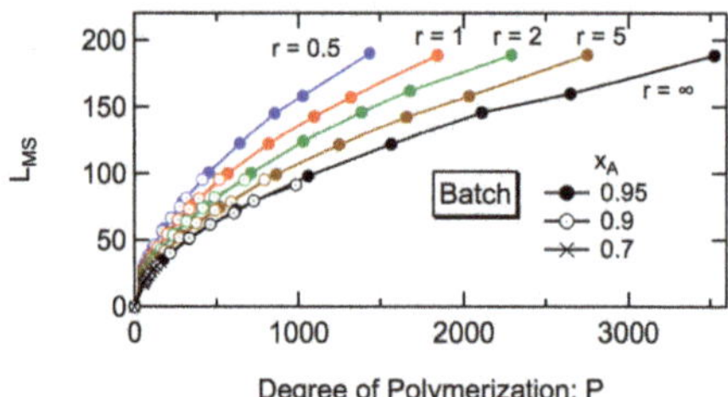

Figure 16. Relationship between L_{MS} and P with various r-values at $x_A = 0.95$ (filled circle), 0.9 (open circle), and 0.7 (cross) for batch polymerization.

Figure 17 shows the expected value of L_{MS}/P for a given $DB_{inf}P/2$. Because the L_{MS}-value for a given P, as well as the magnitude of DB_{inf}, is the same at any conversion level x_A, the calculated results for $x_A = 0.95$ with various r's are shown in Figure 17. Note that the value of $DB_{inf}P/2$ is equal to the average number of branch points per molecule for large polymers. All data points fall nicely on the same universal curve, as in the case of Rg^2.

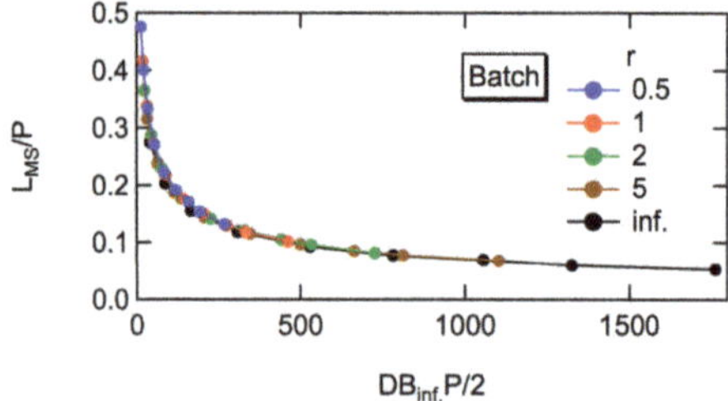

Figure 17. Universal relationship between L_{MS}/P and the number of branch points in a polymer, $DB_{inf}P/2$ for batch polymerization.

Both Rg^2 and L_{MS} show a similar universal relationship, as shown in Figures 13 and 17. Now, consider the relationship between Rg^2 and L_{MS}.

Figure 18a shows the relationship between Rg^2 and L_{MS} for $r = 1$ and $x_A = 0.95$, which shows a linear relationship. The blue line with circular symbols shows the expected Rg^2 for a given L_{MS}.

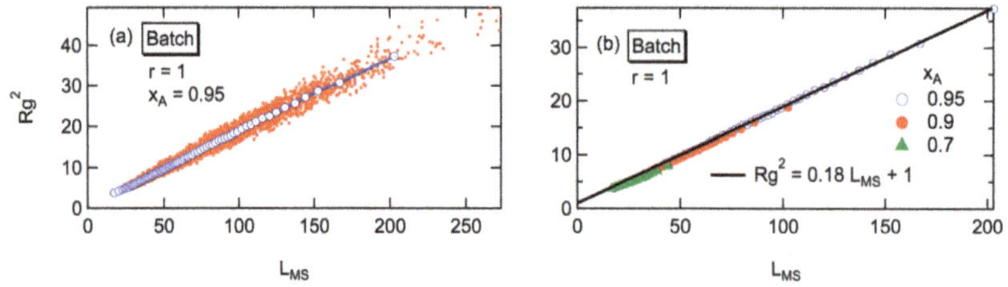

Figure 18. Relationship between Rg^2 and L_{MS} for batch polymerization with $r = 1$. (**a**) Raw data (red dots) and the averages within ΔL_{MS}, i.e., the expected Rg^2-values, at $x_A = 0.95$. (**b**) Expected Rg^2 for a given L_{MS} with various conversion levels.

Figure 18b shows the expected Rg^2 at various conversion levels for $r = 1$. The plotted points shown by the blue circular symbols are the same as those in Figure 18a. The relationship is essentially unchanged by the conversion level, and the relationship fits reasonably well with:

$$Rg^2 = 0.18 L_{MS} + 1. \tag{12}$$

Figure 19 shows the expected Rg^2 for a given L_{MS}, for various combinations of x_A and r. A linear relationship seems to hold for any value of r. The black straight line shows the linear relationship given by Equation (12). Although a slight discrepancy is observed in the cases of $r = 5$ and ∞ for large polymers, the data points are well correlated with Equation (12). Equation (12) is the universal relationship between Rg^2 and L_{MS}, applicable to any combination of r and x_A in batch polymerization.

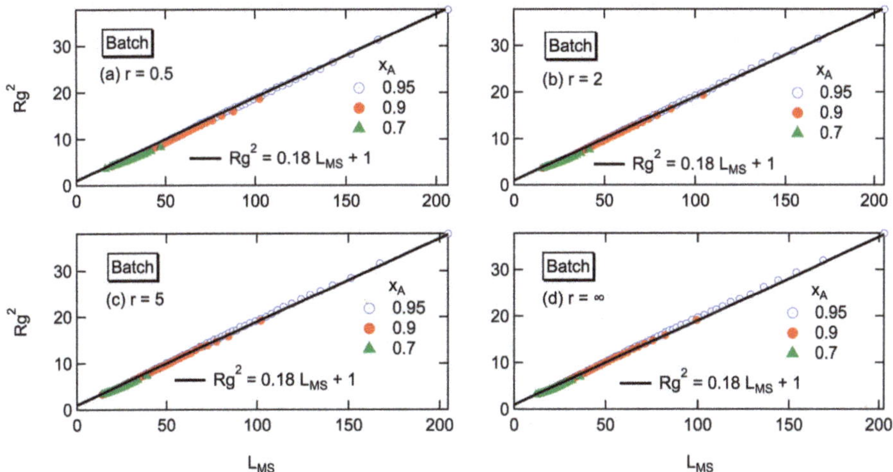

Figure 19. Universal relationship between Rg^2 and L_{MS} for batch polymerization with various combinations of r and x_A. (**a**) $r = 0.5$, (**b**) $r = 2$, (**c**) $r = 5$, and (**d**) $r = \infty$.

For the SCVP, the relationship, $Rg^2 = 0.18 L_{MS} + 0.6$ was reported both for a batch and a CSTR [26]. The proportional coefficient, 0.18 is the same as Equation (12), and the constant term is very close.

For linear polymers, L_{MS} is equal to P, and the following equation is valid for large polymers:

$$Rg^2 = L_{MS}/6 \cong 0.167 L_{MS}. \tag{13}$$

Note that Rg^2 is the mean-square radius of gyration when each monomeric unit is considered as the random walk segment.

In the case of linear polymers, there is no contribution to Rg^2 other than its own chain with $P = L_{MS}$. In the HB polymers, the chains other than the largest span chain can make a contribution to increase the Rg^2-value. The increase in the coefficient from 0.167 to 0.18 could be considered as showing the degree of contribution from the other chains to the magnitude of Rg^2.

For the perfect dendrons, on the other hand, the numerical calculation results are shown in Figure 20, and the relationship for large L_{MS}-values is given by:

$$Rg^2 = 0.5 L_{MS} - 2. \tag{14}$$

The proportionality coefficient changes from 0.167 for linear polymers to 0.5 for perfect dendrons, and the HB polymers comes in between these two extremes. The value of 0.18 for the HP polymers is closer to linear polymers, rather than that for perfect dendrons, which shows that the magnitude of Rg^2

is still mainly determined by the maximum span chain, and the contribution of the other chains is not very significant. The exact physical meaning of the magnitude of coefficient is still an open question. However, the linear relationship found here is of great interest.

For batch polymerization, the value of Rg^2 can be determined analytically without MC simulation, as discussed earlier. By using Equation (12), the magnitude of L_{MS} can also be estimated in a straightforward manner, without relying on the MC simulation.

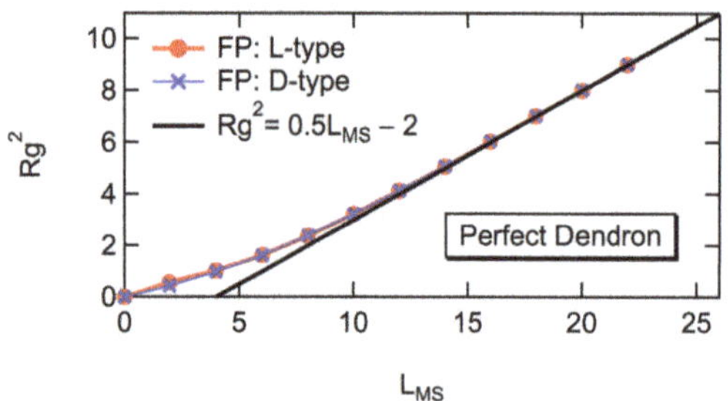

Figure 20. Relationship between Rg^2 and L_{MS} for perfect dendrons, when the focal point is the L-type (red circle) and the D-type (blue cross). For both cases, the relationship is represented by Equation (14) for large polymers.

3.2.2. CSTR

In this section, basic characteristics of Rg^2 and L_{MS} are considered, as was done in the previous section. For a CSTR, however, the variance of Rg^2 for large polymer is quite large [8,9], and it is difficult to determine the statistically valid expected Rg^2-values for large polymers. Inspired by the universal curve shown in Figure 13, the value of g-ratio defined by Equation (9) is plotted with respect to the number of branch points in a polymer molecule, n_b. Figure 21a shows the case with $r = 1$ at $\xi = 0.5$. In the figure, each red dot indicates a set of values for the polymer molecule generated in the MC simulation. With this type of plot, the variance of g for large polymers is rather small, and it is straightforward to determine the expected g-ratio for a given n_b, shown by the blue curve with circular symbols.

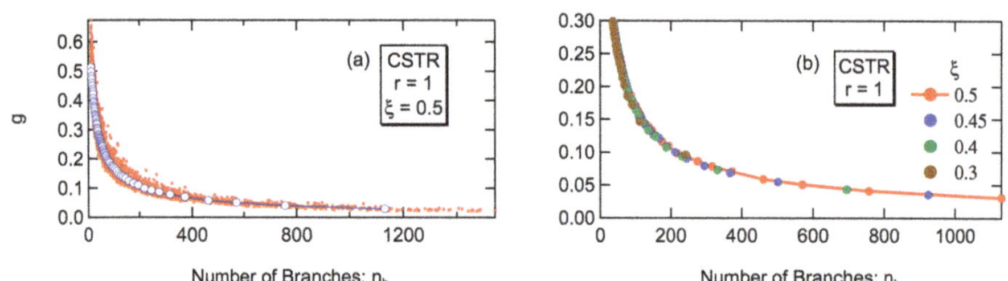

Figure 21. Relationship between the g-ratio and the number n_b of branch points in a polymer for a CSTR with $r = 1$. (**a**) Each data point (red) and the expected g-value (blue), at $\xi = 0.5$. (**b**) The expected g-ratio for $\xi = 0.5$, 0.45, 0.4, and 0.3.

Figure 21b shows the expected g-ratio at various values of ξ for the case with $r = 1$. The data points fall on a single curve irrespective of the values of ξ, i.e., at any steady state conversion level. The universal relationship between g and n_b, unchanged by ξ, is confirmed also for the other r cases, as shown in Figure S1 of Supplementary Materials.

Figure 22 shows how the universal curves, shown in Figure 21b and Figure S1, change with the reactivity ratio r. Because the relationship does not change with ξ for a given r, the expected g-values at $\xi = \xi_{UL}$ are shown in Figure 22. In order to magnify the small differences for smaller g-values, the logarithmic scale plot was used for Figure 22b. Slight differences among curves are observed, and the expected g-ratio is considered a very weak function of r. In particular, the change for $r < 5$ is not very significant.

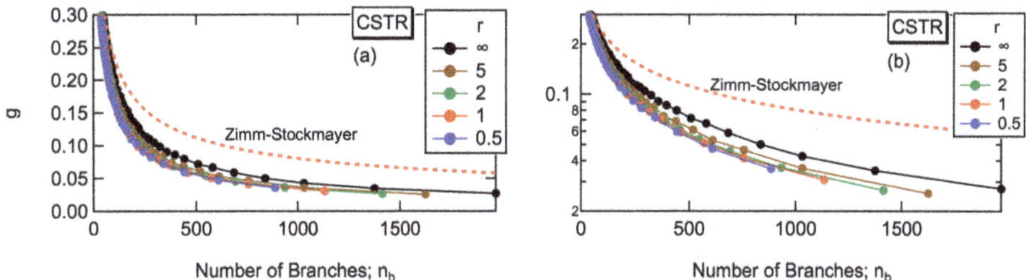

Figure 22. Expected g-ratio and n_b for the HB polymers having n_b branch points in a polymer with various reactivity ratio r for a CSTR. The plotted values are at ξ_{UL} for each r. (**a**) Normal scale plot, and (**b**) logarithmic scale plot for the y-axis.

In Figure 22, the g-ratio for the random branched polymers represented by the Zimm-Stockmayer equation, Equation (11), is also shown by the red broken curve. It is clearly shown that the HB architecture formed in a CSTR is much more compact than for the random branched polymers, i.e., for the HB polymers synthesized in a batch reactor.

Next, consider the maximum span length, L_{MS} for the HB polymers formed in a CSTR. Figure 23a shows the relationship between the weight fraction of the maximum span chain L_{MS}/P and n_b for $r = 1$ with $\xi = 0.5$. Each red dot shows the individual data point, and the blue curve with circular symbols shows the expected value of L_{MS}/P for a given n_b. Figure 23b shows the expected L_{MS}/P for various values of ξ, i.e., for different conversion levels at steady state. The expected values of L_{MS}/P do not change with ξ, and another universal relationship is found. The universal relationship between L_{MS}/P and n_b, unchanged by ξ, is confirmed also for the other r cases, as shown in Figure S2 of Supplementary Materials.

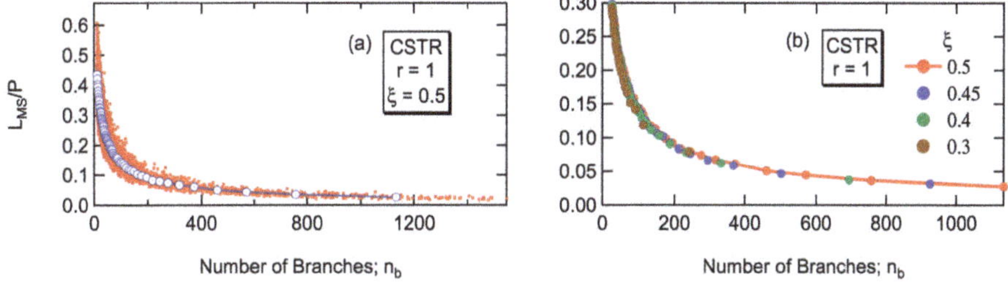

Figure 23. Relationship between L_{MS}/P and n_b for a CSTR with $r = 1$. (**a**) Each data point (red) and the expected L_{MS}/P (blue), at $\xi = 0.5$. (**b**) The expected L_{MS}/P for the HB polymers having n_b branch points in a polymer with $\xi = 0.5, 0.45, 0.4$, and 0.3.

Figure 24 shows how the universal curve changes with the reactivity ratio r. Again, the relationship at ξ_{UL} is shown for each reactivity ratio. Figure 24a is the normal scale plot, which shows the differences among the curves are rather small. To enlarge the differences for smaller L_{MS}/P-values, the logarithmic plot is used for the y-axis of Figure 24b, and it is shown that up to $r = 5$, the differences are rather small, but the curve with $r = \infty$ shows slightly larger L_{MS}/P-values. Similarly with the g-ratio for a CSTR, the value of L_{MS}/P is a very weak function of the reactivity ratio r.

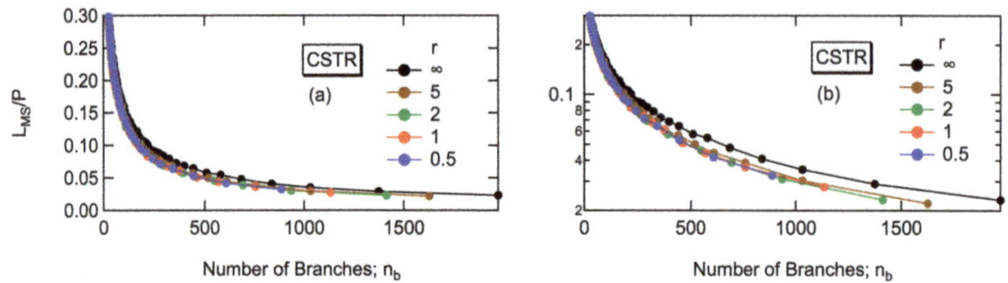

Figure 24. Relationship between L_{MS}/P and n_b for a CSTR with various reactivity ratios. The plotted values are at ξ_{UL} for each r. (**a**) Normal scale plot, and (**b**) logarithmic scale plot.

Finally, consider the relationship between Rg^2 and L_{MS}, as was done for batch polymerization in Figures 18 and 19, which showed a universal relationship, represented by $Rg^2 = 0.18L_{MS} + 1$.

Figure 25a shows the relationship between Rg^2 and L_{MS} for HB polymers formed in a CSTR with $r = 1$ and $\xi = 0.5$, which shows a linear relationship. Figure 25b shows the expected Rg^2 at various ξ-values for $r = 1$. The relationship is essentially unchanged by the steady-state conversion level. The black line represents the relationship given by Equation (11). Although a slight deviation is observed for large values of L_{MS}, overall agreement is satisfactory. Compared with Figure 18, the absolute values of Rg^2 and L_{MS} are smaller for the case of CSTR, because of much more compact architecture formed in a CSTR.

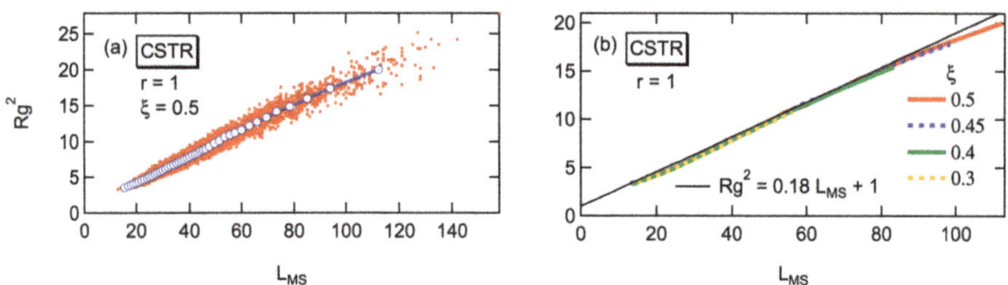

Figure 25. Relationship between Rg^2 and L_{MS} for a CSTR with $r = 1$. (**a**) Raw data and the averages within ΔL_{MS}, i.e., the expected Rg^2-values, at $x_A = 0.95$. (**b**) Expected Rg^2 for various conversion levels.

The expected Rg^2-values for various combinations of ξ and r are shown in Figure S3 of Supplementary Materials. The universal relationship, represented by Equation (12), correlates reasonably well, irrespective of the magnitude of ξ and r, and Equation (12) could be considered as a universal relationship that holds for both batch and CSTR. Even though a CSTR leads to form much more compact HB polymers, such difference in branched structure does not affect the relationship, $Rg^2 = 0.18L_{MS} + 1$.

The physical meaning of the magnitude of proportionality coefficient is not clear at the present stage, however, the proportionality coefficient, 0.18 is closer to that for linear polymers (0.167), rather than for the perfect dendron (0.5). The perfect dendron suffers from the Malthusian packing paradox [27] and cannot fit in the 3D space. Closer value to that for linear polymers may imply that the structure does not suffer from the space dimensionality. In fact, the Rg^2-value of the HB polymers formed with $r = \infty$ in a CSTR is still much larger than that for the perfect dendron [9].

For a CSTR, the value of DB_{inf} for a given r is represented graphically in Figure 4. At least approximately for large polymers, n_b is estimated to be $n_b = DB_{inf}P/2$. Therefore, the magnitude of Rg^2 can be estimated from Figure 22, and the value of L_{MS} could also be estimated by using the relationship, $Rg^2 = 0.18L_{MS} + 1$.

Obviously, various non-idealities, notably the size and structural dependent reaction kinetics, may need to be accounted for in a real system. The information concerning the 3D architecture obtained for the present ideal condition could be used as a starting point for the discussion of such non-ideal reaction kinetics, and the present results would provide a basis for the development of more realistic models for the HB polymer formation.

4. Conclusions

The HB polymer architecture formed in a batch and a CSTR is investigated in detail, by using the MC simulation method, proposed earlier [6–9]. In a CSTR, a highly branched core region consisting of units with large residence times is formed to give much more compact architecture, compared with batch polymerization for large polymers. The branched architecture can be controlled by the residence time distribution.

For batch polymerization, the g-ratio, as well as L_{MS}/P, shows a universal relationship with the average number of branches per molecule, which is independent of conversion x_A and reactivity ratio r. The g-ratio follows the relationship given by the Zimm-Stockmayer equation [15], which shows that the random branched structure is formed in batch polymerization.

For a CSTR, the g-ratio, as well as L_{MS}/P, follows a universal relationship with the number of branches in a polymer molecule, and the relationship is independent of ξ, but is a very weak function of r.

It was found that the Rg^2 is linearly correlated with L_{MS}, represented by $Rg^2 = 0.18L_{MS} + 1$, both for a batch and a CSTR, irrespective of the conversion level and reactivity ratio. The coefficient, 0.18 is essentially the same as for an SCVP [26], and could be considered as a general characteristic of HB polymer architecture. The coefficient is 0.167 for linear polymers, and is 0.5 for perfect dendrons. The physical meaning of the coefficient is still not clear, but the value of 0.18 is closer to that for the linear polymers, rather than the perfect dendron that cannot fit in the 3D space because of the Malthusian packing paradox.

The HB polymer architecture can be controlled by the residence time distribution. Appropriate process control in combination with the chemical control of the reactivity of the second B-group will make it possible to produce HB polymers with well-controlled molecular architecture.

Supplementary Materials: The following are available online at http://www.mdpi.com/2227-9717/7/4/220/s1, Figure S1: Expected g-ratio for the HB polymers having n_b branch points in a polymer for a CSTR with (a) $r = 0.5$, (b) $r = 2$, (c) $r = 5$, and (d) $r = \infty$, for various ξ-values, Figure S2: Relationship between L_{MS}/P and n_b for a CSTR; (a) $r = 0.5$, (b) $r = 2$, (c) $r = 5$, and (d) $r = \infty$, with various ξ-values, Figure S3: Universal relationship between Rg^2 and L_{MS} for a CSTR with various combinations of r and ξ. (a) $r = 0.5$, (b) $r = 2$, (c) $r = 5$, and (d) $r = \infty$.

Funding: This research received no external funding.

Acknowledgments: I would like to express my sincere thanks to Professor Masoud Soroush, Drexel University, for kind invitation to this special issue.

Conflicts of Interest: The author declares no conflict of interest.

References

1. Zheng, Y.; Li, S.; Weng, Z.; Gao, C. Hyperbranched polymers: Advances from synthesis to application. *Chem. Soc. Rev.* **2015**, *44*, 4091–4130. [CrossRef] [PubMed]
2. Lederer, A.; Burchard, W. *Hyperbranched Polymers: Macromolecules in between Deterministic Linear Chains and Dendrimer Structure*; The Royal Society of Chemistry: Cambridge, UK, 2015.
3. Levenspiel, O. *Chemical Reaction Engineering*, 2nd ed.; John Wiley & Sons: New York, NY, USA, 1972.
4. Flory, P.J. Molecular size distribution in three dimensional polymers. VI. Branched polymers containing A-R-B$_{f-1}$ type units. *J. Am. Chem. Soc.* **1952**, *74*, 2718–2723. [CrossRef]
5. Segawa, Y.; Higashihara, T.; Ueda, M. Synthesis of hyperbranched polymers with controlled structure. *Polym. Chem.* **2013**, *4*, 1746–1759. [CrossRef]

6. Tobita, H. Universality in branching frequrncies and molecular dimensions during hyperbranched polymer formation: 1. step polymerization of AB_2 type monomer with equal reactivity. *Macromol. Theory Simul.* **2016**, *25*, 116–122. [CrossRef]

7. Tobita, H. Universality in branching frequrncies and molecular dimensions during hyperbranched polymer formation: 2. step polymerization of AB_2 type monomer with different reactivity for the second B group. *Macromol. Theory Simul.* **2016**, *25*, 123–133. [CrossRef]

8. Tobita, H. Hyperbranched polymers formed through irreversible step polymerization of AB_2 type monomer in a continuous flow stirred-tank reactor (CSTR). *Macromol. Theory Simul.* **2017**, *26*, 1600078. [CrossRef]

9. Tobita, H. Hyperbranched polymers formed through irreversible step polymerization of AB_2 type monomer with substitution effect in a continuous flow stirred-tank reactor (CSTR). *Macromol. Theory Simul.* **2017**, *26*, 1700020. [CrossRef]

10. Kryven, I.; Iedema, P.D. Predicting multidimensional distributive properties of hyperbranched polymer resulting from AB_2 polymerization with substitution, cyclization and shielding. *Polymer* **2013**, *54*, 3472–3484. [CrossRef]

11. Jacobson, H.; Stockmayer, W.H. Intramolecular reaction in polycondensations. I. The theory of linear systems. *J. Chem. Phys.* **1950**, *18*, 1600–1606. [CrossRef]

12. Hawker, C.J.; Lee, R.; Fréchet, M.J. One-step synthesis of hyperbranched dendritic polyesters. *J. Am. Chem. Soc.* **1991**, *113*, 4583–4588. [CrossRef]

13. Wiener, H. Structural determination of paraffin boiling points. *J. Am. Chem. Soc.* **1947**, *69*, 17–20. [CrossRef]

14. Nitta, K. A topological approach to statistics and dynamics of chain molecules. *J. Chem. Phys.* **1994**, *101*, 4222–4228. [CrossRef]

15. Zimm, B.H.; Stockmayer, W.H. The dimensions of chain molecules containing branches and rings. *J. Chem. Phys.* **1949**, *17*, 1301–1314. [CrossRef]

16. Versluis, C.; Souren, F.; Nijenhuis, A. Span-length distributions of star-branched polymers. *Macromol. Theory Simul.* **2002**, *11*, 969–974. [CrossRef]

17. Iedema, P.D.; Hoefsloot, H.C.J. Synthesis of branched polymer architectures from molecular weight and branching distributions for radical polymerization with long-chain branching, accounting for topology-controlled random scission. *Macromol. Theory Simul.* **2001**, *10*, 855–869. [CrossRef]

18. Tobita, H. Random sampling technique to predict the molecular weight distribution in nonlinear polymerization. *Macromol. Theory Simul.* **1996**, *5*, 1167–1194. [CrossRef]

19. Tobita, H. Polymerization Processes, 1. Fundamentals. In *Ullmann's Encyclopedia of Industry Chemistry*; Elvers, B., Ed.; Wiley: Weinheim, Germany, 2015. [CrossRef]

20. Buren, B.D.; Zhao, Y.R.; Puskas, J.E.; McAuley, K.B. Predicting average molecular weight and branching level for self-condensing vinyl copolymerization. *Macromol. React. Eng.* **2018**, *12*, 1700074. [CrossRef]

21. Tobita, H. Hyperbranched polymers formed through self-condensing vinyl polymerization in a continuous stirred-tank reactor (CSTR): 1. molecular weight distribution. *Macromol. Theory Simul.* **2018**, *27*, 1800027. [CrossRef]

22. Hölter, D.; Frey, H. Degree of branching in hyperbranched polymers. 2. Enhancement of the DB: Scope and limitation. *Acta Polym.* **1997**, *48*, 298–309. [CrossRef]

23. Hanselmann, R.; Hölter, D.; Frey, H. Hyperbranched polymers prepared via the core-dilution/slow addition technique: Computer simulation of molecular weight distribution and degree of branching. *Macromolecules* **1998**, *31*, 3790–3801. [CrossRef]

24. Cook, A.B.; Barbey, R.; Burns, J.A.; Perrier, S. Hyperbranched polymers with high degrees of branching and low polydispersity values: Pushing the limits of thiol-yne chemistry. *Macromolecules* **2016**, *49*, 1296–1304. [CrossRef]

25. Tobita, H. Model-based reactor design to control hyperbranched polymer architecture. *Macromol. React. Eng.* **2018**, *12*, 1700065. [CrossRef]

26. Tobita, H. Detailed structural analysis of the hyperbranched polymers formed in self-condensing vinyl polymerization. *Macromol. Theory Simul.* **2019**, *28*, 1800061. [CrossRef]

27. Gordon, M.; Ross-Murphy, S.B. The structure and properties of molecular trees and networks. *Pure Appl. Chem.* **1975**, *43*, 1–26. [CrossRef]

Review

Advances in Mathematical Modeling of Gas-Phase Olefin Polymerization

Mohd Farid Atan [1,2], Mohd Azlan Hussain [1,*], Mohammad Reza Abbasi [1], Mohammad Jakir Hossain Khan [1] and Muhamad Fazly Abdul Patah [1]

[1] Department of Chemical Engineering, Faculty of Engineering, University of Malaya, 50603 Kuala Lumpur, Malaysia; amfarid@unimas.my (M.F.A.); m_abbasi@ifac-mail.org (M.R.A.); jakirkhanbd@gmail.com (M.J.H.K.); fazly.abdulpatah@um.edu.my (M.F.A.P.)

[2] Department of Chemical Engineering and Energy Sustainability, Faculty of Engineering, Universiti Malaysia Sarawak, 94300 Kota Samarahan, Malaysia

[*] Correspondence: mohd_azlan@um.edu.my; Tel.: +60-379-675-214

Received: 31 December 2018; Accepted: 26 January 2019; Published: 30 January 2019

Abstract: Mathematical modeling of olefin polymerization processes has advanced significantly, driven by factors such as the need for higher-quality end products and more environmentally-friendly processes. The modeling studies have had a wide scope, from reactant and catalyst characterization and polymer synthesis to model validation with plant data. This article reviews mathematical models developed for olefin polymerization processes. Coordination and free-radical mechanisms occurring in different types of reactors, such as fluidized bed reactor (FBR), horizontal-stirred-bed reactor (HSBR), vertical-stirred-bed reactor (VSBR), and tubular reactor are reviewed. A guideline for the development of mathematical models of gas-phase olefin polymerization processes is presented.

Keywords: modeling; olefin; gas phase; kinetics

---

1. Introduction

Polyolefin or polyalkene, which is one of the most popular thermoplastic polymers, are formed from the monomer of the alkene group, which possesses a double bond. The most common monomers used to produce this polyolefin are ethylene and propylene. The polyolefin formed through the polymerization of ethylene is either in the form of polyethylene (PE), high-density polyethylene (HDPE), or low-density polyethylene (LDPE), and another type of polyolefin can be produced through a similar procedure from the monomer propylene in the form of polypropylene (PP). Meanwhile, copolymerization of these two monomers produces a polyolefin, which is in the form of ethylene-propylene diene monomer (EPDM). These polyolefins are in high demand and widely used in several sectors, such as automobiles, packaging, construction, and textiles. As reported by Szabó (2015) [1], the annual consumption of polyolefin per capita in the European continent is expected to increase significantly from 88 kg per person in 2015 to 120 kg per person in 2030 with an increment of 36 percent. Figure 1 summarizes the total consumption per capita in three parts of the European continent for the years 2015, 2020, and 2030.

As shown in Figure 1, Western European countries such as France, Italy, and the United Kingdom are considered to be the biggest consumers of this polyolefin compared with their neighbors located in central and Eastern Europe. This is due to the localization of several polyolefin industries, such as SO.T.AC SRL and Montello SPA in Italy, Borgeois in France, Vital Parts Ltd. in the United Kingdom, and Warm-On Gmbh, and Co.Kg in Germany. In regard to this high demand for polyolefins, the performance of the current polymerization process needs to be reviewed and improved, starting from the selection of monomers and catalyst up to the production of end products.

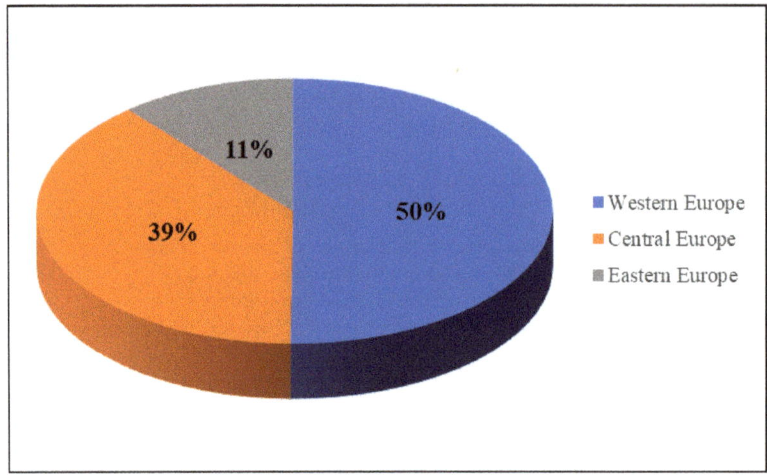

Figure 1. Total consumption per capita in European continent [1].

Much effort has been made by previous researchers to improve this olefin polymerization in the gas phase by reviewing different aspects, such as the thermodynamic properties, the operational conditions, the chemical processing, the reaction mechanism, the catalyst used in the polymerization reaction, and the properties of the end products. Table 1 summarizes the reviews that have been carried out previously on different aspects of gas-phase olefin polymerization.

Table 1. Past reviews analysis.

No.	Review Area	Review Aspect	Ref.
1.	Thermodynamic Properties	Different methods to determine enthalpy and entropy	[2]
2.	Process Design	Design criteria, process condition, protection of instruments against overpressure, instruments for heating up and cooling down, and different types of stirrers for polymerization reactors such as autoclave reactor, high-pressure autoclave reactor, tubular reactor, fluidized bed reactor to improve the process efficiency.	[3–6]
3.	Process Routes	The implementation of the solvent polymerization process, solvent polymerization without deashing, the bulk polymerization process without solvent, the vapor phase polymerization process without deashing and atactic polymer, Unipol I and II, Innovene G, Spherilene S & C, and Borstars in producing the polyolefin.	[5,7]
4.	Olefin Synthesis	The implementation of free-radical methodology, carbine and nitrene methodology, and transition metal C–H bond activation methodology to synthesize the polyolefin.	[8–11]
5.	Catalyst	The utilization of a metallocene catalyst system, Ziegler–Natta catalyst system, Fujita group Invented (FI) catalyst system, and oxide-supported surface organometallic complexes in olefin polymerization synthesis.	[6,12–31]
6.	Process Modeling	The implementation of mathematical models, namely macroscale modeling, mesoscale modeling, microscale modeling, single particle modeling, computational fluids dynamic modeling, microelements modeling, 2D finite element modeling, single pore modeling, and parti-level fragmentation modeling to determine the properties of the polyolefin, and the mass and heat transfer phenomena during the polymerization process.	[8,11,12,22,31–34]
7.	Quality Control	Different types of analysis such as nuclear magnetic resonance (NMR), temperature rising elution fractionation (TREF), gel permeation chromatography (GPC), rheological characterization (zero shear viscosity, zero shear viscosity, shear thinning behavior, dynamic modulus, loss angle, Van-Gurp-Palmen plot, Cole-Cole plot, activation energy, thermorheological complexity, strain-hardening effect, relaxation time, damping function, nonlinear dynamical oscillatory shear, and long-chain branching index), dynamic mechanical analysis, differential scanning calorimeter, neutron scattering, and molecular topology fractionation.	[12,13,22,35]

Table 1. *Cont.*

No.	Review Area	Review Aspect	Ref.
8.	Polyolefin Demand	The analysis of global production and consumption of polypropylene from 1985 until 2000 in the textile industry.	[36]
9.	Physical and Chemical Properties of the Polyolefin	The influence of process conditions on the thermal properties, specific heat capacity, melting point, relative thermal conductivity, density, thermal diffusivity, crystallinity, amorphous phase properties, coefficient of linear thermal expansion, electrical properties, foam structure, shear, and rheological properties.	[37–39]

From the past reviews summarized in Table 1, none of the above review studies reviewed and discussed kinetic modeling together with mass and energy balance modeling in the gas phase. The review article published by Abbasi et al. (2018) [34] only focused on a fluidized bed reactor and did not cover other types of reactors such as tubular reactor or stirred bed reactor. In addition, none of the above review studies proposed simple or proper guidelines to implement and simulate the mathematical model for this olefin polymerization process. Thus, the objective of this review is to review and discuss the past and current development of the mathematical modeling, together with the reaction mechanism of the olefin polymerization in the gas phase. This review will be concluded by proposing general guidelines in implementing and simulating the mathematical model for this olefin polymerization in the gas phase.

2. Mathematical Model Development for Olefin Polymerization

Theoretically, the olefin polymerization reaction occurs when the monomers, which possess reactive functional groups with double or triple bonds, are reacted in the presence of a catalyst under certain conditions of pressure and temperature. The core product formed from this reaction is called a polymer, which possesses a certain of chain length. This produced polymer can be further classified into three main chemical structures, namely, linear chain polymer, branched chain polymer, and network chain polymer. The first is also known as a thermoplastic polymer, which is formed from a monomer that possesses repeat units held by strong covalent bonds. The second is formed by a monomer, which contains the molecules in the form of a linear backbone with branches emanating randomly. Lastly, the third is also known as a gel polymer, which is formed by the extension of a branched chain polymer, which is put under reaction with high conversion [40].

Moreover, with regard to the reactor used for this polymerization process, there are two types of reactors that can be used for this olefin polymerization process, namely homogeneous and heterogeneous reactors. In the former type, the olefin polymerization occurs only in one phase and the reaction can be carried out either in a continuous stirred-tank reactor, in a loop reactor, in a hollow shaft reactor, or in a batch reactor. For the latter type, the reaction occurs in several phases such as emulsion, bubble, cloud, and solid. In this type of reaction, the analysis needs to be in each of the phases to ensure that the predicted data are coherent with the real data plant and it is highly recommended for the gas-phase olefin polymerization process. In addition, this reaction can take place either in a fluidized bed reactor or in a tubular reactor.

Furthermore, in terms of the reaction mechanism, the olefin polymerization is classified under the chain-growth polymerization where the growth of the polyolefin chain "n" occurs when the monomer is reacted with the end of the reactive functional group of the growing polyolefin under certain conditions of temperature (T) and pressure (P). It can be illustrated by using the following reaction mechanism:

$$\text{Polyolefin}(n) + \text{Monomer} \xrightarrow[\text{Catalyst}]{(T,P)} \text{Polyolefin}(n+1)$$

Then, this chain-growth mechanism can be further sub-classified into two different categories, namely coordination and free-radical mechanisms. The former mechanism requires the use of a coordination catalyst such as Phillips catalyst, Ziegler–Natta catalyst, and metallocene-based catalyst during the activation of active sites. Basically, this active site is made of the metal atom (Me) and the ligands (Li), which are covalently bonded. The growth of the polyolefin chain length occurs when this active site reacts with the monomer under certain conditions of temperature and pressure [41,42]. The growth mechanism of this polyolefin can be illustrated by the following reaction mechanism, where M is the monomer:

$$\text{Me} - \text{Li} \overset{(T,P)}{\underset{M}{\rightarrow}} \text{Me} - \text{M} - \text{Li} \overset{(T,P)}{\underset{M}{\rightarrow}} \ldots \overset{(T,P)}{\underset{M}{\rightarrow}} \text{Me} - (\text{M})_n - \text{Li}$$

In contrast, the latter type of mechanism requires the presence of initiators (I) such as oxygen and organic peroxides that can be easily decomposed to form radicals. Later, these radicals will react with the monomer to grow the chain length under certain conditions of temperature and pressure [10,43]. The following free-radical mechanism is used to illustrate the growth of this olefin polymerization process.

$$\text{I} \rightarrow 2\text{R}^\bullet \overset{(T,P)}{\underset{M}{\rightarrow}} \text{M} - \text{R}^\bullet \overset{(T,P)}{\underset{M}{\rightarrow}} \ldots \overset{(T,P)}{\underset{M}{\rightarrow}} (\text{M})_n - \text{R}^\bullet$$

The mathematical model is considered essential in predicting the output of the chemical process, in particular, the olefin polymerization process. It is written in the form of a series of algebraic and differential equations and it comprises three major elements, namely hydrodynamic, kinetic, and transport phenomena. The first phenomenon describes the process outputs such as production rate, temperature, and dynamic poly-dispersity. It allows for observing any changes or transition during the polymerization process. Then, the second phenomenon evaluates the rate or the speed of the reaction. It allows for studying the influence of input parameters or variables such as feed flow rate, type of catalyst, catalyst flow rate, pressure, inlet temperature in increasing the rate of the reaction. The last phenomenon describes the transport of momentum, energy, and chemical species via several media such as liquid, gas, and solid. The first type of transport is called fluid dynamics. The second and third types of transport are called heat and mass transfer, respectively [44].

However, for the olefin polymerization process, the modeling framework is relatively complex due to the high nonlinearity of its process dynamics caused by several factors. The first is due to the complexity of the reaction mechanisms. Theoretically, the olefin polymerization is made up of several stages, such as activation, initiation, propagation, chain transfer reaction, and deactivation. The second is caused by the complexity of the heat transfer phenomena, especially in the gas phase where it theoretically comprises several phases such as cloud, emulsion, bubble, and solid [45]. The third is caused by the physical properties of the flow behavior in both phases; solid and gas. Lastly, the fourth factor is caused by the influence of the reactor used and its operational conditions, such as the volumetric flow rate, the pressure, the reactor temperature, and the size of the catalyst particles on the final properties of the polyolefin. In addition, these final properties are measured by determining the molecular weight distribution, copolymer distribution, sequence length distribution, long chain branching distribution, short chain branching distribution, stereoregularity, and morphology properties, for instance, particle size distribution, pore size distribution, bulk density, and melt index (MI), which are among the most highlighted physicochemical properties, which have been measured frequently. To predict these physicochemical properties by using the mathematical model, there are two available methods, namely population balance modeling and method of moments that can be employed due to their ability in solving a wide range of a complex dynamic polymerization processes in the liquid and gas phases, particularly in characterizing the population of the particles as well as the growth of the polyolefin chain length [46,47].

2.1. Advances in the Polymerization of Olefins in the Gas Phase

Polymerization of olefins in the gas phase is considered as one of the major routes in producing polyolefins such as polypropylene, polyethylene, low-density polyethylene, ethylene-propylene copolymer, and ethylene-1-butene copolymer. A lot of effort has been made previously to model and simulate this polymerization process. Table 2 summarizes the mathematical models that have been implemented previously.

Table 2. Overview of implemented mathematical models.

Model Type/Polyolefin	Process Condition	Model Assumption (s)	Ref.
Two-Phase Model/Ethylene-propylene copolymer	1. Z–N catalyst 2. Fluidized Bed Reactor (FBR) 3. Coordination 4. T = [320, 500] K	1. Emulsion does not at minimum fluidization 2. Well-mixed condition 3. The reaction occurs in the emulsion and bubble phases 4. Bubble and particle size are constant 5. Heat and mass transfer resistance is neglected 6. Solid elutriation is considered	[48]
Two-phase Model/Ethylene-1-butene copolymer	1. Z–N catalyst 2. FBR 3. Coordination 4. T = [310, 317] K 5. P = [1.99, 2] MPa	1. Emulsion does not at minimum fluidization 2. Well-mixed condition 3. The reaction occurs in the emulsion and bubble phases 4. The bubble is in a spherical form with a constant diameter 5. Plug flow condition with constant velocity 6. Heat and mass transfer resistance is neglected 7. Gradient temperature and concentration are neglected 8. Particle size distribution is uniform 9. Solid elutriation is considered	[49]
Two-Phase Model/Polypropylene	1. Z–N catalyst 2. FBR 3. Coordination 4. T = [343, 353] K 5. P = [2, 3] MPa	1. Eulerian–Eulerian approach 2. Immediate consumption of the propylene after injection the catalyst 3. The existence of the interaction between mass, momentum, and energy in emulsion and bubble phases 4. No lifts effects and virtual mass	[50]
Two-Phase Model/Ethylene-propylene copolymer	1. Z–N catalyst 2. FBR 3. Coordination 4. T = 353.15 K 5. P = 2.5 MPa	1. Emulsion does not at minimum fluidization 2. Well-mixed condition 3. The reaction occurs in the emulsion and bubble phases 4. Bubbles are in a spherical form with constant diameters 5. Plug flow condition with constant velocity 6. Heat and mass transfer resistance is neglected 7. Gradient temperature and concentration are neglected 8. Particle size distribution is uniform 9. Solid elutriation is neglected	[51]
Single-phase model/Polyethylene	1. Organic peroxides and oxygen 2. Tubular reactor 3. Free radical 4. T = [403, 574] K 5. P = [152, 304] MPa	1. Formation of a single supercritical phase 2. Plug flow condition 3. Quasi-steady-state assumption 4. The models depend on the ratios of kinetic rate constants 5. Fouling resistances at each tubular zone are uniform 6. No heat transfer model 7. The efficiencies of all initiators are similar	[52–54]

Table 2. *Cont.*

Model Type/Polyolefin	Process Condition		Model Assumption (s)		Ref.
Two-Phase Model/Polypropylene	1. Z–N catalyst 2. FBR 3. Coordination 4. T = [342, 354] K 5. P = [2.2, 2.5] MPa		1. Emulsion does not at minimum fluidization 2. Well-mixed condition 3. The reaction occurs in the emulsion and bubble phases 4. Bubbles are in a spherical form with constant diameters 5. Plug flow condition with constant velocity 6. Heat and mass transfer resistance is neglected 7. Gradient temperature and concentration are neglected 8. Particle size distribution is uniform 9. Solid elutriation is neglected		[55–59]
Single-phase Model/Low-density of Polyethylene	1. Organic peroxides and oxygen 2. Tubular reactor 3. Free radical 4. T = [400, 600] K 5. P = [200, 300] MPa		1. The water temperature in the cooling jacket is constant 2. The grade transition is influenced by the small fraction of the reaction time and is not influenced by the wall heat capacity 3. Steady-state assumption		[60]
Single-phase Model/Low-density of Polyethylene	1. Organic peroxides and oxygen 2. Tubular reactor 3. Free radical 4. T = [400, 600] K 5. P = [200, 300] MPa		1. Plug flow condition and single supercritical phase 2. Molecular weight distribution is in the form of a log-normal distribution shape 3. Jacket temperature and pressure in each section of the tubular reactor are constant 4. The process is isothermal and occurs below the gel point		[61]
Single-phase model/Polypropylene	1. Z–N catalyst 2. FBR 3. Coordination 4. T = 353.15 K 5. P = 2.5 MPa		1. The emulsion is not set at the minimum fluidization 2. Heat and mass transfer resistance is neglected 3. Dynamic monomer internal energy is neglected 4. Pseudo-homogeneous single phase		[62]
Two-Phase Model/Ethylene-1-butene copolymer	1. Z–N catalyst 2. FBR 3. Coordination 4. T = [345, 374] K		1. The reactor comprises four continuous stirred-tank reactors (emulsion phase) and four plug flow reactors (bubble phase) 2. Particles are in a spherical form with a constant dimension 3. The density function of the particles in the outlet stream and in the bed are similar		[63]
Three-Phase Model/Ethylene-1-butene copolymer	1. Z–N catalyst 2. FBR 3. Coordination 4. T = 300 K		1. The reaction occurs in the emulsion and solid phases 2. The emulsion is set at the minimum fluidization and the excess gas to maintain this condition was considered as the bubble phase 3. No gradient of temperature and concentration 4. The existence of resistance to mass transfer between the emulsion and solid phase 5. Rigid and porous catalyst represent the dynamic reaction 6. The mass transfer of emulsion molecules occurs at the surface of the solid catalyst particles		[45,64]
Single-phase Model/Low-density of Polyethylene	1. Organic peroxides and oxygen 2. Tubular reactor 3. Free radical 4. T = [323, 604] K 5. P = [130, 300] MPa		1. Plug flow condition and the supercritical reaction mixture 2. Existence of changes in the physical and transport properties with the axial distance 3. The jacket temperature at each of the reaction zones is constant 4. No pressure pulse 5. The mixture of organic peroxide and transfer agents is considered as one fictitious species 6. The polymer is well separated from other output components of the reactor		[65–67]

Table 2. *Cont.*

Model Type/Polyolefin	Process Condition		Model Assumption (s)		Ref.
Single-phase Model/Ethylene-1-butene Copolymer	1. 2. 3. 4.	Z–N catalyst FBR Coordination T = 345.15 K	1. 2. 3.	One serial of the continuous-stirred-tank reactor (CSTR) model The particles are in a spherical form with constant dimensions The attrition term in the population balance equation is constant	[68]
Single-phase Model/Low-density of Polyethylene	1. 2. 3. 4. 5.	Organic peroxides and oxygen Tubular reactor Free radical T = [423, 574] K P = [101, 355] MPa	1. 2. 3.	The flow regime is situated in the turbulent regime At the point of injection, the pressure of the lateral feed streams and the main reaction mixture are similar No formation of diradical and chain transfer to telogen	[69]
Two-Phase Model/ Ethylene-1-butene copolymer	1. 2. 3. 4.	Chromium oxide catalyst FBR Coordination T = 375 K	1. 2. 3. 4. 5. 6. 7. 8. 9. 10.	Plug flow condition and quasi-steady-state assumption Emulsion phase is homogeneous Crystallinity and swelling are constant The adsorbed gas phase in solid and emulsion phase is in equilibrium The reactor bed porosity and bed porosity are identical at the minimum fluidization Porosity condition for solid discharge is uniform No heat loss through the fluidized bed wall The recycle gas temperatures at the heat exchanger entrance and exit, at the reactor exit, and at the compressor exit are equal Complete fluid back-mixing condition is considered at each of the small sub-sections in the heat exchanger The fluid in each of the sub-sections of the heat exchanger is constant	[70]
Single-phase Model/High-density of Polyethylene	1. 2. 3. 4. 5.	Organic peroxides and oxygen Extruder Free radical T = 443.15 K P = 0.01 MPa	1. 2. 3. 4. 5.	Plug flow condition and the isothermal condition One site kinetic reaction Termination by combination is the only alternative to terminate the polymerization Quasi-state approximation for radicals γ radicals are formed from the decomposition of organic peroxides via the first-order reaction	[71]
Single-phase Model/Low-density of Polyethylene	1. 2. 3. 4. 5.	Organic peroxides and oxygen Tubular reactor Free radical T = [325, 625] K P = [150, 250] MPa	1. 2. 3.	No axial mixing and temperature and concentration gradients in both reactor and jackets The mixture is homogeneous and acts as a supercritical fluid The temperature at each of the jacket zones is uniform	[72]
Two-Phase Model/ Ethylene-1-butene copolymer	1. 2. 3. 4. 5.	Z–N catalyst FBR Coordination T = [310, 317] K P = [1.99, 2] MPa	1. 2. 3.	No radial concentration and temperature gradients and no solid entrainment The injection of the catalyst into the reactor as prepolymer and the mean particle size is uniform The downward direction is considered for the overall flow direction of the polymer	[73]
Two-Phase Model/Polyethylene	1. 2. 3. 4. 5.	Z–N catalyst FBR Coordination T = [298, 701] K P = 0.1 MPa	1. 2. 3.	Simple and dynamic two-phase model and generalized bubbling–turbulent model No variation of bubble diameter and temperature No radial concentration gradient	[74]

Table 2. *Cont.*

Model Type/Polyolefin	Process Condition	Model Assumption (s)	Ref.
Single-phase Model/Polypropylene	1. Z–N catalyst 2. FBR 3. Coordination 4. T = 343.15 K 5. 197.	1. The reaction occurs only in the emulsion phase 2. The emulsion phase is at the minimum fluidization 3. Pseudo-state assumption 4. The catalyst is injected uniformly at each of the emulsion cells and the collection of the products is done at the bottom part of the cells 5. The bubble phase is formed by the excess of the fluidization gas 6. The number of bubble cells over the number of emulsion cells is an integer 7. No mass and heat transfer resistance 8. Solid elutriation is recycled back into the emulsion cells	[75]
Single-phase Model/Polypropylene	1. Z–N catalyst 2. Stirred-bed reactor 3. Coordination 4. T = [330, 351] K 5. P = [1.9, 2.3] MPa	1. Four CSTRs in series approximation 2. All the kinetic sites produce their molecular weight distribution 3. The catalyst possesses a single site scheme to predict the number average molecular weight in the first step and multiple site schemes in the second step	[76]
Two-Phase Model/ Ethylene-1-butene copolymer	1. Z–N catalyst 2. FBR 3. Coordination 4. T = 345.15 K 5. P = 2 MPa	1. Pseudo-homogeneous state 2. The emulsion phase does not remain at the minimum fluidization 3. No mass and heat transfer resistance between solid and emulsion gas 4. No mass and energy transfer resistance between emulsion and bubble phase 5. The reaction is at an isothermal condition 6. No radial temperature gradient 7. No solid entrainment 8. The catalyst is injected continuously 9. Mean size of the particle is uniform	[77]
Single-phase Model/Polypropylene	1. Organic peroxides and oxygen 2. Single screw extruder 3. Free radical 4. T = 480.15 K	1. The isothermal condition 2. One active kinetic site 3. Plug flow condition	[78]
Single-phase Model/High-density Polyethylene	1. Organic peroxides and oxygen 2. Extruder 3. Free radical 4. T = 443.15 K 5. P = 0.01 MPa	1. Only molecules with a vinyl group is present in the untreated polymer 2. The kinetic single site 3. Termination only by combination	[79]
Single-phase Model/Ethylene-1-butene Copolymer	1. Z–N catalyst 2. FBR 3. Coordination 4. T = 360 K 5. P = 2.1 MPa	1. Single-phase CSTR approximation 2. Perfect mixing in the beds 3. No radial temperature and concentration gradient 4. No heat and mass transfer between the solid and gas phase 5. Two-site kinetics scheme	[80]
Single-phase Model/Low-density of Polyethylene	1. Organic peroxides and oxygen 2. Tubular reactor 3. Free radical 4. T = 358.15 K 5. P = 220 MPa	1. Quasi-stationary state approximation 2. Equipment holdups are considered uniform 3. Plug flow condition	[81]

Table 2. *Cont.*

Model Type/Polyolefin	Process Condition	Model Assumption (s)	Ref.
Single-phase Model/Polypropylene	1. Metallocene based catalyst 2. FBR 3. Coordination 4. T = 345.15 K 5. P = 2 MPa	1. The reactor was divided into three main sections called annulus, draft tube and cone 2. No resistance mass and heat transfer between solid and gas phase 3. No energy generated from absorption and desorption of propylene 4. No gradient of velocities for solid and gas 5. The annulus is set at minimum fluidization 6. No heat transfer issued from the annulus to the draft tube 7. Wall and reactor temperature are equal	[82]
Single-phase Model/High-density Polyethylene	1. Organic peroxides and oxygen 2. Extruder 3. Free radical 4. T = 443.15 K 5. P = 0.01 MPa	1. The isothermal condition 2. Quasi-steady-state approximation 3. The single site kinetic scheme 4. No diffusion phenomena	[83]
Single-phase Model/Low-density Polyethylene	1. Organic peroxides and oxygen 2. Tubular reactor 3. Free radical 4. T = 349.15 K 5. P = 228 MPa	1. Quasi-steady-state approximation 2. Feedstock flow rates are constant	[84]
Single-phase Model/Low-density Polyethylene	1. Organic peroxides and oxygen 2. Tubular reactor 3. Free radical 4. T = [323, 604] K 5. P = [130, 300] MPa	1. Decomposition of oxygen into the radicals formed from the initiation is only allowed for the grouping of the activation energy with pre-exponential factor 2. The molecular properties are not considered in the modification of the initial parameter 3. Jacket temperature and pressure are different at each of the jacket zones 4. Plug flow and supercritical reaction mixture condition 5. Radial variation for the physical and transport properties 6. No pressure pulse	[85,86]
Single-phase Model/Low-density Polyethylene and Ethylene-vinyl	1. Organic peroxides and oxygen 2. Tubular reactor 3. Free radical 4. T = [423, 574] K 5. P = [98, 196] MPa	1. The kinetic mechanism is assumed valid for both type of polymer 2. No long-chain branched polymer	[87]
Single-phase Model/Polyethylene	1. Z–N catalyst 2. FBR 3. Coordination 4. T = 300 K	1. Well-mixed and quasi-steady-state approximation 2. The reaction occurs only in the dense phase 3. The removal flow rate is manipulated to ensure the consistency of the bed height 4. The catalyst was injected continuously 5. No influence of inert gas, co-monomer, and hydrogen	[88,89]
Two-Phase Model/Ethylene-1-butene copolymer	1. Z–N catalyst 2. FBR 3. Coordination 4. T = 353.15 K	1. The movement of the emulsion phase follows the plug flow regime 2. Particles dimensions vary 3. Two-site or multiple kinetic schemes 4. No radial gradient of concentration and temperature 5. No heat and mass transfer resistance between the solid and emulsion phases 6. No solid elutriations	[90–92]

Table 2. *Cont.*

Model Type/Polyolefin	Process Condition	Model Assumption (s)	Ref.
Single-phase Model/Ethylene-1-butene Copolymer	1. Z–N catalyst 2. FBR 3. Coordination 4. $T = 353.15$ K	1. Bubble-growth model 2. The emulsion phase is well back-mixed 3. The bubble phase comprises N well-mixed compartments in series 4. No heat and mass transfer resistance between the solid and emulsion phase 5. No reaction occurs in the bubble phase 6. No monomer mass transfer from bubble to emulsion phase 7. The emulsion phase is set at the minimum fluidization 8. Particle size varies	[93]
Single-phase Model/Low-density Polyethylene	1. Organic peroxides and oxygen 2. Tubular reactor 3. Free radical 4. $T = [423, 604]$ K 5. $P = [122, 355]$ MPa	1. No mass accumulation in each volume segments 2. No heat transfer due to initiation, termination and transfer reactions 3. No presence of gel effect	[94]
Single-phase Model/Polyethylene	1. Z–N catalyst 2. FBR 3. Coordination 4. $T = 353.15$ K	1. Well-mixed condition 2. Particle size distribution does not influence the production rate and it is discontinuous 3. The particles are unequally distributed throughout the beds 4. The agglomeration rate is influenced by the operating condition and a function of colliding particle size 5. Particles are in the form of a spherical shape with constant density and no limitation in term of inter- or intraparticle heat and mass transfer	[95]
Single-phase Model/Polyolefin	1. Z–N catalyst 2. FBR 3. Coordination 4. $T = 360$ K	1. The mixture is nonideal 2. No external films resistances 3. CSTR approximation 4. The deactivation has an influence on the rate constant 5. No elutriation and no particle breakage 6. The steady-state assumption	[96]
Single-phase Model/Low-density Polyethylene	1. Organic peroxides and oxygen 2. Tubular reactor 3. Free radical 4. $T = [323, 599]$ K 5. $P = [182, 284]$ MPa	1. Plug flow approximation 2. The generation of high-temperature peroxide 3. The utilization of water or steam as heating fluid in jackets 4. Jacket temperature and pressure are not constant 5. The presence of the friction and pressure pulse which cause the pressure drop 6. The reactivities of the telogen and monomeric radicals are equal 7. Quasi-steady-state approximation	[97]
Single-phase Model/Polyolefin	1. Z–N catalyst 2. FBR, horizontal-stirred-bed reactor (HSBR), vertical-stirred-bed reactor (VSBR) 3. Coordination 4. $T = 343.15$ K 5. $P = 2.5$ MPa	1. Total activation of catalyst since $t = 0$ 2. The rate of initiation and propagation are similar, and higher than the rate of chain transfer 3. The only transformation is from site 1 to 2 4. Quasi-steady-state assumption 5. The occupied site is dominant	[98]
Single-phase Model/Polyethylene	1. Z–N catalyst 2. FBR 3. Coordination 4. $T = 273.15$ K 5. $P = 2.07$ MPa	1. The reactor comprises of bubble and emulsion and the reaction occurs only in the emulsion phase 2. The emulsion phase is set at the minimum fluidization 3. Bubble dimension is constant 4. The emulsion phase is back-mixed 5. No radial concentration and temperature gradients 6. No heat and mass transfer resistance between solid and gas in the emulsion phase 7. No variation in terms of the size of the particles 8. No agglomeration and elutriation of the particles 9. The steady-state approximation	[99]

Table 2. *Cont.*

Model Type/Polyolefin	Process Condition		Model Assumption (s)		Ref.
Single-phase Model/Ethylene-1-butene Copolymer	1. 2. 3. 4. 5.	Z–N catalyst FBR Coordination T = 353 K P = 3.55 MPa	1. 2. 3. 4. 5. 6.	The size of the formed transition metal crystallites are equal Catalyst particle is in a spherical form with a constant dimension The multigrain solid core model The polymer molecular weight is only influenced by the chain transfer reaction to hydrogen under isothermal conditions Implementation of first-order deactivation kinetics for the site deactivation reaction No intraparticle mass transfer resistance for monomers	[100]
Single-phase Model/Low-density Polyethylene	1. 2. 3. 4. 5.	Organic peroxides and oxygen Tubular reactor Free radical T = [300, 617] K P = 294 MPa	1. 2. 3. 4. 5.	Plug flow conditions and quasi-steady-state approximation No variation in velocity, temperature, pressure, physical properties, and initiator efficiency No axial mixing No influence of viscosity on the rate constant No heat of reaction issued from initiation, termination, and chain transfer reaction	[101]
Two-Phase Model/Ethylene-1-butene Copolymer	1. 2. 3. 4. 5.	Z–N catalyst FBR Coordination T = 273.15 K P = 2.07 MPa	1. 2. 3. 4. 5. 6.	The well-mixed condition Amorphous and gas phases are at equilibrium No plasticizing effect of dissolved monomer Terminal monomer or chain do not have any influence on the rate of deactivation No radial or vertical temperature gradient The molecular weight of ethylene and 1-butene are equal	[102]
Single-phase Model/Polypropylene	1. 2. 3. 4. 5.	Z–N catalyst FBR Coordination T = [345, 347] K P = [2.03, 3.55] MPa	1. 2. 3. 4. 5. 6.	The reactor comprises of slide-free gas phase and the ideal gas law Perfect back-mixing of gas and solid Continuous injecting of the catalyst The absence of net accumulation of the monomer No amount of gas in the solid The reactor is adiabatic	[103]
Single-phase Model/Low-density Polyethylene	1. 2. 3. 4. 5.	Organic peroxides and oxygen Tubular reactor Free radical T = [333, 403] K P = [193, 253] MPa	1. 2. 3. 4.	Plug flow condition Supercritical single phase is formed by the reaction mixture Free radicals are not in steady state The gradient of physical properties assumed to be in the axial direction	[104]
Single-phase Model/Polyethylene and Polypropylene	1. 2. 3. 4.	Z–N catalyst FBR Coordination T = 300 K	1. 2. 3. 4. 5.	Emulsion or dense phase was perfectly back-mixed The bubbles have a constant spherical dimension Quasi-steady-state approximation The mass and heat transfer rate between the bubble and emulsion phase is constant No mass and heat transfer resistance between the solid and emulsion phase	[105]

As mentioned in Table 2, the olefin polymerization occurs via two types of mechanism, namely coordination and free-radical mechanisms. The polyolefins formed via the former mechanism in several types of reactors, namely fluidized bed reactor, vertical-stirred reactor, and horizontal-stirred-bed reactor are polypropylene, polyethylene, ethylene-propylene copolymer, and ethylene-1-butene copolymer. Meanwhile, the polyolefins formed via the latter mechanism in several types of reactors such as a tubular reactor, extruder, and autoclave reactor are low-density polyethylene, high-density polyethylene, and polypropylene. Moreover, for this type of olefin polymerization process, the reactions can be considered to occur in several phases, such as emulsion, bubble, and cloud, which differs from the olefin polymerization process in the liquid phase where the reaction can only be considered to occur in a single phase, which is the liquid or slurry phase. Furthermore, reactions that occur via a coordination mechanism requires organometallic-type catalysts such as Ziegler–Natta catalyst, metallocene-based catalyst, or chromium oxide-based catalyst to create and activate the active sites where the olefin reaction occurs. Meanwhile, for the reaction to occur via the free-radical mechanism, it requires the presence of organic peroxide and oxygen as an initiator to create the radicals, which play a role to initiate the growth of the polymer chain.

The details of the reaction mechanisms for the coordination mechanism by using the organometallic catalyst system are tabulated in Table 3.

Table 3. Mechanism of the reaction for the coordination mechanism by using the organometallic catalyst.

Reaction Mechanism	Ref.
Activation of Active Sites $P^*(j) + cCata \xrightarrow{k_{act}(j)} P(0,j)$	[45,48–51,55–59,62,64,70,75,76,80,92,93,95,98,102]
Spontaneous Site Activation $P^* \xrightarrow{k_{actS}(j)} P(0,j)$	[70]
Site Activation by Hydrogen $P^*(j) + H_2 \xrightarrow{k_{actSH_2}(j)} P(0,j)$	[70]
Initiation of Active Sites $P(0,j) + M \xrightarrow{k_{in}(j)} P(1,j)$	[45,48,49,51,55–59,62–64,68,70,73,76,77,80,82,92,93,95,98,100,102]
Propagation $P(n,j) + M \xrightarrow{k_{prop}(j)} P(n+1,j)$	[45,48–51,55–59,62–64,68,70,73,75–77,80,82,88–93,95,96,98–100,102,103,105]
Site Transformation $P(n,j) \xrightarrow{k_{Transf}(j \to k)} P(n,k)$	[75,98]
Chain Transfer to Monomer $P(n,j) + M \xrightarrow{k_{trM}(j)} P(1,j) + P_d(n,j)$	[45,48–51,55–59,62–64,68,70,73,75–77,80,88,89,91,92,95,102]
Chain Transfer to Hydrogen $P(n,j) + H_2 \xrightarrow{k_{fH_2}(j)} P_H(0,j) + P_d(n,j)$ $P_H(0,j) + M \xrightarrow{k_{H_2}(j)} P(1,j)$ $P_H(0,j) + cCata \xrightarrow{k_{H_2C}(j)} P(1,j)$	[45,48,49,51,55–59,62–64,68,70,73,75–77,80,88,89,91,92,95,98,100,102]
Chain Transfer to Co-Catalyst $P(n,j) + cCata \xrightarrow{k_{trCo}(j)} P(1,j) + P_d(n,j)$	[48–51,55–59,62,63,68,73,75–77,88,89,92,102]
Spontaneous Transfer $P(n,j) \xrightarrow{k_{trs}(j)} P_H(0,j) + P_d(n,j)$	[48–51,55–59,62,63,68,70,73,75–77,80,88,89,91,92,95,102]
Deactivation Reaction $P(n,j) \xrightarrow{k_{deac}(j)} P_{deac}(0,j) + P_d(n,j)$ $P(0,j) \xrightarrow{k_{deac}(j)} P_{deac}(0,j)$ $P_H(0,j) \xrightarrow{k_{deac}(j)} P_{deac}(0,j)$	[48–51,55–59,62,70,73,75–77,80,82,93,95,96,98,100,102]
Site Deactivation by Hydrogen $P(n,j) + H_2 \xrightarrow{k_{deacH}(j)} P_{deac}(0,j) + P_d(n,j)$ $P(0,j) + H_2 \xrightarrow{k_{deacH}(j)} P_{deac}(0,j)$	[70]
Site Deactivation by Oxygen $P(n,j) + O_2 \xrightarrow{k_{deacO}(j)} P_{deac}(0,j) + P_d(n,j)$ $P(0,j) + O_2 \xrightarrow{k_{deacO}(j)} P_{deac}(0,j)$	[70]
Oxygen Elimination by Alkyl Aluminum $AL + O_2 \xrightarrow{k_{eO}(j)} SP$	[70]
Reaction with Poisons $P(n,j) + Po \xrightarrow{k_{dP}(j)} P_{dPo}(0,j) + P_d(n,j)$ $P_H(0,j) + Po \xrightarrow{k_{dP}(j)} P_{dPo}(0,j)$ $P(0,j) + Po \xrightarrow{k_{dP}(j)} P_{dPo}(0,j)$	[48–51,55–59,62,95,96,102]

As mentioned in Table 3, the number of stages in the reaction mechanism for this olefin polymerization process in the gas phase via the coordination mechanism by using the organometallic catalyst, namely Ziegler–Natta catalyst, metallocene-based catalyst, or chromium oxide-based catalyst implemented previously differ from one study to another study. Most of the studies considered that the polymerization reaction commences by activating the active site. At this level, the potential active sites are activated by co-catalyst to create a vacant site for the insertion of the monomer during the initiation. In addition, Salau et al. (2008) [70] proposed two additional stages, namely, spontaneous site activation and site activation by hydrogen, which leads toward the same purpose of activating the potential active sites. After creating these vacant sites, most of the studies implemented the initiation stage, where the insertion of monomer in the vacant sites commences. At this stage, the chain length of the polyolefins starts to expand. Later, the growth of this polyolefin chain length continues during the propagation stage. After that, the growing polyolefins undergo several types of chain transfer reaction such as spontaneous transfer, chain transfer to monomer, chain transfer to hydrogen, and chain transfer to co-catalyst with the aim of controlling the molecular weight and the chain length of the polyolefin. In addition, Harshe et al. (2004) [75] and Zacca et al. (1996) [98] proposed a supplementary stage called site transformation reaction, which plays the same role as the chain transfer reactions. However, the vacant site for the insertion of the molecules is altered from the active site j to active site k located in the catalyst. Finally, to terminate the growing of the polyolefin chain length with the main aim to form the dead polyolefin, the active site or the catalyst is deactivated spontaneously or by using hydrogen, oxygen, co-catalyst, and impurities. During this catalyst deactivation, the catalytic activity and selectivity continue to decrease [27].

The mechanisms of the reaction for the olefin polymerization via free radical mechanism by using the organic peroxide and oxygen as the initiator are tabulated in Table 4.

Table 4. Mechanism of the reaction for the free radical mechanism by using the organic peroxide and oxygen.

Reaction Mechanism	Ref.
Initiator Decomposition/Peroxide Initiation $In \xrightarrow{k_{idec}} 2R^\bullet$	[52–54,60,61,65–67,69,72,78,81,84–87,94,97,101]
Peroxide Initiation at High Temperature $O_2 + R^\bullet(n) \xrightarrow{f_0 k_{idhT}} PO_2(n)$	[86,97]
Generation of Peroxide at High Temperature $PO_2(n) \xrightarrow{f_{P0} k_{idP0}} R^\bullet(n)$	[86,97]
Oxygen Initiation $O_2 + M \xrightarrow{k_{in}} 2R^\bullet$	[65–67,72,81,84–86,97,104]
Thermal Initiation $3M \xrightarrow{k_{ith}} R^\bullet(1) + R^\bullet(2)$	[65–67,69,85,86,97]
Generation of Inert $O_2 + R^\bullet(n) \xrightarrow{f_0 k_{ine}} Y$	[65–67,85]
Initiation (Extruder) $In \xrightarrow{k_{ine}} \gamma R^\bullet$ with $\gamma = 4$	[71,79,83]
Chain Initiation $R^\bullet + M \xrightarrow{k_{in}} P(1)$	[52–54,60,61,94,101]
Hydrogen Abstraction (Extruder) $R^\bullet + P(n) \xrightarrow{k_{aH}} P^\bullet(m) + RH$	[83]
Hydrogen Abstraction without a Vinyl Group (Extruder) $R^\bullet + P(n,j) \xrightarrow{k_{aH_2}} R^\bullet(n,j)$	[71,79]
Hydrogen Abstraction with a Vinyl Group (Extruder) $R^\bullet + P(n,j) \xrightarrow{k_{aH_2V}} R^\bullet(n,j-1)$	[71,79]
Propagation $R^\bullet(n) + M \xrightarrow{k_{prop}} R^\bullet(n+1)$	[52–54,60,61,65–67,69,72,81,84–87,94,97,101,104]
Double Bond Propagation $P(m) + R^\bullet(n) \xrightarrow{k_{dbprop}} R^\bullet(n+m)$	[86,97]
Double Bond Propagation (extruder) $P(n,j) + R^\bullet(m,k) \xrightarrow{k_{dbprop}} R^\bullet(n+m,j+k-1)$	[71]
Chain Transfer to Monomer $R^\bullet(n) + M \xrightarrow{k_{trM}} P(m) + R^\bullet(1)$	[52–54,60,61,65–67,69,86,87,94,97,101]
Chain Transfer to Polymer $R^\bullet(m) + P(n) \xrightarrow{k_{trPo}} P(m) + R^\bullet(n)$	[52–54,60,65–67,69,72,78,85–87,94,97,101,104]

Table 4. *Cont.*

Reaction Mechanism	Ref.
Chain Transfer to Polymer (Extruder) $P(m) + P^\bullet(n) \xrightarrow{k_{trPo}} P^\bullet(m) + P(n)$	[83]
Chain Transfer to Polymer without a Vinyl Group (extruder) $R^\bullet(m,k) + P(n,j) \xrightarrow{k_{trP}} P(m,k) + R^\bullet(n,j)$	[71,79]
Chain Transfer to Polymer without a Vinyl Group (extruder) $R^\bullet(m,k) + P(n,j) \xrightarrow{k_{trPV}} P(m,k) + R^\bullet(n,j-1)$	[71,79]
Chain Transfer to Chain Transfer Agent/Solvent $R^\bullet(n) + CTA \xrightarrow{k_{CTA}} P(n) + R^\bullet$	[52–54,60,61,65–67,72,85–87,94,97,101,104]
Incorporation of Chain Transfer Agent $R^\bullet(n) + CTA \xrightarrow{k_{iCTA}} P(n+1)$	[52–54]
Termination by Combination $R^\bullet(n) + R^\bullet(m) \xrightarrow{k_{tcom}} P(n+m)$	[52–54,60,61,65–67,69,72,81,84,86,87,94,97,101,104]
Termination by Combination (extruder) $R^\bullet(n,j) + R^\bullet(m,k) \xrightarrow{k_{tcom}} P(n+m,j+k)$ or $P^\bullet(m) + P^\bullet(n) \xrightarrow{k_{tcom}} P(n+m)$	[71,79,83]
Termination with Initiation Radical (extruder) $R^\bullet(n,j) + R^\bullet \xrightarrow{k_{trad}} P(n,j)$	[71]
Termination by Disproportionation $R^\bullet(n) + R^\bullet(m) \xrightarrow{k_{tdisp}} P(n) + P(m)$	[52–54,60,78,94,101]
Thermal Degradation $R^\bullet(n) \xrightarrow{k_{thd}} P(n) + R^\bullet$	[65–67,72,78,81,84–87,97,104]
Intramolecular Chain Transfer/Backbitting $R^\bullet(n) \xrightarrow{k_{int}} R^\bullet(n)$	[52–54,60,66,67,69,72,85,94,97,101]
β-Scission $R^\bullet(n) + P(m) \xrightarrow{k_{\beta s}} P(n) + R^\bullet(r) + R^\bullet(m-r)$	[53,54,60,78,94]
β-Scission (Extruder) $P^\bullet(m) \xrightarrow{k_{\beta s}} P(n) + P^\bullet(m-n)$	[83]
β-Scission for Sec-radicals $R^\bullet(n) \xrightarrow{k_{\beta s2}} P(n) + R^\bullet$	[52,66,67,69,72,85,97,101]
β-Scission for Tert-radicals $R^\bullet(n) \xrightarrow{k_{\beta s3}} P(n) + R^\bullet$	[52,66,67,72,85,97,101]
Retardation by the Impurities $R^\bullet(n) + Po \xrightarrow{k_{rIm}} P(n)$	[101]
Decomposition of Ethylene $2C_2H_4 \xrightarrow{k_{deco}} 2C + 2CH_4 + Heat$ $C_2H_4 \xrightarrow{k_{deco}} 2C + 2H_2 + Heat$	[101]

As mentioned in Table 4, the number of stages in the reaction mechanism for this olefin polymerization process in the gas phase via a free-radical mechanism by using an organic peroxide and oxygen as the initiator implemented previously differ from one study to another study. Most of the studies considered that the polymerization reaction commences by the initiator decomposition or peroxide initiation to create the vacant site for the monomer insertion by generating the radical. In addition, these radicals can also be generated by implementing other methods, such as (i) initiation and generation of peroxide at high temperature, (ii) initiation of oxygen by reacting the oxygen with the monomer, and (iii) thermal initiation by decomposing the monomer into the radical form. Then, the radicals react with the monomer to produce the growth of the polymer chain. For the olefin reaction that takes place in the extruder, the growth of the polymer chain occurs when the radicals react with polyolefin during the abstraction of the hydrogen. After that, the growth of the polymer continues during the propagation as well as during the double bond propagation. To control the molecular weight and the length of the polyolefin, several chain transfer reactions such as chain transfer to monomer, chain transfer to polymer, chain transfer to chain transfer agent (CTA), and incorporation of CTA are implemented. Finally, to terminate the growing of the polyolefin chain length, with the purpose of forming the dead polyolefin, the radicals are deactivated by using several methods, such as termination by combination and disproportionation, termination by initiating the radicals, thermal degradation, backbiting, β-scission, and retardation by impurities.

2.2. Overview of the Kinetic Model, and the Mass and Energy Balance for Olefin Polymerization in the Gas Phase

Kinetic modeling is used to predict the velocity or the speed of the chemical reaction by determining the reaction rate of the olefin polymerization reaction and it can be determined theoretically by using Equation (1) [106].

$$M + H_2 \overset{k(T)}{\rightarrow} P(0,j) \Rightarrow r = k(T)[M]^a[H_2]^b \tag{1}$$

The value of k depends on the temperature and it can be determined theoretically by using Equation (2) [107].

$$k(T) = k_0 \exp\left(\frac{-E_A}{R}\left[\frac{1}{T} - \frac{1}{T_{ref}}\right]\right) \tag{2}$$

Many studies have been carried out to determine this reaction constant experimentally to ensure the calculated reaction rate is highly accurate. The values of the reaction rate constant together with the values of activation energy (E_A), which are required to be used for the kinetic modeling and simulation for the organometallic-type catalyst system such as Ziegler–Natta catalyst, metallocene-based catalyst, and chromium oxide-based catalyst are mentioned in the following publications [45,49,51,55–59,62,68,73–76,80,82,88,89,91,93,95,98–100,102,103,105,108–110]. For the olefin polymerization via the free-radical mechanism, using organic peroxide and oxygen as the initiator, the values of reaction rate constants, together with their corresponding activation energy, were also published [61,69,71,72,78,83,85,94,97,101,104].

Several mathematical models have been implemented to describe the dynamic behavior of this olefin polymerization process in the gas phase. By referring to Table 2, most of the mathematical models were simulated by using a fluidized bed reactor, followed by the tubular reactor, extruder, autoclave reactor, horizontal-stirred-bed reactor, and vertical-stirred-bed reactor. The fluidized bed reactor, vertical and horizontal-stirred-bed reactors were used to carry out the olefin polymerization reaction by using a Ziegler–Natta catalyst, chromium oxide-based catalyst, and a metallocene-based catalyst system. Meanwhile, the tubular reactor, the extruder, and the autoclave reactor were used to perform the olefin polymerization reaction by using the initiator, namely, organic peroxide and oxygen. To summarize, even though the reaction was carried out in the same type of reactor or by using the same type of catalyst or initiator, the proposed mathematical models slightly differed due to the different considerations in the number of stages in the reaction mechanism and the assumptions to simplify the complexity of the nonlinearity phenomena that occurred during olefin polymerization in the gas phase.

For the olefin synthesis via a coordination mechanism, the following kinetic model can be used to determine theoretically the number of moles of the potential sites and the initiation sites, the population balance for living and dead chains, the moment of the chain length distribution for the living and dead polymer, and the population balance for dead polymer [49,56,58,59,62]. For the number of moles of the potential sites, it is written in the following form:

$$\frac{dP^*(j)}{dt} = F_{in}^*(j) - k_{act}(j)P^*(j) - P^*(j)\frac{R_v}{V_p} \tag{3}$$

Then, the number of moles of the initiation sites can be determined using the following formulas:

$$\frac{dP(0,j)}{dt} = k_{act}(j)P^*(j) - P(0,j)\left\{ k_{deac}(j)[M] + k_{ds}(j) + k_{dP}(j)[Po] + \frac{R_v}{V_p} \right\} \tag{4}$$

$$\frac{dP_H(0,j)}{dt} = Y(0,j)\left\{ k_{fH_2}(j)[H_2] + k_{trs}(j) \right\} - N_H(0,j)\left\{ k_{H_2}(j)[M] + k_{deac}(j) + k_{H_2C}(j)[AL] + k_{dP}[Po] + \frac{R_v}{V_p} \right\} \tag{5}$$

Moreover, to determine the population balance for a living and dead chains polymer for chain length equal to 1 and greater than 2, the following formulas can be used:

$$\begin{aligned}\frac{dP(1,j)}{dt} &= k_{act}(j)P(0,j)[M] + P_H(0,j)\left\{ k_{H_2}(j)[M] + k_{H_2C}(j)[AL] \right\} \\ &+ Y(0,j)\left\{ k_{trM}(j)[M] + k_{trCo}(j)[AL] \right\} - N(1,j)\left\{ k_{prop}(j)[M] \right. \\ &+ k_{trM}(j)[M] + k_{fH_2}(j)[H_2] + k_{trCo}(j)[AL] + k_{trs}(j) + k_{deac}(j) + k_{dP}(j)[Po] + \frac{R_v}{V_p} \left. \right\} \end{aligned} \tag{6}$$

$$\begin{aligned}\frac{dP(n,j)}{dt} &= k_{prop}(j)[M]P(n-1,j) - P(n,j)\left\{ k_{prop}(j)[M] + k_{trM}(j)[M] \right. \\ &+ k_{trCo}(j)[AL] + k_{fH_2}(j)[H_2] + k_{trs}(j) + k_{deac}(j) + k_{dP}(j)[Po] + \frac{R_v}{V_p} \left. \right\} \end{aligned} \tag{7}$$

$$\begin{aligned}\frac{dQ(1,j)}{dt} &= P(1,j)\left\{ [M]k_{trM}(j) + [H_2]k_{fH_2}(j) + [AL]k_{trCo}(j) \right. \\ &+ k_{trs}(j) + k_{deac}(j) + k_{dP}(j)[Po] \left. \right\} - \frac{R_v}{V_p}Q(n,j) \end{aligned} \tag{8}$$

$$\begin{aligned}\frac{dQ(n,j)}{dt} &= P(n,j)\left\{ [M]k_{trM}(j) + [H_2]k_{fH_2}(j) + [AL]k_{trCo}(j) \right. \\ &+ k_{trs}(j) + k_{deac}(j) + k_{dP}(j)[Po] \left. \right\} - \frac{R_v}{V_p}P(n,j) \end{aligned} \tag{9}$$

Lastly, the following equations can be used to calculate the zeroth, first and second moment of the chain length distribution for the living and dead polymer:

$$\begin{aligned}\frac{dY(0,j)}{dt} &= k_{act}(j)P(0,j)[M] + P_H(0,j)\left\{ k_{H_2}(j)[M] + k_{H_2C}(j)[AL] \right\} \\ &- Y(0,j)\left\{ k_{fH_2}(j)[H_2] + k_{trs}(j) + k_{deac}(j) + k_{dP}(j)[Po] + \frac{R_v}{V_p} \right\} \end{aligned} \tag{10}$$

$$\begin{aligned}\frac{dY(1,j)}{dt} &= k_{act}(j)P(0,j)[M] + P_H(0,j)\left\{ k_{H_2}(j)[M] + k_{h_2C}(j)[AL] \right\} \\ &+ Y(0,j)\left\{ k_{trM}(j)[M] + k_{trCo}(j)[AL] \right\} + k_{prop}[M]Y(0,j) \\ &- Y(1,j)\left\{ k_{trM}(j)[M] + k_{fH_2}(j)[H_2] + k_{trCo}(j)[AL] + k_{trs}(j) + k_{deac}(j) \right. \\ &+ k_{dP}(j)[Po] + \frac{R_v}{V_p} \left. \right\} \end{aligned} \tag{11}$$

$$\begin{aligned}\frac{dY(2,j)}{dt} &= k_{act}(j)P(0,j)[M] + P_H(0,j)\left\{ k_{H_2}(j)[M] + k_{H_2C}(j)[AL] \right\} \\ &+ Y(0,j)\left\{ k_{trM}(j)[M] + k_{trCo}(j)[AL] \right\} + k_{prop}[M]\left\{ 2Y(1,j) + Y(0,j) \right\} \\ &- Y(2,j)\left\{ k_{trM}(j)[M] + k_{fH_2}(j)[H_2] + k_{trCo}(j)[AL] + k_{trs}(j) + k_{deac}(j) + k_{dP}(j)[Po] + \frac{R_v}{V_p} \right\} \end{aligned} \tag{12}$$

$$\begin{aligned}\frac{dX(n,j)}{dt} &= Y(n,j)\left\{ [M]k_{trM}(j) + [H_2]k_{fH_2}(j) + [AL]k_{trCo}(j) \right. \\ &+ k_{trs}(j) + k_{deac}(j) + k_{dP}(j)[Po] \left. \right\} - \frac{R_v}{V_p}X(n,j) \end{aligned} \tag{13}$$

Furthermore, there exist several types of heat and mass transfer models implemented previously to describe the dynamic behavior of this polymerization process. For a three-phase model where the reactions were assumed to occur in emulsion, cloud, and bubble phases, the heat and mass transfer models are written in the following form [45]. The mass balance for the potential sites, active sites, and catalyst are written in the following form:

$$\frac{dP^*(j)}{dt} = \frac{Q_{cat}P^*_{in}(j)}{m_s} - \frac{Q_{vprod}P^*(j)\rho_{cat}}{m_s} - R_{iP^*} \tag{14}$$

$$\frac{dP(0,j)}{dt} = \frac{Q_{cat}P(0,j)}{m_s} - \frac{Q_{vprod}P(0,j)\rho_{cat}}{m_s} - R_{iP(0,j)} \tag{15}$$

$$\frac{d[Cat]}{dt} = \frac{Q_{cat}}{m_s} - \frac{Q_{vprod}[Cat]\rho_{cat}}{m_s} \tag{16}$$

Furthermore, for the heat and mass balances from the bubble phase to the cloud phase, it can be determined by using the following formula:

$$u_b\frac{d[R_A]_b}{dz} = -K_{bc}([R_A]_b - [R_A]_c) \text{ with } z = 0, \ [R_A]_b = [R_A]_{b0} \tag{17}$$

$$\frac{d([R_A]_b(T_b - T_{ref}))}{dz} = \frac{H_{bc}}{u_bC_{p,g}}(T_c - T_b) \tag{18}$$

Then, for the heat and mass balance from the cloud phase to the emulsion phase, the equations are written in the following form:

$$u_b\delta\left[\frac{3\left(\frac{u_{mf}}{\varepsilon_{mf}}\right)}{u_b - \frac{u_{mf}}{\varepsilon_{mf}}} + \alpha\right]\frac{d[R_A]_b}{dz} = K_{bc}([R_A]_b - [R_A]_c) - K_{ce}([R_A]_c - [R_A]_e) \tag{19}$$

$$z\frac{d([R_A]_c(T_c - T_{ref}))}{dz} = \frac{H_{ce}}{u_bC_{p,g}}(T_e - T_c) \tag{20}$$

Moreover, for the heat and mass transfer with chemical reaction from the emulsion phase to the catalyst phase, the equations are formulated in the following form:

$$A_{bed}(He)\varepsilon_{mf}\frac{d[R_A]_e}{dt} = K_{ce}([R_A]_c - [R_A]_e)A_{bed}(He)\varepsilon_{mf} + Q_{vm}([R_A]_0 - [R_m]_e) \\ -Q_{vprod}[R_m]_e\varepsilon_{mf} + r_A m_s \tag{21}$$

$$A_{bed}(He)\left[(1 - \varepsilon_{mf})\rho_sC_{p,s} + \varepsilon_{mf}[R_A]_{mf}C_{p,g}\right]\frac{dT_e}{dt} + A_{bed}(He)(T_e - T_{ref})\varepsilon_{mf}C_{p,g}\frac{d[R_m]_e}{dt} = \\ -Q_{vm}[R_m]_eC_{p,g}(T_e - T_f) + A_bH_{be}\int(T_b - T_e)dz + (-\Delta H_r)r_A \\ -Q_{vprod}\varepsilon_{mf}[R_m]_eC_{p,g}(T_e - T_{fs}) - Q_{vprod}\varepsilon_{mf}[R_m]_eC_{p,g}(T_e - T_f) \\ -\pi D(He)(1 - \delta^*)h_w(T_e - T_w) \tag{22}$$

Finally, to determine the population balance for a living and dead polymer for this three-phase model, the following equations can be used:

$$\frac{dY(n,j)}{dt} = R_{Y(n,j)} - \frac{Q_{vprod}Y(n,j)\rho_{cat}}{m_s} \tag{23}$$

$$\frac{dX(n,j)}{dt} = R_{X(n,j)} - \frac{Q_{vprod}X(n,j)\rho_{cat}}{m_s} \tag{24}$$

Then, for a two-phase model where the reactions were assumed to occur in the emulsion and bubble phase, the heat and mass transfer model is written in the following form [48]. For heat and mass balance equations in the emulsion phase, the equations are written as follows:

$$\frac{d(V_e\varepsilon_e[M]_e)}{dt} = [M]_{e,in}U_eA_e - [M]_eU_eA_e - R_{ve}\varepsilon_e[M]_e \\ +K_{be}([M]_b - [M]_e)V_e\left(\frac{\delta}{1-\delta}\right) \\ -(1 - \varepsilon_e)R_{i_e} - \frac{K_eV_e\varepsilon_eA_e[M]_e}{W_e} \tag{25}$$

$$\frac{d(V_e \varepsilon_e [H_2]_e)}{dt} = [H_2]_{e,in} U_e A_e - [H_2]_e U_e A_e - R_{ve} \varepsilon_e [H_2]_e$$
$$+ K_{be}([H_2]_b - [H_2]_e) V_e \left(\frac{\delta}{1-\delta}\right) \tag{26}$$
$$- (1 - \varepsilon_e) R_{i_e} - \frac{K_e V_e \varepsilon_e A_e [H_2]_e}{W_e}$$

$$\left(V_e \left(\varepsilon_e \sum_{i=1}^{m} C_{pi}[M_i]_e + (1-\varepsilon_e)\rho_{poly} C_{p,pol}\right)\right) \frac{d(T_e - T_{ref})}{dt} =$$
$$U_e A_e (T_{e,in} - T_{ref}) \sum_{i=1}^{m} [M_i]_{e,in} C_{pi} - U_e A_e (T_e - T_{ref}) \sum_{i=1}^{m} [M_i]_e C_{pi}$$
$$- R_{ve}(T_e - T_{ref}) \left(\sum_{i=1}^{m} \varepsilon_e C_{pi}[M_i]_e + (1-\varepsilon_e)\rho_{poly} C_{p,pol}\right) + (1-\varepsilon_e) R_{pe} \Delta H_R \tag{27}$$
$$- H_{be} V_e \left(\frac{\delta}{1-\delta}\right)(T_e - T_b) - V_e \varepsilon_e (T_e - T_{ref}) \sum_{i=1}^{m} C_{pi} \frac{d[M_i]_e}{dt}$$
$$- \frac{K_e A_e}{W_e(T_e - T_{ref})} \left(\sum_{i=1}^{m} \varepsilon_e C_{pi}[M_i]_e + (1-\varepsilon_e)\rho_{poly} C_{p,pol}\right)$$

Meanwhile, the equations for heat and mass transfer in the bubble phase are written as follows:

$$\frac{d(V_b \varepsilon_b [M]_b)}{dt} = [M]_{b,in} U_b A_b - [M]_b U_b A_b - R_{vb} \varepsilon_b [M]_b - K_{be}([M]_b - [M]_e) V_b$$
$$- (1 - \varepsilon_b) \frac{A_b}{V_{PFR}} \int R_{i_b} dz - \frac{K_b V_b \varepsilon_b A_b [M]_b}{W_b} \tag{28}$$

$$\frac{d(V_b \varepsilon_b [H_2]_b)}{dt} = [H_2]_{b,in} U_b A_b - [H_2]_b U_b A_b - R_{vb} \varepsilon_b [H_2]_b - K_{be}([H_2]_b - [H_2]_e) V_b$$
$$- (1 - \varepsilon_b) \frac{A_b}{V_{PFR}} \int R_{i_b} dz - \frac{K_b V_b \varepsilon_b A_b [H_2]_b}{W_b} \tag{29}$$

$$\left(V_b \left(\varepsilon_b \sum_{i=1}^{m} C_{pi}[M_i]_b + (1-\varepsilon_b)\rho_{poly} C_{p,pol}\right)\right) \frac{d(T_b - T_{ref})}{dt} =$$
$$U_b A_b (T_{b,in} - T_{ref}) \sum_{i=1}^{m} [M_i]_{b,in} C_{pi} - U_b A_b (T_b - T_{ref}) \sum_{i=1}^{m} [M_i]_b C_{pi}$$
$$- R_{vb}(T_b - T_{ref}) \left(\sum_{i=1}^{m} \varepsilon_b C_{pi}[M_i]_b + (1-\varepsilon_b)\rho_{poly} C_{p,pol}\right) \tag{30}$$
$$+ (1 - \varepsilon_b) \frac{A_b \Delta H_R}{V_{PFR}} \int R_{pb} dz + H_{be} V_b (T_e - T_b) - V_b \varepsilon_b (T_b - T_{ref}) \sum_{i=1}^{m} C_{pi} \frac{d[M_i]_b}{dt}$$
$$- \frac{K_b A_b}{W_b(T_b - T_{ref})} \left(\sum_{i=1}^{m} \varepsilon_b C_{pi}[M_i]_b + (1-\varepsilon_b)\rho_{poly} C_{p,pol}\right)$$

Furthermore, for single-phase known as a well-mixed model where the reactions were assumed to occur in the emulsion phase, the following heat and mass transfer model can be implemented [99,102]. The heat and mass balance equations are formulated as follows:

$$V \varepsilon_{mf} \frac{d[M]}{dt} = U_0 A ([M]_{in} - [M]) - R_v \varepsilon_{mf} [M] - (1 - \varepsilon_{mf}) R_i \tag{31}$$

$$V \varepsilon_{mf} \frac{d[H_2]}{dt} = U_0 A ([H_2]_{in} - [H_2]) - R_v \varepsilon_{mf} [H_2] - (1 - \varepsilon_{mf}) R_i \tag{32}$$

$$\left(\sum_{i=1}^{m} [M_i] C_{pi} V \varepsilon_{mf} + V(1 - \varepsilon_{mf})\rho_{pol} C_{p,pol}\right) \frac{dT}{dt} =$$
$$U_0 A \sum_{i=1}^{m} [M_i]_{in} C_{pi}(T_{in} - T_{ref}) - U_0 A \sum_{i=1}^{m} [M_i] C_{pi}(T - T_{ref}) \tag{33}$$
$$- R_v \left[\sum_{i=1}^{m} [M_i] C_{pi} \varepsilon_{mf} + (1 - \varepsilon_{mf})\rho_{pol} C_{p,pol}\right](T - T_{ref}) + (1 - \varepsilon_{mf}) \Delta H_R R_p$$

Lastly, another single-phase model is known as constant bubble size model, the following heat and mass transfer model was previously implemented [105]. For heat and mass balance in the bubble phase, the equations are written as follows:

$$\overline{[M]}_b = \frac{1}{H}\int_0^{He} [M]_b dh = [M]_e + \left([M]_{e,in} - [M]_e\right)\frac{U_b}{K_{be}He}\left(1 - e^{-\left(\frac{K_{be}He}{U_b}\right)}\right) \tag{34}$$

$$\overline{[H_2]}_b = \frac{1}{H}\int_0^{He} [H_2]_b dh = [H_2]_e + \left([H_2]_{e,in} - [H_2]_e\right)\frac{U_b}{K_{be}He}\left(1 - e^{-\left(\frac{K_{be}He}{U_b}\right)}\right) \tag{35}$$

$$\overline{T}_b = \frac{1}{He}\int_0^{H} T_b dh = T_e + (T_{in} - T_e)\frac{U_b\overline{C}_p}{H_{be}He}\left(1 - e^{-\left(\frac{H_{be}H}{U_b\overline{C}_p}\right)}\right) \tag{36}$$

with

$$\overline{C}_p = \sum_{i=1}^{Nm} \overline{[M_i]}_b C_{pMi} = [M]_b C_{pC_3H_6} + [H_2]_b C_{pH_2} + [N_2]C_{pN_2} \tag{37}$$

For the emulsion phase, the heat and mass transfer equations are written as follows:

$$V_e\varepsilon_{mf}\frac{d[M]_e}{dt} = U_eA_e\varepsilon_{mf}\left([M]_{e,in} - [M]_e\right) + \frac{V_e\delta K_{be}}{(1-\delta)}\left(\overline{[M]}_b - [M]_e\right) - R_{ve}\varepsilon_{mf}[M]_e - (1 - \varepsilon_{mf})R_i \tag{38}$$

$$V_e\varepsilon_{mf}\frac{d[H_2]_e}{dt} = U_eA_e\varepsilon_{mf}\left([H_2]_{e,in} - [H_2]_e\right) + \frac{V_e\delta K_{be}}{(1-\delta)}\left(\overline{[H_2]}_b - [H_2]_e\right) \\ -R_{ve}\varepsilon_{mf}[H_2]_e - (1 - \varepsilon_{mf})R_i \tag{39}$$

$$\left(\sum_{i=1}^{m} V_e\varepsilon_{mf}[M_i]_e C_{pi} + V_e(1 - \varepsilon_{mf})\rho_{pol}C_{p,pol}\right)\frac{dT_e}{dt} = \\ -\sum_{i=1}^{m} V_e\varepsilon_{mf}C_{pi}\frac{d[M_i]_e}{dt}(T_e - T_{ref}) + U_eA_e\varepsilon_{mf}\sum_{i=1}^{m}[M_i]_{e,in}C_{pi}(T_{e,in} - T_{ref}) \\ -U_eA_e\varepsilon_{mf}\sum_{i=1}^{m}[M_i]_e C_{pi}(T_e - T_{ref}) - \frac{V_e\delta H_{be}}{(1-\delta)}(T_e - \overline{T}_b) \\ +R_{ve}\left((1 - \varepsilon_{mf})\rho_{pol}C_{p,pol} + \varepsilon_{mf}\sum_{i=1}^{m}[M_i]_e C_{pi}\right)(T_e - T_{ref}) + (1 - \varepsilon_{mf})\Delta H_R R_{pe} \tag{40}$$

For the olefin synthesis via a free-radical mechanism, the following equations can be used to determine mass and heat transfer models (radicals, polymer, monomer, reactor, and reactor jacket), pressure drop, kinetic model (moment of dead polymer), mass balance for long and short chain branching polymer [61,69]. For the mass balance for the radicals, the equation is written as follows:

$$\frac{d[R(n)]}{dt} = 2fk_{idec}[In] - k_{prop}[M]([R(n)] - [R(n-1)](1 - \delta_{n,1})) \\ -(k_{CTA}[CTA] + k_{trM}[M])\left([R(n)] - \delta_{n,1}\sum_{i=1}^{\infty}[R(i)]\right) - k_{tcom}[R(n)]\sum_{i=1}^{\infty}[R(i)] \tag{41}$$

Then, the mass balance for the polymer is formulated as follows:

$$\frac{d[P(n)]}{dt} = (k_{CTA}[CTA] + k_{trM}[M])[R(n)] + \frac{k_{tcom}}{2}\sum_{m=1}^{n-1}[R(m)][R(n-m)](1 - \delta_{n,1}) \tag{42}$$

Moreover, the mass balance for the monomer used during the synthesis is written as follows:

$$\frac{d[M]}{dt} = -\left(k_{prop} + k_{trM}\right)[M]\sum_{i=1}^{\infty}[R(i)] \tag{43}$$

Meanwhile, for the heat balance for the reactor and the cooling fluid in the reactor jacket, the equations are written as follows:

$$\frac{dT}{dt} = \frac{(-\Delta H_R)LR_{pm}}{\rho u C_p} + \frac{uA_{pipe}(T - T_{cool})}{\rho u A_{sp} C_p} \tag{44}$$

$$\frac{dT_{cool}}{dt} = \pi D_{out} LU \frac{(T_{cool} - T)}{Q_m C_{p,cool}} \tag{45}$$

Furthermore, the pressure drop occurred in the reactor can be determined as follows:

$$\frac{dP}{dz} = -L\left(2f_{ric}\rho\frac{u^2}{D_{in}} + \rho u\frac{du}{dz}\right) \text{ with } f_{ric}^{-1/2} = 4\log\left(f_{ric}^{1/2}Re\right) - 0.4 \tag{46}$$

Then, to determine the zeroth, first and second moments of dead polymer, the equations are formulated as follows:

$$\frac{d(X(0)P(n))}{dz} = \frac{k_{tcom}Y^2(0)}{2} + (k_{trM}[M] + k_{\beta s2})Y(0) \tag{47}$$

$$\begin{aligned}\frac{d(X(1)P(n))}{dz} &= k_{tcom}Y(0)Y(1) + (k_{trM}[M] + k_{\beta s2})Y(1) \\ &+ k_{trPo}\left((Y(1) + Y'(1))X(1) - X(2)\sum_{i=1}^{\infty}[R(i)]\right)\end{aligned} \tag{48}$$

$$\begin{aligned}\frac{d(X(2)P(n))}{dz} &= k_{tcom}\left(Y(0)Y(2) + Y^2(1)\right) + (k_{trM}[M] + k_{\beta s2})Y(2) \\ &+ k_{trPo}\left((Y(2) + Y'(2))X(1) - X(3)\sum_{i=1}^{\infty}[R(i)]\right)\end{aligned} \tag{49}$$

In addition, for the zeroth, first and second moments of temporary dead polymer I and II, the equations are written as follows:

$$\frac{d(X'(0)P(n))}{dz} = k_{tcom}Y(0)Y'(0) - k_{idec}Y'(0) + (k_{trM}[M] + k_{\beta s2})Y'(0) \tag{50}$$

$$\begin{aligned}\frac{d(X'(1)P(n))}{dz} &= k_{tcom}(Y(0)Y'(1) + Y(1)Y'(0)) - k_{idec}Y'(1) + (k_{trM}[M] + k_{\beta s2})Y'(1) \\ &+ k_{trPo}\left((Y(1) + Y'(1))X'(1) - X'(2)\sum_{i=1}^{\infty}[R(i)]\right)\end{aligned} \tag{51}$$

$$\begin{aligned}\frac{d(X'(2)P(n))}{dz} &= k_{tcom}(Y(2)Y'(0) + 2Y(1)Y'(1) + Y(0)Y'(2)) - k_{idec}Y'(2) + (k_{trM}[M] + k_{\beta s2})Y'(2) \\ &+ k_{trPo}\left((Y(2) + Y'(2))X'(1) - X'(3)\sum_{i=1}^{\infty}[R(i)]\right)\end{aligned} \tag{52}$$

$$\frac{d(X''(0)P(n))}{dz} = k_{tcom}Y''(0) - 2k_{idec}Y''(0) \tag{53}$$

$$\frac{d(X''(1)P(n))}{dz} = k_{tcom}Y'(0)Y'(1) - 2k_{idec}Y''(1) \tag{54}$$

$$\frac{d(X''(2)P(n))}{dz} = k_{tcom}\left(Y'(0)Y'(2) + Y''^2(1)\right) - 2k_{idec}Y''(2) \tag{55}$$

Finally, the mass balance for short and long chain branching polymer is defined as follow:

$$\frac{d[P_{scb}(n)]}{dz} = k_{scb}\sum_{i=1}^{\infty}[R(i)] \tag{56}$$

$$\frac{d[P_{lcb}(n)]}{dt} = k_{lcb}(X(1) + X'(1))\sum_{i=1}^{\infty}[R(i)] \tag{57}$$

By referring to Equations (3) until 57, a method of moments and population balances, which are in form of ordinary differential equation (ODEs), were used to build and simulate the kinetics model with the aim of studying the dynamic behavior of the olefin polymerization reaction in a fluidized bed reactor, horizontal-stirred-bed reactor, vertical-stirred-bed reactor, tubular reactor, extruder, and autoclave reactor via two types of reaction mechanisms, namely, coordination and free-radical mechanisms. For the olefin polymerization via a coordination mechanism, Equations (58) to (60) are used to calculate the polydispersity index, number average molecular weight, and the weight average molecular weight, respectively.

$$\mathrm{PDI} = \frac{M_w}{M_n} \tag{58}$$

$$M_n = \frac{\sum\limits_{j=1}^{NS} Y(1,j)X(1,j)}{\sum\limits_{j=1}^{NS} Y(0,j)X(0,j)} MW \tag{59}$$

$$M_w = \frac{\sum\limits_{j=1}^{NS} Y(2,j)X(2,j)}{\sum\limits_{j=1}^{NS} Y(1,j)X(1,j)} MW \tag{60}$$

The polymerization rate is defined as follows [59]:

$$R_p = MW[M]Y(0,j)k_{prop}(j) \tag{61}$$

The melt index or melt flow index (MFI) can be determined by using the following equation [45]:

$$MFI = 3.346 \times 10^{17} M_w^{-3.472} \tag{62}$$

For the olefin polymerization process via a free-radical mechanism, to determine the number and weight average molecular weight, the following equations are used, which are slightly different from the equation used for the coordination mechanism. At this level, the moments of dead and temporary dead polymer, as well as the moments of living polymer, are incorporated to determine these parameters [69]:

$$M_w = MW \frac{X(2) + X'(2) + X''(2) + Y(2) + Y'(2)}{X(1) + X'(1) + X''(1) + Y(1) + Y'(1)} \tag{63}$$

$$M_n = MW \frac{X(1) + X'(1) + X''(1) + Y(1) + Y'(1)}{X(0) + X'(0) + X''(0) + Y(0) + Y'(0)}$$

Lastly, the monomer conversion for this olefin polymerization via a free-radical mechanism is defined by the following equation:

$$y_{mono} = 1 - \frac{[M]u}{[M]_0} \tag{64}$$

To compute the mass and heat balance equations, which represent the olefin reaction via a coordination mechanism, the hydrodynamic correlations that are tabulated in Abbasi et al. (2016) [49] are referred. For the heat and mass balance equations for the olefin polymerization process via a free-radical mechanism, the hydrodynamic correlations that are tabulated in Khazraei and Dhib (2008) [69] are used.

2.3. Numerical Methods for the Simulation of the Mathematical Model

Equation (3) until 57 are either in the form of ODEs or in the form of partial differential equations (PDEs). For the ODEs, they contain only one independent variable, which is generally time (t). The PDEs contains at least two independent variables, which are generally time, height of the reactor, etc. To simulate these types of equations, several methods exist, which are summarized in Table 5.

Table 5. Numerical methods for the process simulation.

Methods	Main Feature	Ref.
Euler's Method	Ability to solve simple and linear ordinary differential equation (ODEs) with the presence of initial values	[111–113]
Monte Carlo Method	Ability to compute the ODEs with random values	[113,114]
Rosenbrock Method	Ability to solve stiff ODEs	[113,115,116]
Backward Euler's Method	Ability to solve stiff ODEs with larger step size	[117]
Finite Difference Method	Ability to solve partial differential equations (PDEs) by approximating the nonlinear system to linear system	[118]
Method of Lines	Ability to solve PDE by approximating the PDE system with an ODE system. In general, the spatial independent variables are substituted by algebraic approximation (as a function of time)	[119]
Finite Element Method	Ability to solve PDEs with the presence of boundary conditions	[120]
Multigrid Methods	Ability to solve high order PDEs, especially parabolic systems.	[121]

By referring to Table 5, because of the stiffness of the ODEs used in this polymerization process, Rosenbrock and backward Euler's method seem to be the most suitable numerical methods to be implemented in simulating these ODEs [58]. Meanwhile, for the PDEs, because the independent variables are not more than two, the finite difference method or method of lines could be applied.

3. A General Guideline to Implement the Mathematical Model

After reviewing the mathematical model, a general guideline can be proposed to ease the implementation of the mathematical model for the olefin polymerization process in the gas phase. The following flowchart in Figure 2 can be used to illustrate the procedure.

For olefin polymerization in the gas phase, two types of major mechanism occur in the process. If the free-radical mechanism (details can be found in Table 4) occurs, the normal choice is to choose a tubular reactor, while for the coordination mechanisms (details can be found in Table 3), the fluidized bed reactors and stirred-bed reactors (horizontal and vertical) are the preferred choices. The next step is to decide on the type of model to be used based on the number of phases assumed in the reactor. For the tubular reactor, the model to be used is normally the single-phase model assuming all the reactions occurs homogeneously in the packed bed tubular reactor. For the fluidized bed reactors (FBR) and stirred-bed reactors (SBR), the model can be either single, two or three phase depending on the assumptions of the location of the sites of the reactions occurring in the process. Examples of single, two, and three phases can be seen in Table 2, with different types of process conditions and assumptions by different researchers.

The final mathematical model for the mass balance (either 1, 2, or 3 phases) will incorporate the kinetic model, transport phenomena process, and the hydrodynamics within the reaction system. The kinetic model includes the method of moments, while the effects of population balance can also be included in the mass balance. A suitable model will also be done for the heat balance incorporating the kinetic model and transport phenomena mechanisms.

Finally, these mathematical models (both the mass and energy balance) can be validated using computational fluid dynamics (CFD) models (ANSYS 6.1, ANSYS Inc., Berkeley, CA, USA) and through experimental pilot plant data [56].

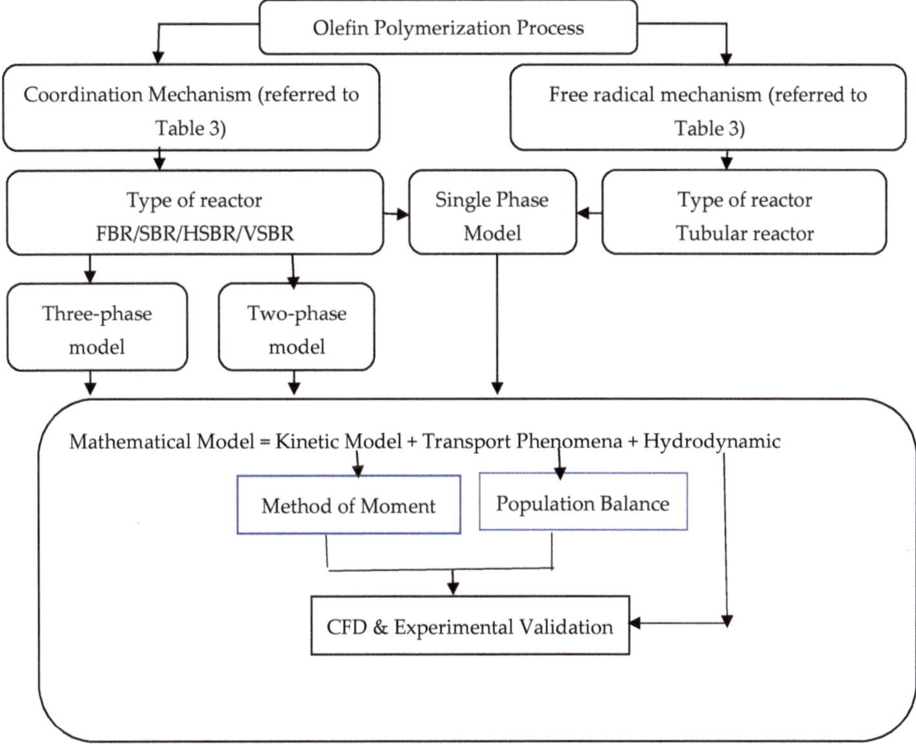

Figure 2. A general flowchart of the guidelines for the implementation of the mathematical model (FBR: fluidized bed reactor; SBR: stirred-bed reactors; HSBR: horizontal-stirred-bed reactor; VSBR: vertical-stirred-bed reactor).

4. Conclusions

Polyolefins such as polyethylene or polypropylene are widely used nowadays in producing several materials for industrial and consumer use. To ensure that the end products are safe, high quality and have an optimized production facility, the processing conditions need to be improved. Thus, a review of the previously implemented gas-phase mathematical models was carried out to scaffold a methodology to build new mathematical models or to improve the existing ones. For the olefin polymerization process in the gas phase, the tubular reactor operates at high pressures while reactors such as a fluidized bed reactor, stirred-bed reactor, vertical-stirred-bed reactor and horizontal-stirred-bed reactor work at moderate pressures. Moreover, most of the modeling in olefin polymerization via a free-radical mechanism considers a single-phase model. Meanwhile, for the olefin polymerization in the gas phase via a coordination mechanism, several models exist, such as the three-phase model, two-phase model, well-mixed (single-phase) model, and constant bubble size model (two-phase with constant bubble size). To simplify this complexity, some researchers ignore extra phases during model implementation. Many studies have been done to improve the precision of the two-phase model as well as single-phase models by adding more details to the model, such as considering solid elutriation. In the future, studies on heat loss through fluidized bed reactors and improvement in the precision of the three-phase model can be carried out. The flowchart in Figure 2 can be used as a guideline to retrofit the available models or to develop new ones. A lot of work has been done on modeling olefin polymerization, but a lot of effort is still needed to validate the models better, which is to minimize the gap between model output and experimental or industrial data with the aim of improving the end products that are environmentally friendly and have a high quality.

Author Contributions: M.F.A. reviewed all the published articles related to olefin polymerization in gas phase and prepared this review manuscript. Meanwhile, M.A.H., M.R.A., M.J.H.K. and M.F.A.P. contribute in reviewing and enriching the content of this review article.

Funding: This research received no external funding.

Acknowledgments: Authors would like to acknowledge the Ministry of Education Malaysia and Universiti Malaysia Sarawak (UNIMAS) in funding this research through public academician scholarship scheme. Last but not least, the authors would like to thank the University of Malaya in providing the research facilities to execute this research.

Conflicts of Interest: The authors declare no conflict of interest

Nomenclature

a	Partial rate constants
A	Area (m^2)
A_{tube}	The cross-sectional area of the cooling jacket tube (m^2)
A_{bed}	The cross-sectional area of the fluidized bed (m^2)
A_{pipe}	The surface area of the pipe (m^2)
A_{sp}	The cross-sectional area of the pipe (m^2)
AL	Alkyl Aluminum (mol/m^3)
b	Partial rate constants
BP	By-product (mol/m^3)
C_i	Concentration of the component i (mol/m^3)
$cCata$	Co – Catalyst (mol/m^3)
C_p	Heat capacity $(J/Kg \cdot K)$
$C_{p,pol}$	Heat capacity of the polyolefin $(J/Kg \cdot K)$
$C_{p,tube}$	Heat capacity in the cooling jacket tube $(J/Kg \cdot K)$
D	Diameter (m)
D_{in}	Inlet diameter (m)
D_{out}	Outer diameter (m)
D_z	Dispersion coefficient (m^2/s)
E_A	Activation energy (J/mol)
F_m	Mass flow rate (kg/s)
$F_{in}^*(j)$	Potential active site flow rate of a site type j injected into the reactor (mol/s)
$F_{m,tube}$	Mass flow rate in the jacket cooling tube (kg/s)
F_{cat}	The fraction of the catalyst in the polyolefin
F_{cata}	Mass fraction of the activated catalyst
F_{dcata}	Mass fraction of the deactivated catalyst
F_{H_2}	Mass fraction of the hydrogen
F_M	Mass fraction of the monomer
FBR	Fluidized bed reactor
$f_0 k_{ine}$	Generation of the inert rate constant $(m^3 mol^{-1} s^{-1})$
He	Height of the reactor (m)
H	Enthalpy of the reactor (J/kg)
H_{be}	Bubble to emulsion heat transfer coefficient $(W/m^3 K)$
H_{bc}	Bubble to cloud heat transfer coefficient $(W/m^3 K)$
H_{ce}	Cloud to emulsion heat transfer coefficient $(W/m^3 K)$

H_{in}	Enthalpy of the inlet feedstock into the reactor (J/kg)
H_2	Hydrogen (mol/m^3)
h_w	Wall heat transfer coefficient (W/m^{2}K)
HSBR	Horizontal-stirred bed reactor
In	Initiators (mol/m^3)
k_0	Pre-exponential reaction rate constant (s^{-1} or m^{3}mol^{-1}s^{-1})
k	Reaction rate constant
$k_{act}(j)$	Activation rate constant for active site type j (m^{3}mol^{-1}s^{-1})
$k_{actH_2}(j)$	Catalyst activation by hydrogen rate constant for a site type j (s^{-1})
$k_{actM}(j)$	Catalyst activation by monomer rate constant for a site type j (s^{-1})
$k_{actS}(j)$	Spontaneous site activation rate constant for active site type j (s^{-1})
$k_{actSA}(j)$	Site activation by alkyl aluminum rate constant for active site type j (s^{-1})
$k_{actSH_2}(j)$	Site activation by hydrogen rate constant for a site type j (s^{-1})
$k_{actSM}(j)$	Site activation by monomer rate constant for a site type j (s^{-1})
k_{aH}	Hydrogen abstraction rate constant (m^{3}mol^{-1}s^{-1})
k_{aH_2}	Hydrogen abstraction without a vinyl group rate constant (m^{3}mol^{-1}s^{-1})
k_{aH_2V}	Hydrogen abstraction with a vinyl group rate constant (m^{3}mol^{-1}s^{-1})
$k_{dP}(j)$	Reaction with poisons rate constant (m^{3}mol^{-1}s^{-1})
$k_{Deac}(j)$	Deactivation rate constant (s^{-1})
$k_{DPo}(j)$	Deactivation by poison rate constant (s^{-1})
$k_{deac}(j)$	Deactivation rate constant for a site type j (s^{-1})
k_{deco}	Decomposition of ethylene rate constant (s^{-1})
$k_{fH_2}(j)$	Chain transfer to hydrogen rate constant for a site type j with terminal monomer M reacting with hydrogen (m^{3}mol^{-1}s^{-1})
k_{idec}	Initiator decomposition rate constant (s^{-1})
$f_0 k_{idhT}$	Peroxide initiator at high-temperature rate constant (m^{3}mol^{-1}s^{-1})
$f_{P0} k_{idP0}$	Peroxide generation at high-temperature rate constant (s^{-1})
k_{in}	Initiation rate constant (m^{3}mol^{-1}s^{-1})
$k_{in}(j)$	Initiation rate constant for a site type j (m^{3}mol^{-1}s^{-1})
k_{ine}	Initiation rate constant in extruder (m^{3}mol^{-1}s^{-1})
k_{ith}	Thermal initiation rate constant (s^{-1})
k_{prop}	Propagation rate constant (m^{3}mol^{-1}s^{-1})
$k_{prop}(j)$	Propagation rate constant for a site type j (m^{3}mol^{-1}s^{-1})
$k_{spont}(j)$	Spontaneous chain transfer rate constant for a site type j (m^{3}mol^{-1}s^{-1})
k_{tcom}	Termination by combination rate constant (m^{3}mol^{-1}s^{-1})
k_{tdisp}	Termination by disproportionation rate constant (m^{3}mol^{-1}s^{-1})
k_{thd}	Thermal degradation rate constant (s^{-1})
$k_{Transf}(j \rightarrow k)$	Site transform from site j to site k rate constant (s^{-1})
k_{trad}	Termination with initiation radical rate constant (m^{3}mol^{-1}s^{-1})
k_{trCTA}	Chain transfer to chain transfer agent rate constant (m^{3}mol^{-1}s^{-1})
k_{trCo}	Chain transfer to co-catalyst rate constant (m^{3}mol^{-1}s^{-1})
$k_{trCo}(j)$	Chain transfer to co-catalyst rate constant for a site type j (m^{3}mol^{-1}s^{-1})
$k_{trH_2}(j)$	Chain transfer to hydrogen rate constants for a site type j (m^{3}mol^{-1}s^{-1})
$k_{trs}(j)$	Chain transfer to solvent rate constants for a site type j (m^{3}mol^{-1}s^{-1})
$k_{trM}(j)$	Chain transfer to monomer rate constant for a site type j (m^{3}mol^{-1}s^{-1})
k_{trM}	Chain transfer to monomer rate constant (m^{3}mol^{-1}s^{-1})
$k_{trs}(j)$	Spontaneous transfer rate constant (m^{3}mol^{-1}s^{-1})
k_{trP}	Chain transfer to the polymer without a vinyl group rate constant (m^{3}mol^{-1}s^{-1})
k_{trPV}	Chain transfer to a polymer with a vinyl group rate constant (m^{3}mol^{-1}s^{-1})
k_{trPo}	Chain transfer to polymer rate constant (m^{3}mol^{-1}s^{-1})
$k_{\beta s}$	β-scission rate constant (m^{3}mol^{-1}s^{-1})
$k_{\beta s2}$	β-scission for secondary radical rate constant (s^{-1})
$k_{\beta s3}$	β-scission for tertiary radical rate constant (s^{-1})

K_z	Heat dispersion coefficient (J/m·s·K)
K_{be}	bubble to emulsion mass transfer coefficient (s^{-1})
K_{bc}	Bubble to cloud mass transfer coefficient (s^{-1})
K_{ce}	Cloud to emulsion mass transfer coefficient (s^{-1})
K_e	Elutriation constant in emulsion phase (kg/m^{2}s)
L	Length of the reactor (m)
M	Monomer used during the polymerization (mol/m^3)
MFI	Melt flow index or Melt index (g/min)
MW	Molecular weight (kg/mol)
M_n	Number average molecular weight (kg/mol)
M_w	Weight average molecular weight (kg/mol)
m	Mass inside the reactor (kg)
m_{poly}	Mass of the polymer inside the reactor (kg)
Me	Metal atoms (mol)
N_s	Number of active sites j
O_2	Oxygen (mol/m^3)
P	Pressure (Pa)
Po	Poison (mol/m^3)
$P^*(j)$	Potential active site of type j (mol)
$P(0,j)$	Uninitiated site of type j produced from activation reaction (mol)
$P_H(0,j)$	Uninitiated site of type j produced from chain transfer to hydrogen reaction (mol)
$P(1)$	Living polymer chain with a chain length one produced by the initiation reaction (mol)
$P(1,j)$	Living polymer chain of type j with a chain length one produced by the initiation reaction (mol)
$P(m)$	Growth living chain with a chain length m with the terminal monomer M (mol)
$P(n)$	Growth living chain with a chain length n with the terminal monomer M (mol)
$P(n+m)$	Growth living chain with a chain length n + m with the terminal monomer M (mol)
$P(n+m,j+k)$	Growth living chain of type j + k with a chain length n + m (mol)
$P(n,j)$	Growth living polymer chain of type j with a chain length n with the terminal monomer M (mol)
$P(m,k)$	Growth living polymer chain of type k with a chain length m with the terminal monomer M (mol)
$P(n+1)$	Growth living chain with a chain length n + 1 with the terminal monomer M (mol)
$P(n+1,j)$	Growth living polymer chain of type j with a chain length n + 1 with the terminal monomer M (mol)
$P_d(n)$	Dead polymer chain (mol)
$P_d(n,j)$	Dead polymer chain of type j (mol)
$P_{dPo}(0,j)$	Impurity site of type j (mol)
$P_{deac}(j)$	Deactivated site of type j (mol)
PDI	Polydispersity index
Po	Impurity (mol/m^3)
q	Heat transfer via the cooling jacket (J/s)
$Q(n,j)$	Dead polymer with n chain length of type j (mol)
Q_{cat}	The mass flow rate of the catalyst (kg/s)
Q_{H_2}	The mass flow rate of the hydrogen (kg/s)
Q_{in}	Inlet flow rate (kg/s)
Q_{Mon}	The mass flow rate of the monomer (kg/s)
Q_m	The mass flow rate (kg/s)
Q_{out}	Outlet flow rate (kg/s)
Q_{outF}	Outflow rate of the polyolefin in the slurry phase (kg/s)
r_A	Rate expression for the active sites (mol/kg catalyst per second)

R_A	Reactant used during polymerization process
$R^\bullet$	Radical (mol)
$R^\bullet(1)$	Living radical with the chain length 1 (mol)
$R^\bullet(2)$	Living radical with the chain length 2 (mol)
$R^\bullet(m)$	Living radical with m chain length (mol)
$R^\bullet(n)$	Living radical with n chain length (mol)
$R^\bullet(n,j)$	Living radical with n chain length of type j (mol)
$R^\bullet(m,k)$	Living radical with n chain length of type k (mol)
$R^\bullet(n+1)$	Living radical with n + 1 chain length (mol)
$R^\bullet(m-r)$	Living radical with m – r chain length (mol)
$R^\bullet(m)$	Living radical with m chain length (mol)
RH	Inert molecule (mol)
R	Gas constant (J/mol·K)
R_d	Deactivation reaction rate (kg/s)
Re	Reynold Number
R_{H_2}	Hydrogen consumption rate (kg/s)
R_p	Polymerization rate (kg/s)
R_{pm}	Polymerization rate (mol/L·s)
R_v	Volumetric production rate of polymer ($m^3 s^{-1}$)
SP	Sub-product (mol/m^3)
T	Temperature (K)
T_{tube}	Cooling jacket tube temperature (K)
T_{ref}	Reference temperature (K)
U	Heat transfer constant or internal energy (W/m^2K)
u	Velocity (m/s)
V	Volume (m^3)
V_p	The volume of polymer phase present in the reactor (m^3)
VSBR	Vertical-stirred bed reactor
V_g	The volume of the gas (m^3)
V_{tube}	The volume of the liquid in the cooling jacket tube (m^3)
WSBR	Well-stirred semi-batch reactor
W	The weight of particle solid (kg)
$X(0)$	Zeroth moment of chain length distribution of the dead polymer (mol)
$X(1)$	The first moment of chain length distribution of the dead polymer (mol)
$X(2)$	The second moment of chain length distribution of the dead polymer (mol)
$X(0,j)$	Zeroth moment of chain length distribution of the dead polymer chain (mol)
$X(1,j)$	The first moment of chain length distribution of the dead polymer chain (mol)
$X(2,j)$	The second moment of chain length distribution of the dead polymer chain (mol)
$X(n,j)$	The n moment of chain length distribution of the dead polymer chain (mol)
Y	Inert molecule (mol)
$Y(0)$	Zeroth moment of chain length distribution of the living polymer (mol)
$Y(1)$	The first moment of chain length distribution of the living polymer (mol)
$Y(2)$	The second moment of chain length distribution of the living polymer (mol)
$Y(0,j)$	Zeroth moment of chain length distribution of the living polymer chain (mol)
$Y(1,j)$	The first moment of chain length distribution of the living polymer chain (mol)
$Y(2,j)$	The second moment of chain length distribution of the living polymer chain (mol)
y_{mono}	Monomer conversion
$Z-N$	Ziegler-Natta

References

1. Szabó, K.P. MOL Group Polyolefin Development in CEE Context. In Proceedings of the 18th Annual CEE & Turkey Refining and Petrochemicals Conference, Budapest, Hungary, 13–15 October 2015.
2. Leute, U.; Dollhopf, W. A Review of the experimental data from the high pressure phase in polyethylene. *Colloid Polym. Sci.* **1980**, *258*, 353–359. [CrossRef]
3. Crossland, B.; Bett, K.E.; Ford, H.; Gardner, A.K. Review of some of the major engineering developments in the high-pressure polyethylene process, 1933–1983. *Proc. Inst. Mech. Eng. Part A J. Power Energy* **1986**, *200*, 237–253. [CrossRef]
4. Romano, U.; Garbassi, F. The environmental issue. A challenge for new generation polyolefins. *Pure Appl. Chem.* **2000**, *72*, 1383–1388. [CrossRef]
5. Sato, H.; Ogawa, H. Review on Development of Polypropylene Manufacturing Process. *RD Rep.* **2009**, *2*, 1–11.
6. Qiao, J.; Guo, M.; Wang, L.; Liu, D.; Zhang, X.; Yu, L.; Liu, Y. Recent advances in polyolefin technology. *Polym. Chem.* **2011**, *2*, 1611–1623. [CrossRef]
7. Wang, D.; Wang, J.; Guo, F.; Gao, Y.; Du, W.; Yang, G. Progress in technology and catalysts for gas phase polyethylene processes. *Adv. Sci. Eng.* **2016**, *8*, 25–31.
8. Gupta, S.K. Low-density polyethylene (LDPE) polymerization—A review. *Curr. Sci.* **1987**, *56*, 979–984.
9. Boaen, N.K.; Hillmyer, M.A. Post-polymerization functionalization of polyolefins. *Chem. Soc. Rev.* **2005**, *34*, 267–275. [CrossRef]
10. Mishra, V.; Kumar, R. Living radical polymerization: A review. *J. Sci. Res.* **2012**, *56*, 141–176.
11. Yu, W.; Liu, J.; Zhou, C. Rheo-chemistry in reactive processing of polyolefin. *Int. Polym. Proc.* **2012**, *27*, 286–298. [CrossRef]
12. Hamielec, A.E.; Soares, J.B.P. Polymerization reaction engineering—Metallocene catalysts. *Prog. Polym. Sci.* **1996**, *21*, 651–706. [CrossRef]
13. Lefebvre, F.; Thivolle-Cazat, J.; Dufaud, V.; Niccolai, G.P.; Basset, J.M. Oxide supported surface organometallic complexes as a new generation of catalysts for carbon-carbon bond activation. *Appl. Catal. A Gen.* **1999**, *182*, 1–8. [CrossRef]
14. Tuchbreiter, A.; Mülhaupt, R. The polyolefin challenges: Catalyst and process design, tailor-made materials, high-throughput development and data mining. *Macromol. Symp.* **2001**, *173*, 1–20. [CrossRef]
15. McKenna, T.F.; Soares, J.B.P. Single particle modelling for olefin polymerization on supported catalysts: A review and proposals for future developments. *Chem. Eng. Sci.* **2001**, *56*, 3931–3949. [CrossRef]
16. Coates, G.W.; Hustad, P.D.; Reinartz, S. Catalysts for the Living Insertion Polymerization of Alkenes: Access to New Polyolefin Architectures Using Ziegler–Natta Chemistry. *Angew. Chem. Int. Ed.* **2002**, *41*, 2236–2257. [CrossRef]
17. Mülhaupt, R. Catalytic Polymerization and Post Polymerization Catalysis Fifty Years after the Discovery of Ziegler's Catalysts. *Macromol. Chem. Phys.* **2003**, *204*, 289–327. [CrossRef]
18. Makio, H.; Kashiwa, N.; Fujita, T. FI Catalysts: A New Family of High Performance Catalysts for Olefin Polymerization. *Adv. Synth. Catal.* **2002**, *344*, 477–493. [CrossRef]
19. Domski, G.J.; Rose, J.M.; Coates, G.W.; Bolig, A.D.; Brookhart, M. Living alkene polymerization: New methods for the precision synthesis of polyolefins. *Prog. Polym. Sci.* **2007**, *32*, 30–92. [CrossRef]
20. Nomura, K.; Kitiyanan, B. Recent progress in precise synthesis of polyolefins containing polar functionalities by transition metal catalysis. *Curr. Org. Synth.* **2008**, *5*, 217–226. [CrossRef]
21. Camacho, D.H.; Guan, Z. Designing late-transition metal catalysts for olefin insertion polymerization and copolymerization. *Chem. Comm.* **2010**, *46*, 7879–7893. [CrossRef] [PubMed]
22. McKenna, T.F.L.; Di Martino, A.; Weickert, G.; Soares, J.B.P. Particle Growth During the Polymerisation of Olefins on Supported Catalysts, 1–Nascent. *Polym. Struc. Macromol. React. Eng.* **2010**, *4*, 40–64. [CrossRef]
23. Heurtefeu, B.; Bouilhac, C.; Cloutet, É.; Taton, D.; Deffieux, A.; Cramail, H. Polymer support of "single-site" catalysts for heterogeneous olefin polymerization. *Prog. Polym. Sci.* **2011**, *36*, 89–126. [CrossRef]
24. Choi, Y.; Soares, J.B.P. Supported single-site catalysts for slurry and gas-phase olefin polymerisation. *Can. J. Chem. Eng.* **2012**, *90*, 646–671. [CrossRef]
25. Baier, M.C.; Zuideveld, M.A.; Mecking, S. Post-Metallocenes in the Industrial Production of Polyolefins. *Angew. Chem. Int. Ed.* **2014**, *53*, 9722–9744. [CrossRef] [PubMed]

26. Shamiri, A.; Chakrabarti, M.; Jahan, S.; Hussain, M.; Kaminsky, W.; Aravind, P.; Yehye, W. The Influence of Ziegler-Natta and Metallocene Catalysts on Polyolefin Structure, Properties, and Processing Ability. *Materials* **2014**, *7*, 5069–5108. [CrossRef] [PubMed]

27. Argyle, M.D.; Bartholomew, C.H. Heterogeneous catalyst deactivation and regeneration: A review. *Catalysts* **2015**, *5*, 145–269. [CrossRef]

28. Pingwei, L.; Weifeng, L.; Wen-Jun, W.; Bo-Geng, L.; Shiping, Z. A Comprehensive Review on Controlled Synthesis of Long-Chain Branched Polyolefins: Part 1, Single Catalyst Systems. *Macromol. React. Eng.* **2016**, *10*, 156–179.

29. Weifeng, L.; Pingwei, L.; Wen-Jun, W.; Bo-Geng, L.; Shiping, Z. A Comprehensive Review on Controlled Synthesis of Long-Chain-Branched Polyolefins: Part 2, Multiple Catalyst Systems and Prepolymer Modification. *Macromol. React. Eng.* **2016**, *10*, 180–200.

30. Sauter, D.; Taoufik, M.; Boisson, C. Polyolefins, a Success Story. *Polymers* **2017**, *9*, 1–13.

31. Alizadeh, A.; McKenna, T.F.L. Particle Growth during the Polymerization of Olefins on Supported Catalysts. Part 2: Current Experimental Understanding and Modeling Progresses on Particle Fragmentation, Growth, and Morphology Development. *Macromol. React. Eng.* **2018**, *12*, 1–24. [CrossRef]

32. Kim, S.D.; Kang, Y. Heat and mass transfer in three-phase fluidized-bed reactors—An overview. *Chem. Eng. Sci.* **1997**, *52*, 3639–3660. [CrossRef]

33. Khan, M.J.H.; Hussain, M.A.; Mansourpour, Z.; Mostoufi, N.; Ghasem, N.M.; Abdullah, E.C. CFD simulation of fluidized bed reactors for polyolefin production—A review. *J. Ind. Eng. Chem.* **2014**, *20*, 3919–3946. [CrossRef]

34. Abbasi, M.R.; Shamiri, A.; Hussain, M. A review on modeling and control of olefin polymerization in fluidized-bed reactors. *Rev. Chem. Eng.* **2018**, 1–23. [CrossRef]

35. Pingwei, L.; Weifeng, L.; Wen-Jun, W.; Bo-Geng, L.; Shiping, Z. A Comprehensive Review on Controlled Synthesis of Long-Chain Branched Polyolefins: Part 3, Characterization of Long-Chain Branched Polymers. *Macromol. React. Eng.* **2017**, *11*, 1–20.

36. Isaeva, V.I.; Aizenshtein, E.M.; Soboleva, O.N. World production and use of polypropylene fibres and thread—A review. *Fibre Chem.* **1997**, *29*, 269–281. [CrossRef]

37. Briskman, B.A. Radiation effects in thermal properties of polymers. An analytical review. I. Polyethylene. *Nucl. Instrum. Methods Phys. Res. Sect. B Beam Interact. Mater.* **2001**, *185*, 116–122. [CrossRef]

38. Gulrez Syed, K.H.; Ali Mohsin, M.E.; Shaikh, H.; Anis, A.; Pulose, A.M.; Yadav, M.K.; Al-Zahrani, S.M. A review on electrically conductive polypropylene and polyethylene. *Polym. Comput.* **2014**, *35*, 900–914. [CrossRef]

39. Mohebbi, A.; Mighri, F.; Ajji, A.; Rodrigue, D. Current issues and challenges in polypropylene foaming: A review. *Cell. Polym.* **2015**, *34*, 299–337. [CrossRef]

40. Kumar, A.; Gupta, R.K. *Fundamentals of Polymer Engineering Second Edition Revised and Expanded*; Marcel Dekker Inc.: New York, NY, USA, 2003; pp. 1–693.

41. Soares, J.B.P.; Pérez, O. Coordination polymerization. In *Handbook of Polymer Synthesis*, 1st ed.; Saldivar-Guerra, E., Vivaldo-Lima, E., Eds.; John Wiley & Sons, Inc.: Hoboken, NJ, USA, 2013; pp. 85–103.

42. Witold, K. Principles of Coordination. In *Polymerisation Heterogenous and Homogenous Catalysys in Polymer Chemistry-Polymerisation of Hydrocarbon, Heterocyclic and Heterounsaturated Monomers*; Witold, K., Ed.; John Wiley & Sons Inc.: Hoboken, NJ, USA, 2001; pp. 1–522.

43. Calhoun, A.; Peacock, A.J. *Polymer Chemistry: Properties and Application*; Hanser, C., Ed.; Hanser: Cincinnati, OH, USA, 2006; pp. 21–31.

44. Bird, R.B.; Stewart, W.E.; Lightfoot, E.N. *Transport Phenomena*, 2nd ed.; Anderson, W., Ed.; John Wiley & Sons, Inc.: Hoboken, NJ, USA, 2002; pp. 1–895.

45. Ibrehem, A.S.; Hussain, M.A.; Ghasem, N.M. Modified mathematical model for gas phase olefin polymerization in fluidized-bed catalytic reactor. *Chem. Eng. J.* **2009**, *149*, 353–362. [CrossRef]

46. Kiparissides, C.; Alexopoulos, A.; Roussos, A.; Dompazis, G.; Kotoulas, C. Population Balance Modeling of Particulate Polymerization Processes. *Ind. Eng. Chem. Res.* **2004**, *43*, 7290–7302. [CrossRef]

47. Ramkrishna, D.; Mahoney, A.W. Population balance modeling. Promise for the future. *Chem. Eng. Sci.* **2002**, *57*, 595–606. [CrossRef]

48. Salahuddin, N.F.; Shamiri, A.; Hussain, M.A.; Mostoufi, N. Fuzzy-GMC Control of Gas-Phase Propylene Copolymerization in Fluidized Bed Reactor. In Proceedings of the 24th Regional Symposium on Chemical Engineering (RSCE 2017), Semarang, Indonesia, 15–16 November 2017.

49. Abbasi, M.R.; Shamiri, A.; Hussain, M.A. Dynamic modeling and Molecular Weight Distribution of ethylene copolymerization in an industrial gas-phase Fluidized-Bed Reactor. *Adv. Powder Technol.* **2016**, *27*, 1526–1538. [CrossRef]

50. Khan, M.J.H.; Hussain, M.A.; Mujtaba, I.M. Developed Hybrid Model for Propylene Polymerisation at Optimum Reaction Conditions. *Polymers* **2016**, *8*, 47. [CrossRef]

51. Shamiri, A.; Wong, S.W.; Zanil, M.F.; Hussain, M.A.; Mostoufi, N. Modified two-phase model with hybrid control for gas phase propylene copolymerization in fluidized bed reactors. *Chem. Eng. J.* **2015**, *264*, 706–719. [CrossRef]

52. Zavala, V.M.; Biegler, L.T. Large-Scale Parameter Estimation in Low-Density Polyethylene Tubular Reactors. *Ind. Eng. Chem. Res.* **2006**, *45*, 7867–7881. [CrossRef]

53. Zavala, V.M.; Laird, C.D.; Biegler, L.T. Interior-point decomposition approaches for parallel solution of large-scale nonlinear parameter estimation problems. *Chem. Eng. Sci.* **2008**, *63*, 4834–4845. [CrossRef]

54. Biegler, L.T. Nonlinear programming strategies for dynamic chemical process optimization. *Theor. Found. Chem. Eng.* **2014**, *48*, 541–554. [CrossRef]

55. Ho, Y.K.; Shamiri, A.; Mjalli, F.S.; Hussain, M.A. Control of industrial gas phase propylene polymerization in fluidized bed reactors. *J. Process Control* **2012**, *22*, 947–958. [CrossRef]

56. Shamiri, A.; Hussain, M.A.; Mjalli, F.S.; Shafeeyan, M.S.; Mostoufi, N. Experimental and Modeling Analysis of Propylene Polymerization in a Pilot-Scale Fluidized Bed Reactor. *Ind. Eng. Chem. Res.* **2014**, *53*, 8694–8705. [CrossRef]

57. Shamiri, A.; Hussain, M.A.; Mjalli, F.S.; Mostoufi, N.; Hajimolana, S. Dynamics and Predictive Control of Gas Phase Propylene Polymerization in Fluidized Bed Reactors. *Chin. J. Chem. Eng.* **2013**, *21*, 1015–1029. [CrossRef]

58. Shamiri, A.; Hussain, M.A.; Mjalli, F.S.; Mostoufi, N.; Shafeeyan, M.S. Dynamic modeling of gas phase propylene homopolymerization in fluidized bed reactors. *Chem. Eng. Sci.* **2011**, *66*, 1189–1199. [CrossRef]

59. Shamiri, A.; Hussain, M.A.; Mjalli, F.S.; Mostoufi, N. Kinetic modeling of propylene homopolymerization in a gas-phase fluidized-bed reactor. *Chem. Eng. J.* **2010**, *161*, 240–249. [CrossRef]

60. Vallerio, M.; Logist, F.; Van Erdeghem, P.; Dittrich, C.; Van Impe, J. Model-Based Optimization of the Cooling System of an Industrial Tubular LDPE Reactor. *Ind. Eng. Chem. Res.* **2013**, *52*, 1656–1666. [CrossRef]

61. Sarmoria, C.; Asteasuain, M.; Brandolin, A. Prediction of molecular weight distributions in polymers using probability generating functions. *Can. J. Chem. Eng.* **2012**, *90*, 263–273. [CrossRef]

62. Shamiri, A.; Hussain, M.A.; Mjalli, F.S.; Mostoufid, N. Improved single phase modeling of propylene polymerization in a fluidized bed reactor. *Comput. Chem. Eng.* **2012**, *36*, 35–47. [CrossRef]

63. Ashrafi, O.; Mostoufi, N.; Sotudeh-Gharebagh, R. Two phase steady-state particle size distribution in a gas-phase fluidized bed ethylene polymerization reactor. *Chem. Eng. Sci.* **2012**, *73*, 1–7. [CrossRef]

64. Ibrehem, A.S.; Hussain, M.A.; Ghasem, N.M. Mathematical model and advanced control for gas-phase olefin polymerization in fluidized-bed catalytic reactors. *Chin. J. Chem. Eng.* **2008**, *16*, 84–89. [CrossRef]

65. Asteasuain, M.; Brandolin, A. Optimal operation of ethylene polymerization reactors for tailored molecular weight distribution. *J. Appl. Polym. Sci.* **2007**, *105*, 2621–2630. [CrossRef]

66. Asteasuain, M.; Brandolin, A. Modeling and optimization of a high-pressure ethylene polymerization reactor using gPROMS. *Comput. Chem. Eng.* **2008**, *32*, 396–408. [CrossRef]

67. Asteasuain, M.; Brandolin, A. High-Pressure Polymerization of Ethylene in Tubular Reactors: A Rigorous Dynamic Model Able to Predict the Full Molecular Weight Distribution. *Macromol. React. Eng.* **2009**, *3*, 398–411. [CrossRef]

68. Ashrafi, O.; Nazari-Pouya, H.; Mostoufi, N.; Sotudeh-Gharebagh, R. Particle size distribution in gas-phase polyethylene reactors. *Adv. Powder Technol.* **2008**, *19*, 321–334. [CrossRef]

69. Khazraei, P.K.F.; Dhib, R. Modeling of ethylene polymerization with difunctional initiators in tubular reactors. *J. Appl. Polym. Sci.* **2008**, *109*, 3908–3922. [CrossRef]

70. Salau, N.P.G.; Neumann, G.A.; Trierweiler, J.O.; Secchi, A.R. Dynamic behavior and control in an industrial fluidized-bed polymerization reactor. *Ind. Eng. Chem. Res.* **2008**, *47*, 6058–6069. [CrossRef]

71. Brandolin, A.; Sarmoria, C.; Failla, M.D.; Vallés, E.M. Mathematical Modeling of the Reactive Modification of High−Density Polyethylene. Effect of Vinyl Content. *Ind. Eng. Chem. Res.* **2007**, *46*, 7561–7570. [CrossRef]

72. Agrawal, N.; Rangaiah, G.P.; Ray, A.K.; Gupta, S.K. Multi-objective optimization of the operation of an industrial low-density polyethylene tubular reactor using genetic algorithm and its jumping gene adaptations. *Ind. Eng. Chem. Res.* **2006**, *45*, 3182–3199. [CrossRef]

73. Kiashemshaki, A.; Mostoufi, N.; Sotudeh-Gharebagh, R. Two-phase modeling of a gas phase polyethylene fluidized bed reactor. *Chem. Eng. Sci.* **2006**, *61*, 3997–4006. [CrossRef]

74. Jafari, R.; Sotudeh-Gharebagh, R.; Mostoufi, N. Performance of the wide-ranging models for fluidized bed reactors. *Adv. Powder Technol.* **2004**, *15*, 533–548. [CrossRef]

75. Harshe, Y.M.; Utikar, R.P.; Ranade, V.V. A computational model for predicting particle size distribution and performance of fluidized bed polypropylene reactor. *Chem. Eng. Sci.* **2004**, *59*, 5145–5156. [CrossRef]

76. Khare, N.P.; Lucas, B.; Seavey, K.C.; Liu, Y.A.; Sirohi, A.; Ramanathan, S.; Chen, C.C. Steady-state and dynamic modeling of gas-phase polypropylene processes using stirred-bed reactors. *Ind. Eng. Chem. Res.* **2004**, *43*, 884–900. [CrossRef]

77. Alizadeh, M.; Mostoufi, N.; Pourmahdian, S.; Sotudeh-Gharebagh, R. Modeling of fluidized bed reactor of ethylene polymerization. *Chem. Eng. J.* **2004**, *97*, 27–35. [CrossRef]

78. Asteasuain, M.; Sarmoria, C.; Brandolin, A. Controlled rheology of polypropylene: Modeling of molecular weight distributions. *J. Appl. Polym. Sci.* **2003**, *88*, 1676–1685. [CrossRef]

79. Asteasuain, M.; P'erez, M.V.; Sarmoria, C.; Brandolin, A. Modelling Molecular Weight Distribution, Vinyl Content and Branching in the Reactive Extrusion of High Density Polyethylene. *Lat. Am. Appl. Res.* **2003**, *33*, 241–249.

80. Chatzidoukas, C.; Perkins, J.D.; Pistikopoulos, E.N.; Kiparissides, C. Optimal grade transition and selection of closed-loop controllers in a gas-phase olefin polymerization fluidized bed reactor. *Chem. Eng. Sci.* **2003**, *58*, 3643–3658. [CrossRef]

81. Cervantes, A.M.; Tonelli, S.; Brandolin, A.; Bandoni, J.A.; Biegler, L.T. Large-scale dynamic optimization for grade transitions in a low density polyethylene plant. *Comput. Chem. Eng.* **2002**, *26*, 227–237. [CrossRef]

82. Meier, G.B.; Weickert, G.; van Swaaij, W.P.M. FBR for catalytic propylene polymerization: Controlled mixing and reactor modeling. *AIChE J.* **2002**, *48*, 1268–1283. [CrossRef]

83. Asteasuain, M.; Sarmoria, C.; Brandolin, A. Peroxide modification of polyethylene. Prediction of molecular weight distributions by probability generating functions. *Polymers* **2002**, *43*, 2363–2373. [CrossRef]

84. Asteasuain, M.; Tonelli, S.M.; Brandolin, A.; Bandoni, J.A. Dynamic simulation and optimisation of tubular polymerisation reactors in gPROMS. *Comput. Chem. Eng.* **2001**, *25*, 509–515. [CrossRef]

85. Asteasuain, M.; Pereda, S.; Lacunza, M.H.; Ugrin, P.E.; Brandolin, A. Industrial high pressure ethylene polymerization initiated by peroxide mixtures: A reduced mathematical model for parameter adjustment. *Polym. Eng. Sci.* **2001**, *41*, 711–726. [CrossRef]

86. Asteasuain, M.; Ugrin, P.E.; Lacunza, M.H.; Brandolin, A. Effect of multiple feedings in the operation of a high-pressure polymerization reactor for ethylene polymerization. *Polym. React. Eng.* **2001**, *9*, 163–182. [CrossRef]

87. Brandolin, A.; Sarmoria, C.; Lótpez-Rodríguez, A.; Whiteley, K.S.; Del Amo Fernandez, B. Prediction of molecular weight distributions by probability generating functions. Application to industrial autoclave reactors for high pressure polymerization of ethylene and ethylene-vinyl acetate. *Polym. Eng. Sci.* **2001**, *41*, 1413–1426. [CrossRef]

88. Ghasem, N.M. Effect of Polymer Particle Size and Inlet Gas Temperature on Industrial Fluidized Bed Polyethylene Reactors. *Chem. Eng. Technol.* **1999**, *22*, 777–783. [CrossRef]

89. Ghasem, N.M. Effect of Polymer Growth Rate and Diffusion Resistance on the Behavior of Industrial Polyethylene Fluidized Bed Reactor. *Chem. Eng. Technol.* **2001**, *24*, 1049–1057. [CrossRef]

90. Fernandes, F.A.N.; Lona, L.M.F. Fluidized bed reactor for polyethylene production. The influence of polyethylene prepolymerization. *Braz. J. Chem. Eng.* **2000**, *17*, 163–170. [CrossRef]

91. Fernandes, F.A.N.; Lona, L.M.F. Fluidized-bed reactor modeling for polyethylene production. *J. Appl. Polym. Sci.* **2001**, *81*, 321–332. [CrossRef]

92. Fernandes, F.A.N.; Lona, L.M.F. Heterogeneous modeling for fluidized-bed polymerization reactor. *Chem. Eng. Sci.* **2001**, *56*, 963–969. [CrossRef]

93. Hatzantonis, H.; Yiannoulakis, H.; Yiagopoulos, A.; Kiparissides, C. Recent developments in modeling gas-phase catalyzed olefin polymerization fluidized-bed reactors: The effect of bubble size variation on the reactor's performance. *Chem. Eng. Sci.* **2000**, *55*, 3237–3259. [CrossRef]

94. Pladis, P.; Kiparissides, C. Dynamic modeling of multizone, multifeed high-pressure LDPE autoclaves. *J. Appl. Polym. Sci.* **1999**, *73*, 2327–2348. [CrossRef]

95. Hatzantonis, H.; Goulas, A.; Kiparissides, C. A comprehensive model for the prediction of particle-size distribution in catalyzed olefin polymerization fluidized-bed reactors. *Chem. Eng. Sci.* **1998**, *53*, 3251–3267. [CrossRef]

96. Khang, D.Y.; Lee, H.H. Particle size distribution in fluidized beds for catalytic polymerization. *Chem. Eng. Sci.* **1997**, *52*, 421–431. [CrossRef]

97. Brandolin, A.; Lacunza, M.H.; Ugrin, P.E.; Capiati, N.J. High Pressure Polymerization of Ethylene. An Improved Mathematical Model for Industrial Tubular Reactors. *Polym. React. Eng.* **1996**, *4*, 193–241. [CrossRef]

98. Zacca, J.J.; Debling, J.A.; Ray, W.H. Reactor residence time distribution effects on the multistage polymerization of olefins. 1. Basic principles and illustrative examples, polypropylene. *Chem. Eng. Sci.* **1996**, *51*, 4859–4886. [CrossRef]

99. McAuley, K.B.; Talbot, J.P.; Harris, T.J. A comparison of two-phase and well-mixed models for fluidized-bed polyethylene reactors. *Chem. Eng. Sci.* **1994**, *49*, 2035–2045. [CrossRef]

100. Choi, K.Y.; Zhao, X.; Tang, S.H. Population Balance Modeling for a Continuous Gas-Phase Olefin Polymerization Reactor. *J. Appl. Polym. Sci.* **1994**, *53*, 1589–1597. [CrossRef]

101. Kiparissides, C.; Verros, G.; Macgregor, J.F. Mathematical-Modeling, Optimization, and Quality-Control of High-Pressure Ethylene Polymerization Reactors. *J. Macromol. Sci.-Rev. Macromol. Chem. Phys.* **1993**, *C33*, 437–527. [CrossRef]

102. McAuley, K.B.; Macgregor, J.F.; Hamielec, A.E. A Kinetic-Model for Industrial Gas-Phase Ethylene Copolymerization. *AIChE J.* **1990**, *36*, 837–850. [CrossRef]

103. Choi, K.Y.; Ray, W.H. The dynamic behavior of continuous stirred-bed reactors for the solid catalyzed gas phase polymerization of propylene. *Chem. Eng. Sci.* **1988**, *43*, 2587–2604. [CrossRef]

104. Brandolin, A.; Capiati, N.J.; Farber, J.N.; Valles, E.M. Mathematical model for high-pressure tubular reactor for ethylene polymerization. *Ind. Eng. Chem. Res.* **1988**, *27*, 784–790. [CrossRef]

105. Choi, K.-Y.; Harmon Ray, W. The dynamic behaviour of fluidized bed reactors for solid catalysed gas phase olefin polymerization. *Chem. Eng. Sci.* **1985**, *40*, 2261–2279. [CrossRef]

106. Zacca, J.J.; Ray, W.H. Modelling of the liquid phase polymerization of olefins in loop reactors. *Chem. Eng. Sci.* **1993**, *48*, 3743–3765. [CrossRef]

107. Yang, Y.; Zou, X.; Xiao, F.; Dong, H. Integrated product-process design approach for polyethylene production. *Chem. Eng. Trans.* **2017**, *61*, 1009–1014.

108. Khare, N.P.; Seavey, K.C.; Liu, Y.A.; Ramanathan, S.; Lingard, S.; Chen, C.-C. Steady-State and Dynamic Modeling of Commercial Slurry High-Density Polyethylene (HDPE) Processes. *Ind. Eng. Chem. Res.* **2002**, *41*, 5601–5618. [CrossRef]

109. Debling, J.A.; Zacca, J.J.; Ray, W.H. Reactor residence-time distribution effects on the multistage polymerization of olefins 3. Multi layered products: Impact polypropylene. *Chem. Eng. Sci.* **1997**, *52*, 1969–2001. [CrossRef]

110. Floyd, S.; Choi, K.Y.; Taylor, T.W.; Ray, W.H. Polymerization of olefins through heterogeneous catalysis III. Polymer particle modelling with an analysis of intraparticle heat and mass transfer effects. *Inst. Syst. Res. Technol. Rep.* **1986**, *32*, 2935–2960. [CrossRef]

111. Busch, J. *Continuous Simulation with Ordinary Differential Equations*; University of Hamburg: Hamburg, Germany, 2013.

112. Lustman, L.; Neta, B.; Grago, W. Solution of ordinary differential initial value problem on an intel hypercube. *Comput. Math. Appl.* **1992**, *23*, 65–72. [CrossRef]

113. Akhtar, M.N.; Durad, M.H.; Ahmed, A. Solving initial value ordinary differential equations by Monte Carlo method. *Proc. Inst. Appl. Math.* **2015**, *4*, 149–174.

114. Bayram, M.; Partal, T.; Buyukoz, G.O. Numerical methods for simulation of stochastic differential equations. *Adv. Diff. Equ.* **2018**, *17*, 1–10. [CrossRef]

115. Lubich, C.; Roche, M. Roesenbrock method for differential algebraic systems with solution-dependet singular matrix multiplying the derivative. *Computing* **1990**, *43*, 325–342. [CrossRef]
116. Roche, M. Rosenbrock methods for differential algebraic equations. *Num. Math.* **1987**, *52*, 45–46. [CrossRef]
117. Bui, T. *Explicit and Implicit Methods in Solving Differential Equations*; University of Connecticut: Storrs, CT, USA, 2009.
118. Causon, D.M.; Mingham, C.G. *Introductory Finite Difference Methods for PDES*; Ventus Publishing ApS: Frederiksberg, Germany, 2010; pp. 1–144.
119. Györi, I. The method of lines for the solution of some nonlinear partial differential equations. *Comput. Math Appl.* **1988**, *15*, 635–658. [CrossRef]
120. Sereno, C.; Rodrigues, A.; Villadsen, J. Solution of partial differential equations systems by the moving finite element method. *Comput. Chem. Eng.* **1992**, *16*, 583–592. [CrossRef]
121. Van Lent, J. Multigrid Methods for Time-Dependent Partial Differential Equations. Ph.D. Thesis, Katholieke Universiteit Leuven, Leuven, Belgium, 2006.

 *processes*

Article

Method of Moments Applied to Most-Likely High-Temperature Free-Radical Polymerization Reactions

Hossein Riazi [1], Ahmad Arabi Shamsabadi [1], Michael C. Grady [2], Andrew M. Rappe [3] and Masoud Soroush [1,*]

[1] Department of Chemical & Biomedical Engineering, Drexel University, Philadelphia, PA 19104, USA; hr339@drexel.edu (H.R.); neginali@sas.upenn.edu (A.A.S.)

[2] Axalta Coating Systems, Philadelphia Navy Yard, PA 19112, USA; mike.grady@axalta.com

[3] Department of Chemistry, University of Pennsylvania, Philadelphia, PA 19104-6323, USA; rappe@sas.upenn.edu

* Correspondence: soroushm@drexel.edu; Tel.: +1-215-895-1710

Received: 13 July 2019; Accepted: 19 September 2019; Published: 26 September 2019

Abstract: Many widely-used polymers are made via free-radical polymerization. Mathematical models of polymerization reactors have many applications such as reactor design, operation, and intensification. The method of moments has been utilized extensively for many decades to derive rate equations needed to predict polymer bulk properties. In this article, for a comprehensive list consisting of more than 40 different reactions that are most likely to occur in high-temperature free-radical homopolymerization, moment rate equations are derived methodically. Three types of radicals—secondary radicals, tertiary radicals formed through backbiting reactions, and tertiary radicals produced by intermolecular chain transfer to polymer reactions—are accounted for. The former tertiary radicals generate short-chain branches, while the latter ones produce long-chain branches. In addition, two types of dead polymer chains, saturated and unsaturated, are considered. Using a step-by-step approach based on the method of moments, this article guides the reader to determine the contributions of each reaction to the production or consumption of each species as well as to the zeroth, first and second moments of chain-length distributions of live and dead polymer chains, in order to derive the overall rate equation for each species, and to derive the rate equations for the leading moments of different chain-length distributions. The closure problems that arise are addressed by assuming chain-length distribution models. As a case study, β-scission and backbiting rate coefficients of methyl acrylate are estimated using the model, and the model is then applied to batch spontaneous thermal polymerization to predict polymer average molecular weights and monomer conversion. These predictions are compared with experimental measurements.

Keywords: method of moments; free-radical polymerization; methyl acrylate; thermal polymerization; high-temperature polymerization

---

1. Introduction

Free-radical polymerization has been used widely to produce a variety of synthetic polymers. The polymerization involves three main (primary) reactions: initiation, propagation and termination [1]. In the initiation reaction, an initiator such as a thermal initiator usually generates free radicals at temperatures above 60 °C [2,3]. The polymerization may also start spontaneously (without adding any known initiator) by a monomer or impurities at high temperatures, typically 140–350 °C [4,5]. In the propagation reactions, free radicals grow in size by reacting with monomers. In the termination reactions, growing macroradicals react with each other to form dead polymer chains. Vinyl, vinylidene, acrylates, and methacrylates are four well-known families of monomers that can participate in free-radical polymerization.

In addition to the primary reactions, free-radical polymerization may involve other reactions, which are known as the secondary reactions. The extent of contribution of each secondary reaction to free-radical polymerization depends on the monomer type, polymerization temperature, and polymerization medium type [6].

Styrene and alkyl acrylates are known to undergo monomer self-initiation, which is a secondary reaction [4,7–9]. Monomer self-initiation is a prevalent secondary reaction in the high-temperature polymerization of many monomers. In case of acrylate monomers, the reaction is significant over 140 °C and potentially obviates the need for a thermal initiator, such as benzoyl peroxide, ammonium persulfate, or azobisisobutyronitrile, to initiate the polymerization [3]. Backbiting and β-scission are two other prevalent secondary reactions in the high-temperature free-radical polymerization of acrylates, while their rates are low around 60–90 °C. Backbiting (intramolecular chain-transfer-to-polymer) reactions produce tertiary radials, which can subsequently propagate and generate short chain branches. The tertiary radical can also participate in a β-scission reaction, which generates a secondary radical and a macromonomer. These tertiary radicals can also undergo a migration reaction along the chain backbone [10]. Intermolecular chain-transfer-to-polymer reactions also produce tertiary radicals, which can subsequently propagate and generate long chain branches. In case of free-radical polymerization in the presence of a solvent, chain transfer to solvent can also occur. Transfer to chain transfer agents is another secondary reaction, which occurs when a transfer agent is present in the reaction medium. Many of these secondary reactions are interdependent. For example, the extent of the β-scission reaction is dependent on the concentration of tertiary radicals formed by backbiting and chain-transfer-to-polymer reactions. Thus, comprehensive models that account for all likely reactions that occur are needed to predict polymer properties reliably.

Given a set of postulated polymerization reactions, one can describe the reaction medium dynamics using one of the four widely-used methods: the method of moments, the tendency modeling, the Monte Carlo simulation, and the Predici simulation package [11]. The tendency modeling can be viewed as a very simplified form of the method of moments [12]; it provides less information than the method of moments. Kinetic Monte Carlo simulations have also been used for modeling polymerization systems. However, it is computationally expensive [13]. Predici is a commercial simulation package used for modeling of polymerization systems [14]. It is able to calculate entire chain-length distributions. Usually, but not always, these methods provide predictions that are in agreement [15].

The method of moments is a powerful and cost-effective method of simulating radical polymerization in different polymerization media [16–19]. It has been used to simulate emulsion copolymerization of vinylidene fluoride and hexafluoropropylene, estimate unavailable kinetic parameters, and predict molecular weight and chain end distributions [18]. Kalfas and Ray used this method to estimate monomer conversion inside particles in a suspension polymerization of vinyl acetate [19]. In the case of running a polymerization in a heterogeneous medium, for which the partitioning coefficient of a monomer between oil and water phases is unknown, the method of moments can be used to estimate the unknown parameter [20]. By dividing an entire range of molecular weights into a limited number of intervals, the method of moments is capable of predicting molecular-weight distributions that can be compared with what is measured experimentally by gel-permeation chromatography [21]. The method of moments was also used to study reaction kinetics in photochemically-mediated atom transfer radical polymerization (ATRP) [22]. In addition, the method of moments enables understanding of the concentrations of free ligands, catalysts, and initiators. Nitroxide-mediated polymerization (NMP) and reversible addition-fragmentation transfer (RAFT) polymerization were also studied by the method of moments approach [23]. In the latter case, one can assess the effect of RAFT agent concentration and rate constants of involved reactions on final polymer properties [24].

Several tutorial and review papers on the method of moments have been published. Mastan and Zhu [25] published a tutorial paper on the method of moments. They considered free-radical polymerization reactions such as initiation, propagation, and termination, but did not account for secondary reactions like monomer self-initiation, backbiting or β-scission [25]. Zhou and Luo [26], in their tutorial paper, focused on method of moments modeling of only living radical polymerization methods such as ATRP, RAFT, and NMP [26]. Bachmann et al. [27] applied the method of moments to a set of polymerization reactions including branching, scission and crosslinking, and concentrated on addressing the closure problems that arise when scission reactions are accounted for. The dependence of the rate of production of a moment of a distribution on a higher-order moment of the same distribution has been referred to as a closure problem. Their formulation allowed for relaxing the assumption of one radical per live chain. However, they did not fully consider backbiting or short-length branching [27]. Other articles on this topic either have not differentiated the radical types (secondary vs. tertiary) or have not fully treated challenging reactions such as β-scission [28–30].

The survey of the literature points to a need for: (i) the derivation of method of moments equations for a comprehensive set of most likely reactions in high-temperature free-radical polymerization; and (ii) a systematic method for the derivation of the contributions of new reactions to moment rate equations when a new reaction must be accounted for. This article was prepared to address these two needs. Based on a comprehensive list of the most likely reactions in high-temperature free-radical homo-polymerization of acrylates, Table S1 (Supporting Information) [31–38], the reader learns how to systematically calculate the contributions of each reaction to different moments and to the production or consumption of different species. Such a comprehensive list consisting of 44 most likely reactions in free-radical polymerization has not been considered elsewhere. β-scission reactions are fully studied herein, and the scissions from both the right-hand side (RHS) and the left-hand side (LHS) of a tertiary radical are considered. Closed-form rate equations are derived for each scission side to describe the contributions of the reaction to the rates of chain-length distribution moments and production/consumption of different chemical species. Three types of radicals and two types of dead polymer chains are considered. The approach presented herein enables the reader to derive easily and systematically new rate equations when a new reaction must be accounted for. As a case study, self-initiated bulk homopolymerization of methyl acrylate (MA) in a batch reactor is considered. Method of moments predictions of conversion and polymer average molecular weights are compared with measurements from polymer sample analyses.

2. Rate Equations

When using the method of moments, it is essential to correctly calculate the contributions of each reaction to the moments of the chain-length distributions of species that are of interest. For example, the jth moment of the chain-length distribution (CLD) of secondary live polymer chains is:

$$\rho_j^{**} = \sum_{n=0}^{\infty} n^j [R_n^{**}], \quad j = 0, 1, 2, \cdots \tag{1}$$

where n is the length (number of monomer units) of a polymer chain, $[R_n^{**}]$ is the molar concentration of live chains that have a secondary radical and n monomer units, and ρ_j^{**} is the jth moment of the chain-length distribution of the secondary radicals. For each class of polymer chains, from the zeroth, first and second moments of the CLD of the chains, one can calculate the number-average and weight-average chain lengths as well as the dispersity of the CLD. The zeroth moment of the CLD of a polymer class is the molar concentration of the polymer class, and the first moment times the monomer molecular weight is the mass concentration of the polymer class [39,40]. Similarly, moments of other CLDs can be defined:

$$\rho_j^{***} = \sum_{n=0}^{\infty} n^j [R_n^{***}], \quad j = 0, 1, 2, \cdots \tag{2}$$

$$\widetilde{\rho}_j^{***} = \sum_{n=0}^{\infty} n^j \left[\widetilde{R}_n^{***}\right] \quad , j = 0,\, 1,\, 2,\, \cdots \tag{3}$$

$$\delta_j = \sum_{n=0}^{\infty} n^j [D_n], \quad j = 0,\, 1,\, 2,\, \cdots \tag{4}$$

$$\epsilon_j = \sum_{n=0}^{\infty} n^j [U_n], \quad j = 0,\, 1,\, 2,\, \cdots \tag{5}$$

where $\left[R_n^{***}\right]$ is the molar concentration of n-monomer-unit live chains that have a tertiary radical produced by intermolecular chain transfer to polymer reactions, $\left[\widetilde{R}_n^{***}\right]$ is the molar concentration of n-monomer-unit live chains that have a tertiary radical produced by backbiting reactions, $[D_n]$ is the molar concentration of n-monomer-unit dead chains without a terminal double bond, and $[U_n]$ is the molar concentration of n-monomer-unit dead chains with a terminal double bond. ρ_j^{***}, $\widetilde{\rho}_j^{***}$, ϵ_j, and δ_j are the jth moments of the chain-length distributions of the R_n^{***}, $\widetilde{R}_n^{***}$, U_n and D_n polymer chains, respectively.

2.1. Initiation with Conventional Thermal Initiators

In this reaction, a molecule of a thermal initiator such as ammonium persulfate or potassium persulfate decomposes into two free radicals:

$$I \xrightarrow{k_d} 2R_0^* \tag{6}$$

where I and R_0^*, respectively, represent an initiator molecule and a monoradical. Contributions of this reaction to the consumption of the initiator and the generation of free radicals with a length of zero (with no monomer units) are given by:

$$r_I = -k_d\,[I] \tag{7}$$

$$r_{R_0^*} = 2f k_d [I] \tag{8}$$

where f is the free-radical generation efficiency of the initiator. An initiator efficiency of less than 100% has been attributed to a phenomenon called the cage effect [41]. It is worth noting that this reaction does not have any monomer-unit contributions to the zeroth, first or second moments of live chains as the length (number of monomer units) of the produced live radicals is zero.

2.2. Self-Initiation of Monomers

Previous studies by our group [42–44] showed that three alkyl acrylate monomers react and form one monoradical with the length of one monomer unit, R_1^{**}, and one monoradical with the length of two monomer units, R_2^{**}:

$$3M \xrightarrow{k_i} R_1^{**} + R_2^{**} \tag{9}$$

Contributions of this reaction to the production/consumption rates of monomer (M), secondary radicals with one monomer unit length (R_1^{**}), secondary radicals with two monomer units length (R_2^{**}), zeroth moment of secondary radicals (ρ_0^{**}), first moment of secondary radicals (ρ_1^{**}), and second moment of the secondary radicals (ρ_2^{**}) are given by:

$$r_M = -3k_i\,[M]^2 \tag{10}$$

$$r_{R_1^{**}} = k_i\,[M]^2 \tag{11}$$

$$r_{R_2^{**}} = k_i\,[M]^2 \tag{12}$$

$$r_{\rho_0^{**}} = \left(1^0 + 2^0\right) k_i \, [M]^2 \tag{13}$$

$$r_{\rho_1^{**}} = \left(1^1 + 2^1\right) k_i \, [M]^2 \tag{14}$$

$$r_{\rho_2^{**}} = \left(1^2 + 2^2\right) k_i \, [M]^2 \tag{15}$$

In the last three (moment) rate equations, the "1" accounts for the single monomer unit of the secondary radical R_1^{**} and "2" for the two monomer units of the secondary radical R_2^{**}. In the case of methacrylates, the monomers undergo the overall (apparent) reactions [42–44]:

$$3M \xrightarrow{k_i} R_1^{**} + R_2^{**} \tag{16}$$

$$2M \xrightarrow{k_{dim}} D_2 \tag{17}$$

Contributions of these reactions to the production rates of monomer (M), secondary radicals with one monomer unit length (R_1^{**}), secondary radicals with two monomer unit length (R_2^{**}), dead polymer chain with two monomer unit length (D_2), zeroth moment of secondary radicals (ρ_0^{**}), first moment of secondary radicals (ρ_1^{**}), and second moment of the secondary radicals (ρ_2^{**}) are given by:

$$r_M = -(3k_i + 2k_{dim}) \, [M]^2 \tag{18}$$

$$r_{R_1^{**}} = k_i \, [M]^2 \tag{19}$$

$$r_{R_2^{**}} = k_i \, [M]^2 \tag{20}$$

$$r_{\rho_0^{**}} = \left(1^0 + 2^0\right) k_i \, [M]^2 \tag{21}$$

$$r_{\rho_1^{**}} = \left(1^1 + 2^1\right) k_i \, [M]^2 \tag{22}$$

$$r_{\rho_2^{**}} = \left(1^2 + 2^2\right) k_i \, [M]^2 \tag{23}$$

$$r_{\delta_0} = \left(2^0\right) k_{dim} [M]^2 \tag{24}$$

$$r_{\delta_1} = \left(2^1\right) k_{dim} [M]^2 \tag{25}$$

$$r_{\delta_2} = \left(2^2\right) k_{dim} [M]^2 \tag{26}$$

2.3. Propagation Reactions

In these reactions, a live chain reacts with a monomer molecule and forms a new live chain that is one monomer unit longer than the reactant live chain. For example, a secondary radical reacts with a monomer molecule and forms a new secondary radical that is one monomer unit longer than the reactant secondary:

$$R_n^{**} + M \xrightarrow{k_p} R_{n+1}^{**} \tag{27}$$

Since there is no change in the population of secondary radicals upon the occurrence of this reaction, the contribution of this reaction to the zeroth moment of the chain-length distribution of the secondary radicals is zero:

$$r_{\rho_0^{**}} = k_p \, [M] \sum (n+1)^0 [R_n^{**}] - k_p \, [M] \sum n^0 [R_n^{**}] = 0 \tag{28}$$

The contributions of this reaction to the production of other species and to other moments are given by:

$$r_M = -k_p \, [M] \sum [R_n^{**}] = -k_p \, [M] \rho_0^{**} \tag{29}$$

$$r_{\rho_1^{**}} = k_p \, [M] \left(\sum (n+1)[R_n^{**}] - \sum n[R_n^{**}] \right) = k_p \, [M] \sum [R_n^{**}] = k_p \, [M] \rho_0^{**} \tag{30}$$

$$r_{\rho_2^{**}} = k_p\,[M]\left(\sum (n+1)^2 [R_n^{**}] - \sum n^2 [R_n^{**}]\right) = k_p\,[M]\sum (2n+1)[R_n^{**}]$$
$$= k_p\,[M]\left(2\rho_1^{**} + \rho_0^{**}\right) \tag{31}$$

$$r_{R_0^*} = -k_p\,[M][R_0^*] \tag{32}$$

$$r_{R_1^{**}} = k_p\,[M]\left([R_0^*] - [R_1^{**}]\right) \tag{33}$$

$$r_{R_2^{**}} = k_p\,[M]\left([R_1^*] - [R_2^{**}]\right) \tag{34}$$

Note that the rate equations satisfy $r_M + r_{\rho_1^{**}} = 0$, confirming that the reaction does not change the total number of monomer units in the system. Tertiary radicals of types R_n^{***} and $\widetilde{R}_n^{***}$ also participate in propagation reactions. As they are more stable than secondary radicals, their propagation reaction rate coefficients are smaller than those of secondary radicals. As a rule of thumb, the propagation rate coefficient of tertiary radicals (k_p^t) is 1000 times less than that of the secondary radicals (k_p) [45]. The product of propagation of a tertiary radical is a secondary radical with one monomer unit longer than the tertiary radical.

$$R_n^{***} + M \xrightarrow{k_p^t} R_{n+1}^{**} \tag{35}$$

$$\widetilde{R}_n^{***} + M \xrightarrow{k_p^t} R_{n+1}^{**} \tag{36}$$

The occurrence of the reaction (35) leads to the formation of long chain branching, while the reaction (36) causes the formation of short chain branches. The contributions of these two reactions to the production rates of different species as well as rates of the moments are presented in the SI (Equations (S1)–(S8) and (S9)–(S16))

2.4. De-Propagation

De-propagation is a reaction that is significant in high-temperature polymerization of methacrylates. In this reaction, a secondary radical generates a monomer molecule and a new secondary radical which is one monomer unit shorter than the reactant secondary radical:

$$R_{n+1}^{**} \xrightarrow{k_{-p}} R_n^{**} + M \tag{37}$$

The contribution of this reaction to the production and consumption of different species as well as to the moments are as follows:

$$r_M = k_{-p}\,\rho_0^{**} \tag{38}$$

$$r_{\rho_0^{**}} = 0 \tag{39}$$

$$r_{\rho_1^{**}} = -k_{-p}\rho_0^{**} \tag{40}$$

$$r_{\rho_2^{**}} = -k_{-p}\left(2\rho_1^{**} - \rho_0^{**}\right) \tag{41}$$

2.5. Backbiting

A backbiting reaction converts a secondary radical to a tertiary counterpart. This reaction is prevalent at high temperatures and involves the abstraction of a hydrogen by a secondary radical from its backbone:

$$R_n^{**} \xrightarrow{k_{bb}} \widetilde{R}_n^{***} \tag{42}$$

To abstract a hydrogen, the chain head with a secondary radical (on carbon atom #1) bends and abstracts a hydrogen usually from carbon #5, or less-likely carbon #7; the radical moves from carbon #1 to carbon #5 or #7. Quantum chemical calculations have shown that carbon #5 is the most probable place for hydrogen abstraction (See Figure 1) [46]. The reaction of a monomer with the produced tertiary radical causes the formation of a short branch in the polymer chain. One challenge here is that this tertiary radical can migrate along the backbone, which is discussed in the next section. Backbiting affects just the rates of moments of secondary and tertiary radicals as follows:

$$r_{\rho_0^{**}} = -r_{\widetilde{\rho}_0^{***}} = -k_{bb} \sum n^0 [R_n^{**}] = -k_{bb}\, \rho_0^{**} \tag{43}$$

$$r_{\rho_1^{**}} = -r_{\widetilde{\rho}_1^{***}} = -k_{bb} \sum n^1 [R_n^{**}] = -k_{bb}\, \rho_1^{**} \tag{44}$$

$$r_{\rho_2^{**}} = -r_{\widetilde{\rho}_2^{***}} = -k_{bb} \sum n^2 [R_n^{**}] = -k_{bb}\, \rho_2^{**} \tag{45}$$

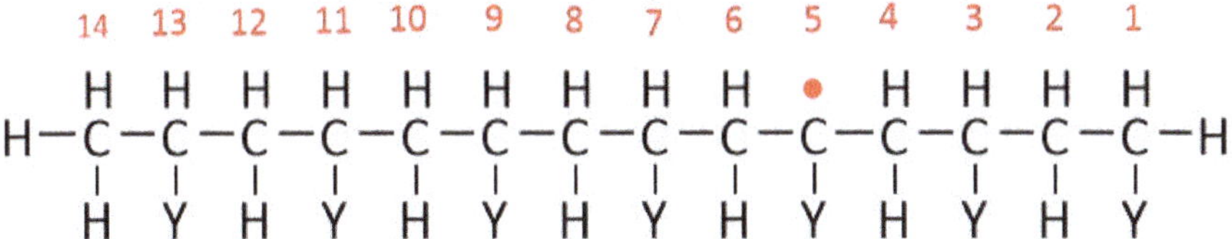

Figure 1. A polymer chain with seven monomer units with a tertiary radical on carbon #5. Y represents a functional group like an alkyl group.

2.6. Mid-Chain Radical Migration

A tertiary radical can migrate along the backbone of the live polymer chain. In the case of a tertiary radical of the type $\widetilde{R}_n^{***}$:

$$\widetilde{R}_n^{***} \xrightarrow{k_{mig}} R_n^{***} \tag{46}$$

This reaction has been reported to occur at high-temperature, low-monomer-concentration polymerization of acrylates [10,47]. Mid-chain radicals that are formed by backbiting usually undergo this migration reaction. Mid-chain radicals formed by macromonomer propagation can also participate in this reaction. Although this reaction does not alter the tertiary nature of radicals, it may change branch sizes (if propagation occurs after migration) or the macromonomer length (if scission happens after migration) [48]. The migration reaction has been reported to be 50 times slower than backbiting but 15 times faster than β-Scission [48]. The occurrence of this reaction causes the generation of low mass macromonomers in polymerizing systems [49]. Migration of mid-chain radicals usually happens between every other monomer (n: $n + 4$); this is why β-scission reaction, which usually occurs after several migration steps, produces macromonomers with different distribution chain length.

Contributions of the reaction (46) to the rates of moments are:

$$r_{\widetilde{\rho}_0^{***}} = -r_{\rho_0^{***}} = -k_{mig}\widetilde{\rho}_0^{***} \tag{47}$$

$$r_{\widetilde{\rho}_1^{***}} = -r_{\rho_1^{***}} = -k_{mig}\widetilde{\rho}_1^{***} \tag{48}$$

$$r_{\widetilde{\rho}_2^{***}} = -r_{\rho_2^{***}} = -k_{mig}\widetilde{\rho}_2^{***} \tag{49}$$

2.7. β-Scission Reaction of $\widetilde{R}_n^{***}$

This β-scission reaction converts a tertiary radical of type $\widetilde{R}_n^{***}$ to a secondary radical R_m^{**} and a macromonomer (unsaturated dead polymer chain) U_{n-m} that possesses a shorter length than its parent chain. The reaction is prevalent at high temperatures and involves a carbon–carbon bond scission, which lowers polymer molecular weight [50]. β-Scission can happen from both RHS and LHS of a tertiary radical where the scission side determines the length of the produced macromonomer as well as the length of the secondary radical. Equations (50) and (51), respectively, show scission from the RHS and LHS of a chain with a free electron on the tertiary carbon #5. In Figure 1, if scission happens from the LHS of the tertiary radical, the covalent bond between carbons #6 and #7 breaks, while the scission from the RHS causes the breakage of the bond between carbons #3 and #4. Regardless of the chain length, scission from the RHS of the tertiary carbon #5 always produces a secondary radical with two-monomer-unit length, while the scission from the LHS of the same tertiary carbon always produces a macromonomer with three-monomer-unit length:

$$\widetilde{R}_n^{***} \xrightarrow{k_\beta} R_2^{**} + U_{n-2} \tag{50}$$

$$\widetilde{R}_n^{***} \xrightarrow{k_\beta} R_{n-3}^{**} + U_3 \tag{51}$$

It has been reported that a secondary radical is most likely to abstract a hydrogen from carbon #5 [51]. Thus, in a subsequent β-Scission reaction, mostly like the C_3-C_4 or C_6-C_7 carbon–carbon bond breaks. Here, we develop equations describing the contribution of this scission site to moment equations. However, the same method is applicable if one needs to account for scission from other tertiary carbons. We start by deriving the contributions of a single chain with seven monomer units (Figure 1) and then generalize them for a chain with n monomer units. Different sets of equations for the RHS and LHS reactions are derived, as the reaction products are different in the two cases.

$$r_{\rho_{0,R}^{**}} = \frac{1}{2}k_\beta[\widetilde{R}_7^{***}] \times 2^0 = \frac{1}{2}k_\beta[\widetilde{R}_7^{***}] \tag{52}$$

where the symbol R in $\rho_{0,R}^{**}$ represents that the scission occurs from the RHS. The pre-factor $\frac{1}{2}$ also emphasizes that the scission is equally likely from the RHS and LHS. Generalizing this, the contribution of an n-monomer-unit chain to the zero moment of secondary radicals is:

$$r_{\rho_{0,R}^{**}} = \frac{1}{2}k_\beta[\widetilde{R}_n^{***}] \times 2^0 = \frac{1}{2}k_\beta[\widetilde{R}_n^{***}] \tag{53}$$

and the contribution of all chains with different lengths is given by:

$$r_{\rho_{0,R}^{**}} = \sum \frac{1}{2}k_\beta[\widetilde{R}_n^{***}] = \frac{1}{2}k_\beta\widetilde{\rho}_0^{***} \tag{54}$$

Similarly, the contributions to the first and second moments of the produced secondary radicals are given by:

$$r_{\rho_{1,R}^{**}} = \sum \frac{1}{2}k_\beta[\widetilde{R}_n^{***}] \times 2 = k_\beta\widetilde{\rho}_0^{***} \tag{55}$$

$$r_{\rho_{2,R}^{**}} = \sum \frac{1}{2}k_\beta[\widetilde{R}_n^{***}] \times 2^2 = 2k_\beta\widetilde{\rho}_0^{***} \tag{56}$$

For the scission from LHS of tertiary carbon #5, Equation (51), the following relations are developed to account the contribution of the reaction to zero, first and second moments of produced secondary radicals:

$$r_{\rho_{0,L}^{**}} = \sum \frac{1}{2}k_\beta[\widetilde{R}_n^{***}] = \frac{1}{2}k_\beta\widetilde{\rho}_0^{***} \tag{57}$$

$$r_{\rho_{1,L}^{**}} = \sum \frac{1}{2}k_\beta[\widetilde{R}_n^{***}] \times (n-3) = \frac{1}{2}k_\beta(\widetilde{\rho}_1^{***} - 3\widetilde{\rho}_0^{***}) \tag{58}$$

$$r_{\rho_{2,L}^{**}} = \sum \frac{1}{2}k_\beta[\widetilde{R}_n^{***}] \times (n-3)^2 = \frac{1}{2}k_\beta(\widetilde{\rho}_2^{***} - 6\widetilde{\rho}_1^{***} + 9\widetilde{\rho}_0^{***}) \tag{59}$$

Following the same concept, contributions of this reaction to the rates of moments of tertiary radicals as well as macromonomers come below:

$$r_{\widetilde{\rho}_{0,L}^{***}} = r_{\widetilde{\rho}_{0,R}^{***}} = -\sum \frac{1}{2}k_\beta[\widetilde{R}_n^{***}] = -\frac{1}{2}k_\beta\widetilde{\rho}_0^{***} \tag{60}$$

$$r_{\widetilde{\rho_{1,L}^{***}}} = r_{\widetilde{\rho_{1,R}^{***}}} = -\sum \frac{1}{2}k_\beta \times n[\widetilde{R_n^{***}}] = -\frac{1}{2}k_\beta\widetilde{\rho_1^{***}} \tag{61}$$

$$r_{\widetilde{\rho_{2,L}^{***}}} = r_{\widetilde{\rho_{2,R}^{***}}} = -\sum \frac{1}{2}k_\beta \times n^2[\widetilde{R_n^{***}}] = -\frac{1}{2}k_\beta\widetilde{\rho_2^{***}} \tag{62}$$

$$r_{\epsilon_{0,R}} = \sum \frac{1}{2}k_\beta[\widetilde{R_n^{***}}] = \frac{1}{2}k_\beta\widetilde{\rho_0^{***}} \tag{63}$$

$$r_{\epsilon_{1,R}} = \sum \frac{1}{2}k_\beta \times (n-2)[\widetilde{R_n^{***}}] = \frac{1}{2}k_\beta(\widetilde{\rho_1^{***}} - 2\widetilde{\rho_0^{***}}) \tag{64}$$

$$r_{\epsilon_{2,R}} = \sum \frac{1}{2}k_\beta \times (n-2)^2[\widetilde{R_n^{***}}] = \frac{1}{2}k_\beta(\widetilde{\rho_2^{***}} - 4\widetilde{\rho_1^{***}} + 4\widetilde{\rho_0^{***}}) \tag{65}$$

$$r_{\epsilon_{0,L}} = \sum \frac{1}{2}k_\beta[\widetilde{R_n^{***}}] = \frac{1}{2}k_\beta\widetilde{\rho_0^{***}} \tag{66}$$

$$r_{\epsilon_{1,L}} = \sum \frac{1}{2}k_\beta \times 3[\widetilde{R_n^{***}}] = \frac{3}{2}k_\beta\widetilde{\rho_0^{***}} \tag{67}$$

$$r_{\epsilon_{2,L}} = \sum \frac{1}{2}k_\beta \times 9[\widetilde{R_n^{***}}] = \frac{9}{2}k_\beta\widetilde{\rho_0^{***}} \tag{68}$$

2.8. β-Scission Reaction of R_n^{***}

Tertiary radicals formed through chain transfer to polymer (R_n^{***}) also undergo scission from RHS and LHS. The most important difference between the β-scission reaction of R_n^{***} and $\widetilde{R_n^{***}}$ is the number of possible sites for scission. As mentioned in Section 2.7, $\widetilde{R_n^{***}}$ most likely undergoes scission from carbon #5 or less likely from carbon #7. It means that just one or two sites are available for scission. On the other hand, R_n^{***} can undergo scission from all available tertiary carbons along its backbone. As in Section 2.7, the β-Scission reaction of a tertiary radical of type R_n^{***} generates a secondary radical and a macromonomer:

$$R_n^{***} \xrightarrow{k_\beta} R_m^{**} + U_{n-m} \tag{69}$$

$$R_n^{***} \xrightarrow{k_\beta} R_{n-m}^{**} + U_m \tag{70}$$

Figure 2 shows the products of both LHS and RHS β-Scission reactions of a chain with seven monomer units. To write the contributions of this reaction to the consumption and generation of different species as well as rates of moments, we first consider scission from RHS and then scission from LHS where there is no chance for carbon #13 to participate in β-scission reaction.

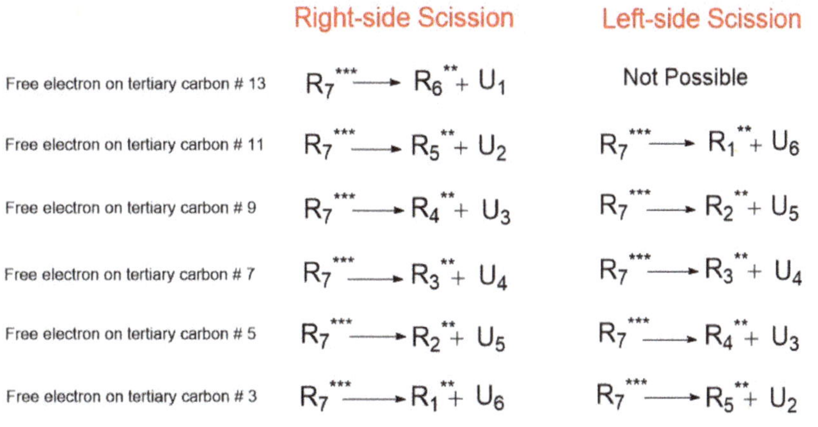

Figure 2. Possible β-scission reactions for a chain with seven monomer units from right-hand side and left-hand side of a tertiary radical formed by transfer to polymer reaction.

In the case of an RHS scission, only one reaction out of the six possible reactions can occur for a chain with seven monomer units every time, Figure 2. Thus, assuming that the reaction sites have the same reactivity to participate in the reaction and equal chances for RHS and LHS scissions for each tertiary radical, the probability of one site to participate in RHS scission is $\frac{1}{2} \times \frac{1}{6}$. The contributions of all possible six reactions of a seven-monomer-unit chain to the zeroth moment of the CLD of the tertiary radicals is:

$$r_{\rho_{0,R}^{***}} = -\frac{1}{6} \times \frac{1}{2} k_\beta [R_7^{***}] \times 6 \tag{71}$$

Generalizing this, the contribution of an n-monomer-unit chain that has $(n-1)$ reaction sites is:

$$r_{\rho_{0,R}^{***}} = -\frac{1}{n-1} \times \frac{1}{2} k_\beta [R_n^{***}] \times (n-1) \tag{72}$$

and the contribution of all chains with different lengths is given by:

$$r_{\rho_{0,R}^{***}} = -\sum \frac{1}{n-1} \times \frac{1}{2} k_\beta [R_n^{***}] \times (n-1) = -\frac{1}{2} k_\beta \rho_0^{***} \tag{73}$$

Similarly, the contributions to the first and second moments of the live chains with tertiary radicals are given by:

$$r_{\rho_{1,R}^{***}} = -\sum \frac{1}{n-1} \times \frac{1}{2} n k_\beta [R_n^{***}] \times (n-1) = -\frac{1}{2} k_\beta \rho_1^{***} \tag{74}$$

$$r_{\rho_{2,R}^{***}} = -\sum \frac{1}{n-1} \times \frac{1}{2} n^2 k_\beta [R_n^{***}] \times (n-1) = -\frac{1}{2} k_\beta \rho_2^{***} \tag{75}$$

The reaction (69) generates secondary radicals as well. As the list of reactions for RHS scission in Figure 2 shows, the contribution of the tertiary chain with seven monomer units to the zeroth moment of the CLD of the R_n^{**} radicals is:

$$r_{\rho_{0,R}^{**}} = +\frac{1}{6} \times \frac{1}{2} k_\beta [R_7^{***}] \times 6 \tag{76}$$

which for an n-monomer-unit tertiary chain that has $(n-1)$ reaction sites becomes:

$$r_{\rho_{0,R}^{**}} = +\frac{1}{n-1} \times \frac{1}{2} k_\beta [R_n^{***}] \times (n-1) \tag{77}$$

and the contribution of all chains with different lengths is given by:

$$r_{\rho_{0,R}^{**}} = \sum \frac{1}{n-1} \times \frac{1}{2} k_\beta [R_n^{***}] \times (n-1) = \frac{1}{2} k_\beta \rho_0^{***} \tag{78}$$

According to Figure 2, the scission reactions from RHS generate secondary live chains with 1, 2, 3, 4, 5, and 6 monomer units. Thus, the contributions of the tertiary chain with seven monomer units to the first moment of the CLD of the R_n^{**} radicals are:

$$\begin{aligned} r_{\rho_{1,R}^{**}} &= \frac{1}{6} \times \frac{1}{2} k_\beta [R_7^{***}]6 + \frac{1}{6} \times \frac{1}{2} k_\beta [R_7^{***}]5 + \frac{1}{6} \times \frac{1}{2} k_\beta [R_7^{***}]4 + \frac{1}{6} \\ &\quad \times \frac{1}{2} k_\beta [R_7^{***}]3 + \frac{1}{6} \times \frac{1}{2} k_\beta [R_7^{***}]2 + \frac{1}{6} \times \frac{1}{2} k_\beta [R_7^{***}]1 \\ &= \frac{1}{6} \times \frac{1}{2} k_\beta [R_7^{***}] (1+2+3+4+5+6) \end{aligned} \tag{79}$$

which is the sum of the multiplication products of the rate of production of each secondary radical and the number of monomer units of the produced secondary radical. Generalizing this, the contribution of a tertiary chain with n monomer units to the first moment of the CLD of the R_n^{**} radicals is:

$$r_{\rho_{1,R}^{**}} = \frac{1}{n-1} \times \frac{1}{2} k_\beta [R_n^{***}](1+2+3+4+\cdots+(n-1)) \tag{80}$$

Using the following identity [52]:

$$1 + 2 + 3 + 4 + \cdots + n = \frac{1}{2}n(n+1) \tag{81}$$

one can write:

$$1 + 2 + 3 + \cdots + (n-1) = \frac{1}{2}(n-1)n \tag{82}$$

Thus, the contribution of all chains with different lengths to the rate of the first moment of the CLD of the secondary radicals is given by:

$$r_{\rho_{1,R}^{**}} = \sum \frac{1}{n-1} \times \frac{1}{2} k_\beta [R_n^{***}](\frac{1}{2}(n-1)n) = \frac{1}{4} k_\beta \sum [R_n^{***}](n) = \frac{1}{4} k_\beta \rho_1^{***} \tag{83}$$

In a similar way, according to Figure 2, we can write the rate of change in the second moment of CLD of the secondary radicals upon RHS of a chain with seven monomer units:

$$\begin{aligned}
r_{\rho_{2,R}^{**}} &= \tfrac{1}{6} \times \tfrac{1}{2} k_\beta [R_7^{***}]6^2 + \tfrac{1}{6} \times \tfrac{1}{2} k_\beta [R_7^{***}]5^2 + \tfrac{1}{6} \times \tfrac{1}{2} k_\beta [R_7^{***}]4^2 + \tfrac{1}{6} \\
&\quad \times \tfrac{1}{2} k_\beta [R_7^{***}]3^2 + \tfrac{1}{6} \times \tfrac{1}{2} k_\beta [R_7^{***}]2^2 + \tfrac{1}{6} \times \tfrac{1}{2} k_\beta [R_7^{***}]1^2 \\
&= \tfrac{1}{6} \times \tfrac{1}{2} k_\beta [R_7^{***}] \,(1^2 + 2^2 + 3^2 + 4^2 + 5^2 + 6^2)
\end{aligned} \tag{84}$$

which is the sum of the products of the production rate of each secondary radical multiplied by the number of monomer units of the secondary radical that the reaction generates, raised to power two. The preceding equation for a chain with n monomer units becomes:

$$r_{\rho_{2,R}^{**}} = \frac{1}{n-1} \times \frac{1}{2} k_\beta [R_n^{***}](1^2 + 2^2 + 3^2 + \cdots + (n-1)^2) \tag{85}$$

Using the following identity [52]:

$$1^2 + 2^2 + 3^2 + \cdots + n^2 = \frac{n(n+1)(2n+1)}{6} \tag{86}$$

one can write:

$$1^2 + 2^2 + 3^2 + \cdots + (n-1)^2 = \frac{(n-1)(n)(2n-1)}{6} \tag{87}$$

Thus, the contribution of all chains with different lengths to the rate of the second moment of the CLD of the secondary radicals is given by:

$$r_{\rho_{2,R}^{**}} = \sum \frac{1}{n-1} \times \frac{1}{2} k_\beta [R_n^{***}] \frac{(n-1)(n)(2n-1)}{6} = \frac{1}{12} k_\beta (2\rho_2^{***} - \rho_1^{***}) \tag{88}$$

The reaction (69) generates unsaturated dead polymers as well. Equations of (76) to (88) are exactly applicable for zeroth, first and second moments of the macromonomers:

$$r_{\epsilon_{0,R}} = \sum \frac{1}{n-1} \times \frac{1}{2} k_\beta [R_n^{***}] \times (n-1) = \frac{1}{2} k_\beta \rho_0^{***} \tag{89}$$

$$r_{\epsilon_{1,R}} = \sum \frac{1}{n-1} \times \frac{1}{2} k_\beta [R_n^{***}] \frac{1}{2}(n-1)(n) = \frac{1}{4} k_\beta (\rho_1^{***}) \tag{90}$$

$$r_{\epsilon_{2,R}} = \sum \frac{1}{n-1} \times \frac{1}{2} k_\beta [R_n^{***}] \frac{(n-1)(n)(2n-1)}{6} = \frac{1}{12} k_\beta (2\rho_2^{***} - \rho_1^{***}) \tag{91}$$

The rate equations derived in this section satisfy $r_{\rho_1^{***}} + r_{\rho_1^{**}} + r_{\epsilon_1} = 0$, implying that the reaction does not change the total number of monomer units in the system.

β-scission from the LHS of a tertiary radical of the type R_n^{***} can also happen, which similarly produces a secondary radical and a macromonomer with less monomer units compared with the parental tertiary radical:

$$R_n^{***} \xrightarrow{k_\beta} R_{n-m}^{**} + U_m \tag{92}$$

As Figure 2 shows, the LHS β-scission reaction produces species with different lengths from those of RHS β-Scission reaction. Thus, a different set of equations is needed to describe the dynamic of this reaction. The details of equations' derivation are mentioned in the Supporting Information (SI, Equations (S17)–(S25)).

2.9. Chain Transfer to a Solvent

A secondary radical (R_n^{**}) can abstract a hydrogen from a solvent molecule (S) to generate a dead polymer chain (D_n) and an active solvent-based radical (R_0^*):

$$R_n^{**} + S \xrightarrow{k_{tr,s}} D_n + R_0^* \tag{93}$$

Solvents are usually added to polymerization media to decrease the viscosity build up upon monomer to polymer conversion and to obviate its consequences like gel and glass effects. Their presence in the reaction media lowers average molecular weights, if the rates of their chain-transfer-to-solvent reactions are appreciable. The extent of chain transfer to solvent varies from one solvent to another, and is usually described with a dimension-less number called the transfer to solvent constant [53]:

$$C_{tr,s} = \frac{k_{tr,s}}{k_p} \tag{94}$$

where $k_{tr,s}$ and k_p are rate coefficients of transfer to solvent and propagation reactions, respectively.

The following rate equations describe the contributions of this reaction to the production and consumption of different species as well as the relevant moments:

$$r_s = -k_{tr,s}[S]\rho_0^{**} \tag{95}$$

$$r_{R_0^*} = k_{tr,s}[S]\rho_0^{**} \tag{96}$$

$$r_{R_1^{**}} = -k_{tr,s}[S][R_1^{**}] \tag{97}$$

$$r_{R_2^{**}} = -k_{tr,s}[S][R_2^{**}] \tag{98}$$

$$r_{\rho_0^{**}} = -k_{tr,s}[S]\rho_0^{**} \tag{99}$$

$$r_{\rho_1^{**}} = -k_{tr,s}[S]\rho_1^{**} \tag{100}$$

$$r_{\rho_2^{**}} = -k_{tr,s}[S]\rho_2^{**} \tag{101}$$

$$r_{\delta_0} = k_{tr,s}[S]\rho_0^{**} \tag{102}$$

$$r_{\delta_1} = k_{tr,s}[S]\rho_1^{**} \tag{103}$$

$$r_{\delta_2} = k_{tr,s}[S]\rho_2^{**} \tag{104}$$

Tertiary radicals of types R_n^{***} and $\widetilde{R}_n^{***}$ can also abstract a hydrogen from a solvent molecule to form a dead polymer chain and an active solvent-based radical. The contributions of these reactions to the production and consumption of different species as well as the relevant moments are shown in SI, Equations (S26)–(S45).

$$R_n^{***} + S \xrightarrow{k_{tr,s}^t} D_n + R_0^* \tag{105}$$

$$\widetilde{R}_n^{***} + S \xrightarrow{k_{tr,s}^t} D_n + R_0^* \tag{106}$$

2.10. Chain Transfer to a Monomer

A secondary radical can abstract a hydrogen from a monomer molecule, leading to the termination of the secondary radical and the generation of a new secondary radical with one monomer unit [54]:

$$R_n^{**} + M \xrightarrow{k_{tr,m}} D_n + R_1^{**} \tag{107}$$

Unlike propagation reactions, this reaction generates a saturated dead polymer chain and a monoradical with one-monomer-unit length. Like transfer to solvent constants, transfer to monomer constants, $C_{tr,m} = k_{tr,m}/k_p$ values, have been reported for different monomers. The following rate equations describe the contributions of this reaction to the production and consumption of different species as well as the relevant moments:

$$r_{R_0^*} = -k_{tr,M}[M][R_0^*] \tag{108}$$

$$r_{R_1^{**}} = k_{tr,M}[M]\rho_0^{**} - k_{tr,M}[M][R_1^{**}] \tag{109}$$

$$r_{R_2^{**}} = -k_{tr,M}[M][R_2^{**}] \tag{110}$$

$$r_M = -k_{tr,M}[M]\rho_0^{**} \tag{111}$$

$$r_{\rho_0^{**}} = 0 \tag{112}$$

$$r_{\rho_1^{**}} = k_{tr,M}[M](\rho_0^{**} - \rho_1^{**}) \tag{113}$$

$$r_{\rho_2^{**}} = k_{tr,M}[M](\rho_0^{**} - \rho_2^{**}) \tag{114}$$

$$r_{\delta_0} = k_{tr,M}[M]\rho_0^{**} \tag{115}$$

$$r_{\delta_1} = k_{tr,M}[M]\rho_1^{**} \tag{116}$$

$$r_{\delta_2} = k_{tr,M}[M]\rho_2^{**} \tag{117}$$

This reaction should does not change the total number of monomer units in the system; $r_M + r_{\rho_1^{**}} + r_{\delta_1} = 0$ holds. As shown below, tertiary radicals can also participate in chain transfer to monomer reactions. The contributions of these reactions to the production and consumption of different species as well as the relevant moments are presented in SI (Equations (S46)–(S65)).

$$R_n^{***} + M \xrightarrow{k_{tr,M}^t} D_n + R_1^{**} \tag{118}$$

$$\widetilde{R}_n^{***} + M \xrightarrow{k_{tr,M}^t} D_n + R_1^{**} \tag{119}$$

2.11. Chain Transfer from a Radical to a Dead Polymer Chain

A secondary radical can abstract a hydrogen from a dead polymer chain, leading to the formation of a new dead polymer chain and a new tertiary radical of type R_n^{***}:

$$R_n^{**} + D_m \xrightarrow{mk_{tr,P}} D_n + R_m^{***} \tag{120}$$

An important point here is that every monomer unit in the dead polymer chain has the chance to be attacked by the macroradical. Thus, we need to multiply the transfer to polymer rate coefficient by the number of monomer units of the reactant dead chain. The following rate equations describe the contributions of this reaction to the production and consumption of different species as well as the relevant moments:

$$r_{R_0^*} = -k_{tr,P}[R_0^*]\delta_1 \tag{121}$$

$$r_{R_1^{**}} = -k_{tr,P}[R_1^{**}]\delta_1 \tag{122}$$

$$r_{R_2^{**}} = -k_{tr,P}[R_2^{**}]\delta_1 \tag{123}$$

$$r_{\rho_0^{**}} = -k_{tr,P}\rho_0^{**}\delta_1 \tag{124}$$

$$r_{\rho_1^{**}} = -k_{tr,P}\rho_1^{**}\delta_1 \tag{125}$$

$$r_{\rho_2^{**}} = -k_{tr,P}\rho_2^{**}\delta_1 \tag{126}$$

$$r_{\rho_0^{***}} = k_{tr,P}\rho_0^{**}\delta_1 \tag{127}$$

$$r_{\rho_1^{***}} = k_{tr,P}\rho_0^{**}\delta_2 \tag{128}$$

$$r_{\rho_2^{***}} = k_{tr,P}\rho_0^{**}\delta_3 \tag{129}$$

$$r_{\delta_0} = 0 \tag{130}$$

$$r_{\delta_1} = -k_{tr,P}\rho_0^{**}\delta_2 + k_{tr,P}\delta_1\rho_1^{**} \tag{131}$$

$$r_{\delta_2} = -k_{tr,P}\rho_0^{**}\delta_3 + k_{tr,P}\delta_1\rho_2^{**} \tag{132}$$

Again, this reaction does not change the total population of monomers in the system, as $r_{\rho_1^{**}} + r_{\rho_1^{***}} + r_{\delta_1} = 0$. Equations (129) and (132) indicate that r_{δ_2} and $r_{\rho_2^{***}}$ depend on δ_3, which implies that the value of δ_3 is needed to calculate δ_2. To address this closure problem, a CLD model should be assumed for the saturated dead polymer chains, which allows one to describe the third moment of the distribution based on the lower moments of the same distribution [55]. For example, based on:

- A re-scaled Gamma distribution [9]:

$$\delta_3 = \left(\frac{\delta_2}{\delta_0\delta_1}\right)(2\delta_0\delta_2 - \delta_1^2) \tag{133}$$

- A log-normal distribution [55]:

$$\delta_3 = \delta_0\left(\frac{\delta_2^{\,3}}{\delta_1}\right) \tag{134}$$

- A Gaussian distribution [9]:

$$\delta_3 = 3\delta_1\delta_2 - \delta_1^3 \tag{135}$$

Similar to the secondary radicals, tertiary radicals of types R_n^{***} and $\widetilde{R}_n^{***}$ can attack dead polymer chains to abstract a hydrogen as shown below. The contributions of these reactions to the production and consumption of different species as well as the relevant moments are presented in SI, Equations (S66)–(S82).

$$R_n^{***} + D_m \xrightarrow{mk_{tr,P}^t} D_n + R_m^{***} \tag{136}$$

$$\widetilde{R}_n^{***} + D_m \xrightarrow{mk_{tr,P}^t} D_n + R_m^{***} \tag{137}$$

2.12. Chain Transfer from a Radical to a Macromonomer

A secondary radical can react with a macromonomer, leading to the formation of a dead saturated chain and a tertiary radical of type R_m^{***}:

$$R_n^{**} + U_m \xrightarrow{(m-1)k_{tr,P}} D_n + R_m^{***} \tag{138}$$

As the abstraction of a hydrogen from an unsaturated carbon double bond is difficult, we consider $(m-1)$ sites for hydrogen abstraction from the macromonomer. This is the reason that the rate coefficient of this reaction is multiplied to $(m-1)$. The following rate equations describe the contributions of this reaction to the production and consumption of different species as well as the relevant moments:

$$r_{R_1^{**}} = -k_{tr,P}[R_1^{**}](\epsilon_1 - \epsilon_0) \tag{139}$$

$$r_{R_2^{**}} = -k_{tr,P}[R_2^{**}](\epsilon_1 - \epsilon_0) \tag{140}$$

$$r_{\rho_0^{**}} = -k_{tr,P}\, \rho_0^{**}(\epsilon_1 - \epsilon_0) \tag{141}$$

$$r_{\rho_1^{**}} = -k_{tr,P}\, \rho_1^{**}(\epsilon_1 - \epsilon_0) \tag{142}$$

$$r_{\rho_2^{**}} = -k_{tr,P}\, \rho_2^{**}(\epsilon_1 - \epsilon_0) \tag{143}$$

$$r_{\rho_0^{***}} = k_{tr,P}\, \rho_0^{**}(\epsilon_1 - \epsilon_0) \tag{144}$$

$$r_{\rho_1^{***}} = k_{tr,P}\, \rho_0^{**}(\epsilon_2 - \epsilon_1) \tag{145}$$

$$r_{\rho_2^{***}} = k_{tr,P}\, \rho_0^{**}(\epsilon_3 - \epsilon_2) \tag{146}$$

$$r_{\delta_0} = k_{tr,P}\, \rho_0^{**}(\epsilon_1 - \epsilon_0) \tag{147}$$

$$r_{\delta_1} = k_{tr,P}\, \rho_1^{**}(\epsilon_1 - \epsilon_0) \tag{148}$$

$$r_{\delta_2} = k_{tr,P}\, \rho_2^{**}(\epsilon_1 - \epsilon_0) \tag{149}$$

$$r_{\epsilon_0} = -k_{tr,P}\, \rho_0^{**}(\epsilon_1 - \epsilon_0) \tag{150}$$

$$r_{\epsilon_1} = -k_{tr,P}\, \rho_0^{**}(\epsilon_2 - \epsilon_1) \tag{151}$$

$$r_{\epsilon_2} = -k_{tr,P}\, \rho_0^{**}(\epsilon_3 - \epsilon_2) \tag{152}$$

The satisfaction of $r_{\rho_1^{**}} + r_{\rho_1^{***}} + r_{\delta_1} + r_{\epsilon_1} = 0$, verifies that this reaction does not change the total number of monomer units in the system. Tertiary radicals are also capable of attacking macromonomers to abstract a hydrogen as shown below. The contributions of these reactions to the production and consumption of different species as well as the relevant moments are presented in the SI, Equations (S83)–(S105).

$$R_n^{***} + U_m \xrightarrow{(m-1)k_{tr,p}^t} D_n + R_m^{***} \tag{153}$$

$$\widetilde{R}_n^{***} + U_m \xrightarrow{(m-1)k_{tr,p}^t} D_n + R_m^{***} \tag{154}$$

2.13. Termination by Combination of a Secondary Radical Type R_n^{**}

A secondary radical can react with another secondary radical, forming a saturated dead polymer chain that includes the monomer units of both radicals:

$$R_n^{**} + R_m^{**} \xrightarrow{k_{tc}} D_{n+m} \tag{155}$$

The reaction creates a carbon–carbon bond between the radical-bearing carbons of the two living chains. The following rate equations describe the contributions of this reaction to the production and consumption of different species as well as the relevant moments:

$$r_{R_0^*} = -k_{tc}[R_0^*]\rho_0^{**} \tag{156}$$

$$r_{R_1^{**}} = -k_{tc}[R_1^{**}]\rho_0^{**} \tag{157}$$

$$r_{R_2^{**}} = -k_{tc}[R_2^{**}]\rho_0^{**} \tag{158}$$

$$r_{\rho_0^{**}} = -k_{tc}\rho_0^{**2} \tag{159}$$

$$r_{\rho_1^{**}} = -k_{tc}\rho_0^{**}\rho_1^{**} \tag{160}$$

$$r_{\rho_2^{**}} = -k_{tc}\rho_0^{**}\rho_2^{**} \tag{161}$$

$$r_{\delta_0} = \frac{1}{2}k_{tc}\rho_0^{**2} \tag{162}$$

$$r_{\delta_1} = k_{tc}\rho_0^{**}\rho_1^{**} \tag{163}$$

$$r_{\delta_2} = k_{tc}\left(\rho_0^{**}\rho_2^{**} + \rho_1^{**2}\right) \tag{164}$$

Here, $r_{\rho_1^{**}} + r_{\delta_1} = 0$, which confirms that this reaction does not change the total number of monomer units. A secondary radical can also react with a tertiary radical of type R_m^{***} or $\widetilde{R}_m^{***}$:

$$R_n^{**} + R_m^{***} \xrightarrow{2k_{tc}^t} D_{n+m} \tag{165}$$

$$R_n^{**} + \widetilde{R}_m^{***} \xrightarrow{2k_{tc}^t} D_{n+m} \tag{166}$$

The contributions of these reactions to the production and consumption of different species as well as the relevant moments are presented in the SI, Equations (S106)–(S131).

2.14. Termination by Combination of a Tertiary Radical Type R_n^{***}

A radical of this kind can react with a secondary radical, which was consider in (165). It can also react with a tertiary radical of the same type R_m^{***} to form a saturated dead polymer chain that includes the monomer units of both radicals:

$$R_n^{***} + R_m^{***} \xrightarrow{k_{tc}^{tt}} D_{n+m} \tag{167}$$

The rate coefficient of this reaction is smaller than that of the reaction of (155) as both reacting centers have tertiary carbons [9]. The following rate equations describe the contributions of this reaction to the production and consumption of different species as well as the relevant moments:

$$r_{\rho_0^{***}} = -k_{tc}^{tt}\rho_0^{***2} \tag{168}$$

$$r_{\rho_1^{***}} = -k_{tc}^{tt}\rho_0^{***}\rho_1^{***} \tag{169}$$

$$r_{\rho_2^{***}} = -k_{tc}^{tt}\rho_0^{***}\rho_2^{***} \tag{170}$$

$$r_{\delta_0} = \frac{1}{2}k_{tc}^{tt}\rho_0^{***2} \tag{171}$$

$$r_{\delta_1} = k_{tc}^{tt}\rho_0^{***}\rho_1^{***} \tag{172}$$

$$r_{\delta_2} = k_{tc}^{tt}\left(\rho_0^{***}\rho_2^{***} + \rho_1^{***2}\right) \tag{173}$$

The sum of the first moments is zero, confirming no change in the total number of monomer units upon the occurrence of this reaction.

A tertiary radical type R_n^{***} can also react with a radical of type $\widetilde{R}_m^{***}$:

$$R_n^{***} + \widetilde{R}_m^{***} \xrightarrow{2k_{tc}^{tt}} D_{n+m} \tag{174}$$

The rate equations describing the contributions of this reaction to the production and consumption of different species as well as the relevant moments are presented in SI, Equations (S132)–(S141).

2.15. Termination by Combination of a Tertiary Radical Type $\widetilde{R}_m^{***}$

For this radical, two possible termination reactions with R_n^{**} and R_n^{***} were discussed in Sections 2.13 and 2.14. The third possible termination by combination reaction is that of two tertiary radicals of the type $\widetilde{R}_n^{***}$:

$$\widetilde{R}_n^{***} + \widetilde{R}_m^{***} \overset{k_{tc}^{tt}}{\rightarrow} D_{n+m} \tag{175}$$

The following rate equations describe the contributions of this reaction to the production and consumption of different species as well as the relevant moments:

$$r_{\widetilde{\rho}_0^{***}} = -k_{tc}^{tt}\widetilde{\rho}_0^{***2} \tag{176}$$

$$r_{\widetilde{\rho}_1^{***}} = -k_{tc}^{tt}\widetilde{\rho}_0^{***}\widetilde{\rho}_1^{***} \tag{177}$$

$$r_{\widetilde{\rho}_2^{***}} = -k_{tc}^{tt}\widetilde{\rho}_0^{***}\widetilde{\rho}_2^{***} \tag{178}$$

$$r_{\delta_0} = \frac{1}{2}k_{tc}^{tt}\widetilde{\rho}_0^{***2} \tag{179}$$

$$r_{\delta_1} = k_{tc}^{tt}(\widetilde{\rho}_1^{***}\widetilde{\rho}_0^{***}) \tag{180}$$

$$r_{\delta_2} = k_{tc}^{tt}(\widetilde{\rho}_2^{***}\widetilde{\rho}_0^{***} + \widetilde{\rho}_1^{***2}) \tag{181}$$

Here, as expected, the sum of the first moments equals to zero.

2.16. Termination by Disproportionation of a Secondary Radical Type R_n^{**}

Two secondary radicals can participate in a disproportionation termination reaction in which one radical abstracts a hydrogen from the other radical, leading to the formation of one saturated and one unsaturated dead chain:

$$R_n^{**} + R_m^{**} \overset{k_{td}}{\rightarrow} D_n + U_m \tag{182}$$

$$R_n^{**} + R_m^{**} \overset{k_{td}}{\rightarrow} D_m + U_n \tag{183}$$

The only difference between these two reactions is in the lengths (number of monomer units) of the produced dead polymer chains, which depends on which radical abstracts the hydrogen. The following rate equations describe the contributions of these two reactions to the production and consumption of different species as well as the relevant moments:

$$r_{R_0^*} = -2k_{td}[R_0^*]\rho_0^{**} \tag{184}$$

$$r_{R_1^{**}} = -2k_{td}[R_1^{**}]\rho_0^{**} \tag{185}$$

$$r_{R_2^{**}} = -2k_{td}[R_2^{**}]\rho_0^{**} \tag{186}$$

$$r_{\rho_0^{**}} = -2k_{td}\rho_0^{**2} \tag{187}$$

$$r_{\rho_1^{**}} = -2k_{td}\rho_0^{**}\rho_1^{**} \tag{188}$$

$$r_{\rho_2^{**}} = -2k_{td}\rho_0^{**}\rho_2^{**} \tag{189}$$

$$r_{\delta_0} = r_{\epsilon_0} = k_{td}\rho_0^{**2} \tag{190}$$

$$r_{\delta_1} = r_{\epsilon_1} = k_{td}\,\rho_0^{**}\rho_1^{**} \tag{191}$$

$$r_{\delta_2} = r_{\epsilon_2} = k_{td}\,\rho_0^{**}\rho_2^{**} \tag{192}$$

The sum of the rates of the first moments, $r_{\delta_1} + r_{\epsilon_1} + r_{\rho_1^{**}}$, equals to zero, confirming that these two reactions do not change the total number of monomer units in the system. Other disproportionation termination pathways for a secondary radical are shown below:

$$R_n^{**} + R_m^{***} \xrightarrow{k_{td}^t} D_n + U_m \tag{193}$$

$$R_n^{**} + R_m^{***} \xrightarrow{k_{td}^t} D_m + U_n \tag{194}$$

$$R_n^{**} + \widetilde{R}_m^{***} \xrightarrow{k_{td}^t} D_m + U_n \tag{195}$$

$$R_n^{**} + \widetilde{R}_m^{***} \xrightarrow{k_{td}^t} D_n + U_m \tag{196}$$

The contributions of each of these reactions to the production and consumption of different species as well as the relevant moments are presented in SI, Equations (S142)–(S185).

2.17. Termination by Disproportionation of a Tertiary Radical Type R_m^{***}

A tertiary radical R_n^{***} can react with a secondary radical, as discussed in the previous section. It can also react with another tertiary radical of the same type. One radical abstracts a hydrogen from the other radical, leading to the formation of a saturated and an unsaturated dead chain:

$$R_n^{***} + R_m^{***} \xrightarrow{k_{td}^{tt}} D_n + U_m \tag{197}$$

$$R_n^{***} + R_m^{***} \xrightarrow{k_{td}^{tt}} D_m + U_n \tag{198}$$

The following rate equations describe the contributions of these reactions to the production and consumption of different species as well as the relevant moments:

$$r_{\rho_0^{***}} = -2k_{td}^{tt}\rho_0^{***^2} \tag{199}$$

$$r_{\rho_1^{***}} = -2k_{td}^{tt}\rho_0^{***}\rho_1^{***} \tag{200}$$

$$r_{\rho_2^{***}} = -2k_{td}^{tt}\rho_0^{***}\rho_2^{***} \tag{201}$$

$$r_{\delta_0} = r_{\epsilon_0} = k_{td}^{tt}\rho_0^{***^2} \tag{202}$$

$$r_{\delta_1} = r_{\epsilon_1} = k_{td}^{tt}\rho_1^{***}\rho_0^{***} \tag{203}$$

$$r_{\delta_2} = r_{\epsilon_2} = k_{td}^{tt}\rho_2^{***}\rho_0^{***} \tag{204}$$

Here, $r_{\rho_1^{***}} + r_{\delta_1} + r_{\epsilon_1} = 0$ confirms that these reactions do not change the total number of monomer units in the system.

A tertiary radical R_n^{***} can also react with a tertiary radical of the type $\widetilde{R}_n^{***}$.

$$\widetilde{R}_n^{***} + R_m^{***} \xrightarrow{k_{td}^{tt}} D_m + U_n \tag{205}$$

$$\widetilde{R}_n^{***} + R_m^{***} \xrightarrow{k_{td}^{tt}} D_n + U_m \tag{206}$$

The contributions of the reaction to the production and consumption of different species as well as the relevant moments are presented in SI, Equations (S186)–(S204).

2.18. Termination by Disproportionation of a Tertiary Radical Type $\widetilde{R}_m^{***}$

A tertiary radical type $\widetilde{R}_m^{***}$ can react with R_n^{**} and R_n^{***}, which were discussed before. It can also react with a tertiary radical of the same type. One radical abstracts a hydrogen from the other radical, leading to the formation of a saturated and an unsaturated dead chain:

$$\widetilde{R}_n^{***} + \widetilde{R}_m^{***} \xrightarrow{k_{td}^{tt}} D_n + U_m \tag{207}$$

$$\widetilde{R}_n^{***} + \widetilde{R}_m^{***} \xrightarrow{k_{td}^{tt}} D_m + U_n \tag{208}$$

The following rate equations describe the contributions of these reactions to the production and consumption of different species as well as the relevant moments:

$$r_{\widetilde{\rho}_0^{***}} = -2k_{td}^{tt}\widetilde{\rho}_0^{***}\widetilde{\rho}_0^{***} \tag{209}$$

$$r_{\widetilde{\rho}_1^{***}} = -2k_{td}^{tt}\widetilde{\rho}_1^{***}\widetilde{\rho}_0^{***} \tag{210}$$

$$r_{\widetilde{\rho}_2^{***}} = -2k_{td}^{tt}\widetilde{\rho}_2^{***}\widetilde{\rho}_0^{***} \tag{211}$$

$$r_{\delta_0} = r_{\epsilon_0} = k_{td}^{tt}\widetilde{\rho}_0^{***}\widetilde{\rho}_0^{***} \tag{212}$$

$$r_{\delta_1} = r_{\epsilon_1} = k_{td}^{tt}\widetilde{\rho}_1^{***}\widetilde{\rho}_0^{***} \tag{213}$$

$$r_{\delta_2} = r_{\epsilon_2} = k_{td}^{tt}\widetilde{\rho}_2^{***}\widetilde{\rho}_0^{***} \tag{214}$$

2.19. Propagation of Radicals by Reacting with a Macromonomer

In this reaction, a secondary radical reacts with a dead unsaturated chain, leading to the production of a tertiary radical of the type R_n^{***} containing all monomer units of the dead polymer chain and the secondary radical:

$$R_n^{**} + U_m \xrightarrow{k_{mac}} R_{n+m}^{***} \tag{215}$$

The following rate equations describe the contributions of this reaction to the production and consumption of different species as well as the relevant moments:

$$r_{R_0^*} = -k_{mac}[R_0^*]\epsilon_0 \tag{216}$$

$$r_{R_1^{**}} = -k_{mac}[R_1^{**}]\epsilon_0 \tag{217}$$

$$r_{R_2^{**}} = -k_{mac}[R_2^{**}]\epsilon_0 \tag{218}$$

$$r_{\rho_0^{**}} = -k_{mac}\rho_0^{**}\epsilon_0 \tag{219}$$

$$r_{\rho_1^{**}} = -k_{mac}\rho_1^{**}\epsilon_0 \tag{220}$$

$$r_{\rho_2^{**}} = -k_{mac}\rho_2^{**}\epsilon_0 \tag{221}$$

$$r_{\rho_0^{***}} = k_{mac}\rho_0^{**}\epsilon_0 \tag{222}$$

$$r_{\rho_1^{***}} = k_{mac}\left(\rho_1^{**}\epsilon_0 + \rho_0^{**}\epsilon_1\right) \tag{223}$$

$$r_{\rho_2^{***}} = k_{mac}\left(\rho_2^{**}\epsilon_0 + \rho_0^{**}\epsilon_2 + 2\rho_1^{**}\epsilon_1\right) \tag{224}$$

$$r_{\epsilon_0} = -k_{mac}\rho_0^{**}\epsilon_0 \tag{225}$$

$$r_{\epsilon_1} = -k_{mac}\rho_0^{**}\epsilon_1 \tag{226}$$

$$r_{\epsilon_2} = -k_{mac}\rho_0^{**}\epsilon_2 \tag{227}$$

These rate equations satisfy $r_{\rho_1^{**}} + r_{\rho_1^{***}} + r_{\epsilon_1} = 0$, confirming that this reaction does not change the total number of monomer units. Tertiary radicals of types R_n^{***} and $\widetilde{R}_n^{***}$ can also attack dead unsaturated chains to propagate and form longer-branched growing chains.

$$R_n^{***} + U_m \xrightarrow{k_{mac}^t} R_{n+m}^{***} \tag{228}$$

$$\widetilde{R}_n^{***} + U_m \xrightarrow{k_{mac}^t} R_{n+m}^{***} \tag{229}$$

The equations describing the contributions of these reactions to the production and consumption of different species as well as the relevant moments are presented in SI, Equations (S205)–(S221).

3. Overall Rate Equations

After considering primary and secondary reactions that are most likely to occur in high-temperature free-radical polymerization and determining the contributions of each reaction to the production of each species and each CLD moment, one needs to obtain all overall rate equations. The overall rate equation for the production of a species (or a CLD moment of a class of polymer chains) is the sum of the rate equations of the production of the species (or the CLD moment of the class of polymer chains in all reactions). For example, the overall rate of production of M is the sum of the individual r_M's given in Equations (10), (18), (29), (38), and (111), Equations (S2), (S10), (S48) and (S58).

4. Case Study: High-Temperature Free-Radical Polymerization of Methyl Acrylate (MA)

The derived rate equations are used in an MA batch polymerization reactor model to predict monomer conversion and polymer average molecular weights at 180 and 200 °C. The details of the experimental study can be found in Ref. [5]. Measurements of monomer conversion and polymer average molecular weights made, respectively, with the gravimetry method and a gel-permeation chromatograph are compared with the same model-predicted polymer properties.

Different rate coefficients for *n*-butyl acrylate (*n*BA) propagation, transfer to polymer, backbiting, β-Scission, transfer to monomer, and termination reactions have been reported in the literature (Table S2) [5,9,45,56–69]. On the other hand, a reliable experimentally determined MA propagation rate coefficient is available [45]. On the basis of the available *n*BA and the reliable experimentally determined MA rate coefficients and using the family-type-behavior approach [45] (explained in Section (20) of the SI), we determined ranges for the activation energies and frequency factors of backbiting and β-scission reactions of MA. The monomer self-initiation rate coefficient reported in Ref. [5] was used here. Other rate coefficients used in this study are reported in Table S4.

After determining the ranges, the *genetic algorithm (ga)* optimizer [70] of MATLAB was used to estimate the activation energies and frequency factors of backbiting and β-scission reactions within the ranges and parameters of a gel effect model [5], from monomer conversion and average molecular weights measurements. The estimated values are presented in Table 1. Figure 3 compares model predictions and measurements of MA conversion, number-average molecular weight, and weight-average molecular weight. It shows a good agreement between the model predictions and measurements.

Table 1. Estimates of methyl acrylate (MA) reaction kinetic parameters obtained in this work (activation energies, kJ mol^{-1}; R, J mol^{-1} K^{-1}).

Reaction	Rate Coefficient
Backbiting (s^{-1})	$k_{bb} = 2.01 \times 10^{10} \exp\left(\frac{-51.03}{RT}\right)$
β-Scission (s^{-1})	$k_\beta = 3.67 \times 10^{14} \exp\left(\frac{-97.90}{RT}\right)$

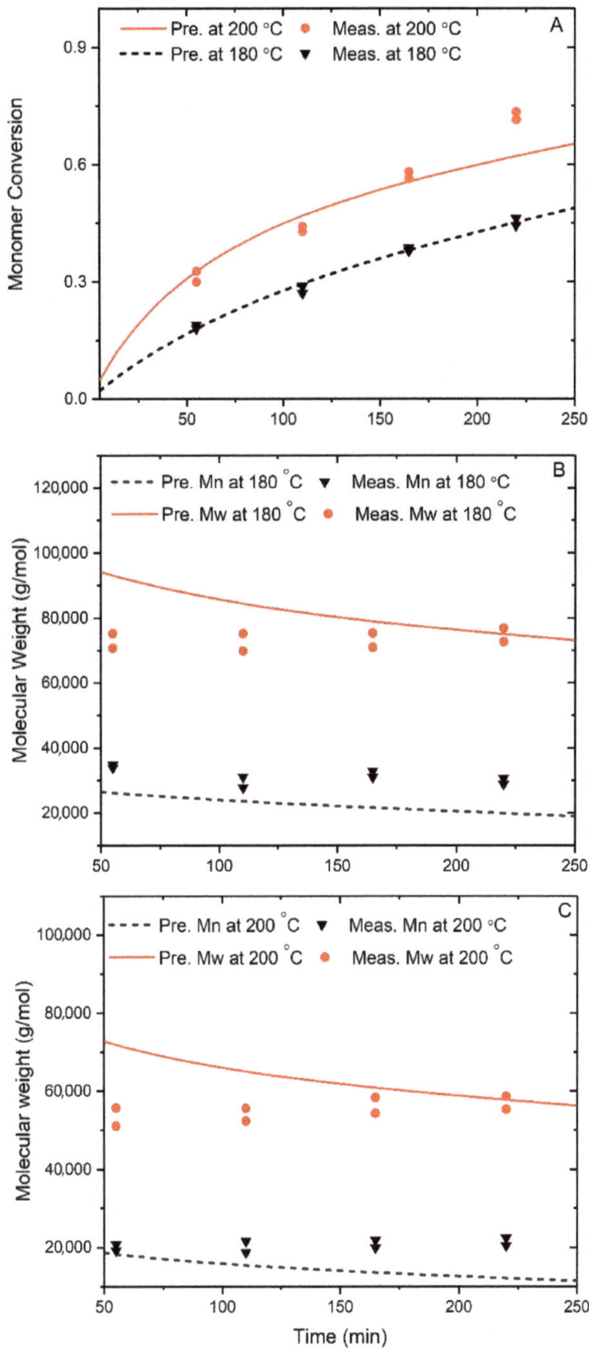

Figure 3. (**A–C**) Comparing model predictions and measurements of monomer conversion, number average molecular weight (Mn) and weight average molecular weight (Mw) at 180 °C and 200 °C.

5. Conclusions

Most-likely reactions that occur in free-radical polymerization of acrylates were considered. This article provides a systematic approach to account for the contributions of each reaction to the production or consumption of each species and to the rates of the moments of chain-length distributions of different polymer types. Three types of free radicals and two types of dead polymers were considered. The closure problems were addressed by assuming a model for each CLD that is affected. Using the systematic approach proposed herein, one can easily derive rate equations that are not considered in this article. Finally, to show the application of the rate equations, they were used in a batch polymerization reactor model that simulates high-temperature bulk free-radical polymerization of MA. The kinetic parameters of the backbiting and β-scission reactions were estimated from measurements of monomer conversion and number- and weight-average molecular weights.

Supplementary Materials: The following are available online at http://www.mdpi.com/2227-9717/7/10/656/s1.

Author Contributions: M.S. and H.R. derived the rate equations and prepared the first draft of this paper. H.R. did the polymerization reactions and A.A.S. measured molecular weights. M.C.G. contributed to the design of the experiments and data analysis. A.M.R. contributed to the design of the experiments and the preparation of this manuscript.

Funding: This material is based upon work supported by the U.S. National Science Foundation under Grant Nos. CBET–1804285 and CBET–1803215. Any opinions, findings, and conclusions or recommendations expressed in this material are those of the authors and do not necessarily reflect the views of the National Science Foundation.

Conflicts of Interest: The authors declare no conflict of interest.

References

1. Matyjaszewski, K.; Davis, T.P. *Handbook of Radical Polymerization*; John Wiley & Sons: Hoboken, NJ, USA, 2003.
2. Riazi, H.; Mohammadi, N.; Mohammadi, H. Emulsion copolymerization of methyl methacrylate/butyl acrylate/iodine system to monosize rubbery nanoparticles containing iodine and triiodide mixture. *Ind. Eng. Chem. Res.* **2013**, *52*, 2449–2456. [CrossRef]
3. Adelnia, H.; Riazi, H.; Saadat, Y.; Hosseinzadeh, S. Synthesis of monodisperse anionic submicron polystyrene particles by stabilizer-free dispersion polymerization in alcoholic media. *Colloid Polym. Sci.* **2013**, *291*, 1741–1748. [CrossRef]
4. Campbell, J.; Teymour, F.; Morbidelli, M. High temperature free radical polymerization. 1. Investigation of continuous styrene polymerization. *Macromolecules* **2003**, *36*, 5491–5501. [CrossRef]
5. Riazi, H.; Shamsabadi, A.A.; Grady, M.C.; Rappe, A.M.; Soroush, M. Experimental and Theoretical Study of the Self-Initiation Reaction of Methyl Acrylate in Free-Radical Polymerization. *Ind. Eng. Chem. Res.* **2018**, *57*, 532–539. [CrossRef]
6. Riazi, H.; Shamsabadi, A.; Corcoran, P.; Grady, M.; Rappe, A.; Soroush, M. On the Thermal Self-Initiation Reaction of n-Butyl Acrylate in Free-Radical Polymerization. *Processes* **2018**, *6*, 3. [CrossRef]
7. Khuong, K.S.; Jones, W.H.; Pryor, W.A.; Houk, K. The mechanism of the self-initiated thermal polymerization of styrene. Theoretical solution of a classic problem. *J. Am. Chem. Soc.* **2005**, *127*, 1265–1277. [CrossRef] [PubMed]
8. Srinivasan, S.; Kalfas, G.; Petkovska, V.I.; Bruni, C.; Grady, M.C.; Soroush, M. Experimental study of the spontaneous thermal homopolymerization of methyl and n-butyl acrylate. *J. Appl. Polym. Sci.* **2010**, *118*, 1898–1909. [CrossRef]
9. Arabi Shamsabadi, A.; Moghadam, N.; Srinivasan, S.; Corcoran, P.; Grady, M.; Rappe, A.; Soroush, M. Study of n-Butyl Acrylate Self-Initiation Reaction Experimentally and via Macroscopic Mechanistic Modeling. *Processes* **2016**, *4*, 15. [CrossRef]
10. Ballard, N.; Veloso, A.; Asua, J. Mid-Chain Radical Migration in the Radical Polymerization of n-Butyl Acrylate. *Polymers* **2018**, *10*, 765. [CrossRef]
11. Ren, S.; Vivaldo-Lima, E.; Dubé, M. Modeling of the Copolymerization Kinetics of n-Butyl Acrylate and D-Limonene Using PREDICI®. *Processes* **2016**, *4*, 1. [CrossRef]

12. Villermaux, J.; Blavier, L. Free radical polymerization engineering—I: A new method for modeling free radical homogeneous polymerization reactions. *Chem. Eng. Sci.* **1984**, *39*, 87–99. [CrossRef]

13. Mastan, E.; Li, X.; Zhu, S. Modeling and theoretical development in controlled radical polymerization. *Prog. Polym. Sci.* **2015**, *45*, 71–101. [CrossRef]

14. Wulkow, M.; Busch, M.; Davis, T.P.; Barner-Kowollik, C. Implementing the reversible addition–fragmentation chain transfer process in PREDICI. *J. Polym. Sci. Part A Polym. Chem* **2004**, *42*, 1441–1448. [CrossRef]

15. Gao, H.; Konstantinov, I.A.; Arturo, S.G.; Broadbelt, L.J. On the modeling of number and weight average molecular weight of polymers. *Chem. Eng. J.* **2017**, *327*, 906–913. [CrossRef]

16. Soroush, M.; Kravaris, C. Nonlinear control of a batch polymerization reactor: An experimental study. *AIChE J.* **1992**, *38*, 1429–1448. [CrossRef]

17. Soroush, M.; Kravaris, C. Optimal design and operation of batch reactors. 2. A case study. *Ind. Eng. Chem. Res.* **1993**, *32*, 882–893. [CrossRef]

18. Apostolo, M.; Arcella, V.; Storti, G.; Morbidelli, M. Kinetics of the emulsion polymerization of vinylidene fluoride and hexafluoropropylene. *Macromolecules* **1999**, *32*, 989–1003. [CrossRef]

19. Kalfas, G.; Ray, W.H. Modeling and experimental studies of aqueous suspension polymerization processes. 1. Modeling and simulations. *Ind. Eng. Chem. Res.* **1993**, *32*, 1822–1830. [CrossRef]

20. Hernández-Barajas, J.; Hunkeler, D.J. Inverse-emulsion polymerization of acrylamide using block copolymeric surfactants: Mechanism, kinetics and modelling. *Polymer* **1997**, *38*, 437–447. [CrossRef]

21. Crowley, T.J.; Choi, K.Y. Calculation of molecular weight distribution from molecular weight moments in free radical polymerization. *Ind. Eng. Chem. Res.* **1997**, *36*, 1419–1423. [CrossRef]

22. Zhou, Y.N.; Luo, Z.H. An old kinetic method for a new polymerization mechanism: Toward photochemically mediated ATRP. *AIChE J.* **2015**, *61*, 1947–1958. [CrossRef]

23. Bonilla, J.; Saldívar, E.; Flores-Tlacuahuac, A.; Vivaldo-Lima, E.; Pfaendner, R.; Tiscareño-Lechuga, F. Detailed modeling, simulation, and parameter estimation of nitroxide mediated living free radical polymerization of styrene. *Polym. React. Eng.* **2002**, *10*, 227–263. [CrossRef]

24. Wang, A.R.; Zhu, S. Modeling the reversible addition–fragmentation transfer polymerization process. *J. Polym. Sci. Part A Polym. Chem.* **2003**, *41*, 1553–1566. [CrossRef]

25. Mastan, E.; Zhu, S. Method of moments: A versatile tool for deterministic modeling of polymerization kinetics. *Eur. Polym. J.* **2015**, *68*, 139–160. [CrossRef]

26. Zhou, Y.N.; Luo, Z.H. State-of-the-art and progress in method of moments for the model-based reversible-deactivation radical polymerization. *Macromol. React. Eng.* **2016**, *10*, 516–534. [CrossRef]

27. Bachmann, R.; Melchiors, M.; Avtomonov, E. Modelling and Optimization of Nonlinear Polymerization Processes. In *Macromolecular Symposia*; Wiley Online Library: Hoboken, NJ, USA, 2016.

28. Zabisky, R.; Chan, W.-M.; Gloor, P.; Hamielec, A. A kinetic model for olefin polymerization in high-pressure tubular reactors: A review and update. *Polymer* **1992**, *33*, 2243–2262. [CrossRef]

29. Ray, W.H. On the mathematical modeling of polymerization reactors. *J. Macromol. Sci. Rev. Macromol. Chem. Phys.* **1972**, *8*, 1–56. [CrossRef]

30. Kiparissides, C. Polymerization reactor modeling: A review of recent developments and future directions. *Chem. Eng. Sci.* **1996**, *51*, 1637–1659. [CrossRef]

31. Tobita, H.; Hamielec, A. Modeling of network formation in free radical polymerization. *Macromolecules* **1989**, *22*, 3098–3105. [CrossRef]

32. Tefera, N.; Weickert, G.; Westerterp, K. Modeling of free radical polymerization up to high conversion. I. A method for the selection of models by simultaneous parameter estimation. *J. Appl. Polym. Sci.* **1997**, *63*, 1649–1661.

33. Zhou, W.; Marshall, E.; Oshinowo, L. Modeling LDPE tubular and autoclave reactors. *Ind. Eng. Chem. Res.* **2001**, *40*, 5533–5542. [CrossRef]

34. Kiparissides, C.; Daskalakis, G.; Achilias, D.; Sidiropoulou, E. Dynamic simulation of industrial poly (vinyl chloride) batch suspension polymerization reactors. *Ind. Eng. Chem. Res.* **1997**, *36*, 1253–1267. [CrossRef]

35. Zhu, S.; Hamielec, A. Modeling of free-radical polymerization with crosslinking: monoradical assumption and stationary-state hypothesis. *Macromolecules* **1993**, *26*, 3131–3136. [CrossRef]

36. Louie, B.M.; Carratt, G.M.; Soong, D.S. Modeling the free radical solution and bulk polymerization of methyl methacrylate. *J. Appl. Polym. Sci.* **1985**, *30*, 3985–4012. [CrossRef]

37. Tefera, N.; Weickert, G.; Westerterp, K. Modeling of free radical polymerization up to high conversion. II. Development of a mathematical model. *J. Appl. Polym. Sci.* **1997**, *63*, 1663–1680. [CrossRef]

38. Brandolin, A.; Lacunza, M.; Ugrin, P.; Capiati, N. High pressure polymerization of ethylene. An improved mathematical model for industrial tubular reactors. *Polym. React. Eng.* **1996**, *4*, 193–241.

39. Prasad, V.; Schley, M.; Russo, L.P.; Bequette, B.W. Product property and production rate control of styrene polymerization. *J. Process Control* **2002**, *12*, 353–372. [CrossRef]

40. Hutchinson, R.A. Modeling of free-radical polymerization kinetics with crosslinking for methyl methacrylate/ethylene glycol dimethacrylate. *Polym. React. Eng.* **1993**, *1*, 521–577. [CrossRef]

41. Barry, J.T.; Berg, D.J.; Tyler, D.R. Radical cage effects: Comparison of solvent bulk viscosity and microviscosity in predicting the recombination efficiencies of radical cage pairs. *J. Am. Chem. Soc.* **2016**, *138*, 9389–9392. [CrossRef]

42. Srinivasan, S.; Lee, M.W.; Grady, M.C.; Soroush, M.; Rappe, A.M. Computational study of the self-initiation mechanism in thermal polymerization of methyl acrylate. *J. Phys. Chem. A* **2009**, *113*, 10787–10794. [CrossRef]

43. Srinivasan, S.; Lee, M.W.; Grady, M.C.; Soroush, M.; Rappe, A.M. Self-initiation mechanism in spontaneous thermal polymerization of ethyl and n-butyl acrylate: A theoretical study. *J. Phys. Chem. A* **2010**, *114*, 7975–7983. [CrossRef] [PubMed]

44. Srinivasan, S.; Rappe, A.M.; Soroush, M. Theoretical Insights Into Thermal Self-Initiation Reactions of Acrylates. In *Computational Quantum Chemistry*; Elsevier: Amsterdam, The Netherlands, 2019; pp. 99–134.

45. Barner-Kowollik, C.; Beuermann, S.; Buback, M.; Castignolles, P.; Charleux, B.; Coote, M.L.; Hutchinson, R.A.; Junkers, T.; Lacík, I.; Russell, G.T. Critically evaluated rate coefficients in radical polymerization–7. Secondary-radical propagation rate coefficients for methyl acrylate in the bulk. *Polym. Chem.* **2014**, *5*, 204–212.

46. Soroush, M.; Rappe, A.M. Theoretical Insights Into Chain Transfer Reactions of Acrylates. In *Computational Quantum Chemistry*; Elsevier: Amsterdam, The Netherlands, 2019; pp. 135–193.

47. Cuccato, D.; Mavroudakis, E.; Dossi, M.; Moscatelli, D. A Density Functional Theory Study of Secondary Reactions in n-Butyl Acrylate Free Radical Polymerization. *Macromol. Theory Simul.* **2013**, *22*, 127–135. [CrossRef]

48. Van Steenberge, P.H.; Vandenbergh, J.; Reyniers, M.-F.; Junkers, T.; D'hooge, D.R.; Marin, G.B. Kinetic Monte Carlo generation of complete electron spray ionization mass spectra for acrylate macromonomer synthesis. *Macromolecules* **2017**, *50*, 2625–2636. [CrossRef]

49. Drache, M.; Stehle, M.; Mätzig, J.; Brandl, K.; Jungbluth, M.; Namyslo, J.C.; Schmidt, A.; Beuermann, S. Identification of β scission products from free radical polymerizations of butyl acrylate at high temperature. *Polym. Chem.* **2019**, *10*, 1956–1967. [CrossRef]

50. Hutchinson, R.A. Modeling of chain length and long-chain branching distributions in free-radical polymerization. *Macromol. Theory Simul.* **2001**, *10*, 144–157. [CrossRef]

51. Liu, S.; Srinivasan, S.; Grady, M.C.; Soroush, M.; Rappe, A.M. Backbiting and β-scission reactions in free-radical polymerization of methyl acrylate. *Int. J. Quantum Chem.* **2014**, *114*, 345–360. [CrossRef]

52. Spiegel, M.R. *Mathematical Handbook of Formulas and Tables*; Schaum's Outline Series; McGraw-Hill: New York, NY, USA, 1968.

53. Brandrup, J.; Immergut, E.H.; Grulke, E.A.; Abe, A.; Bloch, D.R. *Polymer Handbook*; Wiley: New York, NY, USA, 1999; Volume 89.

54. Moghadam, N.; Liu, S.; Srinivasan, S.; Grady, M.C.; Soroush, M.; Rappe, A.M. Computational study of chain transfer to monomer reactions in high-temperature polymerization of alkyl acrylates. *J. Phys. Chem. A* **2013**, *117*, 2605–2618. [CrossRef]

55. Hungenberg, K.-D.; Wulkow, M. *Modeling and Simulation in Polymer Reaction Engineering: A Modular Approach*; John Wiley & Sons: Hoboken, NJ, USA, 2018.

56. Arzamendi, G.; Plessis, C.; Leiza, J.R.; Asua, J.M. Effect of the intramolecular chain transfer to polymer on PLP/SEC experiments of alkyl acrylates. *Macromol. Theory Simul.* **2003**, *12*, 315–324. [CrossRef]

57. Nikitin, A.N.; Hutchinson, R.A.; Buback, M.; Hesse, P. Determination of intramolecular chain transfer and midchain radical propagation rate coefficients for butyl acrylate by pulsed laser polymerization. *Macromolecules* **2007**, *40*, 8631–8641. [CrossRef]

58. Peck, A.N.; Hutchinson, R.A. Secondary reactions in the high-temperature free radical polymerization of butyl acrylate. *Macromolecules* **2004**, *37*, 5944–5951. [CrossRef]

59. Nikitin, A.N.; Hutchinson, R.A.; Wang, W.; Kalfas, G.A.; Richards, J.R.; Bruni, C. Effect of Intramolecular Transfer to Polymer on Stationary Free-Radical Polymerization of Alkyl Acrylates, 5–Consideration of Solution Polymerization up to High Temperatures. *Macromol. React. Eng.* **2010**, *4*, 691–706. [CrossRef]

60. Ballard, N.; Hamzehlou, S.; Asua, J.M. Intermolecular transfer to polymer in the radical polymerization of n-butyl acrylate. *Macromolecules* **2016**, *49*, 5418–5426. [CrossRef]

61. Hamzehlou, S.; Ballard, N.; Reyes, Y.; Aguirre, A.; Asua, J.; Leiza, J. Analyzing the discrepancies in the activation energies of the backbiting and β-scission reactions in the radical polymerization of n-butyl acrylate. *Polym. Chem.* **2016**, *7*, 2069–2077. [CrossRef]

62. Drache, M.; Hosemann, B.; Laba, T.; Beuermann, S. Modeling of Branching Distributions in Butyl Acrylate Polymerization Applying Monte Carlo Methods. *Macromol. Theory Simul.* **2015**, *24*, 301–310. [CrossRef]

63. Hamzehlou, S.; Reyes, Y.; Hutchinson, R.; Leiza, J.R. Copolymerization of n-Butyl Acrylate and Styrene: Terminal vs Penultimate Model. *Macromol. Chem. Phys.* **2014**, *215*, 1668–1678. [CrossRef]

64. Kockler, K.B.; Haehnel, A.P.; Junkers, T.; Barner-Kowollik, C. Determining Free-Radical Propagation Rate Coefficients with High-Frequency Lasers: Current Status and Future Perspectives. *Macromol. Rapid Commun.* **2016**, *37*, 123–134. [CrossRef]

65. Wang, W.; Hutchinson, R.A. High temperature semibatch free radical copolymerization of styrene and butyl acrylate. In *Macromolecular Symposia*; Wiley Online Library: Hoboken, NJ, USA, 2010.

66. Hlalele, L.; Klumperman, B. In situ 1H NMR studies of high-temperature nitroxide-mediated polymerization of n-butyl acrylate. *Macromolecules* **2011**, *44*, 7100–7108. [CrossRef]

67. Barth, J.; Buback, M.; Hesse, P.; Sergeeva, T. Termination and transfer kinetics of butyl acrylate radical polymerization studied via SP-PLP-EPR. *Macromolecules* **2010**, *43*, 4023–4031. [CrossRef]

68. Vir, A.B.; Marien, Y.W.; van Steenberge, P.H.; Barner-Kowollik, C.; Reyniers, M.-F.; Marin, G.B.; D'hooge, D.R. From n-butyl acrylate Arrhenius parameters for backbiting and tertiary propagation to β-scission via stepwise pulsed laser polymerization. *Polym. Chem.* **2019**, *10*, 4116–4125. [CrossRef]

69. Buback, M.; Kuelpmann, A.; Kurz, C. Termination kinetics of methyl acrylate and dodecyl acrylate free-radical homopolymerizations up to high pressure. *Macromol. Chem. Phys.* **2002**, *203*, 1065–1070. [CrossRef]

70. Shahnazari, M.; Samandari-Masouleh, L.; Emami, S. Equipment capacity optimization of an educational building's CCHP system by genetic algorithm and sensitivity analysis. *Energy Equip. Syst.* **2017**, *5*, 375–387.

 *processes*

Article

Separating Electronic from Steric Effects in Ethene/α-Olefin Copolymerization: A Case Study on Octahedral [ONNO] Zr-Catalysts

Francesco Zaccaria [1,2], Roberta Cipullo [1], Andrea Correa [1], Peter H. M. Budzelaar [1], Vincenzo Busico [1] and Christian Ehm [1,*]

[1] Dipartimento di Scienze Chimiche, Università di Napoli Federico II, Via Cintia, 80126 Napoli, Italy; francesco.zaccaria@unipg.it (F.Z.); rcipullo@unina.it (R.C.); andrea.correa@unina.it (A.C.); p.budzelaar@unina.it (P.H.M.B.); busico@unina.it (V.B.)

[2] Dipartimento di Chimica, Biologia e Biotecnologie and CIRCC, Università di Perugia, Via Elce di Sotto 8, 06123 Perugia, Italy

* Correspondence: christian.ehm@unina.it

Received: 30 May 2019; Accepted: 17 June 2019; Published: 20 June 2019

Abstract: Four Cl/Me substituted [ONNO] Zr-catalysts have been tested in ethene/α-olefin polymerization. Replacing electron-donating methyl with isosteric but electron-withdrawing chlorine substituents results in a significant increase of comonomer incorporation. Exploration of steric and electronic properties of the ancillary ligand by DFT confirm that relative reactivity ratios are mainly determined by the electrophilicity of the metal center. Furthermore, quantitative DFT modeling of propagation barriers that determine polymerization kinetics reveals that electronic effects observed in these catalysts affect relative barriers for insertion and a capture-like transition state (TS).

Keywords: olefin copolymerization; reactivity ratios; electronic effects; salan catalysts; post-metallocene; DFT; insertion kinetics; olefin capture

1. Introduction

The production of ethene/α-olefin copolymers is one of the largest-scale processes of the chemical industry [1,2]. Linear low density polyethylene (LLDPE), obtained by the copolymerization of ethene with α-olefins like 1-butene, 1-hexene, or 1-octene, is one of the most representative examples of this class of copolymers, finding broad applications in the packaging industry [3].

The incorporation of α-olefins in a linear polyethylene backbone leads to chain branching, which can dramatically affect polymer crystallization and, subsequently, mechanical and rheological material properties [4–6]. Comonomer content and distribution within the polymer chain are consequently two critical microstructural parameters affecting macroscopic properties of ethene/α-olefin copolymers. The design of catalysts that allow polymer microstructure fine-tuning is therefore especially desirable in this context. Well-defined single-center molecular catalysts based on group IV metallocene and 'post-metallocene' complexes are nowadays often preferred for this kind of copolymerization rather than homologous multi-center, heterogeneous Ziegler–Natta systems [7–9].

Tailoring catalyst properties by ancillary ligand design has been widely exploited to access polymeric materials with desired structural features. Computational chemistry has contributed significantly to the identification of some rationale in the relationships between catalyst structure and properties [10–12]. The demystification of the origins of stereoselectivity is generally considered as a successful example of rational understanding [13–16], although some aspects are still debated [17–22]. Conversely, other catalyst properties, including comonomer affinity, are far less understood.

Typically, the probability of a comonomer to insert in the M–P bond (M = metal, P = polymeryl) depends on the last inserted monomeric unit [23]. In such cases, the tendency of a given catalyst to incorporate ethene or an α-olefin in the growing polymeryl can be described by first-order Markov statistics, employing two simple parameters denoted as r_E and r_C [24–26]:

$$r_E = \frac{k_{EE}}{k_{EC}}; \qquad r_C = \frac{k_{CC}}{k_{CE}} \tag{1}$$

where k is the kinetic constant of the specific propagation reaction indicated by the two subscripts; the first subscript denotes the last *inserted* monomer, while the second indicates the *inserting* one. For instance, k_{EC} refers to the insertion of the comonomer (C) after ethene (E). Under the assumption that the Curtin–Hammett principle applies, reactivity ratios can be expressed in terms of Gibbs free energy differences ($\Delta\Delta G^{\ddagger}$) between different propagation barrier heights:

$$r_E = \frac{k_{EE}}{k_{EC}} = e^{\frac{\Delta G^{\ddagger}_{EC} - \Delta G^{\ddagger}_{EE}}{RT}} = e^{\frac{\Delta\Delta G^{\ddagger}_{(EC-EE)}}{RT}} \; ; \; r_C = \frac{k_{CC}}{k_{CE}} = e^{\frac{\Delta G^{\ddagger}_{CE} - \Delta G^{\ddagger}_{CC}}{RT}} = e^{\frac{\Delta\Delta G^{\ddagger}_{(CE-CC)}}{RT}}. \tag{2}$$

Early polymerization studies have proposed that open coordination geometries might favor coordination and insertion of relatively bulky α-olefins, but this assertion has been challenged [27–30]. While it is generally accepted that electronic effects influence comonomer affinity, no definitive consensus has been reached; electron-withdrawing substituents reportedly lead to higher comonomer contents [31–33], but the opposite has also been observed [8,34]. Uncoupling electronic from steric effects when exploring variations of the ancillary ligands is non-trivial [31,35,36]; furthermore, pronounced electronic effects generally rely on heteroatom (e.g., F, N, O) containing substituents that might interact or even react with the other components of the catalytic pool (e.g., Al-based cocatalysts), leading to in situ catalyst modifications.

Computational analysis of copolymerization mechanisms is challenging [37,38]. Only very recently, an effective DFT protocol for highly accurate predictions of reactivity ratios has been reported, providing excellent mean average deviations (MAD) from experimental $\Delta\Delta G^{\ddagger}$ of only 0.2–0.3 kcal/mol [33,39]. Exploration of reaction pathways revealed the existence of a 'capture-like' transition state (TS) along the olefin approach vector, which can compete with insertion for the rate-limiting step in chain propagation especially for catalysts having additional steric bulk remote from the active pocket [39]. Furthermore, entropic effects were shown to contribute to the good incorporation capability of constrained geometry catalysts (CGC) [33].

To gain deeper insights into the still ill-understood electronic effects determining comonomer affinities, ancillary ligands with comparable steric hindrance but appreciable electronic differences are needed. The four diamine-bis(phenolate) Zr catalysts (from here on catalysts **R₁R₂**, R = Me or Cl) shown in Figure 1 might fulfill these requirements and were therefore selected for a case study. Chlorine and methyl substituents are relatively small, virtually isosteric, but differ in electronic properties—electron-withdrawing (Cl) vs. electron-donating (Me). Furthermore, this class of [ONNO]-type complexes has been widely exploited for mechanistic and kinetic studies due to (1) the ease of modification of *ortho-* and *para-*substituents of the phenolate ligands, allowing effective electronic and steric tuning, and (2) the typical controlled kinetic regimes exhibited by these systems in olefin homo- and co-polymerization [40–51]. Thus, in this work, the reactivity ratios for catalysts **R₁R₂** have been experimentally determined in the copolymerization of ethene with α-olefins, i.e., 1-butene (B) and propene (P). DFT modeling, using the aforementioned optimized protocols [33,39], provides a plausible interpretation for the dependence of reactivity ratios on the substitution pattern of the [ONNO] ligands.

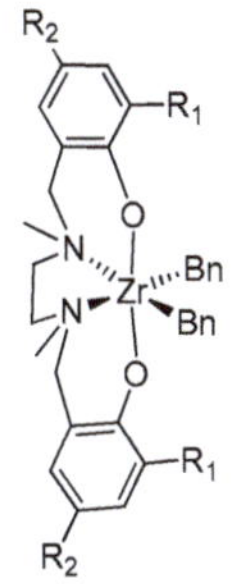

Figure 1. Diamine-bis(phenolate) Zr-complexes **R₁R₂** (Bn = benzyl).

2. Materials and Methods

2.1. Polymerization Experiments

All manipulations were performed under an inert atmosphere of argon or nitrogen, employing Schlenk line techniques or MBRAUN LABmaster 130 gloveboxes. Precatalysts were synthesized according to previously established procedures [44,48]. Methanol, hydrochloric acid, and di-*tert*-butylphenol (TBP) were purchased from Sigma-Aldrich and used as received. All other solvents were purchased from Romil and purified by passing through a mixed-bed activated-Cu/4Å molecular sieves column in an MBRAUN SPS-5 unit (final concentration of O_2 and H_2O < 1 ppm). Ethene and propene were purchased from Rivoira and purified by passing through a mixed-bed activated-Cu/4Å molecular sieves column. Methylalumoxane (MAO) was purchased from Chemtura. All glassware was dried in an oven at 150 °C for at least 16 h prior to use.

Ethene/propene polymerization experiments were performed as reported in [26]. Ethene/1-butene polymerization experiments were performed using a 300 mL magnetically stirred jacketed Pyrex reactor with three necks—one with a 15 mm SVL cap housing a silicone rubber septum, another with a 30 mm SVL cap, and the last with a two-way Rotaflo tap connected to a Schlenk line, and the ethene and 1-butene cylinders. A solution of activator/scavenger was prepared in the reactor by adding TBP (1.6 g) to a solution of MAO (5 mL, [Al] = 1.6 M) in dry toluene (150 mL) and gently stirring at room temperature for 1 h to ensure complete scavenging of "free" trimethylaluminum in MAO by TBP [52,53]. The reactor was thermostated at 25 °C, evacuated to remove argon, and then saturated with 1-butene (1.1 bar) and subsequently with ethene (0.55 or 1.1 bar; see Table S1). Ethene was fed in continuum at a constant operating pressure. Saturation was monitored by gas chromatography (GC) until the gas–liquid equilibrium was established. Concentrations of monomers in the liquid phase were determined using the equations of Kissin [54]. The polymerization was started by injecting a solution of precatalyst in toluene (<1 mL) through the silicone rubber septum with a syringe. The reaction was then stopped by turning off the ethene supply, venting the reactor, and rapidly quenching with a 95/5 *v/v* methanol/HCl (aqueous, concentrated) solution. The resulting polymer was coagulated with 400 mL of the same acidic solution, decanted, and vacuum-dried at 50 °C for 16 h. The reaction time was chosen to guarantee a low 1-butene conversion (<10%), based on GC analysis of the gas phase.

Polymer samples were characterized by nuclear magnetic resonance (NMR) spectroscopy, using a Bruker AVANCE spectrometer (400 MHz for ¹H) equipped with a 5 mm high-temperature cryoprobe on 50 mg mL⁻¹ solutions in tetrachloroethane-1,2-d_2 at 120 °C. Specific parameters for the measurements were as follows: ¹H NMR, 90° pulse, 2 s acquisition time, 10 s relaxation delay, 16 transients, 8.0 kHz spectral width, 32K time domain data points; ¹³C NMR, 45° pulse, 2.3 s acquisition time, 5.0 s relaxation delay, 1K transient, 14 kHz spectral width, 64K time domain data points. Experimental triad distributions have been determined according to the Randall method [25]. Values of reactivity ratio were calculated according to [55] from the conditional probabilities obtained from best-fit calculations of sequence distributions, performed with the Copolstat program (Professor M. Vacatello, University of Naples).

2.2. DFT Calculations

Following the protocol described in [56], all geometries were fully optimized using the Gaussian 09 software package [57] in combination with the OPTIMIZE routine of Baker [58,59] and the BOpt software package [60]. Zr-*n*Pr and Zr-*i*Bu groups have been used to mimic the Zr-polymeryl species after ethene and propene insertion, respectively. The naked cation approximation has been used [38]. All relevant minima and transition states were fully optimized at the TPSSTPSS level [61] of theory employing correlation-consistent polarized valence double-ζ Dunning (DZ) basis sets (cc-pVDZ quality) [62] from the EMSL basis set exchange library [63]. The density fitting approximation was used at the optimization stage (resolution of identity, RI) [64–67]. All calculations were performed with Scf = Tight and Int(Grid = Ultrafine) quality settings. All structures represent either true minima (as indicated by the absence of imaginary frequencies) or transition states (with exactly one imaginary frequency corresponding to the reaction coordinate). Convergence criteria were not loosened with the exception of propene BBRA with Zr-*n*Pr cationic species for **MeMe** (gradient tolerance 0.00015 instead of 0.00010). This approximation should not cause significant error, as all BBRA TSs are located on a plateau of the potential energy surface [39,68]. Final single-point energies and natural bond orbital (NBO) [69] analyses were calculated at the M06-2X level of theory [70] employing triple-ζ Dunning (TZ) basis sets (cc-pVTZ quality) [62]. Enthalpies and Gibbs free energies were then obtained from TZ single-point energies and thermal corrections from the TPSSTPSS/cc-pVDZ-(PP) vibrational analyses; entropy corrections were scaled by a factor of 0.67 to account for decreased entropy in the condensed phase [33,39,71].

3. Results and Discussion

3.1. Ethene/1-Butene Copolymerization

Catalysts R_1R_2 were tested in the copolymerization of ethene and 1-butene at 25 °C and the results are summarized in Table 1. Catalyst performance is significantly influenced by the substitution pattern of the phenolate rings. In particular, replacing Me with Cl leads to a higher comonomer incorporation ($[B]_{cop}$). Catalyst affinity towards 1-butene increases from **MeMe** to **ClCl**; r_E decreases and r_B increases, as shown in Figure 2. Reactivity ratios for **MeCl** and **ClMe** are very similar but the former exhibits a somewhat higher comonomer affinity. This suggests that the effect of Cl vs. Me substitution on comonomer affinity depends weakly on the exact position. Interestingly, despite the evident change of the individual r_x parameters, the $r_E r_B$ ratio is very similar for all four catalysts, as shown in Table 1, and close to unity, indicative of a random comonomer distribution in the polymer chain [23].

Table 1. Experimental results for catalysts R_1R_2 in ethene/1-butene (E/B) copolymerization.

Entry	Catalyst	R_p [a]	$[B]_{cop}$ [b] mol%	P_n [c] $\times 10^{-2}$	r_E [d]	$\Delta\Delta G^{\ddagger}_{(EB\text{-}EE)}$ [e]	r_B [d]	$\Delta\Delta G^{\ddagger}_{(BE\text{-}BB)}$ [e]	$r_E r_B$
1	**MeMe**	1	24	1.00	60(1)	2.4	0.012(1)	−2.6	0.7
2	**MeCl**	7	32	0.53	38(4)	2.2	0.025(4)	−2.2	1.0
3	**ClMe**	31	37	0.24	33(1)	2.1	0.034(4)	−2.0	1.1
4	**ClCl**	360	43	0.22	22(1)	1.8	0.06(2)	−1.7	1.3

In toluene (150 mL); 25 °C; $p(B) = p(E) = 1.1$ bar; cocatalyst = MAO/TBP ($[Al] = 5.4 \times 10^{-2}$ M). E = ethene; B = 1-butene. [a] Productivity in kg(copolymer)·mol(Zr)$^{-1}$·[C$_n$H$_{2n}$]$^{-1}$·h^{-1}; [b] 1-butene content in the copolymer, determined by ^{13}C NMR; [c] average copolymerization degree determined by ^{1}H NMR based on the concentration of chain end groups; [d] average values obtained from two experiments at different $p(E)$ (0.5 and 1.1 bar, see Supporting Information); [e] calculated from experimental r values, according to Equation (2).

Steric effects in the present catalyst class are mainly related to the *ortho*-substituents on the phenolate rings, due to their proximity to the metal center [41]. The Cl and Me substituents considered here are small and isosteric and, in α-olefin homopolymerization, catalysts **ClCl** and **MeMe** exhibit nearly identical, low stereoselectivity [48]. Differences in catalytic performance can therefore be ascribed mainly to electronic effects, resulting from Cl/Me exchange (vide infra).

The marked increase in comonomer affinity is accompanied by an appreciable increase of productivity (R_p) and a decrease of the average copolymerization degree (P_n). Trends in activity are generally difficult to rationalize, since they strongly depend not only on intrinsic barriers for chain propagation, but also on the percentage of active Zr-centers and catalyst decay mechanisms [10,72]. The amount of active metal is often relatively small [73,74], and difficult to determine [75]. Nevertheless, results in Table 1 are in line with general observation that the reactivity of cationic complexes increases with the electrophilicity of the metal center (vide infra) [76], although this can be affected by ion pairing effects in solution [77,78]. Concerning the average P_n, the unsaturated chain ends observed by polymer microstructural analysis indicate a dominant termination pathway by β-H transfer routes; the negligible dependence of P_n on ethene pressure (see Supporting Information) points to β-H transfer to the monomer being favored over β-H transfer to the metal center [79].

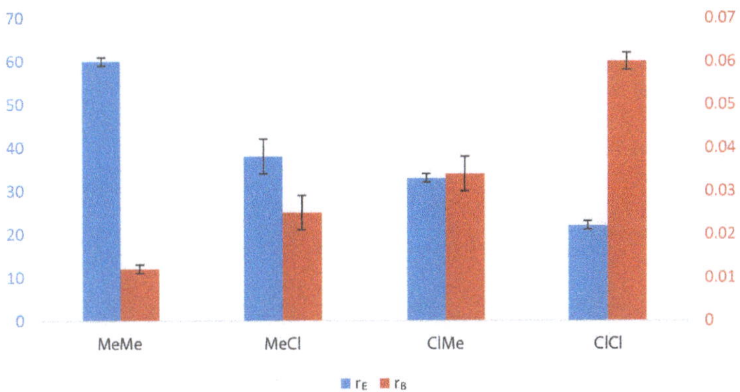

Figure 2. Trends in r_E and r_B for catalysts R_1R_2 (see Table 1).

3.2. Computational Modeling

To unravel the connection between the catalyst structure and comonomer affinity, DFT studies have been carried out on complexes R_1R_2. The computational protocol (see Materials and Methods for details) has been previously benchmarked for group IV precatalysts and TSs relevant to olefin polymerization [54,74,80–82], including specifically those for predicting reactivity ratios in copolymerization [33,39]. A *facfac* geometry for all complexes and TSs has been considered, which is generally accepted to be the most stable for the neutral complexes and the cationic active species; interconversion between active *facfac* and inactive *mermer* configurations of the naked cationic complex is generally assumed to represent a rapid pre-equilibrium with respect to chain propagation [43–45,83].

3.2.1. Connection between Electronic/Steric Properties of R_1R_2 and E/B Copolymerization Performance

As the optimization of TSs for chain propagation is not always straightforward (*vide infra*), a more detailed analysis of the relationship between reactivity ratios and catalyst structure has been initially carried out by considering the neutral precatalysts and two suitable computational parameters to describe the electronic and steric properties of the ancillary ligands. To avoid complications due to the conformational rearrangements of the Zr-benzyl groups, the dichloride analogues of R_1R_2 have been used. Electronic changes of the [ONNO] ligands were estimated by looking at the natural population analysis (NPA) charges of the $ZrCl_2$ fragments, hereinafter denoted as q_{ZrCl_2}. Higher charges are indicative of more electrophilic metal centers and, therefore, of less electron-donating ligands. Indeed,

q_{ZrCl2} increases progressively with the degree of chlorination of the [ONNO]-ligand, going from **MeMe** to **ClCl**, as expected based on the more electron-withdrawing nature of Cl, as shown in Table S6. Figure 3 shows that the experimental $\Delta\Delta G^{\ddagger}$ for r_E and r_B correlate well with q_{ZrCl2} for this catalyst set ($R^2 \geq 0.93$ for linear fitting; see also Table S6). Interestingly, the minor difference in comonomer affinity of **MeCl** and **ClMe** is nicely reproduced. Accordingly, *ortho*-substituent influence on the phenolate ring dominates over *para*-position also from an electronic point of view, in line, for instance, with the higher acidity of *o*-chlorophenol ($pK_a \approx 8.6$) with respect to *p*-chlorophenol ($pK_a \approx 9.4$) [84]. The difference in electronic effects imparted by *ortho*- and *para*-substituents is however usually overshadowed by steric differences [41].

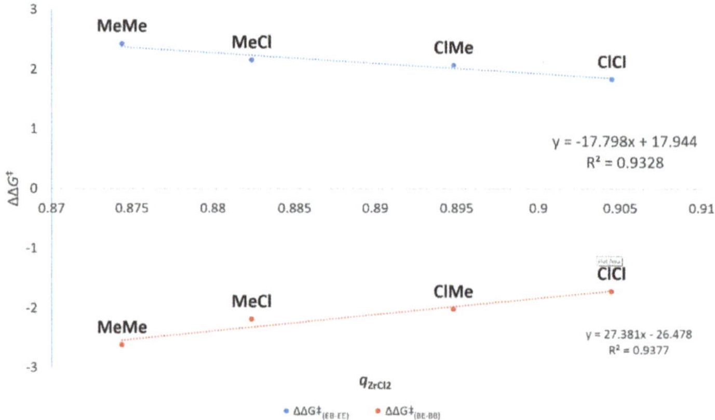

Figure 3. Correlation between $\Delta\Delta G^{\ddagger}$ corresponding to the *r*-parameters for ethene/1-butene copolymerization and q_{ZrCl2} for catalysts **R$_1$R$_2$** (see also Table S6).

To describe the steric hindrance imparted by the ancillary ligand framework, the percentage buried volume in the catalyst active pocket (%V_{bur}, see also Supporting Information) can be used, as proposed by Cavallo and coworkers [85–87]. Scanning spheres having 5 Å radius have been reported to be suitable to explore steric effects in olefin polymerization [19]. Maps of the steric bulk, as shown in Figure S1 and Supporting Information, as well as the %V_{bur} values (total and separately for the four quadrants of the active pocket), as shown in Table 2, clearly indicate that the steric hindrance provided by the ancillary ligand is small and virtually identical for all complexes **R$_1$R$_2$**. This becomes even more evident when compared to a highly stereoselective catalyst of this class with *ortho*-1-adamantyl-*para*-methyl substituent pattern (**AdMe**) [41], providing additional evidence for steric effects at most marginally affecting relative comonomer affinity for this set of catalysts.

Table 2. Results of buried volume analysis for catalysts **R$_1$R$_2$**.

Entry.	Catalyst	Total %V_{bur}	%V_{bur} per Quadrant			
			SW	NW	NE	SE
1	**MeMe**	49.4	48.4	50.5	48.5	50.3
2	**MeCl**	49.5	48.5	50.3	48.5	50.6
3	**ClMe**	49.1	47.7	50.4	47.7	50.5
4	**ClCl**	49.1	47.7	50.5	47.7	50.5
5	**AdMe**	57.8	61.3	53.8	61.7	54.2

Parameters: center of the sphere at the Zr atom; z-axis defined by Zr and the middle point between the two Cl atoms; xz plane defined by Zr and one of the Cl atoms; radius of the sphere = 5 Å; hydrogen atoms included. SW = south-west; NW = north-west; NE = north-east; SE = south-east quadrants (see also Figure S2). **AdMe** provided for comparison.

3.2.2. Quantitative Kinetic Modeling of Reactivity Ratios for $\mathbf{R_1R_2}$ in E/P Copolymerization

Experimental and computational results indicate that differences in comonomer affinity are dictated predominantly by electronic effects in the present case study, yet, they cannot provide a rationale for this observation. Thus, the origins of comonomer affinities have been analyzed more in depth by attempting a direct modelling of the *r*-parameters of catalysts $\mathbf{R_1R_2}$.

The drop of r_E from 60 (**MeMe**), as shown in Table 1, to 22 (**ClCl**) in ethene/1-butene copolymerization is undoubtedly relevant in terms of polymer chemistry, signifying the passage from a very poor to a moderately good comonomer incorporator. However, such decrease corresponds only to a variation of $\Delta\Delta G^{\ddagger}_{(EB\text{-}EE)}$ in the order of about 1 kcal/mol, as shown in Table 1, emphasizing that modeling these relative barriers is very challenging due to the required accuracy. For this reason, only the two extreme cases of **MeMe** and **ClCl** are considered here. From a computational point of view, propene is an appreciably more convenient model α-olefin than 1-butene, especially when complex TSs have to be optimized. For better comparison between experiment and theory, **MeMe** and **ClCl** were re-tested in ethene/propene copolymerization, under reaction conditions analogous to those used for ethene/1-butene. The estimated r_E and r_P values and the corresponding $\Delta\Delta G^{\ddagger}$ are reported in Entries 1–2 of Table 3, showing similar differences in comonomer affinity for propene and 1-butene upon Me/Cl exchange (compare Table 3 to Table 1).

Table 3. Experimental r_E and r_P for **MeMe** and **ClCl** in ethene/propene copolymerization and comparison between corresponding experimental and DFT calculated $\Delta\Delta G^{\ddagger}$.

Entry	Catalyst	r_E	$\Delta\Delta G^{\ddagger}_{(EP\text{-}EE)}$	r_P	$\Delta\Delta G^{\ddagger}_{(PE\text{-}PP)}$	$r_E r_P$
			Experimental			
1	MeMe	41	2.2	0.03	−2.1	1.2
2	ClCl	8.7	1.3	0.19	−1.0	1.7
			DFT–INS only			
3	MeMe	-	1.5 [0.7]	-	−1.6 [−0.5]	
4	ClCl	-	1.3 [0.0]	-	−1.2 [0.2]	
			DFT–INS and BBRA			
5	MeMe	-	1.5 [0.7]	-	−1.6 [−0.5]	
6	ClCl	-	0.6 [0.7]	-	−1.2 [0.2]	

$\Delta\Delta G^{\ddagger}$ in kcal/mol. INS = insertion; BBRA = backbone rearrangement. Numbers in squared brackets are deviations between experimental and DFT calculated $\Delta\Delta G^{\ddagger}_{(EP\text{-}EE)}$ or $\Delta\Delta G^{\ddagger}_{(PE\text{-}PP)}$. $\Delta\Delta G^{\ddagger}$ closer to zero are indicative of better comonomer incorporation.

Under the assumption that insertion is the rate-determining step for chain propagation in all cases, $\Delta\Delta G^{\ddagger}_{DFT}$ indicate no significant difference in comonomer affinity for the two catalysts in terms of r_E, and only a small one in terms of r_P, as shown in Entries 3–4 of Table 3. The better comonomer incorporation capability of **ClCl** for r_P is, although slightly underestimated, reproduced. Experimental errors of r_C are generally higher than those of r_E, due to difficulties in accurate determination [26,37,39,88], and we will therefore focus on the discrepancy of experimental and DFT findings for r_E in the following.

As mentioned earlier, this discrepancy might be traced to a change in the rate-limiting step from insertion to a monomer capture-like process. For *ansa*-metallocenes, a backbone rearrangement (BBRA) corresponding to a breathing motion of backbone to accommodate the incoming olefin in the first coordination sphere of the metal has been identified, potentially affecting all propagation modes [39]. Here, we use the same nomenclature, although the open nature of catalysts $\mathbf{R_1R_2}$ makes this structural rearrangement less obvious, as shown in the Supporting Information. Indeed, as there is no marked steric hurdle to monomer capture in catalysts $\mathbf{R_1R_2}$, BBRA is generally a low energy process with respect to insertion for these catalysts. However, it always remains competitive and becomes kinetically relevant for very low insertion barriers, i.e., in ethene homopolymerization for **ClCl**. Accounting for

BBRA in the computational modelling therefore results in a lower $\Delta\Delta G^{\ddagger}_{(EP\text{-}EE)}$ for **ClCl** (0.6 kcal/mol) compared to competing insertion TSs (1.3 kcal/mol); all the other $\Delta\Delta G^{\ddagger}$ remain unchanged, as can be seen by Entries 3–4 vs. 5–6 in Table 3. The difference in experimental r_E for **MeMe** and **ClCl** is now perfectly reproduced, with a relatively small and constant deviation between experiments and theory of 0.7 kcal/mol that is quite satisfactory considering the complexity of the case study. While absolute rates are, both experimentally and computationally, often hard to determine accurately, relative rates are usually more reliable due to error compensation.

A closer inspection of the optimized geometries and DFT-calculated kinetic profiles provides a plausible interpretation for the different copolymerization performance of **MeMe** and **ClCl**. Relative Gibbs free energies for relevant TSs and reaction intermediates for ethene homo-propagation are summarized in Figure 4, highlighting the different metal-to-olefin distances found for the various geometries.

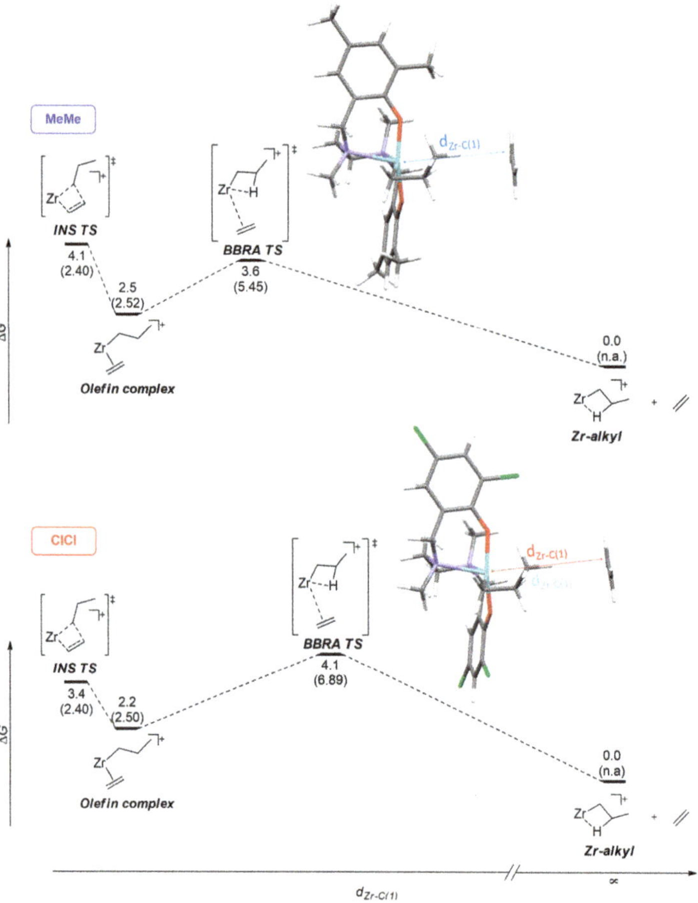

Figure 4. Energy profiles for ethene homopolymerization predicted by DFT for **MeMe** and **ClCl**. INS = insertion, BBRA = backbone rearrangement. Relative Gibbs free energy differences in kcal/mol; numbers in brackets: distances between the Zr and C(1) carbon of ethene in Å ($d_{Zr\text{-}C(1)}$). *n*Pr group used as model for the polymeryl chain after ethene insertion (see Materials and Methods).

ClCl binds ethene slightly more strongly than **MeMe**, likely due to the higher electrophilicity of its metal center (vide supra). The insertion barrier for **ClCl** is lower by 0.7 kcal/mol compared to **MeMe**, suggesting some electronic influence also in this respect [82]. Furthermore, BBRA activation barriers are found to be higher for **ClCl** than for **MeMe**. Higher associative and dissociative $\Delta G^{\ddagger}$ are expected for a more strongly coordinated olefin. In other words, replacing Me with Cl substituents appears to lower insertion but increase BBRA barriers, up to the point that the rate-limiting step can switch. Analogous considerations apply also to propene enchainment after ethene, as reported in the Supporting Information.

It is instructive to analyze the distances between the Zr center and the C(1) carbon of ethene $(d_{\text{Zr-C(1)}})$. Insertion TSs appear to be very similar for the two catalysts, with a $d_{\text{Zr-C(1)}}$ close to 2.4 Å. This structural similarity is reflected by a similar $\Delta S^{\ddagger}_{\text{INS}}$ for the two catalysts; differences in $\Delta G^{\ddagger}_{\text{INS}}$ are mainly due to the $\Delta H^{\ddagger}_{\text{INS}}$ term, as shown by Entries 1 vs. 3 in Table 4. Conversely, $d_{\text{Zr-C(1)}}$ is about 1.5 Å longer in the BBRA TS for **ClCl** than that for **MeMe**, diagnostic of a looser interaction between the metal and the incoming monomer. A lower $\Delta H^{\ddagger}_{\text{BBRA}}$ is found for the former catalyst, as shown by Entries 2 vs. 4 in Table 4, accordingly. Significantly lower $\Delta S^{\ddagger}_{\text{BBRA}}$ than $\Delta S^{\ddagger}_{\text{INS}}$ are estimated by DFT as a smaller fraction of the 2→1 particle entropy loss for monomer capture has been paid at the BBRA rather than the insertion TS.

Table 4. Gibbs free energy, enthalpy, and entropy differences estimated by DFT for **BBRA** and insertion (INS) TS with respect to the β-agostic Zr-alkyl cation (in kcal/mol; see also Figure 4).

Entry	Catalyst	Process	$\Delta G^{\ddagger}$	$\Delta H^{\ddagger}$	$\Delta S^{\ddagger}$
1	MeMe	INS	4.1	−3.7	26
2		BBRA	3.6	−0.7	14
3	ClCl	INS	3.4	−4.6	27
4		BBRA	4.1	−0.4	15

Thus, increased electrophilicity of the active species might lead to lower *insertion* barriers, but not necessarily to lower *propagation* barriers, since BBRA might become the rate-determining step. In particular, in the case of r_{E} for **MeMe** and **ClCl**, the replacement of Me with electron-withdrawing Cl substituents simultaneously lowers propene and raises ethene propagation barriers, as BBRA (and not insertion) becomes kinetically relevant for the latter. Both these effects contribute to increased comonomer incorporation capability.

4. Conclusions

The testing of [ONNO]-type Zr complexes in ethene/α-olefin copolymerization reveals that the replacement of electron-donating Me-substituents of the ancillary ligand with isosteric but electron-withdrawing Cl groups, leads to a significant increase of comonomer affinity in the polymer chain. This indicates that, for this catalyst set, the electronic properties of the active species are the main parameter determining comonomer affinities. This is supported by DFT studies exploring electronic and steric features of the ancillary ligands.

Quantitative kinetic analysis of propagation rates by DFT shows that barriers for olefin insertion are lowered when electron-withdrawing substituents are installed on the catalyst backbone, while those for capture/BBRA increase. Reactivity ratios in copolymerization are therefore determined by the subtle balance between insertion and BBRA kinetics and cooperative effects can simultaneously decrease comonomer propagation barriers but increase those for ethene, when BBRA becomes rate limiting.

Electronic tuning effects in olefin (co)polymerization are still ill understood and the insights presented here allow for the first time a clear separation of steric and electronic effects. The ultimate goal in the design of catalysts for olefin copolymerization is typically to develop systems that prefer α-olefins over the generally more reactive ethene, and to finely tune this preference to access advanced materials for specific applications. The insights presented in this paper provide the means of understanding of electronic effects in copolymerization, which should prove useful for the design of new copolymerization catalysts.

Supplementary Materials: The following are available online at http://www.mdpi.com/2227-9717/7/6/384/s1, Table S1: Polymerization results, Figure S1: Buried volume analysis, Table S2: Electronic charges, Figure S2: Calculated reaction profiles for propene enchainment after ethene, Table S3: Final energies, entropy, and enthalpy corrections in Hartree, Full Gaussian citation.

Author Contributions: Conceptualization, R.C., V.B., C.E.; methodology, R.C., V.B., C.E.; formal analysis, R.C., F.Z., A.C.; investigation, F.Z., R.C., C.E.; resources, V.B.; writing—original draft preparation, F.Z.; writing—review and editing, R.C., A.C., P.H.M.B., V.B., C.E.; visualization, F.Z.; supervision, R.C., P.H.M.B., V.B.

Funding: This research received no external funding.

Conflicts of Interest: The authors declare no conflict of interest.

References

1. Stürzel, M.; Mihan, S.; Mülhaupt, R. From Multisite Polymerization Catalysis to Sustainable Materials and All-Polyolefin Composites. *Chem. Rev.* **2016**, *116*, 1398–1433. [CrossRef] [PubMed]

2. Sauter, D.; Taoufik, M.; Boisson, C. Polyolefins, a Success Story. *Polymers* **2017**, *9*, 185. [CrossRef] [PubMed]

3. Simpson, D.M.; Vaugham, G.A. *Encyclopedia of Polymer Science and Technology*, 4th ed.; John Wiley & Sons, Inc.: Hoboken, NJ, USA, 2014; Volume 5.

4. Bensason, S.; Minick, J.; Moet, A.; Chum, S.; Hiltner, A.; Baer, E. Classification of homogeneous ethylene-octene copolymers based on comonomer content. *J. Polym. Sci. Part B Polym. Phys.* **1996**, *34*, 1301–1315. [CrossRef]

5. Xu, X.; Xu, J.; Feng, L.; Chen, W. Effect of short chain-branching distribution on crystallinity and modulus of metallocene-based ethylene–butene copolymers. *J. Appl. Polym. Sci.* **2000**, *77*, 1709–1715. [CrossRef]

6. Burfield, D.R. Correlation between crystallinity and ethylene content in LLDPE and related ethylene copolymers: Demonstration of the applicability of a simple empirical relationship. *Macromolecules* **1987**, *20*, 3020–3023. [CrossRef]

7. Baier, M.C.; Zuideveld, M.A.; Mecking, S. Post-Metallocenes in the Industrial Production of Polyolefins. *Angew. Chem. Int. Ed.* **2014**, *53*, 9722–9744. [CrossRef] [PubMed]

8. Klosin, J.; Fontaine, P.P.; Figueroa, R. Development of Group IV Molecular Catalysts for High Temperature Ethylene-α-Olefin Copolymerization Reactions. *Acc. Chem. Res.* **2015**, *48*, 2004–2016. [CrossRef] [PubMed]

9. Kaminsky, W. The discovery of metallocene catalysts and their present state of the art. *J. Polym. Sci. Part A Polym. Chem.* **2004**, *42*, 3911–3921. [CrossRef]

10. Ehm, C.; Zaccaria, F.; Cipullo, R. From Mechanistic Investigation to Quantitative Prediction. In *Computational Quantum Chemistry*; Soroush, M., Ed.; Elsevier: Amsterdam, The Netherlands, 2019; pp. 287–326.

11. Hooper, M.S.; Michalak, A. Computational Modeling of Polymerization Catalysts. In *Handbook of Transition Metal Polymerization Catalysts*; Wiley-VCH: Weinheim, Germany, 2018.

12. Rappé, A.K.; Skiff, W.M.; Casewit, C.J. Modeling Metal-Catalyzed Olefin Polymerization. *Chem. Rev.* **2000**, *100*, 1435–1456. [CrossRef]

13. Corradini, P.; Guerra, G.; Cavallo, L. Do New Century Catalysts Unravel the Mechanism of Stereocontrol of Old Ziegler–Natta Catalysts? *Acc. Chem. Res.* **2004**, *37*, 231–241. [CrossRef]

14. Coates, G.W. Precise Control of Polyolefin Stereochemistry Using Single-Site Metal Catalysts. *Chem. Rev.* **2000**, *100*, 1223–1252. [CrossRef] [PubMed]

15. Resconi, L.; Cavallo, L.; Fait, A.; Piemontesi, F. Selectivity in Propene Polymerization with Metallocene Catalysts. *Chem. Rev.* **2000**, *100*, 1253–1346. [CrossRef]

16. Busico, V.; Cipullo, R. Microstructure of polypropylene. *Prog. Polym. Sci.* **2001**, *26*, 443–533. [CrossRef]

17. Talarico, G.; Budzelaar, P.H.M. Analysis of Stereochemistry Control in Homogeneous Olefin Polymerization Catalysis. *Organometallics* **2014**, *33*, 5974–5982. [CrossRef]

18. Castro, L.; Therukauff, G.; Vantomme, A.; Welle, A.; Haspeslagh, L.; Brusson, J.M.; Maron, L.; Carpentier, J.F.; Kirillov, E. A Theoretical Outlook on the Stereoselectivity Origins of Isoselective Zirconocene Propylene Polymerization Catalysts. *Chem. Eur. J.* **2018**, *24*, 10784–10792. [CrossRef] [PubMed]

19. Ehm, C.; Vittoria, A.; Goryunov, G.P.; Kulyabin, P.S.; Budzelaar, P.H.M.; Voskoboynikov, A.Z.; Busico, V.; Uborsky, D.V.; Cipullo, R. Connection of Stereoselectivity, Regioselectivity, and Molecular Weight Capability in *rac*-R′$_2$Si(2-Me-4-R-indenyl)$_2$ZrCl$_2$ Type Catalysts. *Macromolecules* **2018**, *51*, 8073–8083. [CrossRef]

20. Talarico, G.; Budzelaar, P.H.M. Ligand Coordination Driven by Monomer and Polymer Chain: The Intriguing Case of Salalen–Ti Catalyst for Propene Polymerization. *Macromolecules* **2017**, *50*, 5332–5336. [CrossRef]

21. Domski, G.J.; Eagan, J.M.; De Rosa, C.; Di Girolamo, R.; LaPointe, A.M.; Lobkovsky, E.B.; Talarico, G.; Coates, G.W. Combined Experimental and Theoretical Approach for Living and Isoselective Propylene Polymerization. *ACS Catal.* **2017**, *7*, 6930–6937. [CrossRef]

22. Milano, G.; Cavallo, L.; Guerra, G. Site Chirality as a Messenger in Chain-End Stereocontrolled Propene Polymerization. *J. Am. Chem. Soc.* **2002**, *124*, 13368–13369. [CrossRef]

23. Brintzinger, H.H.; Fischer, D.; Mülhaupt, R.; Rieger, B.; Waymouth, R.M. Stereospecific Olefin Polymerization with Chiral Metallocene Catalysts. *Angew. Chem. Int. Ed. Eng.* **1995**, *34*, 1143–1170. [CrossRef]

24. Kakugo, M.; Naito, Y.; Mizunuma, K.; Miyatake, T. Carbon-13 NMR determination of monomer sequence distribution in ethylene-propylene copolymers prepared with δ-titanium trichloride-diethylaluminum chloride. *Macromolecules* **1982**, *15*, 1150–1152. [CrossRef]

25. Randall, J.C. A review of high resolution liquind ^{13}carbon nuclear magnetic resonance characterizations of ethylene-based polymers. *J. Macromol. Sci. Part C* **1989**, *29*, 201–317. [CrossRef]

26. Busico, V.; Cipullo, R.; Segre, A.L. Advances in the ^{13}C NMR characterization of ethene/propene copolymers, 1. *Macromol. Chem. Phys.* **2002**, *203*, 1403–1412. [CrossRef]

27. Dankova, M.; Waymouth, R.M. High Comonomer Selectivity in Ethylene/Hexene Copolymerization by Unbridged Indenyl Metallocenes. *Macromolecules* **2003**, *36*, 3815–3820. [CrossRef]

28. Reybuck, S.E.; Meyer, A.; Waymouth, R.M. Copolymerization Behavior of Unbridged Indenyl Metallocenes: Substituent Effects on the Degree of Comonomer Incorporation. *Macromolecules* **2002**, *35*, 637–643. [CrossRef]

29. Kunz, K.; Erker, G.; Kehr, G.; Fröhlich, R.; Jacobsen, H.; Berke, H.; Blacque, O. Formation of Cyclodimeric (sp^2-C$_1$)-Bridged Cp/-Oxido ("CpC$_1$O"M^{IV}X$_2$) Group 4 Metal Ziegler–Natta Catalyst SystemsHow Important Is the "Constrained Geometry" Effect? *J. Am. Chem. Soc.* **2002**, *124*, 3316–3326. [CrossRef] [PubMed]

30. Möhring, P.C.; Coville, N.J. The influence of cyclopentadienyl ring substituent steric and electronic effects on the ethylene-α-olefin copolymerisation behaviour of (CpR)$_2$ZrCl$_2$ethylalumoxane catalysts. *J. Mol. Catal. A Chem.* **1995**, *96*, 181–195. [CrossRef]

31. Thornberry, M.P.; Reynolds, N.T.; Deck, P.A.; Fronczek, F.R.; Rheingold, A.L.; Liable-Sands, L.M. Synthesis, Structure, and Olefin Polymerization Catalytic Behavior of Aryl-Substituted Zirconocene Dichlorides. *Organometallics* **2004**, *23*, 1333–1339. [CrossRef]

32. Lehmus, P.; Kokko, E.; Härkki, O.; Leino, R.; Luttikhedde, H.J.G.; Näsman, J.H.; Seppälä, J.V. Homo- and Copolymerization of Ethylene and α-Olefins over 1- and 2-Siloxy-Substituted Ethylenebis(indenyl)zirconium and Ethylenebis(tetrahydroindenyl)zirconium Dichlorides. *Macromolecules* **1999**, *32*, 3547–3552. [CrossRef]

33. Zaccaria, F.; Ehm, C.; Budzelaar, P.H.M.; Busico, V. Accurate Prediction of Copolymerization Statistics in Molecular Olefin Polymerization Catalysis: The Role of Entropic, Electronic, and Steric Effects in Catalyst Comonomer Affinity. *ACS Catal.* **2017**, *7*, 1512–1519. [CrossRef]

34. Yano, A.; Hasegawa, S.; Kaneko, T.; Sone, M.; Sato, M.; Akimoto, A. Ethylene/1-hexene copolymerization with Ph$_2$C(Cp)(Flu)ZrCl$_2$ derivatives: Correlation between ligand structure and copolymerization behavior at high temperature. *Macromol. Chem. Phys.* **1999**, *200*, 1542–1553. [CrossRef]

35. Gassman, P.G.; Mickelson, J.W.; Sowa, J.R. 1,2,3,4-Tetramethyl-5-(trifluoromethyl)cyclopentadienide: A unique ligand with the steric properties of pentamethylcyclopentadienide and the electronic properties of cyclopentadienide. *J. Am. Chem. Soc.* **1992**, *114*, 6942–6944. [CrossRef]

36. Nsiri, H.; Belaid, I.; Larini, P.; Thuilliez, J.; Boisson, C.; Perrin, L. Ethylene–Butadiene Copolymerization by Neodymocene Complexes: A Ligand Structure/Activity/Polymer Microstructure Relationship Based on DFT Calculations. *ACS Catal.* **2016**, *6*, 1028–1036. [CrossRef]

37. Friederichs, N.; Wang, B.; Budzelaar, P.H.M.; Coussens, B.B. A combined experimental—molecular modeling approach for ethene–propene copolymerization with C$_2$-symmetric metallocenes. *J. Mol. Catal. A Chem.* **2005**, *242*, 91–104. [CrossRef]

38. Laine, A.; Coussens, B.B.; Hirvi, J.T.; Berthoud, A.; Friederichs, N.; Severn, J.R.; Linnolahti, M. Effect of Ligand Structure on Olefin Polymerization by a Metallocene/Borate Catalyst: A Computational Study. *Organometallics* **2015**, *34*, 2415–2421. [CrossRef]

39. Zaccaria, F.; Cipullo, R.; Budzelaar, P.H.M.; Busico, V.; Ehm, C. Backbone rearrangement during olefin capture as the rate limiting step in molecular olefin polymerization catalysis and its effect on comonomer affinity. *J. Polym. Sci. Part A Polym. Chem.* **2017**, *55*, 2807–2814. [CrossRef]

40. Tshuva, E.Y.; Goldberg, I.; Kol, M. Isospecific Living Polymerization of 1-Hexene by a Readily Available Nonmetallocene C_2-Symmetrical Zirconium Catalyst. *J. Am. Chem. Soc.* **2000**, *122*, 10706–10707. [CrossRef]

41. Busico, V.; Cipullo, R.; Pellecchia, R.; Ronca, S.; Roviello, G.; Talarico, G. Design of stereoselective Ziegler–Natta propene polymerization catalysts. *Proc. Natl. Acad. Sci. USA* **2006**, *103*, 15321–15326. [CrossRef]

42. Ciancaleoni, G.; Fraldi, N.; Cipullo, R.; Busico, V.; Macchioni, A.; Budzelaar, P.H.M. Structure/Properties Relationship for Bis(phenoxyamine)Zr(IV)-Based Olefin Polymerization Catalysts: A Simple DFT Model To Predict Catalytic Activity. *Macromolecules* **2012**, *45*, 4046–4053. [CrossRef]

43. Ciancaleoni, G.; Fraldi, N.; Budzelaar, P.H.M.; Busico, V.; Cipullo, R.; Macchioni, A. Structure−Activity Relationship in Olefin Polymerization Catalysis: Is Entropy the Key? *J. Am. Chem. Soc.* **2010**, *132*, 13651–13653. [CrossRef]

44. Busico, V.; Cipullo, R.; Friederichs, N.; Ronca, S.; Talarico, G.; Togrou, M.; Wang, B. Block Copolymers of Highly Isotactic Polypropylene via Controlled Ziegler–Natta Polymerization. *Macromolecules* **2004**, *37*, 8201–8203. [CrossRef]

45. Busico, V.; Cipullo, R.; Ronca, S.; Budzelaar, P.H.M. Mimicking Ziegler-Natta Catalysts in Homogeneous Phase, 1. C_2-Symmetric Octahedral Zr(IV) Complexes with Tetradentate [ONNO]-Type Ligands. *Macromol. Rapid Commun.* **2001**, *22*, 1405–1410. [CrossRef]

46. Segal, S.; Goldberg, I.; Kol, M. Zirconium and Titanium Diamine Bis(phenolate) Catalysts for α-Olefin Polymerization: From Atactic Oligo(1-hexene) to Ultrahigh-Molecular-Weight Isotactic Poly(1-hexene). *Organometallics* **2005**, *24*, 200–202. [CrossRef]

47. Cohen, A.; Kopilov, J.; Goldberg, I.; Kol, M. C1-Symmetric Zirconium Complexes of [ONNO′]-Type Salan Ligands: Accurate Control of Catalyst Activity, Isospecificity, and Molecular Weight in 1-Hexene Polymerization. *Organometallics* **2009**, *28*, 1391–1405. [CrossRef]

48. Busico, V.; Cipullo, R.; Romanelli, V.; Ronca, S.; Togrou, M. Reactivity of Secondary Metal−Alkyls in Catalytic Propene Polymerization: How Dormant Are "Dormant Chains"? *J. Am. Chem. Soc.* **2005**, *127*, 1608–1609. [CrossRef]

49. Busico, V.; Cipullo, R.; Friederichs, N.; Ronca, S.; Togrou, M. The First Molecularly Characterized Isotactic Polypropylene-block-polyethylene Obtained via "Quasi-Living" Insertion Polymerization. *Macromolecules* **2003**, *36*, 3806–3808. [CrossRef]

50. Preston, A.Z.; Kim, J.; Medvedev, G.A.; Delgass, W.N.; Caruthers, J.M.; Abu-Omar, M.M. Steric and Solvation Effects on Polymerization Kinetics, Dormancy, and Tacticity of Zr-Salan Catalysts. *Organometallics* **2017**, *36*, 2237–2244. [CrossRef]

51. Ciancaleoni, G.; Fraldi, N.; Budzelaar, P.H.M.; Busico, V.; Macchioni, A. Activation of a bis(phenoxy-amine) precatalyst for olefin polymerisation: First evidence for an outer sphere ion pair with the methylborate counterion. *Dalton Trans.* **2009**, 8824–8827. [CrossRef]

52. Busico, V.; Cipullo, R.; Cutillo, F.; Friederichs, N.; Ronca, S.; Wang, B. Improving the Performance of Methylalumoxane: A Facile and Efficient Method to Trap "Free" Trimethylaluminum. *J. Am. Chem. Soc.* **2003**, *125*, 12402–12403. [CrossRef]

53. Zaccaria, F.; Zuccaccia, C.; Cipullo, R.; Budzelaar, P.H.M.; Macchioni, A.; Busico, V.; Ehm, C. BHT-Modified MAO: Cage Size Estimation, Chemical Counting of Strongly Acidic Al Sites, and Activation of a Ti-Phosphinimide Precatalyst. *ACS Catal.* **2019**, *9*, 2996–3010. [CrossRef]

54. Kissin, Y. *Isospecific Polymerization of Olefins: With Heterogeneous Ziegler-Natta Catalysts*; Springer: New York, NY, USA, 1985.

55. Hamm, G.E. In *"Copolymerization"*; Hamm, G.E., Ed.; Wiley-Interscience: New York, NY, USA, 1964; Chapter I.

56. Ehm, C.; Budzelaar, P.H.M.; Busico, V. Calculating accurate barriers for olefin insertion and related reactions. *J. Organomet. Chem.* **2015**, *775*, 39–49. [CrossRef]

57. *Gaussian 09*, Revision B.1; Gaussian, Inc.: Wallingford, CT, USA, 2009; for the full citation see the Supporting Information.

58. Baker, J. *PQS*, version 2.4; Parallel Quantum Solutions: Fayetteville, AR, USA, 2001.

59. Baker, J. An algorithm for the location of transition states. *J. Comput. Chem.* **1986**, *7*, 385–395. [CrossRef]

60. Budzelaar, P.H.M. Geometry optimization using generalized, chemically meaningful constraints. *J. Comput. Chem.* **2007**, *28*, 2226–2236. [CrossRef]

61. Tao, J.; Perdew, J.P.; Staroverov, V.N.; Scuseria, G.E. Climbing the Density Functional Ladder: Nonempirical Meta\char21{}Generalized Gradient Approximation Designed for Molecules and Solids. *Phys. Rev. Lett.* **2003**, *91*, 146401. [CrossRef] [PubMed]

62. Peterson, K.A.; Figgen, D.; Dolg, M.; Stoll, H. Energy-consistent relativistic pseudopotentials and correlation consistent basis sets for the 4d elements Y–Pd. *J. Chem. Phys.* **2007**, *126*, 142101. [CrossRef] [PubMed]

63. Schuchardt, K.L.; Didier, B.T.; Elsethagen, T.; Sun, L.; Gurumoorthi, V.; Chase, J.; Li, J.; Windus, T.L. Basis Set Exchange: A Community Database for Computational Sciences. *J. Chem. Inf. Model.* **2007**, *47*, 1045–1052. [CrossRef] [PubMed]

64. Whitten, J.L. Coulombic potential energy integrals and approximations. *J. Chem. Phys.* **1973**, *58*, 4496–4501. [CrossRef]

65. Baerends, E.J.; Ellis, D.E.; Ros, P. Self-consistent molecular Hartree—Fock—Slater calculations I. The computational procedure. *Chem. Phys.* **1973**, *2*, 41–51. [CrossRef]

66. Feyereisen, M.; Fitzgerald, G.; Komornicki, A. Use of approximate integrals in ab initio theory. An application in MP2 energy calculations. *Chem. Phys. Lett.* **1993**, *208*, 359–363. [CrossRef]

67. Vahtras, O.; Almlöf, J.; Feyereisen, M.W. Integral approximations for LCAO-SCF calculations. *Chem. Phys. Lett.* **1993**, *213*, 514–518. [CrossRef]

68. Flisak, Z.; Ziegler, T. "Dormant" secondary metal-alkyl complexes are not omnipresent. *Proc. Natl. Acad. Sci. USA* **2006**, *103*, 15338–15342. [CrossRef] [PubMed]

69. Foster, J.P.; Weinhold, F. Natural hybrid orbitals. *J. Am. Chem. Soc.* **1980**, *102*, 7211–7218. [CrossRef]

70. Zhao, Y.; Truhlar, D.G. The M06 suite of density functionals for main group thermochemistry, thermochemical kinetics, noncovalent interactions, excited states, and transition elements: Two new functionals and systematic testing of four M06-class functionals and 12 other function. *Theor. Chem. Acc.* **2008**, *120*, 215–241. [CrossRef]

71. Tobisch, S.; Ziegler, T. Catalytic Oligomerization of Ethylene to Higher Linear α-Olefins Promoted by the Cationic Group 4 [(η^5-Cp-(CMe$_2$-bridge)-Ph)M^{II}(ethylene)$_2$]$^+$ (M = Ti, Zr, Hf) Active Catalysts: A Density Functional Investigation of the Influence of the Metal on the Catalyt. *J. Am. Chem. Soc.* **2004**, *126*, 9059–9071. [CrossRef] [PubMed]

72. Crabtree, R.H. Deactivation in Homogeneous Transition Metal Catalysis: Causes, Avoidance, and Cure. *Chem. Rev.* **2015**, *115*, 127–150. [CrossRef] [PubMed]

73. Cipullo, R.; Melone, P.; Yu, Y.; Iannone, D.; Busico, V. Olefin polymerisation catalysts: When perfection is not enough. *Dalton Trans.* **2015**, *44*, 12304–12311. [CrossRef] [PubMed]

74. Zaccaria, F.; Ehm, C.; Budzelaar, P.H.M.; Busico, V.; Cipullo, R. Catalyst Mileage in Olefin Polymerization: The Peculiar Role of Toluene. *Organometallics* **2018**, *37*, 2872–2879. [CrossRef]

75. Desert, X.; Carpentier, J.-F.; Kirillov, E. Quantification of active sites in single-site group 4 metal olefin polymerization catalysis. *Coord. Chem. Rev.* **2019**, *386*, 50–68. [CrossRef]

76. Alameddin, N.G.; Ryan, M.F.; Eyler, J.R.; Siedle, A.R.; Richardson, D.E. Intrinsic Ancillary Ligand Effects in Cationic Zirconium Polymerization Catalysts: Reactions of [L$_2$ZrCH$_3$]$^+$ Cations with H$_2$ and C$_2$H$_4$. *Organometallics* **1995**, *14*, 5005–5007. [CrossRef]

77. Richardson, D.E.; Alameddin, N.G.; Ryan, M.F.; Hayes, T.; Eyler, J.R.; Siedle, A.R. Intrinsic Ancillary Ligand Effects in Cationic Zirconium Polymerization Catalysts: Gas-Phase Reactions of [L$_2$ZrCH$_3$]$^+$ Cations with Alkenes. *J. Am. Chem. Soc.* **1996**, *118*, 11244–11253. [CrossRef]

78. Macchioni, A. Ion Pairing in Transition-Metal Organometallic Chemistry. *Chem. Rev.* **2005**, *105*, 2039–2074. [CrossRef] [PubMed]

79. Resconi, L.; Camurati, I.; Sudmeijer, O. Chain transfer reactions in propylene polymerization with zirconocene catalysts. *Top. Catal.* **1999**, *7*, 145–163. [CrossRef]

80. Ehm, C.; Antinucci, G.; Budzelaar, P.H.M.; Busico, V. Catalyst activation and the dimerization energy of alkylaluminium compounds. *J. Organomet. Chem.* **2014**, *772–773*, 161–171. [CrossRef]

81. Ehm, C.; Budzelaar, P.H.M.; Busico, V. Metal–carbon bond strengths under polymerization conditions: 2,1-insertion as a catalyst stress test. *J. Catal.* **2017**, *351*, 146–152. [CrossRef]

82. Ehm, C.; Budzelaar, P.H.M.; Busico, V. Tuning the Relative Energies of Propagation and Chain Termination Barriers in Polyolefin Catalysis through Electronic and Steric Effects. *Eur. J. Inorg. Chem.* **2017**, 3343–3349. [CrossRef]

83. Ishii, A.; Ikuma, K.; Nakata, N.; Nakamura, K.; Kuribayashi, H.; Takaoki, K. Zirconium and Hafnium Complexes with Cycloheptane- or Cyclononane-Fused [OSSO]-Type Bis(phenolato) Ligands: Synthesis, Structure, and Highly Active 1-Hexene Polymerization and Ring-Size Effects of Fused Cycloalkanes on the Activity. *Organometallics* **2017**, *36*, 3954–3966. [CrossRef]

84. Liptak, M.D.; Gross, K.C.; Seybold, P.G.; Feldgus, S.; Shields, G.C. Absolute pKa Determinations for Substituted Phenols. *J. Am. Chem. Soc.* **2002**, *124*, 6421–6427. [CrossRef]

85. Falivene, L.; Credendino, R.; Poater, A.; Petta, A.; Serra, L.; Oliva, R.; Scarano, V.; Cavallo, L. SambVca 2. A Web Tool for Analyzing Catalytic Pockets with Topographic Steric Maps. *Organometallics* **2016**, *35*, 2286–2293. [CrossRef]

86. Clavier, H.; Correa, A.; Cavallo, L.; Escudero-Adán, E.C.; Benet-Buchholz, J.; Slawin, A.M.Z.; Nolan, S.P. [Pd(NHC)(allyl)Cl] Complexes: Synthesis and Determination of the NHC Percent Buried Volume (%Vbur) Steric Parameter. *Eur. J. Inorg. Chem.* **2009**, *2009*, 1767–1773. [CrossRef]

87. Poater, A.; Cosenza, B.; Correa, A.; Giudice, S.; Ragone, F.; Scarano, V.; Cavallo, L. SambVca: A Web Application for the Calculation of the Buried Volume of N-Heterocyclic Carbene Ligands. *Eur. J. Inorg. Chem.* **2009**, *2009*, 1759–1766. [CrossRef]

88. Galimberti, M.; Mascellani, N.; Piemontesi, F.; Camurati, I. Random ethene/propene copolymerization from a catalyst system based on a "constrained geometry" half-sandwich complex. *Macromol. Rapid Commun.* **1999**, *20*, 214–218. [CrossRef]

Article

A Theoretical and Experimental Study for Screening Inhibitors for Styrene Polymerization

Ali Darvishi, Mohammad Reza Rahimpour * and Sona Raeissi

Chemical Engineering Department, Shiraz University, Shiraz 71345, Iran; ali.darvishi2008@yahoo.com (A.D.); raeissi@shirazu.ac.ir (S.R.)
* Correspondence: rahimpor@shirazu.ac.ir

Received: 23 August 2019; Accepted: 23 September 2019; Published: 1 October 2019

Abstract: Styrene is one of the most important monomers utilized in the synthesis of various polymers. Nevertheless, during distillation, storage, and transportation of ST, undesired polymer (i.e., UP) formation can take place. Thus, the control of undesired polymerization of styrene is a challenging issue facing industry. To tackle the mentioned issue, the antipolymer and antioxidant activity of stable nitroxide radicals (i.e., SNRs) and phenolics in styrene polymerization were studied by density functional theory (DFT) calculation and experimental approach. The electrophilicity index and growth percentage have been determined by DFT calculation and experimental approach, respectively. It is depicted that 2,6-di-tert-butyl-4-methoxyphenol (DTBMP) and 2,6-di-tert-butyl-4-methylphenol (BHT) from phenolics, and 4-hydroxy-2,2,6,6-tetramethyl piperidine 1-Oxyl (4-hydroxy-TEMPO) and 4-oxo-2,2,6,6-tetramethylpiperidine 1-Oxyl (4-oxo-TEMPO) from stable nitroxide radicals were the most effective inhibitors. Also, the growth percentage of DTMBP, BHT, 4-hydroxy-TEMPO, and 4-oxo-TEMPO after 4 h were 16.40, 42.50, 24.85, and 46.8, respectively. In addition, the conversion percentage of DTMBP, BHT, 4-hydroxy-TEMPO, and 4-oxo-TEMPO after 4 h were obtained to be 0.048, 0.111, 0.065, and 0.134, respectively. Furthermore, the synergistic effect of these inhibitors was investigated experimentally, indicating that DTMBP/4-hydroxy-TEMPO exerted the best synergistic effects on the inhibition of polymerization. The optimum inhibition effect was observed at the blend of 4-hydroxy-TEMPO (25 wt.%) and DTMBP (75 wt.%) corresponding to 6.8% polymer growth after 4 h.

Keywords: density functional theory; inhibitors; phenolic; stable nitroxide radicals; styrene; polymerization

---

1. Introduction

Over decades, styrene (ST) monomer has been employed to manufacture important commercial commodities such as acrylonitrile butadiene styrene (ABS), and styrene butadiene rubber (SBR) [1–4]. However, during distillation, storage, and transportation of ST, undesired polymer (UP) formation or organic peroxides fouling can presumably occur in the processing equipment, which reduces the heat transfer rate and increases the required time of cleaning [5,6].

In this respect, the three major mechanisms contributing to thermal self-initiation polymerization of styrene are Mayo, Flory, as well as a mechanism that is initiated through the reaction of oxygen and styrene, which are depicted in Figure 1, and explained briefly in the following.

Firstly, Mayo mechanism involves a Diels–Alder dimerization between two styrene molecules, which is followed by molecule-assisted hemolysis of dimer with a styrene molecule to produce two benzylic radicals. The high reactivity of the produced radicals initiate polymerization provided that they are added to a monomer (see r7).

Secondly, according to Flory mechanism, the initiation of styrene polymerization is attributed to the formation of a biradical, which is followed by biradical propagation through the addition of monomers on both active centers (see r1).

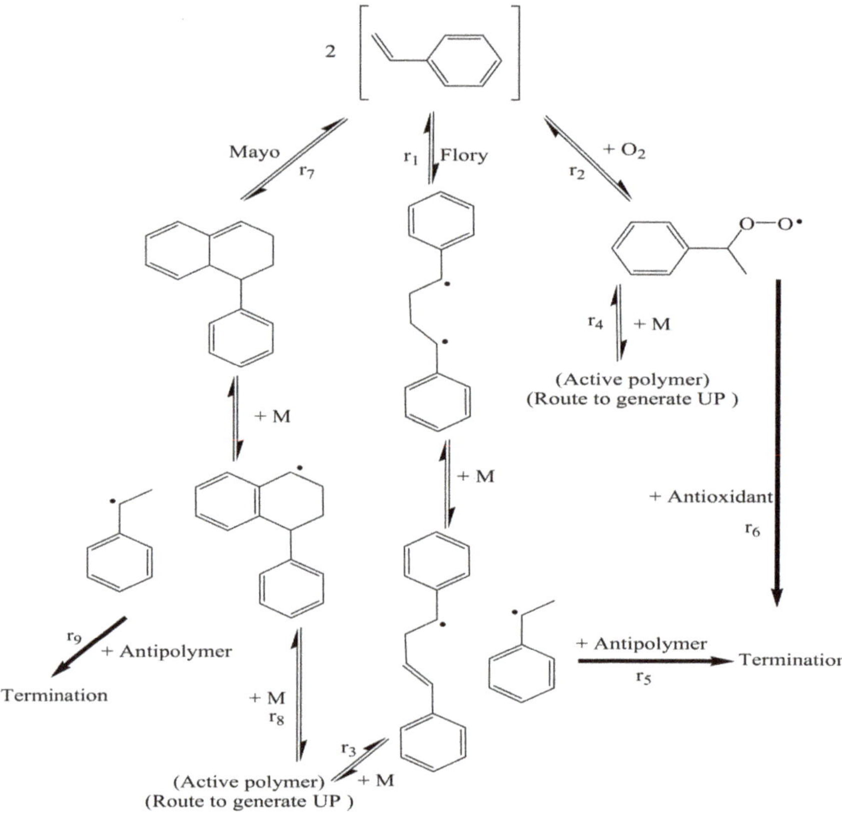

Figure 1. Reaction mechanism for styrene (ST) polymerization. Hydrogen atoms are omitted for clarity.

Thirdly, regarding the oxygen-related mechanism, the oxygen reacts with some of the ST molecules to form an organic peroxide (r2). Following that, considering the fact that organic peroxides are very reactive molecules containing very weak oxygen–oxygen single bonds that can break easily to give free radicals, chain-growth polymerization occurs. Then, these chain-extended polymers react with another monomer, further extending the polymer chains, and subsequently produce very stable compounds. As a result, these stable compounds tend to sediment easily as undesired polymers regarding their low activity in the reaction [7,8].

It should be noted that under some specific conditions, the UP formation can be intensified. By way of illustration, the autocatalytic phenomenon has been reported to take place in the styrene polymerization at high temperatures of around 85 to 130 °C [9–12].

Moreover, the number of methylene groups can play a key role in intensifying polymerization. Indeed, a polymer with greater number of methylene groups exhibit higher activity to react with oxygen, and consequently intensify UP formation.

Furthermore, exposure to heat and light have also been reported as other factors contributing to more sever UP formation. As a matter of fact, when ST monomer is exposed to heat or light, generation of free radicals (ST*) is accelerated, which eventually results in more rapid polymerization. More details of free radical polymerization mechanism can be found elsewhere [13–18].

To talk about the configuration of the UP such as polystyrene, it is in fact highly dependent on the spatial structure of monomer and how it is added to the growing radical chain during polymerization. For example, the polymerized polystyrene formed undesirably is regarded as an atactic polymer;

in other words, it has an amorphous' structure. The production of polystyrene through the radical polymerization leads to formation of atactic polymers in the polymer chains [19].

To address the mentioned issues regarding UP formation, a practical method for the removal of undesired polymers is the mechanical cleaning in which the disassembly of apparatus is inevitable. Nevertheless, this procedure is undeniably accompanied by some drawbacks. As a matter of fact, it is time-consuming, expensive, and also total elimination of UP cannot thoroughly be accomplished through the mechanical cleaning. As a result, the remaining UP could further intensify the growth of UP in the next operation cycles [5,6].

Accordingly, various inhibitors have been suggested to prevent undesired polymer formation [6,20]. Generally, according to mechanism of action, they could be classified into two main types of materials, i.e., acceptor type and donor type, which are called antipolymers and antioxidants, respectively, and are discussed briefly in the following [21].

In the first place, the acceptor radical inhibitors (i.e., antipolymers) are capable of oxidizing the alkyl radicals by accepting hydrogen or even an electron via an addition mechanism through reactions r5 and r9 in Figure 1. The recombination of radicals is an important termination route in any polymerization pathway. As the concentration of radicals increases, the rate of recombination increases as well. These antipolymers directly react with the radical to remove it from the propagation. Acceptor radical inhibitors are efficient in the low-oxygen environments for the deactivation of alkyl radicals [22]. It should be noted that these antipolymers are injected in low concentrations, indicating their low amount of consumption for the purpose of antipolymerization. These inhibitors have quite rapid action on propagator radicals. In fact, they can transform initiator and propagator radicals into either non-radical form or radicals with low reactivity in propagation reaction, and thus inhibiting the radical polymerization [23,24].

By way of illustration, stable nitroxide radicals (SNRs) are typical antipolymers, which are able to terminate the propagation chains through the reaction pathway demonstrating in Figure 2.

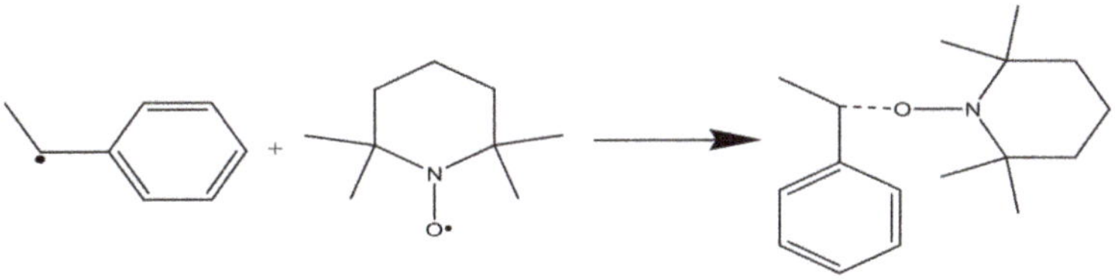

Figure 2. Mechanism of inhibition of styrene polymerization by a typical nitroxide radical. Hydrogen atoms are omitted for clarity.

In the second place, radical donor inhibitors (i.e., antioxidant), as shown in Figure 1, tend to reduce proxy radicals by giving a hydrogen or an electron (see reaction r6). As a result, in contrast with antipolymers, antioxidants exhibit more favorable performance in oxygen-rich environments. Actually, as mentioned before, oxygen can act as the initiator of the polymerization when attacks the styrene molecules. As a consequence, in the absence of antioxidant, ST monomer can react with peroxide radicals to form peroxide chains (see reaction r4), and subsequently lead to formation of UP. This is where the role of using antioxidants becomes noticeable as it can shift the reaction pathway from reaction 4 to reaction 6, which is followed by less polymer formation in the process.

It can generally be claimed that antioxidants are capable of suppressing the formation of new UP seeds. However, they cannot stop UP polymerization thoroughly. To compare the performance of antioxidant with antipolymers, the reaction rate of antioxidants is much lower than that of the antipolymers. Accordingly, they are consumed more slowly, and hence their preventing effects last longer [22,25].

Possible pathways of inhibition in the presence of phenolic antioxidants are shown in Figure 3 [26,27]. The mechanisms of Figure 3a,b occur in the absence of oxygen, while the mechanism in Figure 3c, which is faster than the two former ones, occurs in the presence of oxygen [26–29]. The conventional representative of phenolic antioxidant is 4-tert-butylpyrocatechol (TBC) [5,6,27,30].

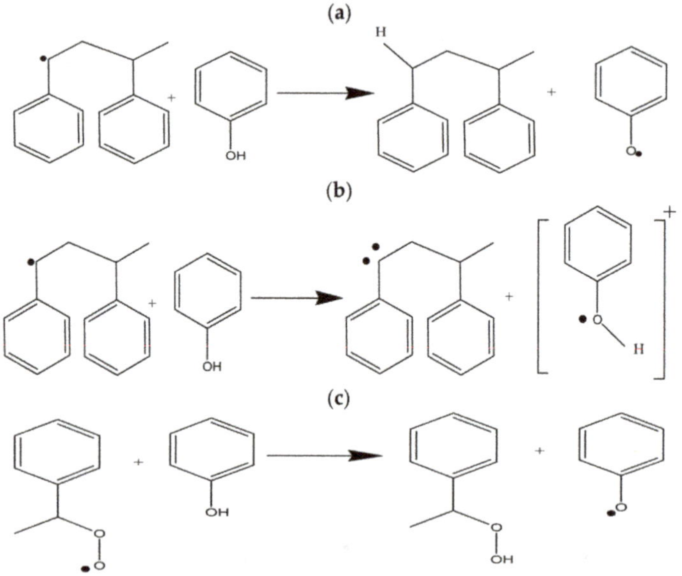

Figure 3. Mechanism of (**a**): H-atom donation, (**b**) electron donation, and (**c**) H-atom donation reaction in the presence of oxygen by typical phenolics. Hydrogen atoms are omitted for clarity.

Nonetheless, in almost all cases, both routes of r1-r3, r7-r8, and r2-r6 successive reactions occur simultaneously to generate UP. Therefore, the main drawback of utilizing antioxidants or antipolymersis that they are not individually capable of inhibiting both routes of UP formation. To address the mentioned issue, the utilization of antipolymer/antioxidant blends has been proposed as the state of the art currently [31–33]. This strategy would result in the simultaneous inhibition of three UP routes.

In order to study the mechanisms and performance of inhibition processes, the two major approaches are experimental evaluation and mathematical calculations such as density functional theory (DFT).

The DFT method demonstrates advantageous features including less demanding computational effort and computing time in conjunction with better agreement with experimental results relative to other procedures. (e.g., Hartree–Fock base method) [34,35]. In fact, the results of calculation through this method are in an acceptable agreement with the experimental chemical properties such as electron-withdrawing of some components. For instance, it is generally known that the reactivity of antioxidant and antipolymer is related to the electron withdrawing tendency of the moieties [36,37]. In addition, Pellecchia and Grassi [38] reported that electron-releasing substituent atom or group enhanced polymerization, whereas the electron-withdrawing substituent atom or group acted as deactivator. Thus, in order to evaluate the electron-withdrawing property of a component, DFT calculations can appropriately be applied.

The main objective of the present study is the investigation of approaches and remedies for eliminating, or at least reducing the blockage of styrene purification units occurred by undesired polymerization. In this regard, a combination of theoretical and experimental approaches is employed to screen and compare different inhibitors including SNRs and phenolics. Furthermore, the synergic effect of antipolymer/antioxidant blends on the undesired ST polymerization is investigated. For the purpose of screening new inhibitors, DFT method is implemented, which can be conducted without experiment. The results of the present study would shed light on the impact of UP inhibitors and pave a reliable path to the optimum utilization of these inhibitors.

2. Method

2.1. Theoretical

In the present study, the global electrophilicity index is utilized as a criterion for the usefulness of different additives in polymerization inhibition. This index can be determined by employing DFT calculations. In fact, in order to calculate the electrophilicity index (i.e., ω) two other parameters including η (i.e., chemical hardness of a radical) and μ (i.e., global chemical potential) indexes are firstly required to be calculated (see Equations (1)–(3)) [39,40]. The η index is the degree of persistence to the charge transfer, while the ω index is the degree of tendency of an atom to attract electrons. The μ index demonstrates the stabilization energy of the components and their electron affinity. In order to calculate μ and η, in the first place, E_{LUMO} and E_{HOMO} should be determined through DFT calculations (UB3LYP/6-311+G level).

$$\eta = (E_{LUMO} - E_{HOMO})/2 \tag{1}$$

$$\mu = (E_{HOMO} + E_{LUMO})/2 \tag{2}$$

$$\omega = \mu^2/2\eta \tag{3}$$

where E_{LUMO} denotes electron energies of the lowest unoccupied molecular orbital (i.e., LUMO) and E_{HOMO} indicates electron energies of the highest occupied molecular orbital for a neutral component (i.e., HOMO). Higher HOMO energy corresponds to stronger electron-donating ability of the molecule [41–46].

2.2. Experimental

2.2.1. Material

2,2,6,6-tetramethylpiperidine 1-Oxyl (TEMPO, 98%), 4-Amino-2,2,6,6-tetramethylpiperidine 1-Oxyl (4-amino-TEMPO, 97%), 4-hydroxy-2,2,6,6-tetramethylpiperidine 1-Oxyl (4-hydroxy-TEMPO, 97%), 4-oxo-2,2,6,6-tetramethylpiperidine 1-Oxyl (4-oxo-TEMPO, 97%), 2,6-di-tert-butyl-4-methylphenol (BHT, (99%), Tert-butyl hydroquinone (TBHQ, 97%), 2,6-di-tert- butyl-4-methoxyphenol (DTBMP, 98%), and 4-Methoxyphenol (MEHQ, 98%) were supplied from Sigma Aldrich. The structures of inhibitors sketched by DFT are summarized in Figure 4. Styrene, styrene polymer, and TBC were supplied from the domestic plant.

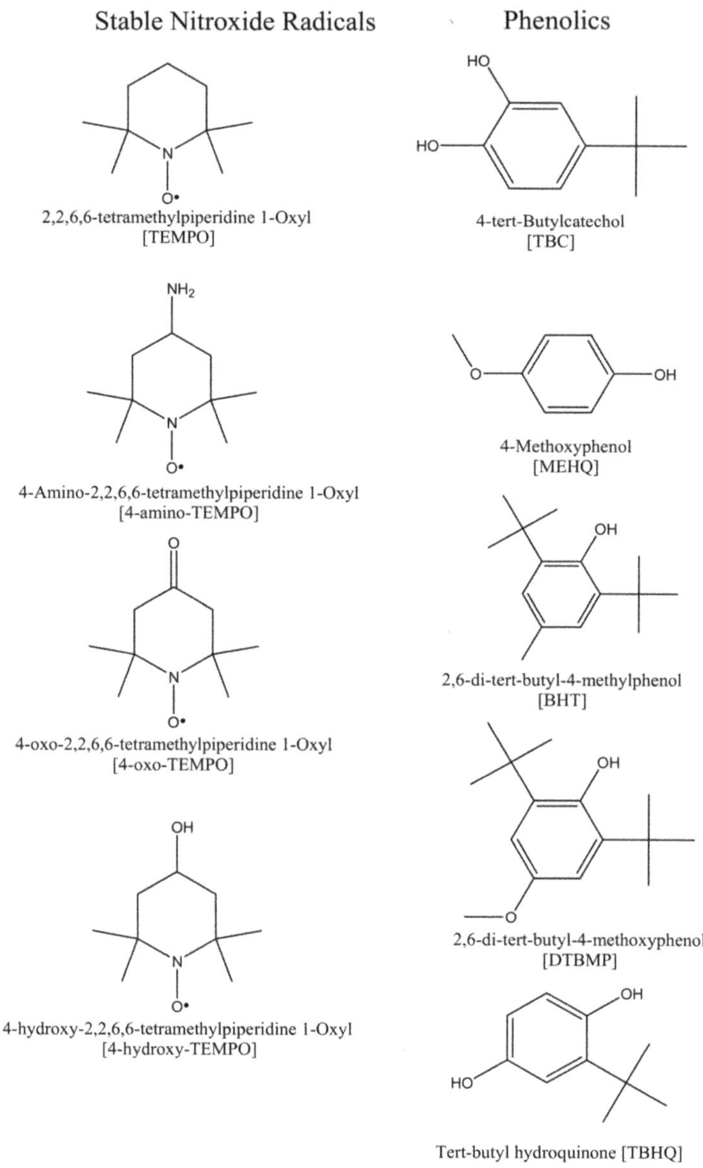

Figure 4. The name and structure of stable nitroxide radicals (SNR) and phenolic components. Hydrogen atoms are omitted for clarity.

The FT-IR spectra were recorded for UP seed (Figure 5). In the FT-IR spectrum of the UP seed, the peaks at 3004 cm^{-1}–3083 cm^{-1} are attributed to the sp^2 C-H stretch, and band at 2923 cm^{-1} and 2850 cm^{-1} corresponds to the sp^3 C–H stretch. Also, signals at 1630 cm^{-1}–1680 cm^{-1} are assigned to the stretching vibrations of aromatic C=C in the UP seed. The signal at 1452 cm^{-1} corresponding to CH$_2$ + C=C bond stretching. Finally, a slight shift of C–O aromatic stretching modes towards lower wavenumbers can be observed.

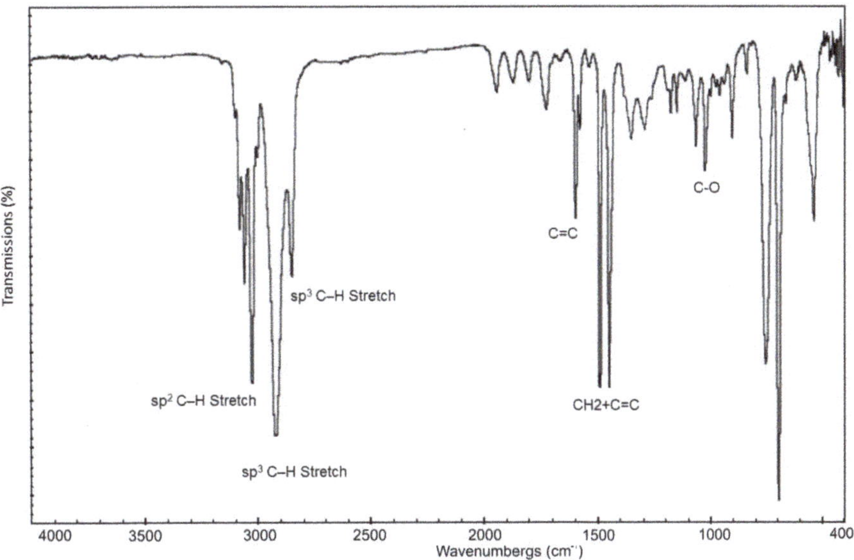

Figure 5. The FT-IR spectrum of undesired polymer seed.

2.2.2. Thermogravimetric Analysis

The initial decomposition temperature is measured via thermogravimetric analysis (TGA) test. The test is performed from ambient temperature up to 350 °C with 10 °C/min temperature ramp and under a nitrogen flow.

2.2.3. Analysis of Product (Styrene, Dimer Styrene, and Trimer Styrene)

The samples were analyzed by GC-MS (model: Agilent 7890A series GC coupled with an Agilent 5975C series MS detector), equipped with a HP-5MS capillary column (inner diameter: 250 μm; film thickness: 0.25 μm; length: 30 m).

2.2.4. FTIR

The structural changes of sample were characterized by an attenuated total reflection Fourier-transformed infrared spectroscopy (ATR/FT-IR).

2.2.5. Inhibition Test

The effect of SNRs, phenolic compounds, and their blends on the polymerization of ST at 115 °C was evaluated by utilizing the apparatus depicted in Figure 6. The system includes an adiabatic cell reactor (ACR), which is made of stainless steel (60 mm length, and 50 mm inside diameter). A K-type thermocouple installed in the middle of ACR is employed to control the ACR temperature. Syringe pumps supply the desired mass of monomer, inhibitor, and water. Before initiation, oxygen is eliminated by nitrogen purge to avoid peroxide formation. The ST polymer was utilized as the nucleation source. The sample was then placed in the ACR, 2 mL of deionized water as an oxygen source was injected into the system, and after adding a 50 ppm inhibitor, 50 mL of monomer was injected. The oxygen dissolved in water acts as initiator. The concentration of dissolved oxygen in water equaled 8.5 mg/L [47]. After 4 h of polymerization, the reacting mixture was cooled, and the polymer was precipitated by adding about 5 mL of methanol. The precipitate was filtered, dried at 100 °C to remove methanol, and weighed. The experimental conditions are tabulated in Table 1.

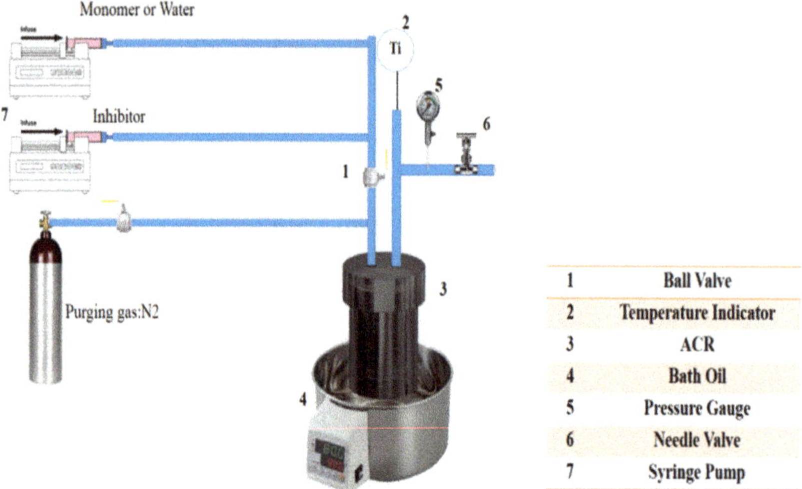

1	Ball Valve
2	Temperature Indicator
3	ACR
4	Bath Oil
5	Pressure Gauge
6	Needle Valve
7	Syringe Pump

Figure 6. The schematic diagram of experimental setup.

Table 1. Experimental conditions.

Properties	Value	Unit
Temperature	115	°C
Pressure	1	bar
Inhibitor dosing	50	ppm
Monomer volume	50	mL
Water dosing	2	mL
Initial weight of ST polymer	0.2	g

3. Result

Different inhibitors and their blends were investigated to examine the effect of the inhibitor on the growth and formation of UP as well as ST conversion. As aforementioned, these studies are based on the DFT calculation and an experimental procedure.

3.1. DFT Calculation Results

The results of DFT calculations applied on SNRs and phenolics are tabulated in Table 2.

Table 2. Calculated values of μ, η, and ω indexes for SNRs and phenolics.

Sample ID	Component	μ	η	ω
		SNRs		
S1	TEMPO	−3.1688	2.4450	2.0170
S2	4-amino-TEMPO	−3.4344	2.5718	2.2675
S3	4-oxo-TEMPO	−3.6544	2.1485	3.1079
S4	4-hydroxy-TEMPO	−3.8239	1.6973	4.3075
		Phenolics		
S5	BHT	−2.6654	2.9900	1.1880
S6	TBC	−3.2072	2.7878	1.8448
S7	TBHQ	−3.2376	2.6937	1.9457
S8	DTBMP	−2.5161	2.8021	1.1297
S9	MEHQ	−3.2282	2.6478	1.9680

As mentioned previously, inhibitors are divided into antipolymers (i.e., S1, S2, S3, and S4 in Table 2) and antioxidants (S5, S6, S7, S8, and S9 in Table 2). The most important difference is that for the antipolymers, higher electrophilicity is desirable, while it is reverse for the antioxidants. In other words, for the purpose of inhibiting polymerization, a suitable antipolymer should possess higher electrophilicity, and inversely, a suitable antioxidant should have lower electrophilicity.

Among the first group (i.e., antipoymers), TEMPO has the lowest electrophilicity index, indicating that this antipolymer has the lowest reactivity, and thus it is more stable. Conversely, the highest electrophilicity index of 4-hydroxy TEMPO indicates that it is less stable and more reactive in comparison with other SNR inhibitors. In addition, it is known that the hardness index is related to molecular stability, and consequently higher hardness means higher stability of molecule [48,49]. In this respect, hard molecules (e.g., TEMPO and 4-amino-TEMPO) possess higher values of chemical hardness, and subsequently, larger gaps between the HOMO and LUMO, whereas soft molecules (e.g., 4-hydroxy-TEMPO and 4-oxo-TEMPO) exhibit lower values of chemical hardness and smaller gaps. The obtained results demonstrate good consistency with those reported in the previous studies [50].

To compare performance of the second group inhibitors (i.e., antioxidants), firstly it should be noted that the distribution of the HOMO energy is a criterion of inhibitory performance of phenolic antioxidants. Thus, considering the fact that the molecules with lower gap energy shows weaker electrophilicity [51,52], among the studied phenolics, DTBMP (1.1297 eV) and BHT (1.1880eV) possess higher electrophilicity in comparison with TBC (1.8448 eV), TBHQ (1.9457 eV), and MEHQ (1.9680 eV).

The electrophilicity of selected SNRs and phenolics that are calculated utilizing DFT are presented in Figure 7.

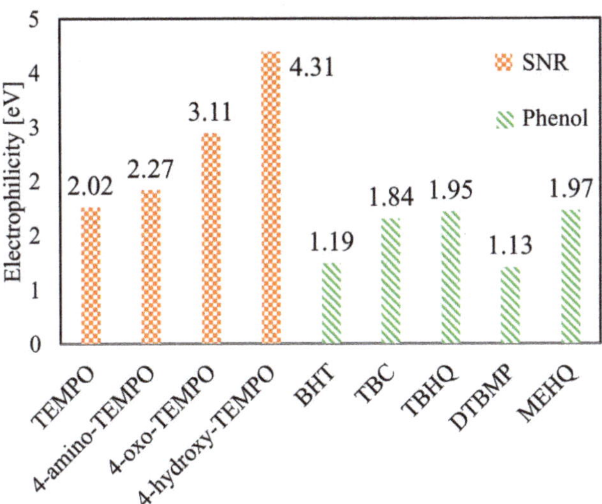

Figure 7. Calculated electrophilicity of SNRs and phenolics utilizing density functional theory (DFT).

For the case of antipolymers, based on the theoretical studies [53], the capability of hydrogen bonding by SNRs affects the properties and proton affinities. In fact, the present oxygen in the N–O bond of SNRs is a suitable H-bond acceptor. The mentioned oxygen can compete with other oxygen atoms, which are present within the structure, even in the solid state. It is observedthat the order of electrophilicity of SNRs is: 4-hydroxy-TEMPO > 4-oxo-TEMPO > 4-amino-TEMPO > TEMPO. Hence, it is expected that 4-oxo-TEMPO shows the best inhibition effect in the UP formation and styrene conversion.

For the case of antioxidants, however, it can be observed that electrophilicity of phenolic antioxidants are in the following order: MEHQ > TBHQ > TBC > BHT > DTBMP. Hence, DTBMP is expected to show the best inhibition effect in the UP formation.

In order to acquire a deeper understanding of the performance of different inhibitors and to confirm the findings of DFT calculations, the experimental results are presented and compared in the following section.

3.2. Experimental Results

The growth percentage, which is calculated by Equation (4), is considered as a criterion to compare the performance of inhibitors.

$$\text{Growth \%} = \frac{\text{weight}_{\text{Final}} - \text{weight}_{\text{Initial}}}{\text{weight}_{\text{Initial}}} \times 100 \tag{4}$$

The initial weight equals to nucleation source, and the final weight equals to the summation of UP formation weight added to the initial weight. Firstly, it should be noted that aside from the UP formation, dimer and oligomer formation is inevitable as well. However, the number of undesired dimers and oligomers production is very small because the radical mechanism is prevailing, which mainly leads to the polymerization. However, these quantities must be controlled in a specified range because dimers and oligomers are regarded as impurities, which do not participate in further reactions in the polymerization unit. As a result, they must be separated in the styrene purification unit.

In this set of experiments, styrene conversion as well as the formation of dimers and oligomers are investigated after 4 and 8 h of operation to ensure those small quantities of dimer and oligomer are generated. In this respect, Tables 3 and 4 tabulate the effects of inhibitors on the weight and growth of undesired polymer, the quantities of generated dimer and oligomer, and conversion of styrene. The obtained results reveal that inhibition effects do not change significantly when the reaction time increases from 4 to 8 h, and they are approximately identical.

Table 3. Measured weights of polymer and the growth percentage in the presence of inhibitors after 4 h.

Sample ID	Component	Weight [g]	Growth Percentage	Outlet Mass Fraction [wt.%]			Conversion [%]
		SNRs		Styrene	Dimer	Trimer	
S1	TEMPO	0.395	97.35	99.719	0.034	0.015	0.231
S2	4-amino-TEMPO	0.320	60.05	99.808	0.021	0.009	0.142
S3	4-oxo-TEMPO	0.313	56.30	99.811	0.020	0.009	0.134
S4	4-hydroxy-TEMPO	0.250	24.85	99.885	0.013	0.006	0.065
		Phenolics		Styrene	Dimer	Trimer	
S5	BHT	0.285	42.50	99.839	0.022	0.010	0.111
S6	Commercial TBC	0.305	52.65	99.811	0.028	0.012	0.139
S7	TBHQ	0.363	81.25	99.749	0.034	0.015	0.201
S8	DTBMP	0.233	16.40	99.902	0.012	0.005	0.048
S9	MEHQ	0.387	93.35	99.730	0.053	0.023	0.251

Table 4. Measured weights of polymer and the growth percentage in the presence of inhibitors after 8 h.

Sample ID	Component	Weight [g]	Growth Percentage	Outlet Mass Fraction [wt.%]			Conversion [%]
		SNRs		Styrene	Dimer	Trimer	
S1	TEMPO	0.474	136.82	99.630	0.045	0.019	0.320
S2	4-amino-TEMPO	0.464	132.07	99.654	0.034	0.015	0.296
S3	4-oxo-TEMPO	0.422	111.00	99.703	0.035	0.015	0.257
S4	4-hydroxy-TEMPO	0.350	74.79	99.783	0.019	0.008	0.167
		Phenolics		Styrene	Dimer	Trimer	
S5	BHT	0.399	99.50	99.713	0.036	0.015	0.237
S6	Commercial TBC	0.470	135.08	99.643	0.038	0.016	0.307
S7	TBHQ	0.526	162.81	99.568	0.054	0.023	0.382
S8	DTBMP	0.319	59.47	99.812	0.019	0.008	0.138
S9	MEHQ	0.630	215.16	99.493	0.062	0.027	0.491

The obtained results regarding styrene conversion is proportionally in agreement with that of UP growth. Hence, only one of these two parameters is required to be evaluated as the inhibition criterion. It should also be mentioned that the styrene dimer and trimer only refer to 2,4-diphenyl-1-butene and 2,4,6-triphenyl-1-hexene, respectively.

The lower weight of polymer, growth percentage, and conversion all imply better inhibition effect. Hence, 4-hydroxy-TEMPO and DTBMP offer better inhibition effect among the SNRs and phenolics, respectively. It can be understood from the result that the substituent in the 4′-position of the cyclic structure has a significant effect on the ability of TEMPO in inhibiting polymerization. Consequently, among the SNRs, 4-hydroxy TEMPO (24.85% growth and 0.065% conversion) and 4-oxo-TEMPO (56.3% growth and 0.134% conversion) have the greatest effectiveness, respectively, after 4 h. As presented in Tables 3 and 4, the styrene conversion and polymerization inhibition performance of phenolics are in the following order: DTBMP > BHT > commercial TBC > TBHQ > MEHQ. Results show that DTBMP and BHT are the most effective phenolics.

The experimental results of growth percentage are plotted against the calculated electrophilicity from DFT calculations in Figure 8. The observed trends show that in SNRs, increase in electrophilicity leads to lower growth percentage, while in the phenolics, higher growth percentage can be seen with the increase in electrophilicity. Consequently, phenolics with lower electrophilicity and SNRs with higher electrophilicity are preferred. Such a difference could be explained by different mechanisms presented in Figure 1 (see r5, r6, and r9).

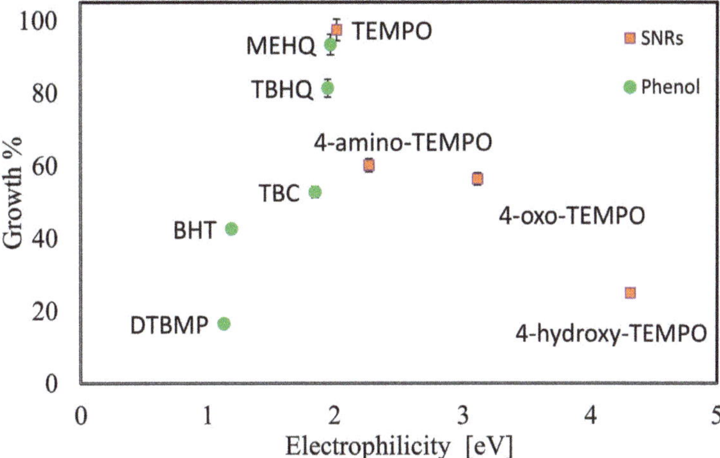

Figure 8. Calculated electrophilicity and measured growth percentage after 4 h of SNRs and phenolics.

A comparison of electrophilicity and growth percentage of 4-amino-TEMPO and 4-oxo-TMEPO reveals that in spite of the considerable difference in electrophilicity, their growth percentage differs slightly. This observation can be attributed to their different thermal behavior at various experimental condition. Actually, the key parameter that plays a very influential role in the performance of inhibitors is the operating temperature, which is around 115 °C in the styrene purification process. To clarify the thermal behavior of inhibitors toward temperature, TGA patterns, which determine the initial decomposition temperature, are shown in Figure 9a,b.

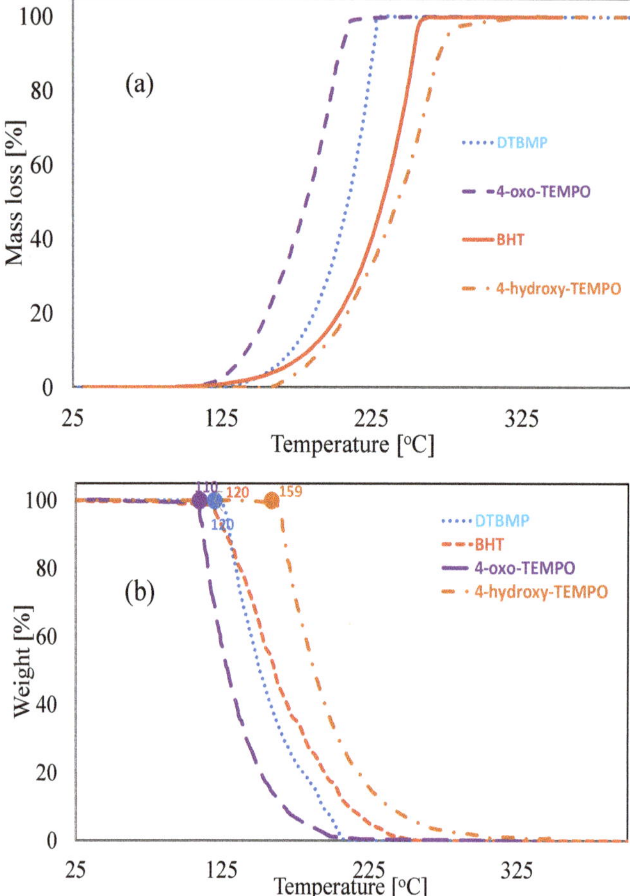

Figure 9. Thermogravimetric analysis of BHT, DTBMP, 4-hydroxy-TEMPO, and 4-oxo-TEMPO. (**a**) mass loss % and; (**b**) weight %.

In fact, as the temperature reaches the initial decomposition temperature, the inhibitors start to break down, and thereby lose their capability of inhibiting polymerization. As displayed in Figure 9, the initial decomposition temperature of BHT, DTMBP, and 4-hydroxy-TEMPO are higher than 115 °C, while that of 4-oxo-TEMPO starts from 110 °C. These results imply that at the operating temperature of 115 °C, 4-oxo-TEMPO molecules decompose and lose their ability in UP inhibition.

The mechanism of 4-oxo-TEMPO decomposition, which is called keto-enol tautomerism, is well-explained in the literature [54,55]. Through the decomposition of 4-oxo-TEMPO, nitroso and hydroxylamine compounds are formed, which is followed by further decomposition of nitroso and hydroxylamine compounds, leading to production of TEMPOH and TEMPO with low electrophilicity. The formation of less electrophilic TEMPOH and TEMPO justifies the unexpected performance of 4-oxo-TEMPO molecule, considering its high electrophilicity [54,55].

The functional groups of the 4-oxo-TEMPO and its decomposition products are further characterized by FT-IR to investigate the chemical structure changes during 4-oxo-TEMPO heating. The FT-IR spectra in the range of 400–4000 cm^{-1} are shown in Figure 10. The peaks at 1700 cm^{-1} in the FT-IR spectra of 4-oxo-TEMPO were assigned to the C=O stretching mode. For decomposition product of 4-oxo-TEMPO, no peak at 1700 cm^{-1} (main peak for 4-oxo-TEMPO) is observed, which means the breakage of C=O bond in 4-oxo-TEMPO, and absence of 4-oxo-TEMPO in the decomposition product. This finding provides another strong evidence which indicates that the mass losses observed in the TGA is merely attributed to the decomposition of 4-oxo-TEMPO, rather than other reasons (e.g., boiling or evaporation).

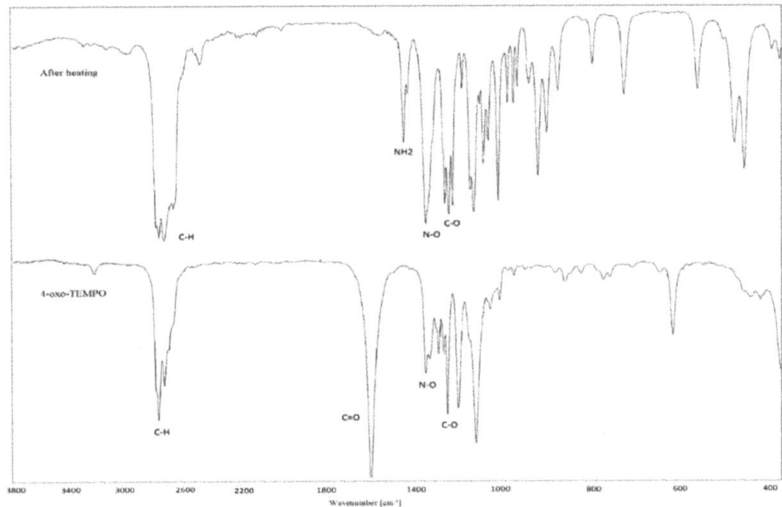

Figure 10. The FTIR spectrum of 4-oxo-TEMPO and decomposition products of 4-oxo-TEMPO.

As previously mentioned, 4-hydroxy-TEMPO and 4-oxo-TEMPO from SNRs and DTBMP and BHT from phenolics show the best performance in the inhibition of UP and styrene conversion. Hence, their synergistic effect in the inhibition process is worth investigating.

3.3. Synergic effect of Antioxidant and Antipolymer

In order to determine the synergistic effect of the optimum blend of antioxidants and antipolymers in the inhibition process, a mixture of antioxidants and antipolymers (50 wt.%), which is injected to the process with the overall concentration of 50 ppm, was prepared and evaluated at the experimental conditions explained previously.

As tabulated in Table 5, the blends of optimum antioxidants and antipolymers improve the inhibition process in comparison with the commercial TBC sample. Moreover, the results show that the blends illustrate superior performance relative to single-component inhibitors, which arises from the fact that antipolymers are responsible for the inhibition of two polymerization pathways (i.e., r1–r3 and r7–r8 shown in Figure 1) and antioxidants just terminate the other polymerization pathway (i.e., r2–r6). The growth percentage of DTMBP/4-hydroxy-TEMPO, DTMBP/4-oxo-TEMPO, and BHT/4-hydroxy-TEMPO blends are 8.60, 10.25, and 12.65, respectively.

Table 5. Measured weights of polymer and the growth percentage in the presence of inhibitor blends after 4 h.

Sample ID	Component	Weight [g]	Growth Percentage
S6	Commercial TBC	0.3053	52.65
S10	DTMBP/4-hydroxy-TEMPO	0.2042	8.60
S11	DTMBP/4-oxo-TEMPO	0.2185	10.25
S12	BHT/4-hydroxy-TEMPO	0.2093	12.65
S13	BHT/4-oxo-TEMPO	0.2847	42.35

The growth percentage of all the single and blended inhibitors are compared in Figure 11. As can be seen, the inhibitor composed of DTMBP/4-hydroxy-TEMPO provides the best inhibition performance, while TEMPO shows the worst one.

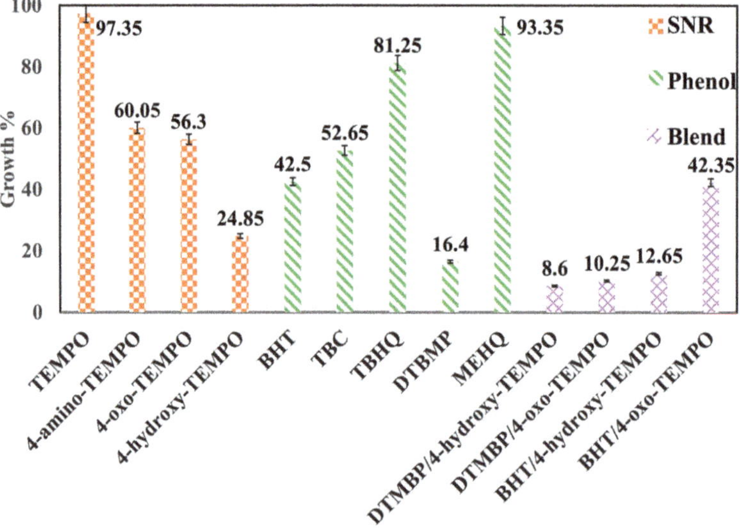

Figure 11. Growth percentage of SNRs, phenolics, and blends.

The synergism of SNRs and phenolics blends depends on the reversible hydrogen atom transfer between them. In fact, by utilizing the blends of SNRs and phenolics, hydrogen acceptance by SNRs and hydrogen donation by phenolics are enhanced, which is followed by improvement in inhibition performance.

Hydrogen atom hopping between 4-hydroxy-TEMPO and DTMBP results in a dynamic equilibrium of phenoxyl radical and 4-hydroxy-TEMPO. The phenoxyl radical can abstract the labile hydrogen from the product of dimerization of styrene Diels-Alder self-reaction [56,57]. This reaction yields a high flux of alkyl radicals and accelerates oxygen consumption. Hydrogen abstraction from DTMBP by TEMPO is not favorable due to the bond dissociation energies of TEMPOH and MEHQ [58]. The reaction is driven by hydrogen abstraction from styrene dimer by the phenoxyl radical to form an aromatic compound.

The blend of BHT/4-oxo-TEMPO did improve the inhibition of UP. This is, in fact, due to low abstraction of BHT hydrogen by the 4-oxo-TEMPO and decomposition of 4-oxo-TEMPO.

The effect of concentration of DTMBP/4-hydroxy-TEMPO blend on the inhibition process is depicted in Figure 12. The obtained results reveal that the blend of 4-hydroxy-TEMPO (25 wt.%) and DTMBP (75 wt.%), causes the lowest growth percentage, and thereby higher inhibition efficiency is obtained. However, other concentrations are also more effective than single SNR or phenolic compounds.

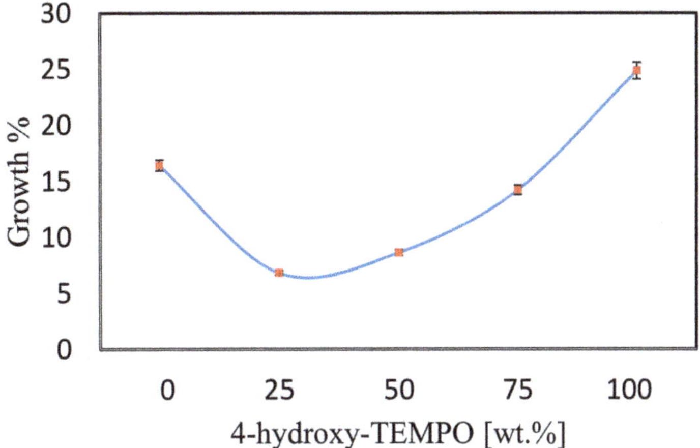

Figure 12. The effect of DTMBP/4-hydroxy-TEMPO concentration on the growth percentage.

4. Conclusions

The DFT calculation and an experimental procedure were employed to analyze the inhibition performance of various antioxidants and antipolymer inhibitors regarding the undesired styrene polymerization during styrene purification process. The electrophilicity index and growth percentage (or styrene conversion) were obtained by DFT calculation and experiments, respectively. Furthermore, the relation between the calculated electrophilicity and the measured growth percentage and styrene conversion was determined. In addition, in order to find the optimum inhibitor, both growth percentage and electrophilicity were taken into consideration. Accordingly, 4-oxo-TEMP and 4-hydroxy-TEMPO were the best antipolymers, while DTMBP and BHT were the best antioxidant inhibitors. After four hours of operation, the lowest growth percentage of 16.40 and conversion percentage of 0.048 were obtained in the case of DTBMP. Besides, the synergistic effect of antioxidant and antipolymer inhibitors were also investigated by the experimental procedure. Results showed that a combination of 4-hydroxy-TEMPO (25 wt.%) and DTMBP (75 wt.%) provided the best synergistic inhibition effect with the lowest growth percentage of 8.60 after 4 h of operation.

Author Contributions: A.D.: Conceived and designed the analysis, Collected the data, Contributed data or analysis tools, Performed the analysis, Wrote the paper; M.R.R.: Conceived and designed the analysis, Contributed data or analysis tools, Performed the analysis, Wrote the paper; S.R.: Conceived and designed the analysis, Performed the analysis, Wrote the paper.

Funding: This research received no external funding.

Conflicts of Interest: The authors confirm that there are no known conflict of interest associated with this publication and there has been no significant financial support for this work that could have influenced its outcome.

References

1. Cai, S.; Zhang, B.; Cremaschi, L. Review of moisture behavior and thermal performance of polystyrene insulation in building applications. *Build. Environ.* **2017**, *123*, 50–65. [CrossRef]
2. Ghosh, J.; Ghorai, S.; Bhunia, S.; Roy, M.; De, D. The role of devulcanizing agent for mechanochemical devulcanization of styrene butadiene rubber vulcanizate. *Polym. Eng. Sci.* **2018**, *58*, 74–85. [CrossRef]
3. Kamelian, F.S.; Saljoughi, E.; Nasirabadi, P.S.; Mousavi, S.M. Modifications and research potentials of acrylonitrile/butadiene/styrene (ABS) membranes: A review. *Polym. Compos.* **2018**, *39*, 2835–2846. [CrossRef]
4. Dimian, A.C.; Bildea, C.S. Energy Efficient Styrene Process: Design and Plantwide Control. *Ind. Eng. Chem. Res.* **2019**, *58*, 4890–4905. [CrossRef]

5. Ovejero, G.; Romero, M.D.; Díaz, I.; Mestanza, M.; Díez, E. Bentonite as an Alternative Adsorbent for the Purification of Styrene Monomer: Adsorption Kinetics, Equilibrium and Process Design. *Adsorpt. Sci. Technol.* **2010**, *28*, 101–123. [CrossRef]

6. Díaz, I.; Langston, P.; Ovejero, G.; Romero, M.D.; Díez, E. Purification process design in the production of styrene monomer. *Chem. Eng. Process Process Intensif.* **2010**, *49*, 367–375. [CrossRef]

7. Miller, G.H.; Perizzolo, A.F. Styrene popcorn polymers. *J. Polym. Sci.* **1955**, *18*, 411–416. [CrossRef]

8. Hsieh, H.; Quirk, R.P. *Anionic Polymerization: Principles and Practical Applications*; CRC Press: Boca Raton, FL, USA, 1996.

9. Liao, C.-C.; Wu, S.-H.; Su, T.-S.; Shyu, M.-L.; Shu, C.-M. Thermokinetics evaluation and simulations for the polymerization of styrene in the presence of various inhibitor concentrations. *J. Therm. Anal. Calorim.* **2006**, *85*, 65–71. [CrossRef]

10. Chen, C.-C.; Duh, Y.-S.; Shu, C.-M. Thermal polymerization of uninhibited styrene investigated by using microcalorimetry. *J. Hazard. Mater.* **2009**, *163*, 1385–1390. [CrossRef]

11. Nanda, A.; Kishore, K. Autocatalytic oxidative polymerization of indene by cobalt porphyrin complex and kinetic investigation of the polymerization of styrene. *Macromolecules* **2001**, *34*, 1600–1605. [CrossRef]

12. Cui, J.; Ni, L.; Jiang, J.; Pan, Y.; Wu, H.; Chen, Q. Computational Fluid Dynamics Simulation of Thermal Runaway Reaction of Styrene Polymerization. *Org. Process Res. Dev.* **2019**, *23*, 389–396. [CrossRef]

13. Mohammadi, Y.; Pakdel, A.S.; Saeb, M.R.; Boodhoo, K. Monte Carlo simulation of free radical polymerization of styrene in a spinning disc reactor. *Chem. Eng. J.* **2014**, *247*, 231–240. [CrossRef]

14. Sobani, M.; Haddadi-Asl, V.; Mirshafiei-Langari, S.-A.; Salami-Kalajahi, M.; Roghani-Mamaqani, H.; Khezri, K. A kinetics study on the in situ reversible addition–fragmentation chain transfer and free radical polymerization of styrene in presence of silica aerogel nanoporous particles. *Des. Monomers Polym.* **2014**, *17*, 245–254. [CrossRef]

15. Sütekin, S.D.; Atıcı, A.B.; Güven, O.; Hoffman, A.S. Controlling of free radical copolymerization of styrene and maleic anhydride via RAFT process for the preparation of acetaminophen drug conjugates. *Radiat. Phys. Chem.* **2018**, *148*, 5–12. [CrossRef]

16. Matyjaszewski, K.; Davis, T.P. *Handbook of Radical Polymerization*; John Wiley & Sons: Hoboken, NJ, USA, 2003.

17. Chiefari, J.; Chong, Y.; Ercole, F.; Krstina, J.; Jeffery, J.; Le, T.P.; Mayadunne, R.T.A.; Meijs, G.F.; Moad, C.L.; Moad, G. Living free-radical polymerization by reversible addition—Fragmentation chain transfer: the RAFT process. *Macromolecules* **1998**, *31*, 5559–5562. [CrossRef]

18. Encinas, M.V.; Lissi, E.A.; Norambuena, E. Inhibition of Styrene Polymerization by β-Nitrostyrene. A Novel Inhibition Mechanism. *Macromolecules* **1998**, *31*, 5171–5174. [CrossRef]

19. Harper, C.A.; Petrie, E.M. *Plastics Materials and Processes: A Concise Encyclopedia*; John Wiley & Sons: Hoboken, NJ, USA, 2003.

20. Luo, K.; Zheng, W.; Zhao, X.; Wang, X.; Wu, S. Effects of antioxidant functionalized silica on reinforcement and anti-aging for solution-polymerized styrene butadiene rubber: Experimental and molecular simulation study. *Mater. Des.* **2018**, *154*, 312–325. [CrossRef]

21. Gomes, G.d.P.; Loginova, Y.; Vatsadze, S.Z.; Alabugin, I.V. Isonitriles as Stereoelectronic Chameleons: The Donor–Acceptor Dichotomy in Radical Additions. *J. Am. Chem. Soc.* **2018**, *140*, 14272–14288. [CrossRef]

22. Kolthoff, I.; Bovey, F. Studies of Retarders and Inhibitors in the Emulsion Polymerization of Styrene. I. Retarders1a. *J. Am. Chem. Soc.* **1948**, *70*, 791–799. [CrossRef]

23. Bevington, J.; Ebdon, J.; Huckerby, T. An appraisal of NMR methods for study of end-groups derived from initiators in radical polymerizations. *Eur. Polym. J.* **1985**, *21*, 685–694. [CrossRef]

24. Yokota, H.; Kawakatsu, T. Modeling induction period of polymer crystallization. *Polymer* **2017**, *129*, 189–200. [CrossRef]

25. Cohen, S.G. Inhibition and Retardation of the Peroxide Initiated Polymerization of Styrene. *J. Am. Chem. Soc.* **1947**, *69*, 1057–1064. [CrossRef]

26. Wright, J.S.; Carpenter, D.J.; McKay, D.J.; Ingold, K. Theoretical calculation of substituent effects on the O—H bond strength of phenolic antioxidants related to vitamin E. *J. Am. Chem. Soc.* **1997**, *119*, 4245–4252. [CrossRef]

27. Wright, J.S.; Johnson, E.R.; DiLabio, G.A. Predicting the activity of phenolic antioxidants: Theoretical method, analysis of substituent effects, and application to major families of antioxidants. *J. Am. Chem. Soc.* **2001**, *123*, 1173–1183. [CrossRef] [PubMed]

28. Tikhonov, I.; Roginsky, V.; Pliss, E. The chain-breaking antioxidant activity of phenolic compounds with different numbers of O-H groups as determined during the oxidation of styrene. *Int. J. Chem. Kinet.* **2009**, *41*, 92–100. [CrossRef]

29. Mardare, D.; Matyjaszewski, K. *Thermal Polymerization of Styrene in the Presence of Stable Radicals and Inhibitors*; Carnegie-Mellon Univ. Pittsburgh Pa. Dept of Chemistry: Pittsburgh, PA, USA, 1994.

30. Kemmere, M.; Mayer, M.; Meuldijk, J.; Drinkenburg, A. The influence of 4-tert-butylcatechol on the emulsion polymerization of styrene. *J. Appl. Polym. Sci.* **1999**, *71*, 2419–2422. [CrossRef]

31. Finson, S.; Jury, J.M.; Crivello, J.V. Azodioxides as Inhibitors and Retarders in Photoinitiated Cationic Polymerization. *Macromol. Chem. Phys.* **2013**, *214*, 1806–1816. [CrossRef]

32. Moghadam, N.; Liu, S.; Srinivasan, S.; Grady, M.C.; Soroush, M.; Rappe, A.M. Computational Study of Chain Transfer to Monomer Reactions in High-Temperature Polymerization of Alkyl Acrylates. *J. Phys. Chem. A* **2013**, *117*, 2605–2618. [CrossRef]

33. Bartlett, P.D.; Kwart, H. Some Inhibitors and Retarders in the Polymerization of Liquid Vinyl Acetate. II. 1, 3, 5-Trinitrobenzene and Sulfur1. *J. Am. Chem. Soc.* **1952**, *74*, 3969–3973. [CrossRef]

34. Anbazhakan, K.; Sadasivam, K.; Praveena, R.; Dhandapani, M. Target prediction and antioxidant analysis on isoflavones of demethyltexasin: A DFT study. *J. Mol. Model.* **2019**, *25*, 169. [CrossRef]

35. Sharma, V.; Arora, E.K.; Cardoza, S. 4-Hydroxy-benzoic acid (4-diethylamino-2-hydroxy-benzylidene) hydrazide: DFT, antioxidant, spectroscopic and molecular docking studies with BSA. *Luminescence* **2016**, *31*, 738–745. [CrossRef] [PubMed]

36. Tüdős, F.; Kende, I.; Azori, M. Inhibition kinetics of the polymerization of styrene. II. Investigations on the effect of s-trinitrobenzene. *J. Polym. Sci. Part A Gen. Pap.* **1963**, *1*, 1353–1368.

37. Tüdős, F.; Kende, I.; Azori, M. Inhibition kinetics of polymerization of styrene. III. Investigations on the effect of substituted trinitrobenzenes. *J. Polym. Sci. Part A Gen. Pap.* **1963**, *1*, 1369–1381.

38. Pellecchia, C.; Grassi, A. Syndiotactic-specific polymerization of styrene: Catalyst structure and polymerization mechanism. *Top. Catal.* **1999**, *7*, 125–132. [CrossRef]

39. Domingo, L.R.; Pérez, P. Global and local reactivity indices for electrophilic/nucleophilic free radicals. *Org. Biomol. Chem.* **2013**, *11*, 4350–4358. [CrossRef] [PubMed]

40. De Vleeschouwer, F.; van Speybroeck, V.; Waroquier, M.; Geerlings, P.; de Proft, F. Electrophilicity and nucleophilicity index for radicals. *Org. Lett.* **2007**, *9*, 2721–2724. [CrossRef] [PubMed]

41. Zheng, Y.-Z.; Zhou, Y.; Liang, Q.; Chen, D.-F.; Guo, R.; Xiong, C.-L.; Xu, X.J.; Zhang, Z.N.; Huang, Z.J.; Xu, X.J.; et al. Solvent effects on the intramolecular hydrogen-bond and anti-oxidative properties of apigenin: A DFT approach. *Dyes Pigments* **2017**, *141*, 179–187. [CrossRef]

42. Zheng, Y.-Z.; Chen, D.-F.; Deng, G.; Guo, R.; Fu, Z.-M. The antioxidative activity of piceatannol and its different derivatives: Antioxidative mechanism analysis. *Phytochemistry* **2018**, *156*, 184–192. [CrossRef]

43. Zheng, Y.-Z.; Chen, D.-F.; Deng, G.; Guo, R. The Substituent Effect on the Radical Scavenging Activity of Apigenin. *Molecules* **2018**, *23*, 1989. [CrossRef] [PubMed]

44. Zheng, Y.-Z.; Deng, G.; Chen, D.-F.; Liang, Q.; Guo, R.; Fu, Z.-M. Theoretical studies on the antioxidant activity of pinobanksin and its ester derivatives: Effects of the chain length and solvent. *Food Chem.* **2018**, *240*, 323–329. [CrossRef]

45. Wang, X.; Lin, F.; Qu, J.; Hou, Z.; Luo, Y. DFT Studies on Styrene Polymerization Catalyzed by Cationic Rare-Earth-Metal Complexes: Origin of Ligand-Dependent Activities. *Organometallics* **2016**, *35*, 3205–3214. [CrossRef]

46. Coote, M.L.; Henry, D.J. Computer-Aided Design of a Destabilized RAFT Adduct Radical: Toward Improved RAFT Agents for Styrene-block-Vinyl Acetate Copolymers. *Macromolecules* **2005**, *38*, 5774–5779. [CrossRef]

47. Kauffman, G. The Cost of Clean Water in the Delaware River Basin (USA). *Water* **2018**, *10*, 95. [CrossRef]

48. Cinar, Z. The role of molecular modeling in TiO$_2$ photocatalysis. *Molecules* **2017**, *22*, 556. [CrossRef] [PubMed]

49. Del Castillo, R.M.; Salcedo, R.; Martínez, A.; Ramos, E.; Sansores, L.E. Electronic Peculiarities of a Self-Assembled M12L24 Nanoball (M = Pd+ 2, Cr, or Mo). *Molecules* **2019**, *24*, 771. [CrossRef] [PubMed]

50. Galván, J.E.; Gil, D.M.; Lanús, H.E.; Altabef, A.B. Theoretical study on the molecular structure and vibrational properties, NBO and HOMO–LUMO analysis of the POX3 (X = F, Cl, Br, I) series of molecules. *J. Mol. Struct.* **2015**, *1081*, 536–542. [CrossRef]

51. Al-Majedy, Y.; Al-Duhaidahawi, D.; Al-Azawi, K.; Al-Amiery, A.; Kadhum, A.; Mohamad, A. Coumarins as potential antioxidant agents complemented with suggested mechanisms and approved by molecular modeling studies. *Molecules* **2016**, *21*, 135. [CrossRef]

52. Rao, P.S.; Puyad, A.L.; Bhosale, S.V.; Bhosale, S.V. Triphenylamine-Merocyanine-Based D1-A1-π-A2/A3-D2 Chromophore System: Synthesis, Optoelectronic, and Theoretical Studies. *Int. J. Mol. Sci.* **2019**, *20*, 1621.

53. Ichikawa, K.; Sasada, R.; Chiba, K.; Gotoh, H. Effect of Side Chain Functional Groups on the DPPH Radical Scavenging Activity of Bisabolane-Type Phenols. *Antioxidants* **2019**, *8*, 65. [CrossRef]

54. Murayama, K.; Yoshioka, T. Studies on stable free radicals. IV. Decomposition of stable Nitroxide radicals. *Bull. Chem. Soc. Jpn.* **1969**, *42*, 2307–2309. [CrossRef]

55. Ma, Y.; Loyns, C.; Price, P.; Chechik, V. Thermal decay of TEMPO in acidic media via an N-oxoammonium salt intermediate. *Org. Biomol. Chem.* **2011**, *9*, 5573–5578. [CrossRef] [PubMed]

56. Boutevin, B.; Bertin, D. Controlled free radical polymerization of styrene in the presence of nitroxide radicals I. Thermal initiation. *Eur. Polym. J.* **1999**, *35*, 815–825. [CrossRef]

57. Conte, M.; Ma, Y.; Loyns, C.; Price, P.; Rippon, D.; Chechik, V. Mechanistic insight into TEMPO-inhibited polymerisation: Simultaneous determination of oxygen and inhibitor concentrations by EPR. *Org. Biomol. Chem.* **2009**, *7*, 2685–2687. [CrossRef] [PubMed]

58. Ma, Y. Nitroxides in Mechanistic Studies: Ageing of Gold Nanoparticles and Nitroxide Transformation in Acids. Ph.D. Thesis, University of York, York, UK, 2010.

 *processes*

Article

Modeling and Observer-Based Monitoring of RAFT Homopolymerization Reactions

Patrick M. Lathrop [†], Zhaoyang Duan [†], Chen Ling , Yossef A. Elabd * and Costas Kravaris *

Artie McFerrin Department of Chemical Engineering, Texas A&M University, College Station, TX 77843, USA; pml25@tamu.edu (P.M.L.); danjoyer@tamu.edu (Z.D.); ling0408@tamu.edu (C.L.)
* Correspondence: elabd@tamu.edu (Y.A.E.); kravaris@tamu.edu (C.K.);
  Tel.: +1-979-845-7506 (Y.A.E.); +1-979-458-4514 (C.K.)
† These authors contributed equally to this work.

Received: 13 September 2019; Accepted: 16 October 2019; Published: 20 October 2019

Abstract: Reversible addition–fragmentation chain–transfer (RAFT) polymerization of methyl methacrylate (MMA) is modeled and monitored using a multi-rate multi-delay observer in this work. First, to fit the RAFT reaction rate coefficients and the initiator efficiency in the model, in situ ^{1}H nuclear magnetic resonance (NMR) experimental data from small-scale (<2 mL) NMR tube reactions is obtained and a least squares optimization is performed. ^{1}H NMR and size exclusion chromatography (SEC) experimental data from large-scale (>400 mL) reflux reactions is then used to validate the fitted model. The fitted model accurately predicts the polymer properties of the large-scale reactions with slight discordance at late reaction times. Based on the fitted model, a multi-rate multi-delay observer coupled with an inter-sample predictor and dead time compensator is designed, to account for the asynchronous multi-rate measurements with non-constant delays. The multi-rate multi-delay observer shows perfect convergence after a few sampling times when tested against the fitted model, and is in fair agreement with the real data at late reaction times when implemented based on the experimental measurements.

Keywords: RAFT polymerization; multi-rate observer; nonlinear sampled-data system; measurements with delay; parameter fitting

---

1. Introduction

Reversible addition–fragmentation chain–transfer (RAFT) polymerization is a living polymerization that is based on free-radical polymerization [1]. Free-radical polymerization reactions are made living by the addition of a RAFT agent, which controls the polymerization via chain equilibration, in which radicals are shared between growing polymer chains. Since RAFT polymerization was first reported by Krstina et al. [2], it has been further investigated by many researchers to show that it works with a wide variety of monomers and RAFT agents [2–9]. RAFT polymerization has been used to create polymers in many fields, including drug delivery [10], electrochemical applications such as fuel cells [11] and batteries [12], and surface modification [13], showing its versatility and applicability.

Key advantages of RAFT polymerization include the synthesis of polymers with narrow polydispersities [1], the ability to form polymers of complex architecture (i.e., block, star), and the ability to work with multiple monomer types—allowing for synthesis of homopolymers and block co-polymers with controlled molecular weights and dispersities. Control of properties such as dispersity and molecular weight is especially important during block copolymer synthesis, in which polymer morphology and micro-phase structure is determined by the polymer block composition [14]. However, no information about these properties is available from online sensors. Offline analysis techniques like ^{1}H NMR and SEC, which give more insight into the process, require time to analyze samples, and the results are usually obtained after significant delays or even after the whole process is completed. The lack of timely information, which is critical to the progress of polymerization, brings challenges to the quality control of RAFT polymerization. A live model, which accounts for these measurement delays with observers, would be useful for monitoring RAFT polymerization for quality control.

RAFT polymerization was first modeled by Zhang and Ray [15]. They modeled RAFT polymerization of methyl methacrylate in batch, semi-batch, and continuous stirred tank reactors, providing insight into how the mechanism of RAFT polymerization behaved throughout a typical reaction. Many more models have been developed to further explore the fundamentals of RAFT polymerization as more experimental data has become available [16–23]. Few of these models are validated against experimental data at higher monomer conversions, and those that are have few experimental data points above 60% monomer conversion. Additionally, the RAFT agent kinetic rate constants found in these models as fitting parameters are expected to be different for each monomer and RAFT agent pair.

The monitoring of polymerization processes with state estimation techniques based on mathematical models would help target desired polymer compositions and molecular weights during large scale reactions by predicting reaction termination times, and ensuring controlled polymerization by predicting expected dispersities of reactions. However, it is not an easy task to develop the state estimation due to the lack of online measurements and the significant nonlinearity of the systems. There have been a number of studies on state estimation for polymerization applications, with a focus on free radical polymerization, since the early 1980s. The extended Kalman filter (EKF) has been popularly applied to industry with good performance [24–31]. This kind of estimator is designed based on a local linearization approximation and is thereby less effective in the presence of strong system nonlinearities. The state observer, which reconstructs the missing state variables based on the system dynamics with certain feedback terms from the measurement, is an alternative way to design a soft sensor for polymerization processes. For example, Astorga implemented a continuous-discrete observer to an emulsion copolymerization reactor [32]. Appelhaus et al. developed an extended Luenberger observer for a batch polycondensation reactor, in order to get the estimation for concentrations of ethylene glycol and hydroxyl end groups as well as a mass transfer parameter [33]. Several high-gain observers have also been designed for polymerization processes [34,35]. In the observer approach, the convergence properties were mathematically proved under most circumstances, and the computational cost is lowered compared to EKF. Although a significant number of studies have been completed for monitoring different polymerization applications, no state estimation study has been conducted on RAFT polymerizations. Tatiraju and Soroush used a nonlinear open-loop reduced-order observer to estimate the unobservable states in a homopolymerization reactor, and obtained accurate estimation results [36,37]. In our work, a similar idea will be adopted as the basis for the design of a multi-rate multi-delay observer.

One key challenge for the state estimation problem for polymerization is that the measurements are not available at the same rate and usually come with delays. For example, in polymerization processes, the sampling may not be done in a uniform frequency and the measurement information only becomes available after analysis. Therefore, there is a need for the design of a multi-rate multi-delay observer to address this issue. For the multi-rate estimation, the problem has been widely studied based on EKF framework [38]. Moving horizon estimation (MHE) [39–41] and other state observer design [37,42,43] have also been investigated. However, most of the EKF and MHE-based methods will not account for the inter-sample dynamic behavior and are incapable of giving estimation for asynchronously sampled measurements. As for the delay effects accompanied with sampling measurements, some observer design approaches were proposed accordingly [44–51], but few of them have been proved stable through a rigorous approach. In some recent works by Ling and Kravaris [52–57], a multi-rate multi-delay observer based on a continuous-time design, coupled with inter-sample predictors and dead time compensators, is proposed. Irrespective of perturbations in the sampling schedule, the input-to-output stability property for this observer was first well established for linear systems [53,57], and extended to nonlinear systems with noises [55,56], in the presence of non-uniform and asynchronous measurement with non-constant, arbitrarily large measurement delays, as long as the maximum sampling period does not exceed a certain limit. This observer will be used to estimate the missing information between the sampling measurements for the RAFT polymerization process in the current work.

In this study, small-scale experiments are performed and characterized with in situ ^{1}H NMR analysis to more accurately monitor RAFT polymerization. Accurate prediction of RAFT specific kinetic rate constants (specifically, k_{raftf}, k_{raftr}, k_{raftfR} and k_{raftrR}) is obtained, allowing for an improved RAFT polymerization model that can be used to monitor larger scale reactions to high monomer conversion values. The improved model is used to test multi-rate, multi-delay observers, which improve predictive modeling of larger scale RAFT reactions allowing for real-time monitoring of reactions and targeted reaction termination times.

2. Model and Observer Design

2.1. Modeling RAFT Polymerization—Improving the Model by Zhang and Ray

The RAFT polymerization model used in this work is based on a model presented by Zhang and Ray [15]. This subsection will provide a brief review of their model as well as the specific changes and additions made to their model to help the reader follow subsequent sections of this paper.

RAFT polymerization is modeled by including standard free radical polymerization kinetics (i.e., initiation, propagation, chain transfer and termination) with the RAFT reaction steps in the presence of a RAFT agent. The RAFT reactions are shown in Scheme 1. These RAFT steps create three dormant polymer populations based on the state of the RAFT agent end group. In total, five polymer populations exist within the model: growing polymer chains, dead polymer chains and three types of dormant polymer chains. To simplify the model, these populations are modeled using moments, for which the mathematical definitions are given in Table 1. The terms brackets are as follows: $[P_n]$, concentration of growing polymer chain of degree of polymerization n; $[P_nAR]$, concentration of RAFT agent bonded to a primary radical and a polymer chain of degree of polymerization n; $[P_nA]$, concentration of RAFT agent bonded to a polymer chain of degree of polymerization n; $[P_nAP_m]$, concentration of RAFT agent bonded to two polymer chains of degree of polymerization n and m; and $[D_n]$, concentration of dead polymer chains of degree of polymerization n. When using polymer moments, it is not possible to apply different kinetic rate constants based on polymer chain length, and we must make the assumption that the kinetic rate constants are independent of polymer chain length. This assumption, although it may introduce some error, is made out of necessity for model simplicity.

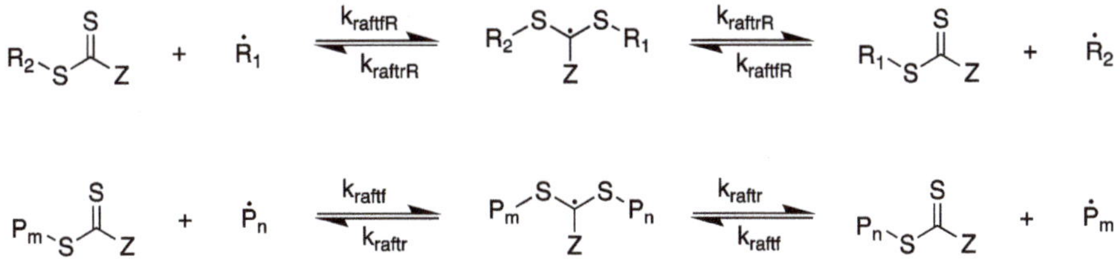

Scheme 1. Kinetic scheme of reactions involving the RAFT agent.

Using these polymer moment definitions and the monomer concentration (C_M), the monomer conversion (X_p), number average degree of polymerization (DP_n), weight average degree of polymerization (DP_w) and dispersity ($Đ$) can be calculated as shown in Equation (1):

$$
\begin{aligned}
X_p &= \frac{\mu_1 + \nu_1 + \xi_1 + \lambda_1 + \zeta_{10}}{\mu_1 + \nu_1 + \xi_1 + \lambda_1 + \zeta_{10} + C_M}, \\
DP_n &= \frac{\mu_1 + \nu_1 + \xi_1 + \zeta_{10} + \lambda_1}{\mu_0 + \nu_0 + \xi_0 + 0.5\zeta_{00} + \lambda_0}, \\
DP_w &= \frac{\mu_2 + \nu_2 + \xi_2 + \zeta_{20} + 0.5\zeta_{11} + \lambda_2}{\mu_1 + \nu_1 + \xi_1 + \zeta_{10} + \lambda_1}, \\
Đ &= \frac{DP_w}{DP_n}.
\end{aligned}
\tag{1}
$$

Table 1. Polymer moment definitions.

Chain Type	Moment Definition
Growing Polymer	$\mu_i = \sum_{n=1}^{\infty} n^i [P_n]$
Dormant Polymer Type 1	$\nu_i = \sum_{n=1}^{\infty} n^i [P_n AR]$
Dormant Polymer Type 2	$\xi_i = \sum_{n=1}^{\infty} n^i [P_n A]$
Dormant Polymer Type 3	$\zeta_{ij} = \sum_{n=1}^{\infty} \sum_{m=1}^{\infty} n^i m^i [P_n A P_m]$
Dead Polymer	$\lambda_i = \sum_{n=1}^{\infty} n^i [D_n]$

Using the kinetic scheme from Zhang and Ray and Scheme 1, the 23 balance equations given in Equation (2) are used to represent each of the species in the model. The definitions for the state variables are listed in Table 2, and the definitions and values for the reaction rate constants are given in Table 3.

This model differs from the Zhang and Ray model in two ways: first, in the way termination is modeled, and second, in the rate constants. In the Zhang and Ray model, termination is modeled accounting for both combination and disproportionation reactions. In this model, termination is modeled accounting for just one general termination reaction based on more recent work in the literature [58]. With this change in how termination reactions are modeled, the corresponding rate constants are also changed. In the Zhang and Ray model, k_{tc} and k_{td} are used to represent the rates for combination and disproportionation reactions, respectively. In this work, a single termination reaction rate constant k_t is used instead to represent all termination reactions. Additionally, the propagation rate constant has been updated from the Zhang and Ray model based on a more recent work in the literature [59]. Finally, two additional kinetic rate constants for the RAFT agent pre-equilibrium reactions are included (k_{raftfR} and k_{raftrR}) based on literature [23,60,61] showing a difference for reactions between the RAFT agent and primary radicals when compared to reactions between the RAFT agent and growing polymer chains.

Most of the reaction rate constants are based on free radical polymerization. The exceptions to this are k_{trCTA}, k_{raftf}, k_{raftr}, k_{raftfR} and k_{raftrR}. These five reaction rate constants are dependent on the choice of RAFT agent used, and are expected to change for each RAFT agent. In this model, the effect of chain transfer to CTA is considered negligible and k_{trCTA} is set to zero:

$$
\begin{aligned}
\frac{d\mu_0}{dt} =\ & k_p C_R C_M + k_{trm}\mu_0 C_M - k_t(C_R + \mu_0)\mu_0 - k_{trm}\mu_0 C_M - k_{trs}\mu_0 C_S \\
& - k_{trCTA}\mu_0 C_{CTA} - k_{raftf}\mu_0(C_{AR} + \xi_0) + k_{raftr}(\nu_0 + \zeta_{00}), \\[4pt]
\frac{d\mu_1}{dt} =\ & k_p C_R C_M + k_{trm}\mu_0 C_M + k_p C_M \mu_0 - k_t(C_R + \mu_0)\mu_1 - k_{trm}\mu_1 C_M - k_{trs}\mu_1 C_S \\
& - k_{trCTA}\mu_1 C_{CTA} - k_{raftf}\mu_1(C_{AR} + \xi_0) + k_{raftr}(\nu_1 + \zeta_{10}), \\[4pt]
\frac{d\mu_2}{dt} =\ & k_p C_R C_M + k_p C_M(\mu_0 + 2\mu_1) - k_t(C_R + \mu_0)\mu_2 - k_{trm}\mu_2 C_M - k_{trs}\mu_2 C_S \\
& k_{trCTA}\mu_2 C_{CTA} - k_{raftf}\mu_2(C_{AR} + \xi_0) + k_{raftr}(\nu_2 + \zeta_{20}), \\[4pt]
\frac{d\nu_0}{dt} =\ & k_{raftf}\mu_0 C_{AR} + k_{raftfR} C_R \xi_0 - k_{raftr}\nu_0 - k_{raftrR}\nu_0, \\[4pt]
\frac{d\nu_1}{dt} =\ & k_{raftf}\mu_1 C_{AR} + k_{raftfR} C_R \xi_1 - k_{raftr}\nu_1 - k_{raftrR}\nu_1, \\[4pt]
\frac{d\nu_2}{dt} =\ & k_{raftf}\mu_2 C_{AR} + k_{raftfR} C_R \xi_2 - k_{raftr}\nu_2 - k_{raftrR}\nu_2, \\[4pt]
\frac{d\xi_0}{dt} =\ & -k_{raftf}\mu_0 \xi_0 - k_{raftfR} C_R \xi_0 + k_{raftr}\zeta_{00} + k_{raftrR}\nu_0, \\[4pt]
\frac{d\xi_1}{dt} =\ & -k_{raftf}\mu_0 \xi_1 - k_{raftfR} C_R \xi_1 + k_{raftr}\zeta_{10} + k_{raftrR}\nu_1, \\[4pt]
\frac{d\xi_2}{dt} =\ & -k_{raftf}\mu_0 \xi_2 - k_{raftfR} C_R \xi_2 + k_{raftr}\zeta_{20} + k_{raftrR}\nu_2, \\[4pt]
\frac{d\zeta_{00}}{dt} =\ & 2k_{raftf}\mu_0 \xi_0 - 2k_{raftr}\zeta_{00}, \\[4pt]
\frac{d\zeta_{10}}{dt} =\ & k_{raftf}(\mu_1 \xi_0 + \mu_0 \xi_1) - 2k_{raftr}\zeta_{10}, \\[4pt]
\frac{d\zeta_{20}}{dt} =\ & k_{raftf}(\mu_2 \xi_0 + \mu_0 \xi_2) - 2k_{raftr}\zeta_{20}, \\[4pt]
\frac{d\zeta_{11}}{dt} =\ & 2k_{raftf}\mu_1 \xi_1 - 2k_{raftr}\zeta_{11}, \\[4pt]
\frac{d\lambda_0}{dt} =\ & k_t(C_R + \mu_0)\mu_0 + k_{trm}\mu_0 C_M + k_{trs}\mu_0 C_S + k_{trCTA}\mu_0 C_{CTA}, \\[4pt]
\frac{d\lambda_1}{dt} =\ & k_t(C_R + \mu_0)\mu_1 + k_{trm}\mu_1 C_M + k_{trs}\mu_1 C_S + k_{trCTA}\mu_1 C_{CTA}, \\[4pt]
\frac{d\lambda_2}{dt} =\ & k_t(C_R + \mu_0)\mu_2 + k_{trm}\mu_2 C_M + k_{trs}\mu_2 C_S + k_{trCTA}\mu_2 C_{CTA}, \\[4pt]
\frac{dC_R}{dt} =\ & 2f k_d C_I - k_p C_R C_M - k_t C_R(C_R + \mu_0) + k_{trs}\mu_0 C_S + k_{trCTA}\mu_0 C_{CTA} \\
& - k_{raftfR} C_R(C_{AR} + \xi_0) + k_{raftrR}(2C_{RAR} + \nu_0), \\[4pt]
\frac{dC_{AR}}{dt} =\ & -k_{raftf} C_{AR}\mu_0 - k_{raftfR} C_{AR} C_R + k_{raftr}\nu_0 + 2k_{raftrR} C_{RAR}, \\[4pt]
\frac{dC_{RAR}}{dt} =\ & k_{raftfR} C_R C_{AR} - 2k_{raftrR} C_{RAR}, \\[4pt]
\frac{dC_I}{dt} =\ & -k_d C_I, \\[4pt]
\frac{dC_M}{dt} =\ & -k_p C_M(C_R + \mu_0) - k_{trm}\mu_0 C_M, \\[4pt]
\frac{dC_S}{dt} =\ & -k_{trs}\mu_0 C_S, \\[4pt]
\frac{dC_{CTA}}{dt} =\ & -k_{trCTA}\mu_0 C_{CTA}.
\end{aligned}
\tag{2}
$$

Table 2. State variables.

Variables	Description
μ_0	Growing polymer chain zeroth moment
μ_1	Growing polymer chain first moment
μ_2	Growing polymer chain second moment
ν_0	Dormant polymer chain type 1 zeroth moment
ν_1	Dormant polymer chain type 1 first moment
ν_2	Dormant polymer chain type 1 second moment
ζ_0	Dormant polymer chain type 2 zeroth moment
ζ_1	Dormant polymer chain type 2 first moment
ζ_2	Dormant polymer chain type 2 second moment
ζ_{00}	Dormant polymer chain type 3 zeroth moment
ζ_{10}	Dormant polymer chain type 3 first moment
ζ_{20}	Dormant polymer chain type 3 second moment part one
ζ_{11}	Dormant polymer chain type 3 second moment part two
λ_0	Dead polymer chain zeroth moment
λ_1	Dead polymer chain first moment
λ_2	Dead polymer chain second moment
C_R	Concentration of primary radicals
C_{AR}	Concentration of RAFT agent
C_{RAR}	Concentration of primary intermediate
C_I	Concentration of initiator
C_M	Concentration of monomer
C_S	Concentration of solvent
C_{CTA}	Concentration of chain transfer agent

Table 3. Reaction rate constant definitions.

Reaction Rate Constant	Definition	Equation/Value	Units	Reference
k_d	Initiator formation	$1.0533 \times 10^{15} e^{\frac{-30704}{RT}}$	s^{-1}	[62]
k_p	Live chain propagation	$10^{6.427} e^{\frac{-5344}{RT}}$	$L\ mol^{-1}\ s^{-1}$	[59]
k_{trM}	Radical transfer to monomer	$k_p \times 10^{-5}$	$L\ mol^{-1}\ s^{-1}$	[63]
k_{trS}	Radical transfer to solvent	0, assumed negligible	$L\ mol^{-1}\ s^{-1}$	
k_{trCTA}	Radical transfer to CTA	0, assumed negligible	$L\ mol^{-1}\ s^{-1}$	
k_t	Termination	$10^{9.586} e^{\frac{-3106}{RT}}$	$L\ mol^{-1}\ s^{-1}$	[58]
k_{raftf}	Forward RAFT main equilibrium	Unknown	$L\ mol^{-1}\ s^{-1}$	
k_{raftr}	Reverse RAFT main equilibrium	Unknown	s^{-1}	
k_{raftfR}	Forward RAFT pre-equilibrium	Unknown	$L\ mol^{-1}\ s^{-1}$	
k_{raftrR}	Reverse RAFT pre-equilibrium	Unknown	s^{-1}	
f	Initiator efficiency	Unknown		

Methyl methacrylate is known for undergoing the Trommsdorff–Norrish effect (i.e., the gel effect) during polymerization, in which the propagation rate accelerates, and the termination rate decelerates, at high viscosities as growing polymer chains entangle. This effect is included in the model by Zhang and Ray, and was considered in this work. In this work, high solvent to monomer ratios dilute the polymer concentration and prevent the viscosity from increasing high enough to induce the Trommsdorff–Norrish effect, and therefore this effect was not observed.

The large-scale reaction experiments in this work are performed under reflux and do not occur at a constant temperature. As monomer is consumed, the composition, and therefore the bubble point of the volatile components in the reactor change. Temperature of the reactor for large-scale experiments is estimated by calculating the bubble point of a binary mixture based on the volatile components of the reacting mixture (i.e., the solvent and the monomer). Data for the bubble point temperature of the binary mixture was calculated using UNIFAQ in ASPEN, and a polynomial fit was applied to the data and the corresponding equation is shown in Equation (3), where x_{THF} denotes the mole fraction composition of tetrahydrofuran (THF) in the reacting mixture. This fit was used within the model to calculate reaction temperature as a function of reacting mixture composition for the large-scale experiments:

$$T = 9.4893x_{THF}^3 - 4.032x_{THF}^2 - 39.482x_{THF} + 100.13. \tag{3}$$

The RAFT polymerization model includes the following assumptions. The reactions are performed under ideal mixing. Reactions involving polymers and their corresponding kinetic rate constants are independent of polymer chain length, allowing for polymer moments to be used. All primary radicals are treated as identical, regardless of the reactions leading to their formation (i.e., a primary radical formed from initiator decomposition is identical to a primary radical from chain transfer reactions). Kinetic rate constants k_{raftf}, k_{raftr}, k_{raftfR} and k_{raftrR} are not dependent on reaction temperature.

2.2. Observer Design

2.2.1. Multi-Rate Multi-Delay Observer—Background

This part gives the basic background for the design method for the multi-rate multi-delay observer used in this work. The state estimation problem for the RAFT polymerization process in the current work will be a special case of the observer presented in this section.

The starting point for the design of the multi-rate multi-delay observer will be a continuous time observer. A reduced order observer that simulates a subsystem driven by the measurements can be built following Soroush [64]. In the following part, the particular observer driven by the measurement is introduced.

For simplicity, consider a continuous system with a part of the state vector to be directly measured. Note that a measurement variable could always be included as a state variable through appropriate coordinate transformation:

$$\dot{x}_R(t) = f_R(x_R(t), x_M(t)), \tag{4a}$$
$$\dot{x}_M(t) = f_M(x_R(t), x_M(t)), \tag{4b}$$
$$y(t) = x_M(t), \tag{4c}$$

where $x_R \in \mathbb{R}^{n-m}$ and $x_M \in \mathbb{R}^m$ are the unmeasured and measured state vectors; y is the continuous output. It is assumed that the unmeasured subsystem (4a) is locally asymptotically stable.

A reduced order observer that is driven by the measurable outputs is of the form

$$\dot{\hat{x}}_R(t) = f_R(\hat{x}_R(t), y(t)), \tag{5}$$

where $\hat{x}_R \in \mathbb{R}^{n-m}$ is the vector of estimated states that guarantee the estimation error $e(t) = \hat{x}_R(t) - x_R$ converges to 0 as $t \to \infty$ [64]. This reduced order observer is based on a different philosophy than the traditional closed-loop observers that involve a correction term multiplied by an observer gain. However, it still incorporates the feedback effect in the sense that it uses the measurement to drive the simulation; and the stability of system dynamics would imply the stability of the observer error dynamics.

When the sampling is performed at a slow rate with asynchronous intervals, the missing inter-sample information needs to be accurately predicted by a multi-rate observer. Now, consider the dynamic system (4a) and (4b) with multi-rate slow-sampled measurements

$$\dot{x}_R(t) = f_R(x_R(t), x_M(t)),$$
$$\dot{x}_M(t) = f_M(x_R(t), x_M(t)), \tag{6}$$
$$y^i(t_j^i) = x_M^i(t_j^i), \qquad j \in \mathbb{Z}_+, i = 1, 2, \ldots, m,$$

where t^i_j denotes the j-th sample time for the i-th measured component of x_M, at some sequence of time instants $S = \{t_k\}^\infty_{k=0}$. The intervals between each sampling time are not necessarily uniform.

A multi-rate observer is designed based on a continuous-time observer, coupled with inter-sample predictors to handle the inter-sample behavior. For the predictor, the dynamic system (6) is simulated to obtain a prediction for the missing output information in between two consecutive measurements. This prediction will then be given to the continuous-time observer (5) functioning as the continuous outputs. For $t \in [t^+_k, t_{k+1}]$, the multi-rate sampled-data observer for the multi-rate system (6) is [56]:

$$
\begin{aligned}
\dot{\hat{x}}_R(t) &= f_R(\hat{x}_R(t), w(t)) \quad t \in [t_k, t_{k+1}), \\
\dot{w}(t) &= f_M(\hat{x}_R(t), w(t)), \quad t \in [t_k, t_{k+1}), \\
w^i(t_{k+1}) &= y^i(t_{k+1}),
\end{aligned}
\tag{7}
$$

where $w \in \mathbb{R}^m$ is the predicted output. The predictors will give continuous estimates of the sampled outputs between two sampling times, but note that these two sampling times t_k, t_{k+1} are not necessarily from the same output. The i-th component $w^i(t)$ will be reset to $y^i(t_{k+1})$ once the new measurement obtained, and the other predictor states will not change until the corresponding measurements become available.

Now, consider a system with possible delays in the multi-rate sampled measurement $y^i(t^i_j)$

$$
\begin{aligned}
\dot{x}_R(t) &= f_R(x_R(t), x_M(t)), \\
\dot{x}_M(t) &= f_M(x_R(t), x_M(t)), \\
y^i(t^i_j) &= x^i_M(t^i_j - \delta^i_j), \qquad j \in \mathbb{Z}_+, i = 1, 2, \ldots, m,
\end{aligned}
\tag{8}
$$

where t^i_j is the time when j-th measurement of x^i_M is obtained after certain delay $\delta^i_j \geqslant 0$. That is to say, the measurement $y^i(t^i_j)$ arrived at t^i_j reflects the value of x^i_M at time $t^i_j - \delta^i_j$.

The proposed observer for the system (6) with multiple measurement delays is based on the multi-rate observer design (7) combined with dead time compensation. When the sampled measurement arrives at t^i_j after a delay δ^i_j, a dead time compensator would be initiated to recalculate the past estimates for $t \in [t^i_j - \delta^i_j, t^i_j]$ with the following observer design [55]:

$$
\dot{\hat{x}}_R(t) = f_R(\hat{x}_R(t), w(t)), \tag{9a}
$$
$$
\dot{w}(t) = f_M(\hat{x}_R(t), w(t)), \tag{9b}
$$
$$
w^i(t^i_j - \delta^i_j) = y^i(t^i_j), \tag{9c}
$$
$$
w^{i'}(t^{i'}_{j'} - \delta^{i'}_{j'}) = y^{i'}(t^{i'}_{j'}), \qquad \forall t^{i'}_{j'}, (t^{i'}_{j'} - \delta^{i'}_{j'}) \in [t^i_j - \delta^i_j, t^i_j), \tag{9d}
$$

where $w \in \mathbb{R}^m$ denotes the corrected state prediction that generates the past estimates for $x_M(t)$. Equation (9c) formulates the reinitialization of the i-th dead time compensator for $x^i_M(t)$ by using the delayed measurement $y^i(t^i_j)$ corresponding to the sampling time $t^i_j - \delta^i_j$. The other measurement $y^{i'}(t^{i'}_{j'})$ that is sampled between $t^i_j - \delta^i_j$ and t^i_j can be used to reset other corrected state prediction $w^{i'}(t)$ at their respective sampling times, as is represented by Equation (9d).

After the dead time compensation has updated the past estimates, the inter-sample prediction will be applied for the missing points between two measurements for $t \in [t^i_j, t^{i'}_{j'}]$ in the same spirit as the delay-free multi-rate observer (7). If a measurement is available at t^i_j without any delay, then no dead time compensation is needed and the inter-sample prediction runs immediately after reset. A detailed explanation for the estimation process will be given in the following part based on the RAFT polymerization process that is studied in this paper.

2.2.2. Prediction of RAFT Polymerization

Throughout the polymerization, it is important to keep track of monomer conversion (X_p), degree of polymerization (DP_n), and dispersity ($Đ$) of the polymer formed in the reacting mixture. Monomer conversion shows the reaction is proceeding, and is used to gauge how quickly polymerization occurs. Degree of polymerization is used to target the reaction end point when a desired degree of polymerization is reached. Dispersity is used to gauge the control of the RAFT agent on the polymerization; high dispersities indicate loss of control of the polymerization.

^{1}H NMR and SEC are used to characterize these properties from the reaction mixture aliquots. However, this data is not available in real time. Data from these analysis techniques become available only after analysis, and the time to perform analysis for each technique differs. In general, ^{1}H NMR provides monomer conversion (X_p) data after a typical delay of approximately 30 min. SEC provides molecular weight data (M_n, M_w), degree of polymerization data (DP), dispersities ($Đ$), and monomer conversion data (X_p) after a typical delay of 4 h. The delays for each technique are inherent to proper sample preparation and analysis procedure for each technique. As a result, reaction data for updating the observer comes with a multi-rate, multi-delay nature.

Note that the concerned measured variables X_p, DP_n and $Đ$ are not part of the state variables. However, through a coordinate transformation via (1), we can introduce them as three state variables into the dynamic system (2) and substitute three existing states. The RAFT process can be described in the form of Equation (8) with $x_M = [X_p \, DP_n \, Đ]^T$ and x_R to be a column vector with 20 remaining state variables out of the 23 variables listed in Table 2. The measurement for the problem could be formulated as

$$
\begin{aligned}
y^1(t_j^{NMR}) &= X_p(t_j^{NMR} - \delta_j^{NMR}), \\
y^2(t_{j'}^{SEC}) &= DP_n(t_{j'}^{SEC} - \delta_{j'}^{SEC}), \qquad j, j' \in \mathbb{Z}_+, \\
y^3(t_{j'}^{SEC}) &= Đ(t_{j'}^{SEC} - \delta_{j'}^{SEC}).
\end{aligned}
\tag{10}
$$

As these measurements are sampled with delay, no timely information is available in between measurements. Using observers to monitor the polymerization allows for an estimation of the missing information between each measurement, and thus provides an accurate estimation near the end point of the reaction to guide the operation of processes. Therefore, in this case, instead of the unmeasured states x_R, the accurate estimation for x_M is expected from the observer through:

$$
\hat{x}_M(t) = w(t),
\tag{11}
$$

where $w(t)$ is calculated from both the dead time compensator (9) and the inter-sample predictor (7).

Figure 1 depicts the two-step estimation process from time t_0. We will use the workflow after the n-th sample is taken, to demonstrate how the multi-rate multi-delay observer works. As a first step, when the delayed ^{1}H NMR measurement (which gives X_p) for the n-th sample becomes available at t_n^{NMR}, the dead time compensation is triggered. Past estimation is generated by integrating the observer and compensator equations for the time range $[t_n^{NMR} - \delta_n^{NMR}, t_n^{NMR}]$. In this compensator, the new value for X_p is used as a delay-free measurement to reset the corresponding compensator w^1. The compensator consequently updates the estimation for all states at t_n^{NMR}, the ending time of compensation. This step makes use of the available measurements in the delay-free observer, in the same order as they are sampled. As a second step, the updated estimates are used as the initial condition for the observer coupled with inter-sample output predictors (7) at t_n^{NMR}. In addition, this multi-rate multi-delay observer now works like a delay-free multi-rate observer with no delayed measurement, until the next measurement from SEC becomes available at t_{n+1}^{NMR}.

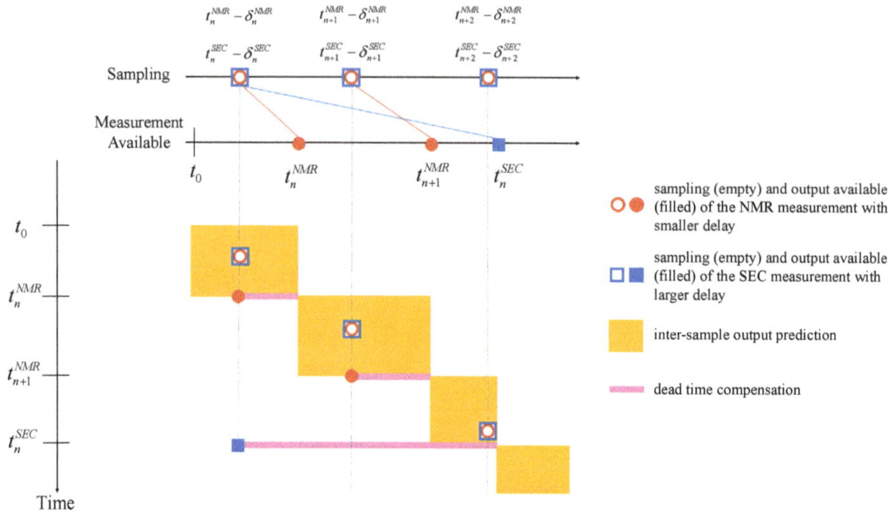

Figure 1. An illustration of the two-step estimation process of a multi-rate multi-delay observer for the RAFT polymerization process starting from t_0.

Note that the proposed multi-rate multi-delay observer is based on the continuous reduced-order observer (5). It has been proved that, as long as the maximum sampling period satisfies a certain limit, the stability of error dynamics for the continuous observer of (5) implies the validity of the multi-rate multi-delay observer [56]. However, as the continuous observer of (5) does not have a feedback correction term led by an observer gain, the rate of error dynamics is not adjustable.

3. Experimental Methods and Results

3.1. Materials and Methods

Materials. 4-cyano-4-(((dodecylthio)carbonothioyl)thio)pentanoic acid (chain transfer agent (CTA), 97%, Boron Molecular, Raleigh, NC, USA), tetrahydrofuran (THF, anhydrous, $\leq$99.9%, inhibitor-free, Sigma-Aldrich, Milwaukee, WI, USA), tetrahydrofuran (HPLC THF, inhibitor-free, for HPLC, $\leq$99.9%, Sigma-Aldrich, Milwaukee, WI, USA), tetrahydrofuran-d$_8$ (THF-d$_8$, $\leq$99.5% atom % D, Sigma-Aldrich, Milwaukee, WI, USA), chloroform-d (CDCl$_3$, 99.96 atom % D, contains 0.03% (v/v) TMS, Sigma-Aldrich, Milwaukee, WI, USA), and methanol (MeOH, ACS reagent, $\leq$99.8%, Sigma-Aldrich, Milwaukee, WI, USA) were used as received. Azobis(isobutyronitrile) (AIBN, 98%, Sigma-Aldrich, Milwaukee, WI, USA) was purified via recrystallization twice from methanol. Methyl methacrylate (MMA, 99%, contains $\leq$30 ppm MEHQ as inhibitor, Sigma-Aldrich, Milwaukee, WI, USA) was purified by distillation over CaH$_2$ at a reduced pressure.

Characterization. The molecular weights and dispersitys of all polymers and reacting mixture aliquots were determined by size exclusion chromatography (SEC) using a Waters GPC system (Milford, MA, USA) equipped with a THF Styragel column (Styragel@HR 5E, effective separation of molecular weight range: 2–4000 kg mol^{-1}) and a 2414 reflective index (RI) detector. All measurements were performed at 40 °C, where THF was used as the mobile phase at a flow rate of 1.0 mL min^{-1}. Poly(styrene) standards (Shodex, Tokyo, Japan) with molecular weights ranging from 2.97 to 983 kg mol^{-1} were used for calibration. Molecular weight values determined against the poly(styrene) standards were converted to the true poly(methyl methacrylate) values using the Universal Calibration procedure. Chemical structures of all polymers and reacting mixture aliquots from large-scale reactions were characterized by ^{1}H NMR (Nuclear Magnetic Resonance) Spectroscopy using a Bruker Avance 500 MHz spectrometer (Billerica, MA, USA) at 23 °C with CDCl$_3$ as the solvent. The chemical shifts were referenced to chloroform at 7.27 ppm. Small-scale PMMA macro-CTA reactions were characterized by ^{1}H NMR spectroscopy using a Varian NMRS 500 MHz spectrometer (Palo Alto, CA, USA) at 60 °C with THF-d$_8$ as the solvent. The chemical shifts were referenced to THF at 1.73 ppm.

Polymer Synthesis Procedures. Two synthesis scales were executed in this work: large-scale and small-scale reactions. Large-scale reactions (>400 mL) consisted of a reflux reactor, which allowed for the removal of reacting mixture via aliquots during the reaction with negligible effect on the reacting volume. All aliquots were analyzed by ^{1}H NMR spectroscopy and SEC. Small-scale reactions (<2 mL) consisted of reactions occurring in sealable temperature-controlled NMR tubes, where frequent in situ NMR analysis of the reacting mixture is possible without collecting aliquots from the reactor. SEC analysis of small-scale reactions were only conducted on the final product.

Large-Scale Reaction Procedure. A general procedure for large-scale polymer synthesis in a reflux reactor is outlined here with reaction details for each reaction listed in Table 4. Monomer and CTA were mixed with solvent in a 2000 mL three-neck round-bottom flask, where the central neck of the flask was connected to a reflux condenser (connected to bubbler and a nitrogen source from a Schlenk line) and the other two necks of the flask were sealed with rubber septa. The reacting mixture was degassed by bubbling nitrogen gas through the reacting mixture for 2 h. After degassing, the reactor was placed into an oil bath, covered in aluminum foil, and heated to reflux. In a separate 10 mL vial sealed with a septum, the initiator was dissolved in solvent and degassed by bubbling nitrogen gas through the solution for 5 min. At the first sign of reflux, the degassed initiator solution was injected into the reacting mixture, and a 1 mL aliquot was collected from the reactor for ^{1}H NMR and SEC analysis. The reaction was carried out under reflux for seven days. Additional aliquots were collected throughout the reaction for analysis. The resulting polymer was twice precipitated dropwise in methanol, filtered, and then dried under dynamic vacuum in an oven at room temperature for 24 h.

Small-Scale Reaction Procedure. A general procedure for small-scale polymer synthesis in an NMR tube reactor is outlined here with reaction details for each reaction listed in Table 4. Monomer, CTA and initiator were mixed with solvent in a sealable low pressure/vacuum NMR tube, connected to a nitrogen Schlenk line, and subjected to four freeze-pump-thaw degassing cycles. After degassing, the NMR tube was sealed under a nitrogen environment. The NMR tube reactor was inserted into a Varian NMRS 500 MHz NMR spectrometer at 60 °C for 23 h, where NMR spectra were collected every 10 min. The resulting polymer was twice precipitated in methanol, filtered, and then dried under dynamic vacuum in an oven at room temperature for 24 h.

Synthesis of Poly(methyl methacrylate). The synthesis of poly(methyl methacrylate) (PMMA) macro-CTA is shown in Scheme 2. In total, four PMMA synthesis reactions were performed: two large-scale and two small-scale reactions. Reaction details for each reaction are listed in Table 4. Product details for each reaction are listed in Table 5. Monomer: MMA; solvent: THF; initiator: AIBN. ^{1}H NMR (500 MHz, CDCl$_3$, 23 °C) δ (ppm): 3.57 (s, 3H, O-CH$_3$), 1.84-1.76 (d, 2H, CH$_2$-C(CH$_3$)), 0.94-0.74 (d, 3H, CH$_2$-C(CH$_3$)).

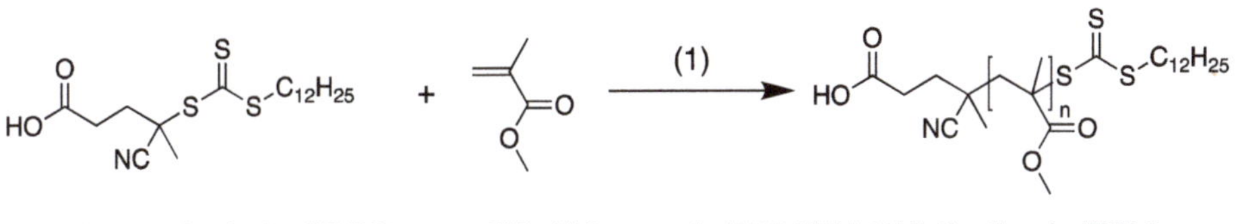

Scheme 2. Synthesis of PMMA macro-CTA. (1) Large-scale: THF, AIBN, 161 h; Small-scale: THF-d$_8$, AIBN, 23 h.

Table 4. Reaction conditions for all polymerization reactions.

Reaction [a]	Scale	Recipe [b]	Monomer (g)	CTA (g)	Initiator (g)	Solvent (g) [c]
PMMA-1	Large	215:1:0.1	250.63	4.7	0.191	626.58
PMMA-2	Large	70:1:0.1	173.62	10.0	0.407	714.48
PMMA-3	Small	215:1:0.1	0.5333	0.01	0.0004	1.3331
PMMA-4	Small	70:1:0.1	0.3472	0.02	0.0008	0.8681

[a] Polymer-reaction number. [b] A:B:C = Monomer:CTA:Initiator (in mol). [c] THF-d_8 was used for PMMA small-scale reactions.

Table 5. Final product properties for all polymerization reactions.

Reaction [a]	Scale	Recipe [b]	Yield (g)	M_n (g mol^{-1}) [c]	M_w (g mol^{-1}) [c]	Đ [c]
PMMA-1	Large	215:1:0.1	240 (94%)	21121	26347	1.247
PMMA-2	Large	70:1:0.1	158 (86%)	7877	10025	1.273
PMMA-3	Small	215:1:0.1	-	11971	14698	1.228
PMMA-4	Small	70:1:0.1	-	6669	8101	1.215

[a] Polymer-reaction number. [b] A:B:C = Monomer:CTA:Initiator (in mol). [c] Determined by SEC.

3.2. Experimental Results

Figure 2 shows the SEC profiles for large-scale reaction of PMMA (PMMA-1) as a function of time. Data only after 240 min is shown as aliquots at earlier times did not precipitate due to low monomer conversion. Molecular weights and dispersity data were calculated from the SEC data for all aliquots (all time points) and are listed in Appendix A (Table A1) for PMMA-1 and the other large-scale polymerization reaction (PMMA-2: Table A2). The SEC profiles clearly shift to earlier elution volumes at later reaction times, indicating the growth of polymer chains throughout the reaction. Molecular weight increases from an M_n of 6,324 Da at 240 min to an M_n of 21,121 Da at 8580 min. Dispersity is well controlled throughout the reaction as shown by the minimal change in the breadth of the profiles and dispersities less than 1.3 for all profiles except for the profiles at reaction times of 240 min and 300 min. High dispersities at low monomer conversions (i.e., early reaction times) are expected in RAFT polymerization reactions. Chromatogram profiles form a tail at high elution volumes at later reaction times signifying the formation of dead polymer chains due to termination reactions.

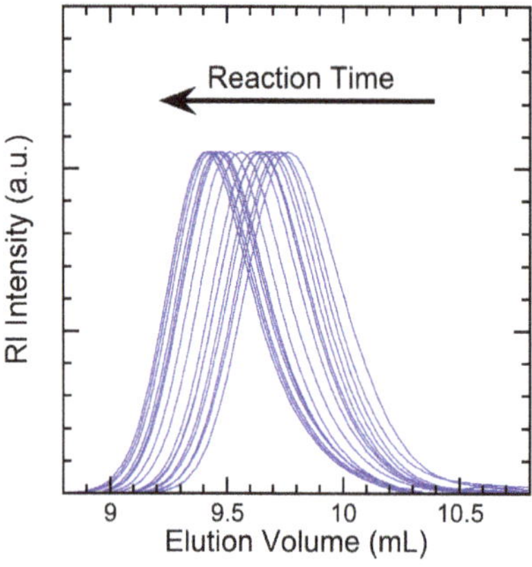

Figure 2. SEC profiles for PMMA-1 vs. time (normalized intensity bands).

Figure 3 shows the corresponding ¹H NMR spectra for all aliquots for the large-scale PMMA reaction (PMMA-1). Analysis of these spectra provides monomer conversion data, which is listed for each time point in Appendix A (Table A1) for PMMA-1 and the other large-scale polymerization reaction (PMMA-2: Table A2). Four primary bands are visible (a, b, c and d) in each spectrum and were subsequently integrated. Monomer conversion was calculated from comparing integral of band b at specified time to integral of band b at t = 0, where integral of band a was set as the reference. Monomer conversion ranges from 0% at 0 min to 95% at 8,580 min.

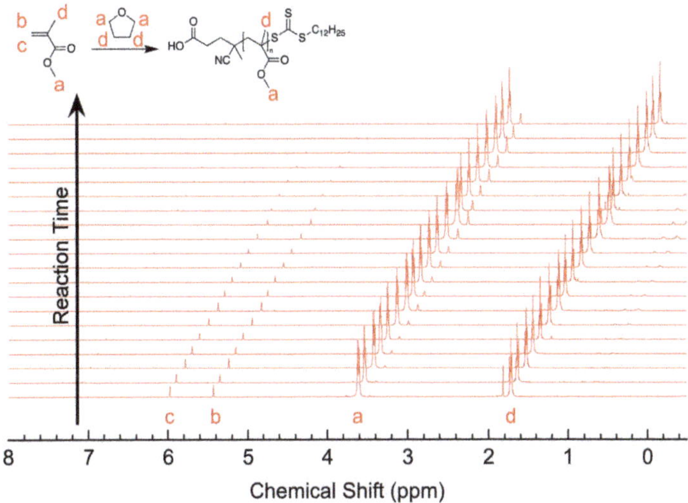

Figure 3. ¹H NMR spectra of PMMA-1 vs. time (scaled to reference band a).

Figure 4 shows the polymerization kinetics data as calculated by the ¹H NMR analysis in Figure 3. Tabulated results are listed in Appendix A (Table A1) for PMMA-1 and the other large-scale polymerization reaction (PMMA-2: Table A2).

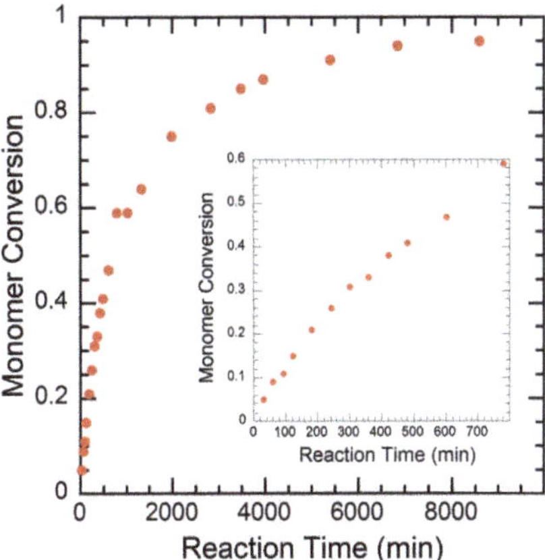

Figure 4. Polymerization kinetics (monomer conversion) of PMMA-1 as determined by ¹H NMR. The inset window highlights magnified early reaction time results.

Figure 5 shows the polymerization kinetics as calculated by NMR analysis for the small-scale PMMA polymerization (PMMA-3). The NMR spectra for PMMA-3 were processed similar to the procedure described for PMMA-1. Using THF-d$_8$ in PMMA-3 instead of THF in PMMA-1 did not affect the NMR spectra processing procedure. The data appears to follow first order reaction kinetics similar to the large-scale reaction results.

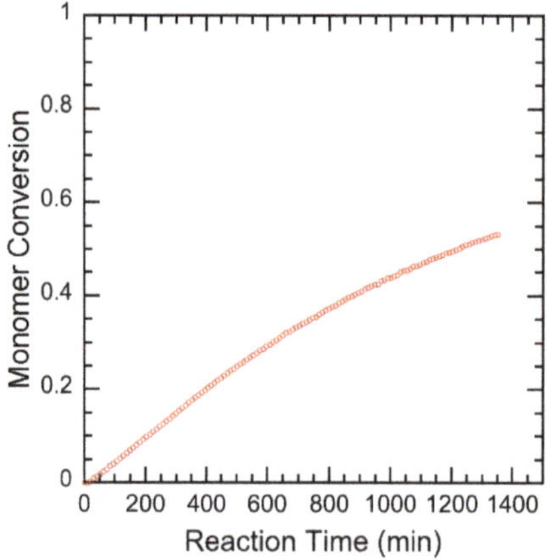

Figure 5. Polymerization kinetics (monomer conversion) of PMMA-3 as determined by in situ ^{1}H NMR analysis.

Data for the other polymerization reactions (PMMA-2, PMMA-4) is available in Appendix A.

4. Results and Discussion

In this part, the results for both model fitting and the multi-rate multi-delay observer will be demonstrated. The small-scale experiment results, which contain 127 conversion data points measured using ^{1}H NMR and 1 ending data point measured with SEC for both PMMA-3 and PMMA-4 sets, are used for parameter fitting purposes. The two sets of large-scale experimental results, with fewer but more comprehensive and practical sampled data points, are to be used to validate the calculated parameters. The observer will be designed based on the fitted model and tested on one of the large-scale reaction data sets.

4.1. Parameter Fitting

For the model (2) introduced in Section 2, values of the parameters for some specific experiments have been identified by other researchers as shown in Table 3. With regards to the experimental setting introduced in Section 3, a RAFT agent is used for which no kinetics data was found in literature. Therefore, its kinetic characteristics remain unknown. Additionally, initiator efficiency varies in different environments and must be identified. Therefore, there are five parameters for the current system to be fitted: the forward RAFT main equilibrium reaction rate coefficient k_{raftf}, the reverse RAFT main equilibrium reaction rate coefficient k_{raftr}, the forward RAFT pre-equilibrium reaction rate coefficient k_{raftfR}, the reverse RAFT pre-equilibrium reaction rate coefficient k_{raftrR}, and the initiator efficiency f. While the small scale reaction is operated under a constant temperature of 60 °C and the large scale reaction at the bubble point of the reaction mixture (65–80 °C), it is assumed that the effect of temperature on these parameters is negligible over this range. This assumption, while not ideal, is necessary due to limitations in the reaction techniques. Small-scale reactions cannot be performed at reflux, and large-scale reactions must be performed at reflux. The small-scale reactions were performed as close to reflux temperature as possible without risking evaporation of solvent, and damage to the NMR.

Because the least squares parameter fitting problem is highly non-convex and non-smooth, the PatternSearch solver in MATLAB (version 9.7.0.1190202 (R2019b), MathWorks, Natick, MA, USA) with multiple random initial points is used to solve for the global optimal solution. In order to set up the optimization problem, an appropriate objective function is needed. It is preferable to use the variable that has a large number of sampled data in the objective function, which is the conversion X_p in this case. However, it is observed that, when minimizing the error of conversion, in the basin region near the global optimum, the error becomes less sensitive to the fitting parameters. Additionally, the optimal solution could not guarantee whether the model also matches the SEC data. Thereby, the dispersity ($Đ$) and the number average degree of polymerization (DP_n) measured from SEC for the ending point are also used as a part of the objective function, and normalized to the proper scale with a weighting coefficient of 5 and $1/25$, respectively. The global optimal parameters are calculated as: $k_{raftf} = 3.74 \times 10^4$ L/(mol $\cdot$ s), $k_{raftr} = 1.82$ /s, $k_{raftfR} = 9.34 \times 10^6$ L/(mol $\cdot$ s), $k_{raftrR} = 8.75$ /s and $f = 0.323$. These values match the typical values found in the literature [60].

Figure 6 shows the experimental data obtained from the small-scale in situ ^{1}H NMR experiments, as well as the obtained fits which give the best estimation of the fitting parameters using least squares. It can be observed that the fitting model shows excellent agreement with the X_p measurement. As for the single sampled $Đ$ and DP_n points on the right two plots, the estimated values from the model also match the experimental data well, especially for the PMMA-3 set.

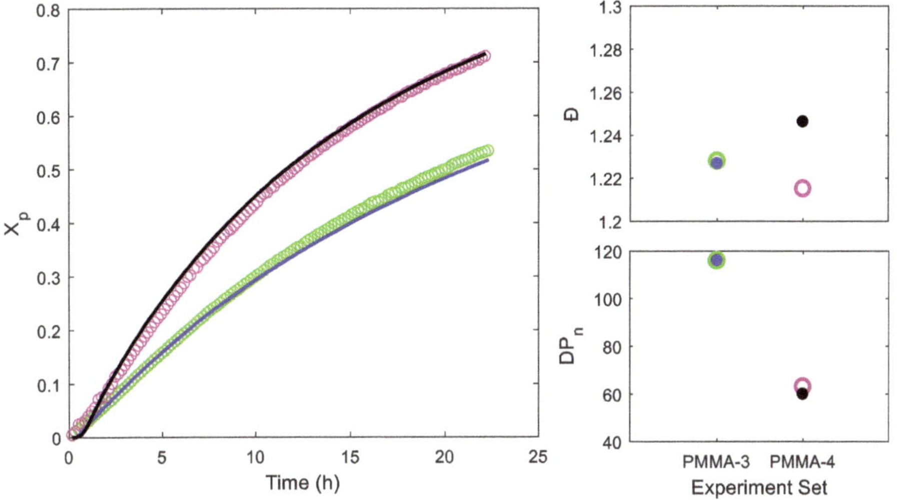

Figure 6. Experimental data obtained from small-scale reactions versus optimal fitting results. (Green circles are the data collected from PMMA-3 set and magenta ones are from PMMA-4 set. The solid symbols represent the $Đ$ and DP_n estimated from the model.)

For validation purposes, we test the model against two large-scale reaction sets (PMMA-1 and PMMA-2) in Figure 7. The measurement of conversion X_p is obtained from ^{1}H NMR measurement. The DP_n and $Đ$ data are collected from SEC measurement. The large-scale reactions are operated under reflux. The temperature is simulated as a function of the composition and used for the calculation of reaction kinetics coefficients except for k_{raftf}, k_{raftr}, k_{raftfR} and k_{raftrR}, which are assumed to be independent of temperature.

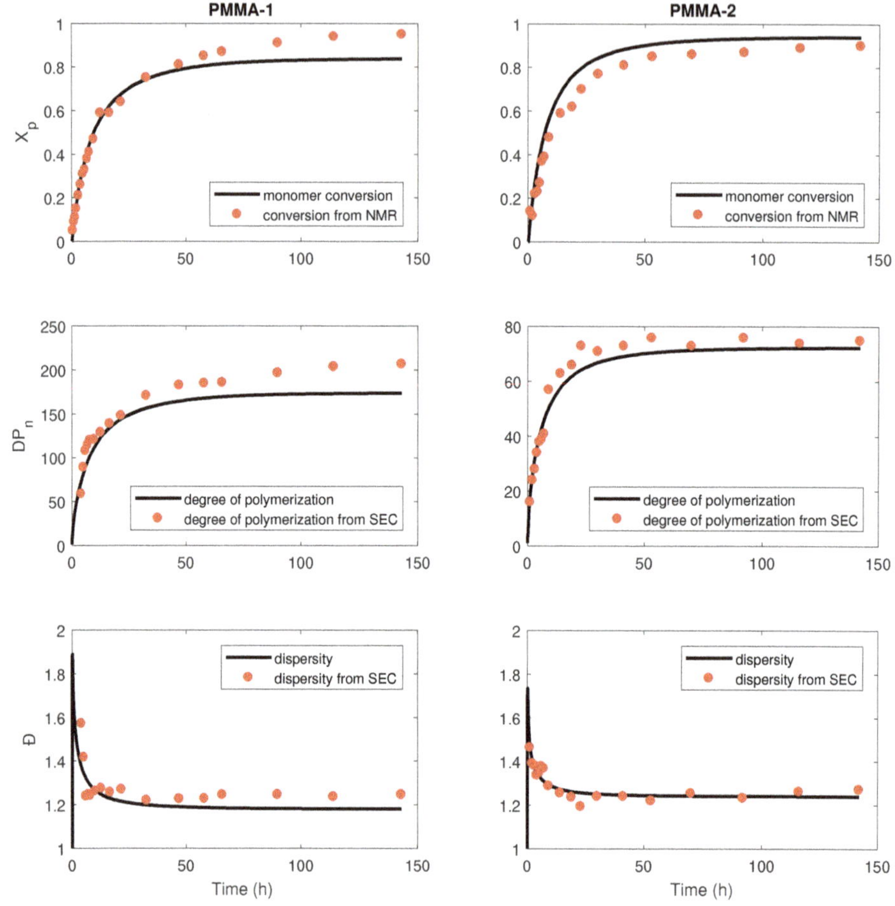

Figure 7. First row: conversion of MMA into polymer with respect to time; second row: evolution of number average degree of polymerization with respect to time; third row: evolution of dispersity with respect to time; Left: PMMA-1; Right: PMMA-2.

Overall, the model satisfactorily matches the experimental data. Monomer conversion (X_p) and degree of polymerization (DP_n) match well at early reaction times, with some discordance appearing at late reaction times, especially for PMMA-1. The discordance may be due to temperature effects which are unaccounted for in the model due to the assumption that k_{raftf}, k_{raftr}, k_{raftfR} and k_{raftrR} are not functions of temperature. The large-scale and small-scale experimental reactions were performed at temperatures as close to each other as experimentally possible, but there was still a 10 to 15 °C difference between the two reaction scales. Dispersities ($Đ$) match well, with the sharp peak around $Đ = 2$ in the model at early times expected from typical RAFT polymerization. SEC data at early reaction times is not available for comparison because the corresponding reaction aliquots did not precipitate due to low monomer conversion.

4.2. Multi-Rate Multi-Delay Observer

In this part, we test the multi-rate multi-delay observer proposed in Section 2.2.2 against both the fitting model and the experimental data, to check its fidelity to the model and its performance under real circumstances in the presence of errors.

The data set collected from the PMMA-2 large-scale experiment is used to test the design of the observer, as this data set shows a better consistence with the fitting model than the data set from the PMMA-1 large-scale experiment. During the reaction, aliquots were collected approximately every 5% monomer conversion, which resulted in a non-uniform sampling schedule with respect to time. The sampling schedule for the reaction and the delays in measurement for those samples are shown in Table 6.

Table 6. Actual sampling schedule and measurement delays in the large-scale reactor.

Sampling (h)	1	2	3	4	5	6	7	9	14	19	23	30	41	53	70	92	116	142
SEC delay (h)	2	3	3	4	4	5	5	6	6	4	4	6	4	3	6	8	5	4
NMR delay (h)	0.5	0.5	1	1	1	1	0.5	1	1.5	1	0.5	0.5	2.5	1	0.5	0.5	1.5	1

In Figure 8, simulated data generated from the model is used as the measurement to test the observer's performance against the fitting model with the same sampling schedule as shown in Table 6. The observer shows good prediction for monomer conversion, as the measurement from NMR has less delay than the measurement from SEC. The estimated DP_n undergoes a period of oscillation and Đ shows a higher peak than the actual data, due to the fact that the first measurement from SEC only becomes available 3 h after the reaction starts. However, the prediction shows higher accuracy after a few samplings.

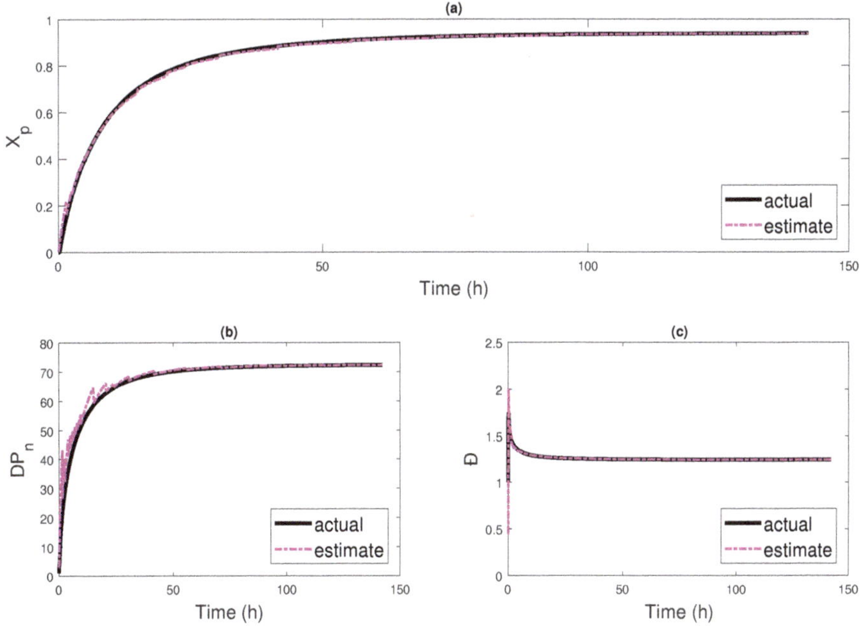

Figure 8. Test of observer performance against the model for (**a**) monomer conversion X_p; (**b**) number average molecular weight DP_n, (**c**) dispersity Đ.

In Figure 9, the observer is designed based on real experimental data, and the comparison between estimated values from this observer and the sampled points is demonstrated. In these figures, the plotted estimation only shows the real-time information at each moment, and does not reflect the updated historical estimation for the past delayed time period that is corrected by the dead-time compensator later on. As a result, the magenta line does not go through the black measured points, but leaps to the levels corresponding to the newest sampling data after certain delays, with the dead-time compensator correction effects applied. Because the experimental data points themselves are fluctuating with measurement noises and the fitted model already shows deviation from the real samples, the performance of the observer is not as good as in Figure 8, in terms of a slower convergence rate and fluctuation at early reaction times. However, the observer still predicts conversion and dispersity with high accuracy after more sampling points become available.

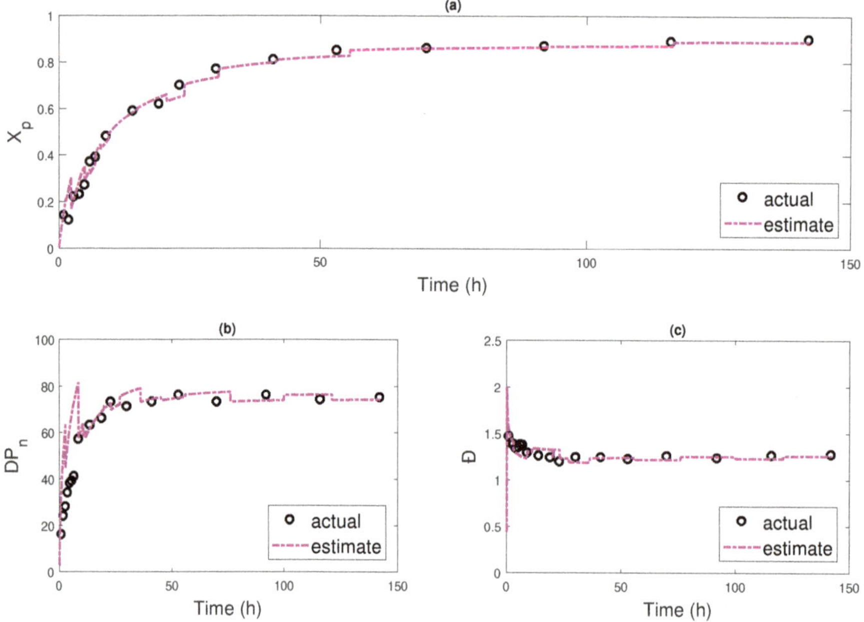

Figure 9. Observer performance based on experimental data for (**a**) monomer conversion X_p; (**b**) number average molecular weight DP_n; (**c**) dispersity $Đ$.

5. Conclusions

In this work, a RAFT polymerization model was improved using small-scale reactions with in situ ^{1}H NMR analysis that allowed for more accurate prediction of RAFT reaction kinetic rate parameters (specifically, k_{raftf}, k_{raftr}, k_{raftfR} and k_{raftrR}). The resulting model with the fitted parameters was used to predict RAFT reactions on a larger scale with differing initial conditions to high values of monomer conversion, X_p. The fitted model accurately predicted the polymer properties of the large-scale reactions with slight discordance at late reaction times. Finally, multi-rate, multi-delay observers were used to more accurately monitor a reaction and its product's quality in real time by incorporating information from the sampled and delayed measurements. The observer was tested both against the fitted model and the experimental data. At early reaction times, both observers show some deviations due to limited characterization data. However, at later reaction times, the observer against the fitted model shows perfect convergence, while the one based on the experimental data matches well with slight shifts upon receiving new measurements, owing to the inherent noises of the analysis techniques. Ultimately, accurate monitoring allowed for targeted termination, making it easier to synthesize polymers of desired properties.

RAFT polymerization is very versatile, being compatible with a wide array of monomer and RAFT agent types. In this work, a trithiocarbonate RAFT agent was used; however, other types, such as dithiocarbonates, are also common. Certain RAFT agents work better with certain monomer types; and now some RAFT agents, such as pyrazole based RAFT agents, even work effectively with all monomer types viable for RAFT polymerization. The methodology presented in this work for monitoring RAFT polymerization reactions could be applied to different monomer-RAFT agent combinations, allowing for monitoring of RAFT polymerizations in general.

In this work, the small-scale reactions used for the parameter fitting occurred at a different temperature than the large-scale reactions, due to limitations in the experimental apparatus and procedures. The temperature difference leads to some discordance when comparing the model to the large-scale reaction results. This discordance is attributed to the temperature difference. Large-scale reactions occur at a slightly higher temperature than the small-scale reactions, and would be expected to exhibit larger reaction rate constants. The methodology to determine the values of the RAFT reaction rate coefficients can be improved by determining temperature dependent reaction rate constant values. Further studies with multiple small-scale reactions at various temperatures can be performed to determine the temperature dependence of the RAFT reaction rate constants.

Additionally, as a living polymerization, polymers synthesized via RAFT polymerization are able to be chain extended into block copolymers. The block copolymers can be synthesized with different monomers with controlled block composition, while maintaining low dispersity. Block copolymers synthesized in this fashion have many applications due to the multiple properties. Many of these properties, such as morphology, depend on the block composition ratios, giving importance to targeting precise block composition during synthesis. Future work in this area would include modeling and monitoring these chain-extension polymerizations using a polymer reactant synthesized via RAFT polymerization. The resulting block copolymer compositions can be targeted, based on desired block copolymer properties, by using the observer based RAFT polymerization model to accurately predict reaction termination times and conditions.

Author Contributions: All authors contributed to the work according to the following categories: conceptualization, P.M.L. and C.L.; methodology, P.M.L. and C.L.; software, Z.D. and C.L.; validation, Z.D., C.L., and P.M.L.; formal analysis, P.M.L., Z.D. and C.L.; investigation, P.M.L. and C.L.; resources, Y.A.E.; writing—original draft preparation, Z.D. and P.M.L; writing—review and editing, P.M.L., Z.D., C.K., Y.A.E. and C.L.; visualization, Z.D., P.M.L. and C.L.; supervision, C.K. and Y.A.E.; project administration, C.K. and Y.A.E.; funding acquisition, C.K. and Y.A.E.

Funding: This research was funded by the National Science Foundation through Grant Nos. CBET-1706201 and CBET-1703645.

Conflicts of Interest: The authors declare no conflict of interest.

Abbreviations

The following abbreviations are used in this manuscript:

^{1}H NMR	Proton Nuclear Magnetic Resonance
AIBN	Azobis(isobutyronitrile)
CTA	Chain Transfer Agent
EKF	Extended Kalman Filter
MHE	Moving Horizon Estimation
MMA	Methyl Methacrylate
PMMA	Poly(methyl methacrylate)
RAFT	Reversible Addition–Fragmentation Chain–Transfer
SEC	Size Exclusion Chromatography
THF	Tetrahydrofuran

Appendix A

Table A1. Summary of ^{1}H NMR and SEC data for PMMA-1.

Time (h)	Time (min)	NMR Conv. (%)	SEC Conv. (%)	M_n (g mol^{-1})	M_w (g mol^{-1})	Đ [a]	DP [b]
0	0	0	-	-	-	-	-
0.5	30	5	-	-	-	-	-
1	60	9	-	-	-	-	-
1.5	90	11	-	-	-	-	-
2	120	15	-	-	-	-	-
3	180	21	42.9	9637	12,409	1.288	92
4	240	26	27.5	6324	9949	1.573	59
5	300	31	41.5	9334	13,227	1.417	89
6	360	33	50.4	11,261	13,954	1.239	108
7	420	38	53.5	11,927	14,858	1.246	115
8	480	41	55.6	12,383	15,382	1.242	120
10	600	47	56.3	12,523	15,800	1.262	121
13	780	59	60.1	13,346	17,018	1.275	129
17	1020	59	64.5	14,293	17,964	1.257	139
22	1320	64	68.9	15,240	19,357	1.270	148
33	1980	75	79.7	17,560	21,400	1.219	171
47	2820	81	85.1	18,724	22,974	1.227	183
58	3480	85	86.2	18,957	23,296	1.229	185
66	3960	87	86.6	19,052	23,744	1.246	186
90	5400	91	91.5	20,107	25,067	1.247	197
114	6840	94	94.9	20,824	25,766	1.237	204
143	8580	95	96.2	21,121	26,347	1.247	207

[a] Đ = dispersity; [b] DP = degree of polymerization.

Table A2. Summary of ^{1}H NMR and SEC data for PMMA-2.

Time (h)	Time (min)	NMR Conv. (%)	SEC Conv. (%)	M_n (g mol^{-1})	M_w (g mol^{-1})	Đ [a]	DP [b]
0	0	0	-	-	-	-	-
1	60	14	23.4	2042 [c]	2990 [c]	1.464	16
2	120	12	34.4	2814 [c]	3918 [c]	1.392	24
3	180	22	40.0	3208 [c]	4433 [c]	1.382	28
4	240	23	48.1	3777	5055	1.338	34
5	300	27	53.6	4162	5604	1.346	38
6	360	37	55.5	4292	5910	1.377	39
7	420	39	58.1	4472	6122	1.369	41
9	540	48	81.5	6112	7877	1.289	57
14	840	59	90.7	6759	8496	1.257	63
19	1140	62	94.2	7003	8672	1.238	66
23	1380	70	104.1	7698	9203	1.195	73
30	1800	77	101.2	7497	9304	1.241	71
41	2460	81	104.9	7754	9620	1.241	73
53	3180	85	108.9	8035	9809	1.221	76
70	4200	86	104.2	7704	9671	1.255	73
92	5520	87	109.1	8051	9933	1.234	76
116	6960	89	106.0	7834	9893	1.263	74
142	8520	90	106.6	7877	10,025	1.273	75

[a] Đ = dispersity; [b] DP = degree of polymerization; [c] These values are below the lower calibration limit of the SEC and determined by extrapolating the calibration curve to the corresponding elution volumes.

PMMA-2. Figure A1 shows the SEC profiles for PMMA-2 as a function of time. Data only after 60 min is shown as aliquots at earlier times did not precipitate due to low monomer conversion. Molecular weights and dispersity data were calculated from SEC data for all aliquots (all time points) and are listed in Table A2. The SEC profiles clearly shift to earlier elution volumes at later reaction times, indicating the growth of polymer chains throughout the reaction. Molecular weight increases from an M_n of 2042 Da at 60 min to an M_n of 7877 Da at 8520 min. Early reaction time profiles show a second peak near an elution volume of 10.7 mL, which is attributed to unreacted CTA present at early reaction times. The height of this CTA peak relative to the polymer peak decreases as the reaction progresses, disappearing after 540 min of reaction time. Dispersity is well controlled throughout the reaction as shown by the lack of change in the breadth of the chromatogram profiles, with dispersities less than 1.3 for all profiles at reacting times greater than 540 min. This is partially due to the presence of the CTA peak. Additionally, high dispersities at low monomer conversion (i.e., early reaction times) are expected in RAFT polymerization reactions. Chromatogram profiles form a tail at high elution volumes at later reaction times signifying the formation of dead polymer chains due to termination reactions.

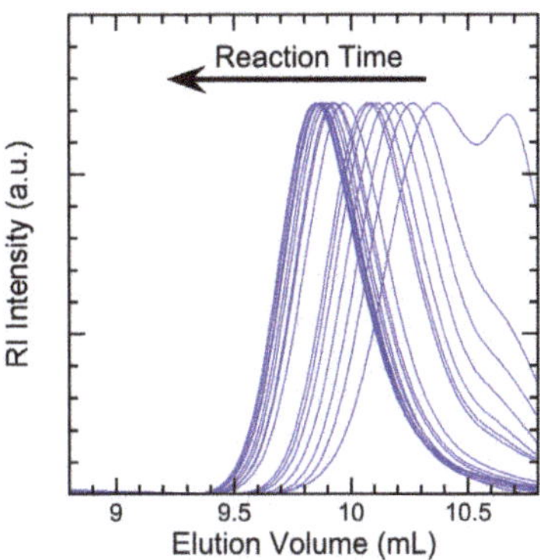

Figure A1. SEC profiles for PMMA-2 vs. time (normalized intensity bands).

Figure A2 shows the corresponding ^{1}H NMR spectra for all aliquots for the large-scale PMMA reaction (PMMA-2). Analysis of these spectra provide monomer conversion data, which is listed for each time point in Table A2. Four primary bands are visible (a, b, c and d) in each spectrum and were subsequently integrated. Monomer conversion was calculated from comparing the integral of band b at specified time to the integral of band b at t = 0, where the integral of band a was set as the reference. Monomer conversion ranges from 0% at 0 min to 90% at 8520 min.

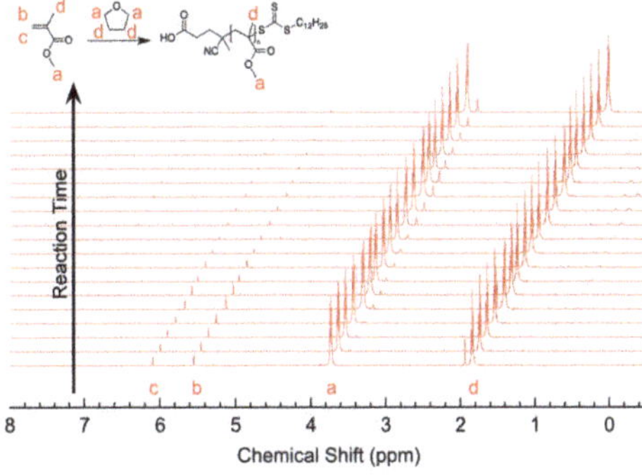

Figure A2. ^{1}H NMR spectra of PMMA-2 vs. time (scaled to reference band a).

Figure A3 shows the polymerization kinetics data as calculated by the ^{1}H NMR analysis in Figure A2. Tabulated results are listed in Table A2 for PMMA-2. The reaction appears to follow first order reaction kinetics.

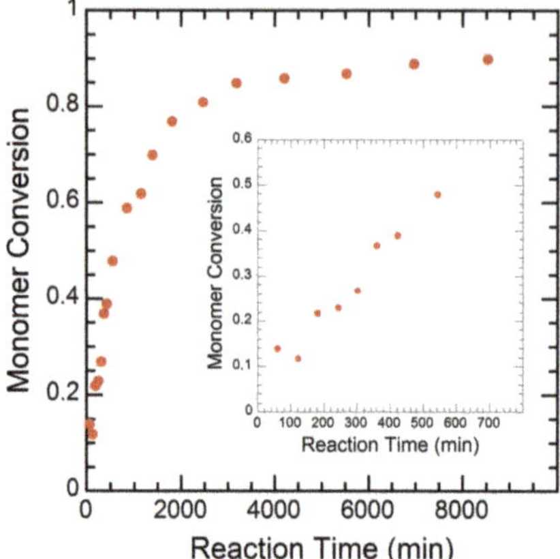

Figure A3. Polymerization kinetics (monomer conversion) of PMMA-2 as determined by ^{1}H NMR. The inset window highlights magnified early reaction time results.

PMMA-4. Figure A4 shows the polymerization kinetics as calculated by ^{1}H NMR analysis for the small-scale PMMA polymerization (PMMA-4). The ^{1}H NMR spectra for PMMA-4 were processed similar to the procedure used for PMMA-2. Using THF-d_8 in PMMA-4 instead of THF in PMMA-2 did not affect the NMR spectra processing procedure. The data appears to follow first order reaction kinetics similar to the large-scale reaction results.

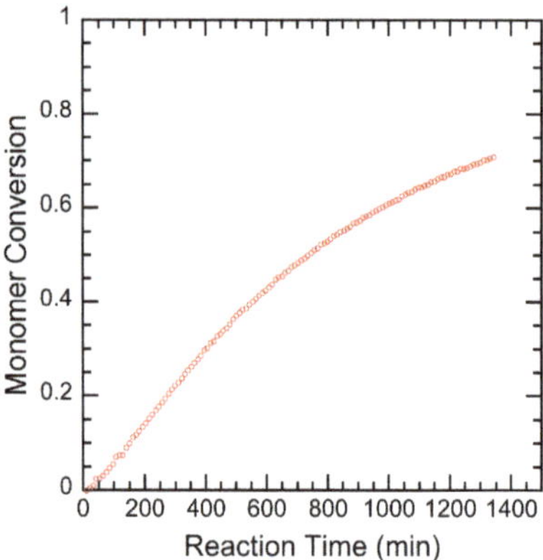

Figure A4. Polymerization kinetics (monomer conversion) of PMMA-4 as determined by in situ ^{1}H NMR analysis.

References

1. Moad, G.; Chiefari, J.; Chong, Y.K.; Krstina, J.; Mayadunne, R.T.A.; Postma, A.; Rizzardo, E.; Thang, S.H. Living free radical polymerization with reversible addition–fragmentation chain transfer (the life of RAFT). *Polym. Int.* **2000**, *49*, 993–1001. [CrossRef]
2. Krstina, J.; Moad, C.L.; Moad, G.; Rizzardo, E.; Berge, C.T.; Fryd, M. A new form of controlled growth free radical polymerization. In *Macromolecular Symposia*; Hüthig & Wepf Verlag: Basel, Switzerland, 1996; Volume 111, pp. 13–23.
3. Gardiner, J.; Martinez-Botella, I.; Tsanaktsidis, J.; Moad, G. Dithiocarbamate RAFT agents with broad applicability - the 3,5-dimethyl-1H-pyrazole-1-carbodithioates. *Polym. Chem.* **2016**, *7*, 481–492. [CrossRef]

4. Gardiner, J.; Martinez-Botella, I.; Kohl, T.M.; Krstina, J.; Moad, G.; Tyrell, J.H.; Coote, M.L.; Tsanaktsidis, J. 4-Halogeno-3,5-dimethyl-1H-pyrazole-1-carbodithioates: Versatile reversible addition fragmentation chain transfer agents with broad applicability. *Polym. Int.* **2017**, *66*, 1438–1447. [CrossRef]

5. Moad, G. RAFT Polymerization—Then, and Now. In *Controlled Radical Polymerization: Mechanisms*; American Chemical Society: Akron, OH, USA, 2015, Volume 1187, pp. 211–246.

6. Ye, Y.S.; Choi, J.H.; Winey, K.I.; Elabd, Y.A. Polymerized Ionic Liquid Block and Random Copolymers: Effect of Weak Microphase Separation on Ion Transport. *Macromolecules* **2012**, *45*, 7027–7035. [CrossRef]

7. Stace, S.J.; Fellows, C.M.; Moad, G.; Keddie, D.J. Effect of the Z- and Macro-R-Group on the Thermal Desulfurization of Polymers Synthesized with Acid/Base "Switchable" Dithiocarbamate RAFT Agents. *Macromol. Rapid Commun.* **2018**, *39*, e1800228. [CrossRef]

8. Chiefari, J.; Mayadunne, R.T.A.; Moad, C.L.; Moad, G.; Rizzardo, E.; Postma, A.; Skidmore, M.A.; Thang, S.H. Thiocarbonylthio compounds (S=C(Z)S-R) in free radical polymerization with reversible addition–fragmentation chain transfer (RAFT polymerization). Effect of the activating group Z. *Macromolecules* **2003**, *36*, 2273–2283. [CrossRef]

9. Chiefari, J.; Chong, Y.; Ercole, F.; Krstina, J.; Jeffery, J.; Le, T.P.; Mayadunne, R.T.; Meijs, G.F.; Moad, C.L.; Moad, G.; et al. Living free-radical polymerization by reversible addition- fragmentation chain transfer: The RAFT process. *Macromolecules* **1998**, *31*, 5559–5562. [CrossRef]

10. Montgomery, K.S.; Davidson, R.W.M.; Cao, B.; Williams, B.; Simpson, G.W.; Nilsson, S.K.; Chiefari, J.; Fuchter, M.J. Effective macrophage delivery using RAFT copolymer derived nanoparticles. *Polym. Chem.* **2018**, *9*, 131–137. [CrossRef]

11. Nykaza, J.R.; Benjamin, R.; Meek, K.M.; Elabd, Y.A. Polymerized ionic liquid diblock copolymer as an ionomer and anion exchange membrane for alkaline fuel cells. *Chem. Eng. Sci.* **2016**, *154*, 119–127. [CrossRef]

12. Nykaza, J.R.; Savage, A.M.; Pan, Q.W.; Wang, S.J.; Beyer, F.L.; Tang, M.H.; Li, C.Y.; Elabd, Y.A. Polymerized ionic liquid diblock copolymer as solid-state electrolyte and separator in lithium-ion battery. *Polymer* **2016**, *101*, 311–318. [CrossRef]

13. Moehrke, J.; Vana, P. The Kinetics of Surface-Initiated RAFT Polymerization of Butyl acrylate Mediated by Trithiocarbonates. *Macromol. Chem. Phys.* **2017**, *218*, 1600506. [CrossRef]

14. Bates, F.S.; Fredrickson, G.H. Block copolymer thermodynamics: Theory and experiment. *Annu. Rev. Phys. Chem.* **1990**, *41*, 525–557. [CrossRef] [PubMed]

15. Zhang, M.; Ray, W.H. Modeling of "living" free-radical polymerization with RAFT chemistry. *Ind. Eng. Chem. Res.* **2001**, *40*, 4336–4352. [CrossRef]

16. Zetterlund, P.B.; Gody, G.; Perrier, S. Sequence-Controlled Multiblock Copolymers via RAFT Polymerization: Modeling and Simulations. *Macromol. Theory Simul.* **2014**, *23*, 331–339. [CrossRef]

17. Hernandez-Ortiz, J.C.; Jaramillo-Soto, G.; Palacios-Alquisira, J.; Vivaldo-Lima, E. Modeling of Polymerization Kinetics and Molecular Weight Development in the Microwave-Activated RAFT Polymerization of Styrene. *Macromol. React. Eng.* **2010**, *4*, 210–221. [CrossRef]

18. Barner-Kowollik, C.; Quinn, J.F.; Morsley, D.R.; Davis, T.P. Modeling the reversible addition–fragmentation chain transfer process in cumyl dithiobenzoate-mediated styrene homopolymerizations: Assessing rate coefficients for the addition–fragmentation equilibrium. *J. Polym. Sci. Part A Polym. Chem.* **2001**, *39*, 1353–1365. [CrossRef]

19. Wang, A.R.; Zhu, S.P. Effects of diffusion-controlled radical reactions on RAFT polymerization. *Macromol. Theory Simul.* **2003**, *12*, 196–208. [CrossRef]

20. Chaffey-Millar, H.; Busch, M.; Davis, T.P.; Stenzel, M.H.; Barner-Kowollik, C. Advanced computational strategies for modelling the evolution of full molecular weight distributions formed during multiarmed (Star) polymerisations. *Macromol. Theory Simul.* **2005**, *14*, 143–157. [CrossRef]

21. Tobita, H. Modeling Controlled/Living Radical Polymerization Kinetics: Bulk and Miniemulsion. *Macromol. React. Eng.* **2010**, *4*, 643–662. [CrossRef]

22. Lopez-Dominguez, P.; Hernandez-Ortiz, J.C.; Vivaldo-Lima, E. Modeling of RAFT Copolymerization with Crosslinking of Styrene/Divinylbenzene in Supercritical Carbon Dioxide. *Macromol. Theory Simul.* **2018**, *27*, 1700064. [CrossRef]

23. Vana, P.; Davis, T.P.; Barner-Kowollik, C. Kinetic analysis of reversible addition fragmentation chain transfer (RAFT) polymerizations: Conditions for inhibition, retardation, and optimum living polymerization. *Macromol. Theory Simul.* **2002**, *11*, 823–835. [CrossRef]

24. Adebekun, D.K.; Schork, F.J. Continuous solution polymerization reactor control. 2. Estimation and nonlinear reference control during methyl methacrylate polymerization. *Ind. Eng. Chem. Res.* **1989**, *28*, 1846–1861. [CrossRef]

25. Jo, J.H.; Bankoff, S.G. Digital monitoring and estimation of polymerization reactors. *AIChE J.* **1976**, *22*, 361–369. [CrossRef]

26. Kim, K.J.; Choi, K.Y. On-line estimation and control of a continuous stirred tank polymerization reactor. *J. Process Control* **1991**, *1*, 96–110. [CrossRef]

27. Ellis, M.F.; Taylor, T.W.; Jensen, K.F. On-line molecular weight distribution estimation and control in batch polymerization. *AIChE J.* **1994**, *40*, 445–462. [CrossRef]

28. Crowley, T.J.; Choi, K.Y. On-line monitoring and control of a batch polymerization reactor. *J. Process Control* **1996**, *6*, 119–127. [CrossRef]

29. Mutha, R.K.; Cluett, W.R.; Penlidis, A. On-line nonlinear model-based estimation and control of a polymer reactor. *AIChE J.* **1997**, *43*, 3042–3058. [CrossRef]

30. Dimitratos, J.; Georgakis, C.; El-Aasser, M.S.; Klein, A. Dynamic modeling and state estimation for an emulsion copolymerization reactor. *Comput. Chem. Eng.* **1989**, *13*, 21–33. [CrossRef]

31. Kozub, D.J.; MacGregor, J.F. State estimation for semi-batch polymerization reactors. *Chem. Eng. Sci.* **1992**, *47*, 1047–1062. [CrossRef]

32. Astorga, C.M.; Sheibat-Othman, N.; Othman, S.; Hammouri, H.; McKenna, T.F. Nonlinear continuous-discrete observers: Application to emulsion polymerization reactors. *Control Eng. Pract.* **2002**, *10*, 3–13. [CrossRef]

33. Appelhaus, P.; Engell, S. Design and implementation of an extended observer for the polymerization of polyethylenterephthalate. *Chem. Eng. Sci.* **1996**, *51*, 1919–1926. [CrossRef]

34. Van Dootingh, M.; Viel, F.; Rakotopara, D.; Gauthier, J.P.; Hobbes, P. Nonlinear deterministic observer for state estimation: Application to a continuous free radical polymerization reactor. *Comput. Chem. Eng.* **1992**, *16*, 777–791. [CrossRef]

35. Viel, F.; Busvelle, E.; Gauthier, J.P. Stability of polymerization reactors using I/O linearization and a high-gain observer. *Automatica* **1995**, *31*, 971–984. [CrossRef]

36. Tatiraju, S.; Soroush, M. Nonlinear state estimation in a polymerization reactor. *Ind. Eng. Chem. Res.* **1997**, *36*, 2679–2690. [CrossRef]

37. Tatiraju, S.; Soroush, M.; Ogunnaike, B.A. Multirate nonlinear state estimation with application to a polymerization reactor. *AIChE J.* **1999**, *45*, 769–780. [CrossRef]

38. Soroush, M. State and parameter estimations and their applications in process control. *Comput. Chem. Eng.* **1998**, *23*, 229–245. [CrossRef]

39. López-Negrete, R.; Biegler, L.T. A Moving Horizon Estimator for processes with multi-rate measurements: A Nonlinear Programming sensitivity approach. *J. Process Control* **2012**, *22*, 677–688. [CrossRef]

40. Krämer, S.; Gesthuisen, R. Multirate state estimation using moving horizon estimation. In Proceedings of the 16th IFAC World Conference (IFAC 2005), Prague, Czech Republic, 3–8 July 2005; pp. 1–6.

41. Liu, A.; Zhang, W.; Yu, L.; Chen, J. Moving horizon estimation for multi-rate systems. In Proceedings of the 54th IEEE Conference on Decision and Control, Osaka, Japan, 15–18 December 2015; pp. 6850–6855.

42. Zambare, N.; Soroush, M.; Grady, M.C. Real-time multirate state estimation in a pilot-scale polymerization reactor. *AIChE J.* **2002**, *48*, 1022–1033. [CrossRef]

43. Ling, C.; Kravaris, C. State observer design for monitoring the degree of polymerization in a series of melt polycondensation reactors. *Processes* **2016**, *4*, 4. [CrossRef]

44. Kahelras, M.; Ahmed-Ali, T.; Giri, F.; Lamnabhi-Lagarrigue, F. Sampled-data chain-observer design for a class of delayed nonlinear systems. *Int. J. Control* **2018**, *91*, 1076–1090. [CrossRef]

45. Ahmed-Ali, T.; Karafyllis, I.; Lamnabhi-Lagarrigue, F. Global exponential sampled-data observers for nonlinear systems with delayed measurements. *Syst. Control Lett.* **2013**, *62*, 539–549. [CrossRef]

46. Nadri, M.; Hammouri, H. Design of a continuous-discrete observer for state affine systems. *Appl. Math. Lett.* **2003**, *16*, 967–974. [CrossRef]

47. Karafyllis, I.; Kravaris, C. From continuous-time design to sampled-data design of observers. *IEEE Trans. Autom. Control* **2009**, *54*, 2169–2174. [CrossRef]

48. Zambare, N.; Soroush, M.; Ogunnaike, B.A. A method of robust multi-rate state estimation. *J. Process Control* **2003**, *13*, 337–355. [CrossRef]

49. Antoniades, C.; Christofides, P.D. Feedback control of nonlinear differential difference equation systems. *Chem. Eng. Sci.* **1999**, *54*, 5677–5709. [CrossRef]

50. Liu, J.; Muñoz de la Peña, D.; Christofides, P.D.; Davis, J.F. Lyapunov-based model predictive control of nonlinear systems subject to time-varying measurement delays. *Int. J. Adapt. Control Signal Process.* **2009**, *23*, 788–807. [CrossRef]

51. Liu, J.; Muñoz de la Peña, D.; Christofides, P.D. Distributed model predictive control of nonlinear systems subject to asynchronous and delayed measurements. *Automatica* **2010**, *46*, 52–61. [CrossRef]

52. Ling, C.; Kravaris, C. Multi-rate observer design using asynchronous inter-sample output predictions. In Proceedings of the 2017 American Control Conference (ACC), Seattle, WA, USA, 24–26 May 2017; pp. 376–381.

53. Ling, C.; Kravaris, C. Multi-rate observer design for process monitoring using asynchronous inter-sample output predictions. *AIChE J.* **2017**, *63*, 3384–3394. [CrossRef]

54. Ling, C.; Kravaris, C. Multi-rate sampled-data observers based on a continuous-time design. In Proceedings of the 2017 IEEE 56th Annual Conference on Decision and Control (CDC), Melbourne, Australia, 12–15 December 2017; pp. 3664–3669.

55. Ling, C.; Kravaris, C. Multi-rate Sampled-data Observer Design for Nonlinear Systems with Asynchronous and Delayed Measurements. In Proceedings of the 2019 American Control Conference (ACC), Philadelphia, PA, USA, 10–12 July 2019; pp. 1128–1133.

56. Ling, C.; Kravaris, C. Multi-Rate Sampled-Data Observer Design Based on a Continuous-Time Design. *IEEE Trans. Autom. Control.* **2019**. [CrossRef]

57. Ling, C.; Kravaris, C. A dead time compensation approach for multirate observer design with large measurement delays. *AIChE J.* **2019**, *65*, 562–570. [CrossRef]

58. Buback, M.; Kowollik, C. Termination kinetics of methyl methacrylate free-radical polymerization studied by time-resolved pulsed laser experiments. *Macromolecules* **1998**, *31*, 3211–3215. [CrossRef]

59. Beuermann, S.; Buback, M.; Davis, T.P.; Gilbert, R.G.; Hutchinson, R.A.; Olaj, O.F.; Russell, G.T.; Schweer, J.; Van Herk, A.M. Critically evaluated rate coefficients for free-radical polymerization, 2. Propagation rate coefficients for methyl methacrylate. *Macromol. Chem. Phys.* **1997**, *198*, 1545–1560. [CrossRef]

60. Moad, G.; Rizzardo, E.; Thang, S.H. Radical addition–fragmentation chemistry in polymer synthesis. *Polymer* **2008**, *49*, 1079–1131. [CrossRef]

61. Perrier, S.; Barner-Kowollik, C.; Quinn, J.F.; Vana, P.; Davis, T.P. Origin of inhibition effects in the reversible addition fragmentation chain transfer (RAFT) polymerization of methyl acrylate. *Macromolecules* **2002**, *35*, 8300–8306. [CrossRef]

62. MARTEN, F.L.; HAMIELEC, A.E. High Conversion Diffusion-Controlled Polymerization. In *Polymerization Reactors and Processes*; ACS Symposium Series; American Chemical Society: Akron, OH, USA, 1978; Volume 104, pp. 3–43.

63. Hiorns, R. Polymer Handbook. *Polym. Int.* **2000**, *49*, 807. [CrossRef]

64. Soroush, M. Nonlinear state-observer design with application to reactors. *Chem. Eng. Sci.* **1997**, *52*, 387–404. [CrossRef]

 *processes*

Article

Data-Driven Estimation of Significant Kinetic Parameters Applied to the Synthesis of Polyolefins

Santiago D. Salas [1,2]**, Amanda L. T. Brandão** [3]**, João B. P. Soares** [4] **and José A. Romagnoli** [1,]*****

[1]   Department of Chemical Engineering, Louisiana State University, Baton Rouge, LA 70809, USA;
     ssalas3@lsu.edu

[2]   Escuela Superior Politécnica del Litoral, ESPOL, Facultad de Ciencias Naturales y Matemáticas,
     Campus Gustavo Galindo Km. 30.5 Vía Perimetral, P.O. Box 09-01-5863, Guayaquil, Ecuador

[3]   Engenharia Química, Pontifícia Universidade Católica do Rio de Janeiro, Rio de Janeiro 22451-900, Brazil;
     amanda.lemette@puc-rio.br

[4]   Department of Chemical and Materials Engineering, University of Alberta, Edmonton, AB T6G 1H9, Canada;
     jsoares@ualberta.ca

*   Correspondence: jose@lsu.edu; Tel.: +1-225-578-1377

Received: 23 April 2019; Accepted: 16 May 2019; Published: 22 May 2019

Abstract: A data-driven strategy for the online estimation of important kinetic parameters was assessed for the copolymerization of ethylene with 1,9-decadiene using a metallocene catalyst at different diene concentrations and reaction temperatures. An initial global sensitivity analysis selected the significant kinetic parameters of the system. The retrospective cost model refinement (RCMR) algorithm was adapted and implemented to estimate the significant kinetic parameters of the model in real time. After verifying stability and robustness, experimental data validated the algorithm performance. Results demonstrate the estimated kinetic parameters converge close to theoretical values without requiring prior knowledge of the polymerization model and the original kinetic values.

Keywords: data-driven parameter estimation; retrospective cost model refinement algorithm; global sensitivity analysis; polyolefin synthesis

---

1. Introduction

Polyolefins, mainly polypropylene and polyethylene, are the most common plastics worldwide. The annual growth rate projected for such materials is estimated to be around 3–5% in the next decade [1], which makes polyolefins a continuously growing and attractive product. Metallocene catalysts such as Dow Chemical's constrained-geometry catalyst (CGC), produce polyolefins with narrow molecular weight distributions (MWD), while allowing the easy addition of α-olefins, dienes, and macromonomers into the growing chains [2]. The incorporation of macromonomers generates copolymers with long-chain branches (LCB), which, besides enhancing physical and mechanical properties, improves the processability of the final plastic materials [3–5]. The reaction pathways that lead to the formation of LCBs in ethylene/α-olefins/diene copolymers are complex. Various experimental investigations have studied their polymerization kinetics [6–8], while others have focused on the development of mechanistic models to explain their microstructures and to predict properties of interest [9–12].

Brandão et al. (2017) [13] proposed a mechanistic model for the semi-batch copolymerization of ethylene and 1,9-decadiene with a metallocene catalyst, which was validated using experimental measurements including the ethylene flow rate (F_M), the number-average molecular weight (M_n), and the weight-average molecular weight (M_w). The model assumed that LCBs were formed by incorporating macromonomers through pendant unsaturations resulting from the copolymerization of 1,9-decadiene. In addition, two methodologies to calculate the polyolefin MWD including the adaptive orthogonal collocation method and Monte-Carlo simulation, were compared [14]. Both methodologies could describe the MWD of polymers made under different experimental conditions, such as reaction temperature and catalyst concentration. For verification, the computed distributions were contrasted with experimental measurements obtained from a high-temperature gel permeation chromatography (GPC) [15].

Even though fundamental models and experimental measurements provide important information to understand the dynamic evolution of a system, neither of them is exact. In real operations, experimental uncertainties may affect previous validations and prevent the model to predict relevant properties accurately. Nonlinear state or parameter estimation strategies, on the other hand, are elegant ways to combine both the experimental data and complex mathematical models [16], which may improve the description of the polymerization system under study. Online estimation techniques are powerful computational tools that can generate immediate knowledge of a particular system of interest, enhance the control actions, and monitor continuously relevant properties [17].

Recent applications of nonlinear estimation methods in polymerization have focused on the reconstruction of the state vector [17–22] rather than on the estimation of polymerization kinetic parameters. Some of the main aims when using state estimators include noise reduction for control purposes, and the prediction of relevant polymer properties such as the MWD, M_n and M_w. Typically, the kinetic parameters of a polymerization model must be estimated before the model can be utilized. The maximum likelihood or the least squares approach are common criteria that aim to match a semi-empirical or fundamental model with the available experimental data. In this context, common strategies include the use of computational packages such as gEST, available in the platform gPROMS [23,24], or metaheuristic/machine-learning algorithms, such as particle swarm optimization or the differential evolution algorithm [13,25,26]. These methods, however, achieve a general description of the model, without considering process disturbances, impurities, experimental errors, and side reactions that may occur during the polymerization.

Few authors have studied algorithms for the online estimation of kinetic parameters in polymerization processes. An initial study in this topic was introduced by Sirohi and Choi (1996) [27]. They implemented an extended Kalman (EKF) filter to estimate kinetic parameters and heat transfer coefficients using a computational experiment. Li et al. (2004) [28] employed an EKF for the simultaneous estimation of states and parameters in a continuous reactor for the ethylene-propylene-diene polymerization. Chen et al. (2005) [29] investigated a particle filter strategy in batch polymerization for joint state and parameter estimation. Finally, Sheibat-Othman et al. (2008) [30] compared different online parameter estimation strategies, including the minimization-based approach, EKF, high gain, and adaptive observer. The results were evaluated qualitatively for the solution homopolymerization of acrylic acid using measurements from infrared spectroscopy. However, all of these methods required an adjoint model, or relied on the explicit knowledge of the parameter dependences, which implied that the mathematical model had to be computable and known by the estimation algorithm (e.g., the EKF requires the Jacobian of the states dynamics). In addition, initial information on the parameters was required; otherwise, the algorithm failed to converge [31].

In this contribution, a data-driven online parameter estimation strategy was assessed for the copolymerization of ethylene with 1,9-decadiene using a metallocene catalyst in a semi-batch reactor. The initial phase corresponded to the selection of the significant kinetic parameters of the polymerization model. A global sensitivity analysis that used an improved version of the Sobol method permitted the identification of the most important kinetic parameters of the system [32]. Subsequently, the retrospective cost model refinement (RCMR) algorithm [33–35], a data-driven method that does not require knowledge of the nonlinear model and the initial values of the estimated parameters, was adapted and implemented for the online estimation of the significant kinetic parameters. Different channels allowed the reconstruction of each parameter of interest towards its real value. Finally, the RCMR algorithm was tested and validated with experimental data to verify its applicability.

2. Process Modelling

Fundamental models in polymerization are advantageous tools that contribute to product development and process troubleshooting. The mathematical model described in this work was proposed by Brandão et al. (2017) [13,14]. The reaction mechanism adopted for the copolymerization of ethylene with 1,9-decadiene used the catalyst dimethylsilyl (*N*-tert-butylamido) (tetramethylcyclopentadienyl) titanium dichloride (CGC)/MAO. The reaction mechanism is described as follows:

Catalyst activation

$$C \xrightarrow{k_a} C^* \tag{1}$$

Initiation

$$C^* + M \xrightarrow{k_{p_{11}}} P_1^* \; (+m) \tag{2}$$

Propagation (ethylene)

$$P_q^* + M \xrightarrow{k_{p_{11}}} P_{q+1}^* \; (+m) \tag{3}$$

Propagation (diene)

$$P_q^* + D \xrightarrow{k_{p_{12}}} P_{q+1}^* (+d) \tag{4}$$

Transfer to monomer and β-hydride elimination

$$P_q^* \xrightarrow{k_t} L_q^= + C^* \tag{5}$$

Living chain deactivation

$$P_q^* + P_r^* \xrightarrow{k_{dP}} L_q + L_r + 2\,DC \tag{6}$$

Macromonomer reincorporation

$$P_q^* + L_r^= \xrightarrow{k_b K} P_{q+r}^* (+2\,lcb) \tag{7}$$

where C is the catalyst precursor, C^* is an active catalyst site, M is the ethylene, D is the diene, DC is a dead catalyst site, m and d are the total number of ethylene and diene units inserted into the growing polymer chains, P_q^* is a living polymer chain of size q, $L_q^=$ is a dead polymer chain of size q containing a terminal unsaturation, and L_q is a dead polymer chain of size q without a terminal unsaturation.

Under assumptions such as constant ethylene concentration, excess co-catalyst concentration, well-mixed reactor, initiation rate equal to propagation rate for ethylene ($k_{p_{11}}$), and propagation controlled by the chemical nature of the monomer species, the set of differential equations that describe the system are shown in Equations (8)–(21).

$$\frac{dC}{dt} = -k_a\left(\frac{C}{V}\right)V \tag{8}$$

$$\frac{dC^*}{dt} = k_a\left(\frac{C}{V}\right)V - k_{p_{11}}\left(\frac{C^*}{V}\right)\left(\frac{M}{V}\right)V + k_t\left(\frac{\mu_0}{V}\right)V \tag{9}$$

$$\frac{dDC}{dt} = k_{dP}\left(\frac{\mu_0}{V}\right)^2 V \tag{10}$$

$$\frac{dD}{dt} = -k_{p_{12}}\left(\frac{\mu_0}{V}\right)\left(\frac{D}{V}\right)V \tag{11}$$

$$\frac{d(m)}{dt} = k_{p_{11}}\left[\left(\frac{C^*}{V}\right) + \left(\frac{\mu_0}{V}\right)\right]\left(\frac{M}{V}\right)V \tag{12}$$

$$\frac{d(d)}{dt} = k_{p_{12}}\left(\frac{\mu_0}{V}\right)\left(\frac{D}{V}\right)V \tag{13}$$

$$\frac{d(lcb)}{dt} = 2k_b\varphi\left(\frac{\mu_0}{V}\right)\left(\frac{\lambda_1}{V}\right)V \tag{14}$$

$$\frac{d\mu_0}{dt} = -k_t\left(\frac{\mu_0}{V}\right)V - k_{dP}\left(\frac{\mu_0}{V}\right)^2 V + k_{p_{11}}\left(\frac{C^*}{V}\right)\left(\frac{M}{V}\right)V \tag{15}$$

$$\frac{d\mu_1}{dt} = -k_t\left(\frac{\mu_1}{V}\right)V - k_{dP}\left(\frac{\mu_0}{V}\right)\left(\frac{\mu_1}{V}\right)V + k_{p_{11}}\left[\left(\frac{C^*}{V}\right) + \left(\frac{\mu_0}{V}\right)\right]\left(\frac{M}{V}\right)V + k_{p_{12}}\left(\frac{\mu_0}{V}\right)\left(\frac{D}{V}\right)V \\ + k_b\varphi\left(\frac{\mu_0}{V}\right)\left(\frac{\lambda_2}{V}\right)V \tag{16}$$

$$\frac{d\mu_2}{dt} = -k_t\left(\frac{\mu_2}{V}\right)V - k_{dP}\left(\frac{\mu_0}{V}\right)\left(\frac{\mu_2}{V}\right)V + k_{p_{11}}\left[\left(\frac{C^*}{V}\right) + \left(\frac{\mu_0}{V}\right) + 2\left(\frac{\mu_1}{V}\right)\right]\left(\frac{M}{V}\right)V \\ + k_{p_{12}}\left[\left(\frac{\mu_0}{V}\right) + 2\left(\frac{\mu_1}{V}\right)\right]\left(\frac{D}{V}\right)V + k_b\varphi\left[\left(\frac{\mu_0}{V}\right)\left(\frac{\lambda_3}{V}\right) + 2\left(\frac{\mu_1}{V}\right)\left(\frac{\lambda_2}{V}\right)\right]V \tag{17}$$

$$\frac{d\lambda_0}{dt} = k_t\left(\frac{\mu_0}{V}\right)V + k_{dP}\left(\frac{\mu_0}{V}\right)^2 V - k_b\varphi\left(\frac{\mu_0}{V}\right)\left(\frac{\lambda_1}{V}\right)V \tag{18}$$

$$\frac{d\lambda_1}{dt} = k_t\left(\frac{\mu_1}{V}\right)V + k_{dP}\left(\frac{\mu_0}{V}\right)\left(\frac{\mu_1}{V}\right)V - k_b\varphi\left(\frac{\mu_0}{V}\right)\left(\frac{\lambda_2}{V}\right)V \tag{19}$$

$$\frac{d\lambda_2}{dt} = k_t\left(\frac{\mu_2}{V}\right)V + k_{dP}\left(\frac{\mu_0}{V}\right)\left(\frac{\mu_2}{V}\right)V - k_b\varphi\left(\frac{\mu_0}{V}\right)\left(\frac{\lambda_3}{V}\right)V \tag{20}$$

$$\frac{dV}{dt} = \left(\frac{d\lambda_1}{dt} + \frac{d\mu_1}{dt}\right)\frac{MM}{\rho_{PE}} \tag{21}$$

where,

$$\varphi = \frac{d}{d+m} \tag{22}$$

$$\lambda_3 = \frac{\lambda_2}{\lambda_0\lambda_1}\left(2\lambda_2\lambda_0 - \lambda_1{}^2\right) \tag{23}$$

$$MM = \varphi MM_D + (1 - \varphi)MM_M \tag{24}$$

When the polymerization mechanism leads to moment closure problems, the qth-moment balance equation requires the definition of the $(q+1)^{th}$ moments. Otherwise, the balance equation cannot be solved. Hulburt and Katz (1964) [36] developed a closure method that can be written in the form of algebraic expressions, using a distribution approximation procedure. The closure expression for λ_3 was then obtained as approximate algebraic equation in the form of Equation (23).

The concentration of ethylene was kept constant during the polymerization, $\frac{dM}{dt} = 0$. Thus, the inlet flow rate of ethylene (F_M), which represents the continuous demand of ethylene during the polymerization, was approximated to the expected demand of monomer during the reaction, as denoted in Equation (25).

$$F_M \approx k_{p_{11}}\left(\frac{C^*}{V}\right)\left(\frac{M}{V}\right)V + k_{p_{11}}\left(\frac{\mu_0}{V}\right)\left(\frac{M}{V}\right)V \tag{25}$$

The average properties of the resultant polymers were computed as written in Equations (26)–(28).

$$M_n = \frac{\lambda_1 + \mu_1}{\lambda_0 + \mu_0}MM \tag{26}$$

$$M_w = \frac{\lambda_2 + \mu_2}{\lambda_1 + \mu_1} MM \tag{27}$$

$$PDI = \frac{M_w}{M_n} = \frac{\lambda_2 + \mu_2}{\lambda_0 + \mu_0} \tag{28}$$

The fundamental model relied on both the reparametrized and the classical Arrhenius law to compute the rate constants as listed in Table 1.

Table 1. Kinetic rate constants for the copolymerization of ethylene and 1,9-decadiene.

Rate Constant	Arrhenius Equation
Catalyst activation	$k_a = exp\left[k_1 + k_2\left(\frac{T-T_r}{T}\right)\right]$
Propagation	$k_{p_{11}} = k_{0p} exp\left(-\frac{E_{ap}}{RT}\right), \quad k_{0p} = 10^{k_7}$
Monomer transfer & β-hydride elimination	$k_t = exp\left[k_3 + k_4\left(\frac{T-T_r}{T}\right)\right]$
Living chain deactivation	$k_{dP} = exp\left[k_5 + k_6\left(\frac{T-T_r}{T}\right)\right]$

T: Temperature inside the reactor, T_r : reference temperature set to 130 °C.

The molar concentration and total amount of monomer, listed in Table 2, were obtained using the Peng-Robinson equation to calculate fugacity, and UNIQUAC model to determine the activity coefficients in the liquid phase.

Table 2. Ethylene concentration and total moles of ethylene in toluene at different temperatures.

T, [C°]	$[C_2H_4]$, [mol L^{-1}]	$[M]$, [mol]
120	0.49472	0.07420
130	0.43732	0.06560
140	0.38141	0.05721

The parameters k_{1-7} had their values determined stochastically by using particle swarm optimization (PSO) [37] in the homopolymerization experiments. After this procedure, the identifiability analysis indicated that only four parameters (k_3, k_5, k_6 and k_7) could be estimated simultaneously. The remaining parameters (k_1, k_2 and k_4), although important for model computations, could not be estimated independently with the available data; therefore, their values were kept constant and equal to the values provided by the PSO. Then, the four selected parameters were estimated using the computational package ESTIMA [25]. The experimental data used to estimate the parameters included the average properties M_n and M_w, and the ethylene feed rates (F_M), which was the only continuous measurement. The copolymerization kinetic parameters $k_{p_{12}}$ and k_b were estimated by ESTIMA considering the experimental data of M_n and M_w only.

Table 3 lists the parameters of the system, including the kinetic parameters in the reparametrized Arrhenius equations, pre-exponential constants, activation energies, as well as other relevant thermodynamic properties and constants. It is important to remind the reader that the kinetic parameters (k_{1-7}) are needed to describe the actual rate constants. Table 4 provides the initial conditions considered in the current investigation. The interested reader is encouraged to consult the original publication for more details [13]. A complete explanation of the variables and kinetic parameters is included in the Nomenclature.

Table 3. Parameters of the copolymerization of ethylene with 1,9-decadiene using dimethylsilyl (*N*-tert-butylamido) (tetramethylcyclopentadienyl) titanium dichloride (CGC)/MAO.

Parameter	Value	Units	Original Rate Constants [1]
k_1	-2.92	-	$k_a = \alpha(k_1, k_2)$
k_2	25.00	-	
k_3	2.58 ± 0.08	-	$k_t = \alpha(k_3, k_4)$
k_4	17.2	-	
k_5	10.91 ± 0.95	-	$k_{dP} = \alpha(k_5, k_6)$
k_6	30.12 ± 2.10	-	
k_7	7.56 ± 0.07	-	$k_{p_{11}} = \alpha(k_7)$
$k_{p_{12}}$	2039.8 ± 54.7	$\mathrm{L\ mol^{-1}s^{-1}}$	-
k_b	908.7 ± 69.0	$\mathrm{L\ mol^{-1}s^{-1}}$	-
E_{ap}	20520.0	$\mathrm{J\ mol^{-1}}$	-
MM_M	28.05	$\mathrm{g\ mol^{-1}}$	-
MM_D	138.254	$\mathrm{g\ mol^{-1}}$	-
ρ_{PE}	940	$\mathrm{g\ L^{-1}}$	-
R	8.31451	$\mathrm{J\ mol^{-1}\ K^{-1}}$	-

[1] $k_i = \alpha(k_\beta, k_\gamma)$ represents k_i as a function of k_β and k_γ.

Table 4. Initial conditions for simulations and experiments.

Variable	Homopolymerization	Copolymerization		Units
		A	B	
$C(0)$	$0.767 \times 10^{-6} \times V(0)$	$0.271 \times 10^{-6} \times V(0)$	$0.271 \times 10^{-6} \times V(0)$	mol
$C^*(0)$	0	0	0	mol
$DC(0)$	0	0	0	mol
$D(0)$	0	$0.3 \div MM_D$	$0.4 \div MM_D$	mol
$(M)(0)$	0	0	0	mol
$(d)(0)$	0	0	0	mol
$(lcb)(0)$	0	0	0	mol
$\mu_0(0)$	0	0	0	mol
$\mu_1(0)$	0	0	0	mol
$\mu_2(0)$	0	0	0	mol
$\lambda_0(0)$	10^{-14}	10^{-14}	10^{-14}	mol
$\lambda_1(0)$	10^{-14}	10^{-14}	10^{-14}	mol
$\lambda_2(0)$	10^{-14}	10^{-14}	10^{-14}	mol
$V(0)$	0.15	0.15	0.15	L

3. Experimental Equipment and Setup

3.1. Materials

The materials utilized in the experimental runs were methylaluminoxane (MAO, 10 wt. % in toluene), anhydrous ethyl alcohol ($\geq$99.5%), toluene anhydrous (99.8%), 1,9-Decadiene (98%), triisobutylaluminum (TIBA) (25 wt. % in toluene), n-butyllithium solution (2.5 M in hexane), sodium ($\geq$99%, stored in mineral oil), which were provided by Sigma-Aldrich (USA). Moreover, dimethylsily (n-tert-butylamido) (tetramethylcyclopentadienyl) titanium dichloride (CGC) (85.0–99.8%) was acquired from Boulder Scientific Company (USA), and nitrogen (>99.998%) and ethylene were provided by Praxair Technology (USA).

3.2. Process Description

Prior polymerization, six cycles of nitrogen venting and vacuuming at 125 °C were applied in the reactor to remove oxygen. Then, the reactor received 150 mL of toluene and 0.5 g of TIBA (impurity scavenger), and the temperature was increased to 120 °C and kept constant for 20 min.

For homopolymerization, after the reactor purging, 150 mL of toluene was charged at ambient temperature. A solution of MAO was added into the reactor through a cannula under nitrogen pressure. The reactor was then heated until reaching the reaction temperature (120, 130 or 140 °C). Then, ethylene was injected into the reactor until the solvent was saturated. After stabilizing the temperature, the catalyst solution was added into the reactor under nitrogen pressure. During polymerization, the reactor temperature remained constant, with variations of ±0.15 °C from the set point. Ethylene was supplied on demand, maintaining a constant reactor pressure (120 psig). When the final reaction time was achieved, the ethylene supply valve was closed, and the reactor contents were immediately transported into a 1 L beaker with 100–250 mL of ethanol. Afterwards, the polymer was kept overnight under constant stirring, then filtered and dried in an oven. The copolymerization procedure was analogous to the homopolymerization procedure. The unique difference is that after adding MAO, the co-monomer solution was injected into the reactor following the same procedure used to feed MAO.

The average properties and the molecular weight distributions of the polymer samples were measured using a Polymer Char High-Temperature Gel Permeation Chromatographer (GPC) calibrated with polystyrene narrow standards and using a universal calibration curve in accordance with the methodology described by Soares and McKenna (2013) [15].

4. Data-Driven Estimation of Significant Kinetic Parameters

4.1. Parameter Selection: Global Sensitivity Analysis

A global sensitivity analysis shows how significant inputs are with respect to one or various outputs. A robust and widely used variance-based sensitivity analysis is the Sobol method [38]. This method proposes the expansion of a function $G = g(z_1, \ldots, z_j, \ldots, z_J)$ into terms of increasing dimensions with mutually independent input parameters such that all summands are mutually orthogonal, as explained in Equation (29).

$$G = g_0 + \sum_{j=1}^{J} g_j(z_j) + \sum_{1 \leq j < b \leq J} g_{jb}(z_j, z_b) + \cdots + g_{1,2,\cdots,J}(z_1, \ldots, z_J) \tag{29}$$

where the index j denotes a parameter of interest, b another parameter, and J is the total number of evaluated parameters. Each term in Equation (29) has quadratic integrability over the domain of existence, where g_0 is a constant, $g_j = g_j(z_j)$, $g_{jb} = g_{jb}(z_j, z_b)$, and so forth. Equation (30) shows the decomposition of the variance of G.

$$V(G) = \sum_{j=1}^{J} V_j + \sum_{1 \leq j < b \leq J} V_{jb} + \ldots + V_{1,\ldots,j,\ldots,J} \tag{30}$$

where $V_j, V_{jb}, V_{1,\cdots,j,\cdots J}$ are the individual variances of functions $g_j, g_{jb}, g_{1,\cdots,j,\cdots,J}$.

Sensitivity indices help understand the variance decomposition from Equation (30). First-order sensitivity indices $(\hat{S}_j)$ permit the selection and classification of the most sensitive parameters, depending on the individual importance of their contribution in changing the variance of the function of interest. The main effect of varying parameter z_j on the output value G is measured by $\hat{S}_j$, as presented in Equation (31). In addition, the total sensitivity index $(\hat{S}_{Tj})$ incorporates the sum of all the effects that involve the parameter z_j. The total sensitivity index for parameter z_j is computed as indicated in Equation (32).

$$\hat{S}_j = \frac{\hat{V}_j}{\hat{V}} \tag{31}$$

$$\hat{S}_{Tj} = 1 - \frac{\hat{V}_{-j}}{\hat{V}} \tag{32}$$

where $\hat{V}_{-j}$ is the sum of all variance terms that exclude z_j.

$\hat{S}_j$ and $\hat{S}_{Tj}$ can be compared to evaluate whether a model is additive or not. For non-additive models, $\hat{S}_j < \hat{S}_{Tj}$; for additive models, $\hat{S}_j = \hat{S}_{Tj}$. Additive models are those in which no interactions between evaluated parameters occur [39].

The Sobol standard method may be improved by introducing sampling and resampling matrices [40,41], and even better performance is achieved when the results of the evaluated functions are averaged, creating extra data points [32]. In this study, we used an improved version of Sobol's method, as implemented by Salas et al. (2017) [42], including a third sampling matrix to avoid unfeasible scenarios. The method follows the steps below:

(1) Define an objective function, and the dimension (D) for a sample of input parameters. For each parameter, define an uncertainty index. In this case, we adopted 4% of change with respect to the mean value.

(2) Build three random matrices M_1, M_2 and M_3—Equations (33a)–(33c), respectively—of dimension $D \times J$ based on the defined uncertainty: M_1 is the sampling matrix, M_2 is the resampling matrix, and M_3 is the backup matrix.

$$M_1 = \begin{bmatrix} z_{11} \cdots & z_{1j} \cdots & z_{1J} \\ \vdots & \vdots & \vdots \\ z_{D1} \cdots & z_{Dj} \cdots & z_{DJ} \end{bmatrix} \tag{33a}$$

$$M_2 = \begin{bmatrix} z'_{11} \cdots & z'_{1j} \cdots & z'_{1J} \\ \vdots & \vdots & \vdots \\ z'_{D1} \cdots & z'_{Dj} \cdots & z'_{DJ} \end{bmatrix} \tag{33b}$$

$$M_3 = \begin{bmatrix} z''_{11} \cdots & z''_{1j} \cdots & z''_{1J} \\ \vdots & \vdots & \vdots \\ z''_{D1} \cdots & z''_{Dj} \cdots & z''_{DJ} \end{bmatrix} \tag{33c}$$

(3) Evaluate the row vectors of matrices M_1 and M_2. If the output is unfeasible, meaning that the combination of inputs in a vector caused the simulation to break or other related problems, use the next available feasible row of the matrix M_3, and update the matrices to $M_1{'}$ and $M_2{'}$, which denote the improved sampling and resampling matrices, respectively. Then, calculate the total average ($\hat{g}_0$) of both evaluations as described in Equation (34).

$$g_S = g(M_1{'}), \qquad g_R = g(M_2{'})$$

$$\hat{g}_0 = \frac{1}{2D} \sum_{d=1}^{D} \left(g_S + g_R \right) \tag{34}$$

where g_S represents the output vector of $M_1{'}$ and g_R is the output vector of $M_2{'}$.

(4) Generate a matrix N_q formed by all columns of matrix $M_2{'}$, except the column of the z_q parameter, which is pulled from $M_1{'}$, as explained in Equation (35a). Subsequently, generate another matrix N_{Tj} formed with all columns of $M_1{'}$ and with the column of the z'_j parameter, pulled from $M_2{'}$ as denoted in Equation (35b).

$$N_j = \begin{bmatrix} z'_{11} \cdots & z_{1j} \cdots & z'_{1J} \\ \vdots & \vdots & \vdots \\ z'_{D1} \cdots & z_{Dj} \cdots & z'_{DJ} \end{bmatrix} \tag{35a}$$

$$N_{Tj} = \begin{bmatrix} z_{11} \cdots & z'_{1j} \cdots & z_{1J} \\ \vdots & \vdots & \vdots \\ z_{D1} \cdots & z'_{Dj} \cdots & z_{DJ} \end{bmatrix} \tag{35b}$$

(5) Evaluate the row vectors of matrices N_j and N_{Tj}. If an evaluated function is unfeasible, the output is replaced by $\hat{g}_0$. The outputs are obtained in column vectors.

$$g'_j = g(N_j), \quad g'_{Rj} = g(N_{Tj})$$

where g'_j is the output vector of matrix N_j, and g'_{Rj} is the output vector of matrix N_{Tj}.

(6) A sample generates the following estimates, which are calculated based on scalar products of the vectors from above.

$$\gamma_q^2 = \frac{1}{2D} \sum_{d=1}^{D} \left(g_S \cdot g_R + g'_j \cdot g'_{Rj} \right) \tag{36}$$

$$\hat{V} = \frac{1}{2D} \sum_{d=1}^{D} \left(g_S^2 + g_R^2 \right) - \hat{g}_0^2 \tag{37}$$

$$\hat{V}_q = \frac{1}{2D} \sum_{d=1}^{D} \left(g_S \cdot g'_{Rj} + g_R \cdot g'_j \right) - \gamma_j^2 \tag{38}$$

$$\hat{V}_{-q} = \frac{1}{2D} \sum_{d=1}^{D} \left(g_S \cdot g'_j + g_R \cdot g'_{Rj} \right) - \gamma_j^2 \tag{39}$$

where γ_j^2 is the squared mean value of the outputs for each parameter z_j. The selection of sensitive parameters relies on the first and total sensitivity indices. Equation (40) introduces the objective function, defined in this case as:

$$G = \sum_{i=1}^{i} \frac{y_i - h_i(z)}{\sigma_{y_i}^2} \tag{40}$$

where y_i is the measurement at each time interval i, $h_i(z)$ is the calculated measurement, and $\sigma_{y_i}^2$ is the variance of the experimental fluctuations.

4.2. Data-Driven Parameter Estimation

4.2.1. Estimation Problem

The polymerization model described in Equations (8)–(21) can be written in compact discrete-time form as portrayed in Equation (41).

$$\begin{cases} x_{i+1} = f(x_i, u_i, z) + v_i \\ y_i = h(x_i, u_i, z) + w_i \end{cases} \tag{41}$$

where $x \in \mathbb{R}^{l_x}$ is the state vector, $u \in \mathbb{R}^{l_u}$ is the vector of inputs, $z \in \mathbb{R}^{l_\mu}$ is the unknown parameter vector, $y \in \mathbb{R}^{l_y}$ is the vector of measurements, and $v \in \mathbb{R}^{l_x}$ and $w \in \mathbb{R}^{l_y}$ are the model and measurement errors, respectively.

For estimation purposes, the compact model is considered to be as shown in Equation (42).

$$\begin{cases} \hat{x}_{i+1} = f(\hat{x}_i, u_i, \hat{z}) + v_i \\ \hat{y}_i = h(\hat{x}_i, u_i, \hat{z}) + w_i \end{cases} \tag{42}$$

where $\hat{x}$ is the estimated state vector, $\hat{y}$ is the vector of estimated measurements, and $\hat{z}$ is the output of the parameter estimator. The parameter estimator is updated by minimizing a cost function based on the performance variable (e).

Considering an ARMA model with a built-in integrator, $\hat{z}$ is given by:

$$\hat{z}_i = \sum_{l=1}^{n_c} P_{li}\hat{z}_{i-1} + \sum_{l=1}^{n_c} Q_{li}e_{i-l} + R_i g_i \tag{43}$$

where,

$$g_i = g_{i-1} + e_{i-1} \tag{44a}$$

$$e_i = \hat{y}_i - y_i \tag{44b}$$

and $P_i \in \mathbb{R}^{l_z \times l_z}$, $Q_i, R \in \mathbb{R}^{l_z \times l_y}$ are the coefficient matrices that are updated recursively by the algorithm. The integrator is combined with the estimator to guarantee that the performance variable approaches to zero as the iterations approach to infinity. Rewriting Equation (43), the following is obtained:

$$\hat{z}_i = \Phi_i \theta_i \tag{45}$$

where,

$$\Phi_i = I_{l_z} \otimes \phi_i^T \in \mathbb{R}^{l_z \times l_\theta}, \tag{46a}$$

$$\phi_i = \left[\hat{z}_{i-1}^T \ldots \hat{z}_{i-n_c}^T \ e_{i-1}^T \ldots e_{i-n_c}^T \ g_i^T \right]^T \tag{46b}$$

$$\theta_i = vec\left[P_{1i} \ldots P_{n_c i} \ Q_{1i} \ldots Q_{n_c i} \ R_i \right] \in \mathbb{R}^{l_\theta} \tag{46c}$$

$$l_\theta = l_z^2 n_c + l_z l_y (n_c + 1) \tag{46d}$$

and Φ is the regressor matrix, θ contains the estimator coefficients computed by the RCMR algorithm. The operator "$\otimes$" is the Kronecker product, and vec represents the column-stacking operator.

It is assumed that z is identifiable, which is guaranteed from the global sensitivity analysis (in the absence of an observability/detectability analysis), and that the input signal u_k is persistently exciting.

4.2.2. Retrospective Cost Model Refinement (RCMR) Algorithm

The retrospective performance variable is defined as follows:

$$\hat{e}_i = e_i - G_f(q)\left(\Phi_i \hat{\theta} - \hat{z}_i \right) \tag{47}$$

where q is the forward-shift operator, and $\hat{\theta}$ has the parameter estimation coefficients to be optimized.

$$G_f(q) = \sum_{n=1}^{n_f} \frac{N_n}{q^n} \tag{48}$$

For all n, $N_n \in \mathbb{R}^{l_y \times l_z}$. G_f is a finite impulse response filter of order n_f. Equation (47) is then rewritten as follows:

$$\hat{e}_i = e_i + N\Phi_{bi}\hat{\theta} - NZ_{bi} \tag{49}$$

where,

$$N = \left[N_1 \ldots N_{n_f} \right] \in \mathbb{R}^{l_y \times l_z n_f},$$

$$\Phi_{bi} = \begin{bmatrix} \phi_{i-1} \\ \vdots \\ \phi_{i-n_f} \end{bmatrix} \epsilon \, \mathbb{R}^{l_z n_f \times l_\theta},$$

$$\mathbf{Z}_{bi} = \begin{bmatrix} \hat{z}_{i-1} \\ \vdots \\ \hat{z}_{i-n_f} \end{bmatrix} \epsilon \, \mathbb{R}^{l_z n_f}.$$

The retrospective cost function, defined by Goel and Bernstein (2018) [34,35], is minimized by making use of recursive least squares. Let $\mathbf{P}_0 = \mathbf{R}_\theta^{-1}$ and $\theta_i = 0$.

The algorithm that updates the estimator coefficients is as follows:

$$\theta_i = \theta_{i-1} - \mathbf{P}_i \Phi_{bi-1}{}^\mathrm{T} \mathbf{N}^\mathrm{T} \mathbf{R}_{ei-1}(\mathbf{N}\Phi_{bi-1}\theta_{i-1} + e_{i-1} - \mathbf{N}\mathbf{Z}_{bi-1}) \tag{50}$$

$$\mathbf{P}_i = \lambda^{-1}\mathbf{P}_{i-1} - \lambda^{-1}\mathbf{P}_{i-1}\Phi_{bi-1}{}^\mathrm{T}\mathbf{N}^\mathrm{T}\Gamma_i^{-1}\mathbf{N}\Phi_{bi-1}\mathbf{P}_{i-1} \tag{51}$$

$$\Gamma_i = \lambda\mathbf{R}_{ei-1}{}^{-1} + \mathbf{N}\Phi_{bi-1}\mathbf{P}_{i-1}\Phi_{bi-1}{}^\mathrm{T}\mathbf{N}^\mathrm{T} \tag{52}$$

where $\mathbf{R}_e$ and $\mathbf{R}_\theta$ are positive-definite matrices, and $\lambda \leq 1$ is the forgetting factor.

4.3. Framework Implementation

The assembly of the implemented strategies is summarized in Figure 1. Initially, the global sensitivity analysis provides information on the most important parameters of the polymerization system. Once these parameters are identified, the proposed framework tries to estimate their values asynchronously, updating/estimating the parameters whenever measurements are available. Monitoring and signal processing are other challenges of the proposed methodology. The estimated properties are expected to be close to the experimental and theoretical values, and noise reduction of the measurements is anticipated to occur.

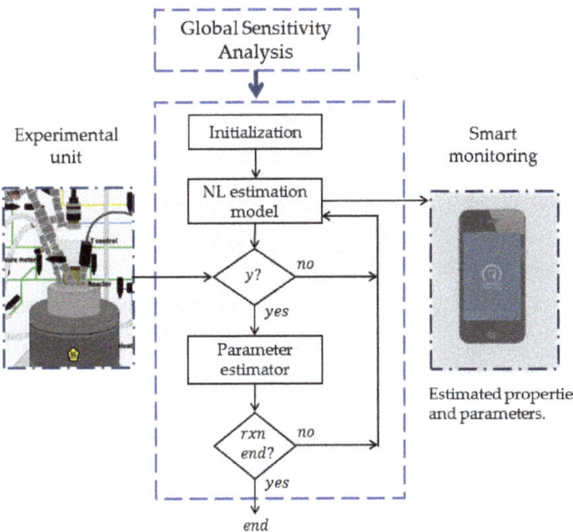

Figure 1. Framework for the online estimation of significant kinetic parameters.

Figure 2 portrays the architecture of the data-driven estimation strategy. The experimental unit generates measurements y, which are assumed to be driven by the inputs u. The data-driven adaptive estimator consists of the nonlinear estimation model, which is also driven by the inputs u, and the RCMR algorithm. Although the nonlinear estimation model is required to generate the estimated measurements $\hat{y}$, it does not provide knowledge for the parameter updates. The estimated parameter $\hat{z}$ is updated by the estimator, which seeks the minimization of the error signal e.

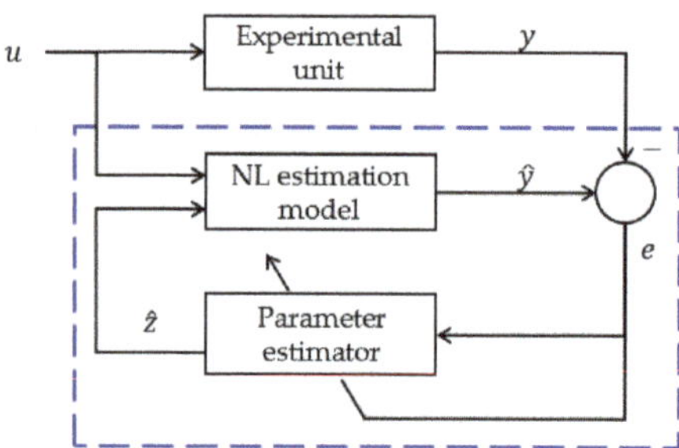

Figure 2. Data-driven parameter estimation architecture.

5. Results

F_M is the only measurement obtained continuously. Thus, the global sensitivity analysis and the data-driven estimation are performed using F_M as the input signal. The other available measurements (M_w, M_n) are employed as a reference for comparing the accuracy of the estimated properties.

The system was simulated in MATLAB R2015a, running on a PC Intel Core™ i7-4790K CPU @ 4.00 GHz with 16.0 Gb of installed RAM. The ODEs of the system were solved using ode23s [43], based on a modified Rosenbrock formula of order 2. The sampling time of F_M was every 1 s, and the RCMR algorithm together with the nonlinear model run in approximately 0.0035 s, which makes feasible the online deployment of the proposed strategy. Computational methods can be used for online applications if they are faster than the real process by a factor of 100 [44].

5.1. Global Sensitivity Analysis

From the global sensitivity analysis of the seven kinetic parameters (k_{1-7}), the fifth and seventh show the highest overall sensitivity, as portrayed in Figure 3. This result is consistent because k_7 is the exponent in the pre-exponential propagation rate constant ($k_{0p} = 10^{k_7}$) of the Arrhenius equation. The propagation rate constant determines the monomer consumption rate; thus, it strongly influences the value of F_M (ethylene flow rate to the reactor), seeking to maintain the ethylene concentration in the reactor constant during the polymerization. The rate constant for the living chain (catalyst) deactivation (k_{dP}) is a function of k_5 (as defined in Table 1). Since k_{dP} influences the moment equations, it is expected that this parameter is influential as well. These results are in agreement with the work of Brandão et al. (2017) [13], because k_5 and k_7 belong to the parameter set classified as significant when an identifiability analysis was applied over the seven parameters. Figure 3a illustrates the first-order, and Figure 3b the total sensitivity indices when using 100 samples.

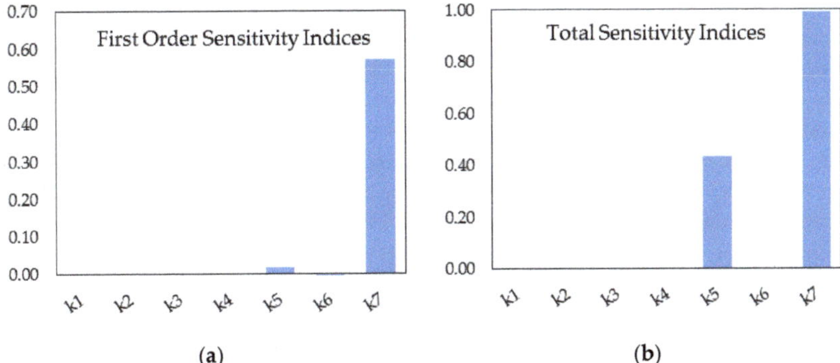

(a)

(b)

Figure 3. Global sensitivity indices: (**a**) first-order sensitivity index; (**b**) total sensitivity index.

5.2. Homopolymerization

Homopolymerization experiments at different reaction temperatures were used to test the RCMR algorithm on its kinetic parameter estimation capabilities. The RCMR algorithm was implemented considering: $\mathbf{P}_0 = \mathbf{R}_\theta{}^{-1}$, $\theta_0 = 0$, and $n_c = 0$.

Initially, only k_7, and consequently $k_{\mathrm{P}_{11}}$, were estimated, using the initial guess $k_7 = 0$. To improve the convergence of the algorithm it was assumed that $l_z = 2$, meaning that two parameters were estimated rather than one. For the non-estimated parameter, a constant value of $\hat{z} = 11.3$ provided a satisfactory response and tradeoff. With these considerations, the architecture selected was: $n_f = 2$ so that $l_\theta = 2$, $\lambda = 0.999$, and

$$G_f(q) = \frac{\begin{bmatrix} 1 & 0 \end{bmatrix}}{q} + \frac{\begin{bmatrix} 0 & 1 \end{bmatrix}}{q^2} \tag{53}$$

For estimating a single parameter, the algorithm considered $\mathbf{R}_\theta = 0.01 I_{l_\theta}$, and $\mathbf{R}_e = 0.1$.

Figure 4 shows the estimates for k_7 and $k_{\mathrm{P}_{11}}$ at different polymerization temperatures. The estimated parameters converge, in all cases, close to theoretical values without requiring prior knowledge of the initial value or range of the parameter.

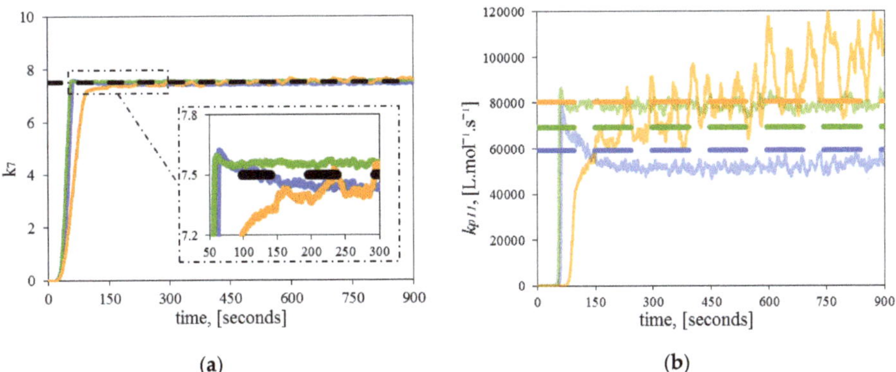

(a)

(b)

Figure 4. Estimation of a single significant parameter. Comparison between theoretical (dashed line), and estimated kinetic parameters (continuous line) at 120 °C (blue), 130 °C (green) 140 °C (orange): (**a**) Dynamic estimation of k_7; (**b**) Dynamic estimation of $k_{\mathrm{p}11}$.

Furthermore, the most significant kinetic parameters of the system, k_5 and k_7, and consequently k_{dP} and $k_{\mathrm{P}_{11}}$, were estimated simultaneously. The same architecture implemented for single parameter estimation was used in the simultaneous estimation, with the distinction that the parameters were

153

estimated in separate channels using the same error signal. The non-estimated parameters were different: $\hat{z}_{k_5} = 23.5$, and $\hat{z}_{k_7} = 15.0$, and everything else remained the same.

Figure 5 shows the results of the estimated k_5, k_{dP}, k_7, and $k_{P_{11}}$ at different polymerization temperatures. In all cases, the estimated parameters (k_5, k_7) approached their theoretical values, converging from an initial value of 0 in both cases. A noisy response was observed, as in Figure 4, which can be attributed to the presence of impurities that could not be removed during the experiments, or to the occurrence of side reactions not included in the fundamental model. In addition, as the polymerization temperature increased, the estimated parameters became more sensitive to noise, which provided the insight that temperature is proportional to the noise/uncertainty of the experimental data.

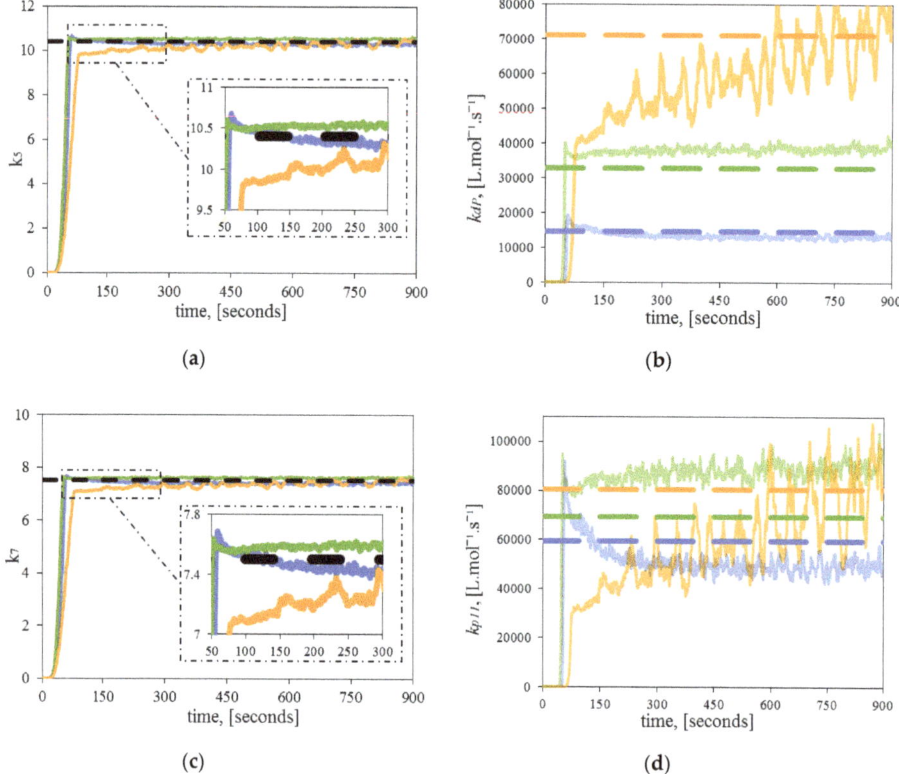

Figure 5. Estimation of two significant parameters. Comparison between theoretical (dashed line), and estimated kinetic parameters (continuous line) at 120 °C (blue), 130 °C (green) 140 °C (orange): (a) Dynamic estimation of k_5; (b) Dynamic estimation of k_{dP}; (c) Dynamic estimation of k_7, (d) Dynamic estimation of k_{p11}.

Figure 6 illustrates how the RCMR can both estimate F_M and reduce the measurement's noise. The reader should note that it takes some time for the estimated F_M to achieve its expected value. Goel and Bernstein (2018) [13] explained that the unknown parameter moves towards different subspaces until it tends to the subspace spanned by $N_1{}^T$. In addition, there is a delay time difference between the estimated $\hat{F}_M$ with a single and two parameters, which is mostly related to the tuning. Finally, as stated before, the $\hat{F}_M$ at the highest temperature (140 °C) shows an oscillatory response.

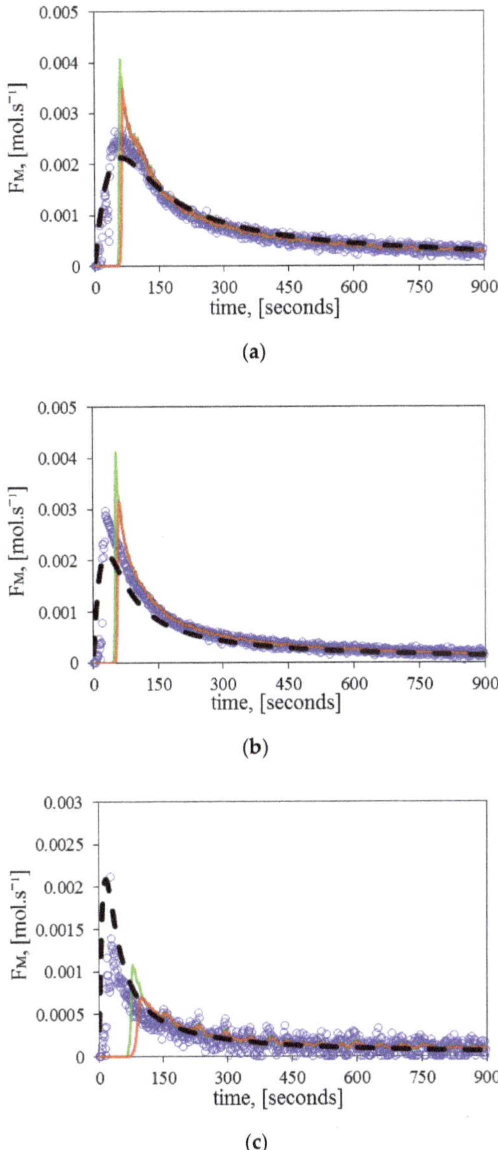

Figure 6. Comparison of the monomer flow rate (F_M) between the experimental values (circles), fundamental model (dashed line), estimated with a single parameter (continuous red line), and estimated with two parameters (continuous green line) at: (**a**) 120 °C, (**b**) 130 °C, (**c**) 140 °C.

Figure 7 compares the results computed by the nonlinear model, the measured results, and the estimated average properties at different polymerization temperatures. The results at 120 °C of both M_w and M_n are very close to their theoretical and experimental values, but as the temperature increases, the uncertainty also increases. The estimation of two parameters simultaneously appears to provoke less reliable results when compared to the estimated properties with a single parameter, especially at higher temperatures.

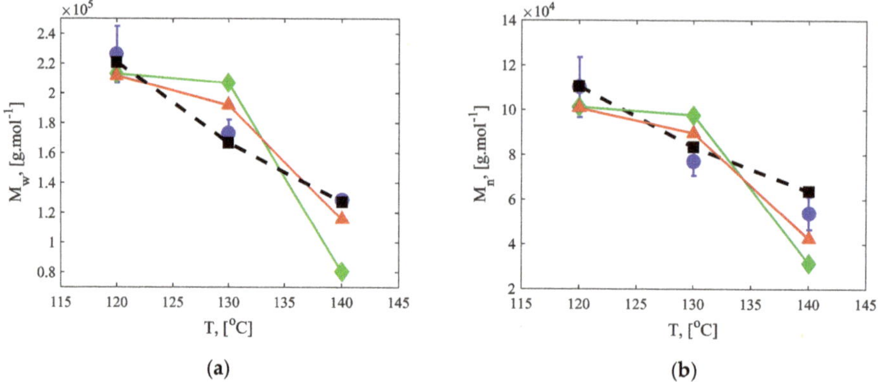

Figure 7. Comparison of the final average properties at different temperatures for experimental values (circles), fundamental model (squares–dashed line), estimated with a single parameter (triangles–continuous line), and estimated with two parameters (diamonds–continuous line): (**a**) weight-average molecular weight (M_w); (**b**) number-average molecular weight (M_n).

5.3. Copolymerization

Following the criterion used for the homopolymerization experiments, the copolymerizations used the same RCMR architecture for the estimation of significant kinetic parameters and important polymer properties. Indeed, the criterion extends to the application of the same channels for estimating one and two significant kinetic parameters and their resulting properties.

The copolymerization experiments considered only one temperature. Initially, k_7 and consequently $k_{P_{11}}$ were estimated for the copolymerizations described in Table 4a,b. Figure 8 illustrates the results of the estimated k_7 and $k_{P_{11}}$ at 120 °C and different initial diene concentrations. The unknown parameter converges towards the theoretical value without requiring prior knowledge at both initial diene concentrations.

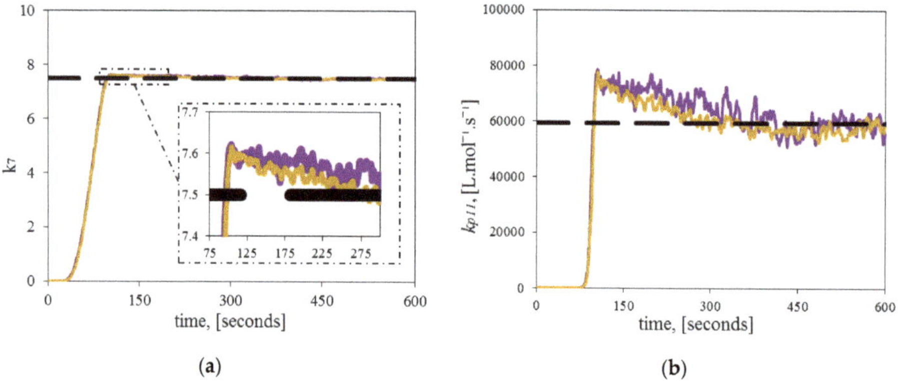

Figure 8. Estimation of a single significant parameter. Comparison between theoretical (dashed line), and estimated kinetic parameters (continuous line) at different initial diene concentrations for the copolymerization experiment A (purple), and copolymerization experiment B (gold); (**a**) Dynamic estimation of k_7; (**b**) Dynamic estimation of k_{p11}.

Moreover, the parameters k_5 and k_7, and, consequently, k_{dP} and $k_{P_{11}}$, were estimated simultaneously, using the same architecture and tuning used in the homopolymerizations. Figure 9 shows the results of the estimated k_5, k_{dP}, k_7, and $k_{P_{11}}$ at different initial diene concentrations at 120 °C. In all cases, the unknown parameters k_5 and k_7 approach the theoretical values, starting from an initial guess of 0. An interesting observation in these experiments is the slight decreasing trend (negative slope) of the parameters, made clearer in the dynamic evolution of k_{dP} and $k_{P_{11}}$. The reason for this behavior might be related to LCBs formation during the copolymerization. The presence of LCBs in the living chains might cause a steric hindrance to the incorporation of ethylene molecules, which disfavors deactivation and propagation reactions. However, this hypothesis must be proved by additional experiments and further simulations.

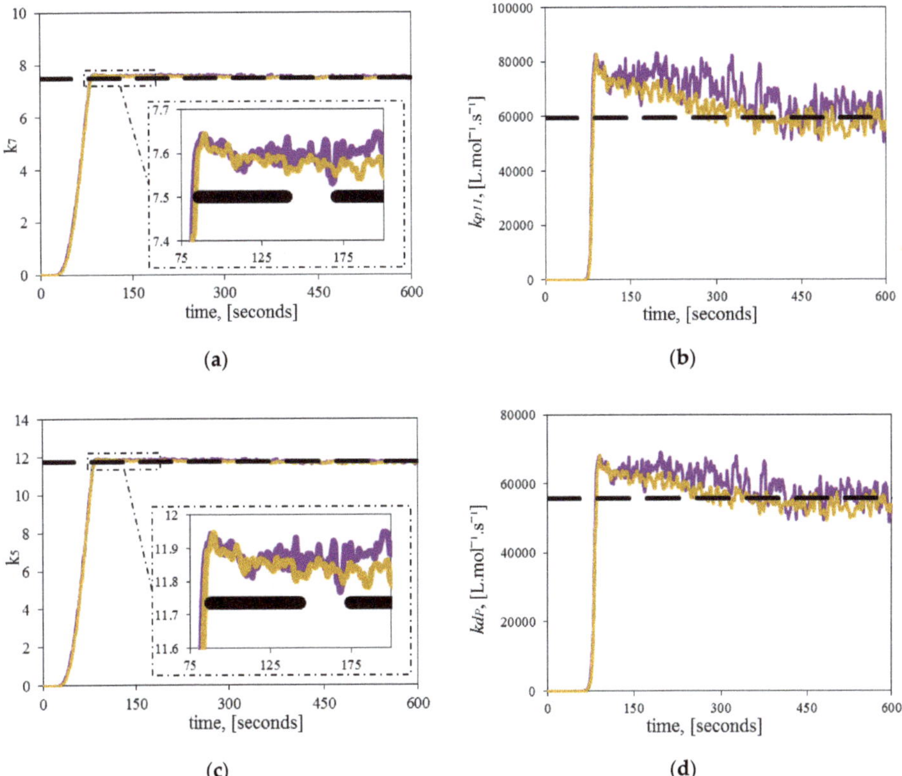

Figure 9. Estimation of two significant parameters. Comparison between theoretical (dashed line), and estimated kinetic parameters (continuous line) at different initial diene concentrations for the copolymerization experiment A (purple), and copolymerization experiment B (gold): (**a**) Dynamic estimation of k_7; (**b**) Dynamic estimation of k_{P11}; (**c**) Dynamic estimation of k_5; (**d**) Dynamic estimation of k_{dP}.

Figure 10 demonstrates how the RCMR can estimate F_M effectively and reduce the measurement's noise. As before, it takes some time for the estimated F_M to achieve values close to experimental and theoretical values. There are no visible differences observed when the initial concentration of diene varies, which makes it possible to conclude that temperature is more influential on the reaction behavior. In contrast to the homopolymerization results (Figure 6), the estimated F_M in the copolymerization experiments shows a higher delay time of convergence. It could be argued that it should be influenced by the increase in the complexity of modelling, but because the RCMR algorithm is a purely data-driven strategy, which does not require information on the nonlinear model, the reasons must be totally related to the nature of the experiment.

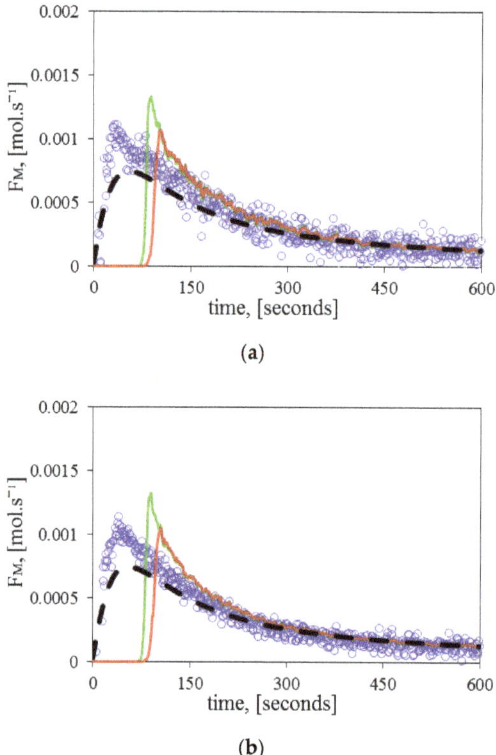

(a)

(b)

Figure 10. Comparison of the monomer flow rate (F_M) during copolymerization between the experimental values (circles), fundamental model (dashed line), estimated with a single parameter (continuous red line), and estimated with two parameters (continuous green line) at different initial diene concentrations: (**a**) copolymerization experiment A, (**b**) copolymerization experiment B.

Finally, Figure 11 compares the results computed by the fundamental model, the measured results, and the estimated average properties of the copolymerization experiments at different initial diene concentrations. Data on the average properties was obtained during the polymerization experiments. The results show that the RCMR algorithm, besides computing the unknown parameters, can estimate M_n satisfactorily using the error signal obtained as the difference between the estimated and measured F_M. This signal allows the estimator to gain enough information on the system to estimate M_n. Of course, the estimation additionally relies on the correctness of the model and the non-estimated parameters as well, but the obtained results are comparable with theoretical and experimental values. On the other hand, M_w achieves a similar dynamic when compared to the fundamental model, but the estimates fail to attain perfect values close to the experimental measurements.

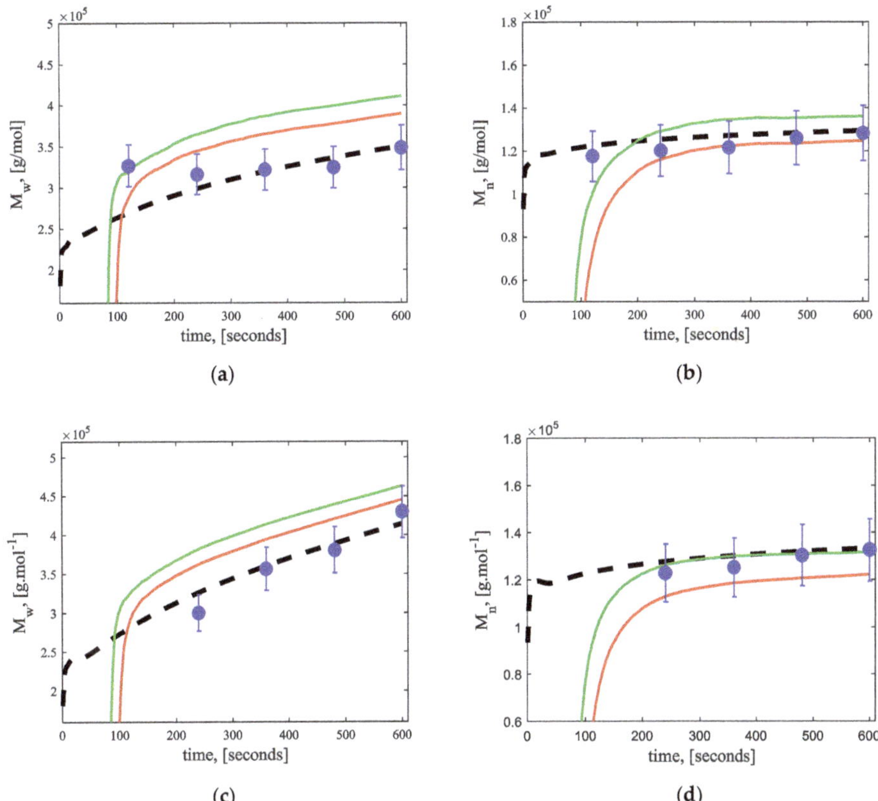

Figure 11. Comparison of average polymer properties at different initial diene concentrations between experimental values (circles), fundamental model (dashed line), estimated with a single parameter (red continuous line), and estimated with two parameters (green continuous line). (**a**) Weight-average molecular weight (M_w) for copolymerization A; (**b**) number-average molecular weight (M_n) for copolymerization A; (**c**) weight-average molecular weight (M_w) for copolymerization B; (**d**) number-average molecular weight (M_n) for copolymerization B.

6. Conclusions

In this contribution, a data-driven strategy for the online estimation of important kinetic parameters was assessed and implemented for the copolymerization of ethylene with 1,9-decadiene using dimethylsilyl (*N*-tert-butylamido) (tetramethylcyclopentadienyl) titanium dichloride (CGC)/MAO as catalyst. A global sensitivity analysis was performed initially to all polymerization kinetic parameters. The first and total sensitivity indices made it possible to choose the significant parameters of the model. Thereafter, the RCMR algorithm, a strategy never implemented in polymerization applications, permitted the estimation of the significant kinetic parameters, which were assumed to be unknown. After verifying consistency, the proposed strategy was tested in the copolymerization of ethylene with 1,9-decadiene at different diene concentrations. Overall, results were satisfactory, showing not only adequacy in signal processing, but also in parameter and property estimation.

The usage of data-driven algorithms such as the RCMR represents a paradigm that could permit easier estimation of parameters of nonlinear systems, such as those observed in polymer synthesis. Disturbances in the experimental data (e.g., impurities, experimental errors, and less frequent side reactions) that might not be captured by the fundamental model could be overcome by applying this strategy.

The proposed strategy also holds the advantage that it is capable of being adapted to the conditions of the experiment, and it can estimate the important kinetic parameters of the system while the reaction is running. This could be beneficial for processes where the major kinetic parameters are unknown for all the different types of catalysts, or the composition of the catalysts might be slightly different from each other, resulting in important differences. In the particular case of the RCMR algorithm, it has the additional advantage of estimating the parameters from an initial value of zero without requiring knowledge of the model.

Future work in this field will include the evaluation of the proposed framework and estimation algorithm in other chemical processes and polymerization systems.

Author Contributions: S.D.S. designed the framework that combines process modelling, global sensitivity analysis, and the RCMR algorithm under the supervision of J.A.R.; A.L.T.B. collected the experimental data and reviewed the reaction mechanism and the process modelling in collaboration with J.B.P.S.; S.D.S. wrote the article, while everyone contributed in reviewing and enriching the content.

Funding: This research received no external funding.

Acknowledgments: The authors acknowledge Ankit Goel from the University of Michigan for providing guidelines that facilitated the implementation of the RCMR algorithm.

Conflicts of Interest: The authors declare no conflict of interest.

Nomenclature

Notation

C	catalyst precursor, [mol]
C^*	active catalyst site, [mol]
D	1,9-decadiene, [mol]
DC	dead catalyst site, [mol]
d	total amount of 1,9-decadiene inserted into the growing polymer chains
E_{ap}	activation energy for ethylene propagation reaction rate, [J·mol^{-1}]
F_M	ethylene feed flow rate, [mol·s^{-1}]
K	total number of pendant unsaturation's present in the dead chains
k_a	catalyst activation constant, [s^{-1}]
k_b	macromonomer reincorporation rate constant, [L·mol^{-1}·s^{-1}]
k_{dP}	living chain deactivation rate constant, [L·mol^{-1}·s^{-1}]
$k_{p_{11}}$	propagation rate constant for ethylene, [L·mol^{-1}·s^{-1}]
$k_{p_{12}}$	propagation rate constant for 1,9-decadiene, [L·mol^{-1}·s^{-1}]
k_t	termination rate constant, [s^{-1}]
k_{0p}	pre-exponential factor for ethylene propagation reaction rate, [L·mol^{-1}·s^{-1}]
lcb	long chain branching
$L_i^=$	dead polymer chain that contains a terminal unsaturation and has chain size i, [mol]
L_i	dead polymer chain with chain size i and without a terminal unsaturation, [mol]
M	ethylene, [mol]
m	total amount of ethylene inserted into the growing polymer chains
MM	average molar mass of the repeating unit, [g·mol^{-1}]
MM_D	molar mass of 1,9-decadiene, [g·mol^{-1}]
MM_M	molar mass of ethylene, [g·mol^{-1}]
M_n	number average molecular weight, [g·mol^{-1}]
M_w	weight average molecular weight, [g·mol^{-1}]
PDI	polydispersity index
P_i^*	living polymer chain with chain size i, [mol]
R	ideal gas constant, [J·mol^{-1}·K^{-1}]
T	reaction temperature, [K or °C]
T_r	reference temperature, [K or °C]
V	reaction mixture volume, [L]

Greek letters

λ_k	k^{th} moment for the dead chain
μ_k	k^{th} moment for the living chain
ρ_{PE}	Polyethylene density, $[g \cdot L^{-1}]$
φ	Average frequency of pendant double bonds in the polymer chains

References

1. Liu, P.; Liu, W.; Wang, W.J.; Li, B.G.; Zhu, S. A Comprehensive Review on Controlled Synthesis of Long-Chain Branched Polyolefins: Part 1, Single Catalyst Systems. *Macromol. React. Eng.* **2016**, *10*, 156–179. [CrossRef]
2. Stadler, F.J.; Piel, C.; Klimke, K.; Kaschta, J.; Parkinson, M.; Wilhelm, M.; Münstedt, H. Influence of type and content of various comonomers on long-chain branching of ethene/α-olefin copolymers. *Macromolecules* **2006**, *39*, 1474–1482. [CrossRef]
3. Soares, J.B.; Hamielec, A.E. Bivariate chain length and long chain branching distribution for copolymerization of olefins and polyolefin chains containing terminal double-bonds. *Macromol. Theory Simul.* **1996**, *5*, 547–572. [CrossRef]
4. Wang, W.J.; Yan, D.; Zhu, S.; Hamielec, A.E. Kinetics of long chain branching in continuous solution polymerization of ethylene using constrained geometry metallocene. *Macromolecules* **1998**, *31*, 8677–8683. [CrossRef]
5. Chum, P.S.; Kruper, W.J.; Guest, M.J. Materials properties derived from INSITE metallocene catalysts. *Adv. Mater.* **2000**, *12*, 1759–1767. [CrossRef]
6. Choo, T.N.; Waymouth, R.M. Cyclocopolymerization: A mechanistic probe for dual-site alternating copolymerization of ethylene and α-olefins. *J. Am. Chem. Soc.* **2002**, *124*, 4188–4189. [CrossRef]
7. Naga, N.; Toyota, A. Unique Insertion Mode of 1,7-Octadiene in Copolymerization with Ethylene by a Constrained-Geometry Catalyst. *Macromol. Rapid Commun.* **2004**, *25*, 1623–1627. [CrossRef]
8. Mehdiabadi, S.; Soares, J.B. Production of Ethylene/α-Olefin/1, 9-Decadiene Copolymers with Complex Microstructures Using a Two-Stage Polymerization Process. *Macromolecules* **2011**, *44*, 7926–7939. [CrossRef]
9. Soares, J.B. Mathematical Modeling of the Long-Chain Branch Structure of Polyolefins Made with Two Metallocene Catalysts: An Algebraic Solution. *Macromol. Theory Simul.* **2002**, *11*, 184–198. [CrossRef]
10. Ferreira, L.C., Jr.; Nele, M.; Costa, M.A.; Pinto, J.C. Mathematical Modeling of MWD and CBD in Polymerizations with Macromonomer Reincorporation and Chain Running. *Macromol. Theory Simul.* **2010**, *19*, 496–513. [CrossRef]
11. Mogilicharla, A.; Mitra, K.; Majumdar, S. Modeling of propylene polymerization with long chain branching. *Chem. Eng. J.* **2014**, *246*, 175–183. [CrossRef]
12. Albeladi, A.; Mehdiabadi, S.; Soares, J.B. Modeling Possible Long Chain Branching Reactions for Polyethylene in a Semi-Batch Reactor. 2018. Available online: http://dc.engconfintl.org/prex/49 (accessed on 26 February 2019).
13. Brandão, A.L.; Alberton, A.L.; Pinto, J.C.; Soares, J.B. Copolymerization of Ethylene with 1, 9-Decadiene: Part I–Prediction of Average Molecular Weights and Long-Chain Branching Frequencies. *Macromol. Theory Simul.* **2017**, *26*, 1600059. [CrossRef]
14. Brandão, A.L.; Alberton, A.L.; Pinto, J.C.; Soares, J.B. Copolymerization of Ethylene with 1, 9-Decadiene: Part II—Prediction of Molecular Weight Distributions. *Macromol. Theory Simul.* **2017**, *26*, 1700040. [CrossRef]
15. Soares, J.B.; McKenna, T.F. *Polyolefin Reaction Engineering*; John Wiley & Sons: Hoboken, NJ, USA, 2013.
16. Qin, S.J. Process data analytics in the era of big data. *AIChE J.* **2014**, *60*, 3092–3100. [CrossRef]
17. Salas, S.D.; Ghadipasha, N.; Zhu, W.; Mcafee, T.; Zekoski, T.; Reed, W.F.; Romagnoli, J.A. Framework Design for Weight-Average Molecular Weight Control in Semi-Batch Polymerization. *Control Eng. Pract.* **2018**, *78*, 12–23. [CrossRef]
18. Kozub, D.J.; MacGregor, J.F. State estimation for semi-batch polymerization reactors. *Chem. Eng. Sci.* **1992**, *47*, 1047–1062. [CrossRef]
19. Tatiraju, S.; Soroush, M. Nonlinear state estimation in a polymerization reactor. *Ind. Eng. Chem. Res.* **1997**, *36*, 2679–2690. [CrossRef]
20. Lopez, T.; Alvarez, J. On the effect of the estimation structure in the functioning of a nonlinear copolymer reactor estimator. *J. Process Control* **2004**, *14*, 99–109. [CrossRef]

21. Galdeano, R.; Asteasuain, M.; Sánchez, M.C. Unscented transformation-based filters: Performance comparison analysis for the state estimation in polymerization processes with delayed measurements. *Macromol. React. Eng.* **2011**, *5*, 278–293. [CrossRef]

22. Hashemi, R.; Kohlmann, D.; Engell, S. Optimizing control and state estimation of a continuous polymerization process in a tubular reactor with multiple side-streams. *Macromol. React. Eng.* **2016**, *10*, 415–434. [CrossRef]

23. Zoellner, K.; Reichert, K.H. Gas phase polymerization of butadiene—Kinetics, particle size distribution, modeling. *Chem. Eng. Sci.* **2001**, *56*, 4099–4106. [CrossRef]

24. Ghadipasha, N.; Geraili, A.; Romagnoli, J.A.; Castor, C.A.; Drenski, M.F.; Reed, W.F. Combining on-line characterization tools with modern software environments for optimal operation of polymerization processes. *Processes* **2016**, *4*, 5. [CrossRef]

25. Schwaab, M.; Biscaia, E.C., Jr.; Monteiro, J.L.; Pinto, J.C. Nonlinear parameter estimation through particle swarm optimization. *Chem. Eng. Sci.* **2008**, *63*, 1542–1552. [CrossRef]

26. Salas, S.D.; Romagnoli, J.A.; Tronci, S.; Baratti, R. A geometric observer design for a semi-batch free-radical polymerization system. *Comput. Chem. Eng.* **2019**, *126*, 391–402. [CrossRef]

27. Sirohi, A.; Choi, K.Y. On-line parameter estimation in a continuous polymerization process. *Ind. Eng. Chem. Res.* **1996**, *35*, 1332–1343. [CrossRef]

28. Li, R.; Corripio, A.B.; Henson, M.A.; Kurtz, M.J. On-line state and parameter estimation of EPDM polymerization reactors using a hierarchical extended Kalman filter. *J. Process Control* **2004**, *14*, 837–852. [CrossRef]

29. Chen, T.; Morris, J.; Martin, E. Particle filters for state and parameter estimation in batch processes. *J. Process Control* **2005**, *15*, 665–673. [CrossRef]

30. Sheibat-Othman, N.; Peycelon, D.; Othman, S.; Suau, J.M.; Fevotte, G. Nonlinear observers for parameter estimation in a solution polymerization process using infrared spectroscopy. *Chem. Eng. J.* **2008**, *140*, 529–538. [CrossRef]

31. Salas Ortiz, S.D. A Model-Based Framework for the Smart Manufacturing of Polymers. Ph.D. Thesis, Louisiana State University and Agricultural and Mechanical College, Baton Rouge, LA, USA, 2019; p. 4897.

32. Wu, Q.L.; Cournede, P.H.; Mathieu, A. An efficient computational method for global sensitivity analysis and its application to tree growth modelling. *Reliab. Eng. Syst. Saf.* **2012**, *107*, 35–43. [CrossRef]

33. Goel, A.; Duraisamy, K.; Bernstein, D.S. Parameter estimation in the burgers equation using retrospective-cost model refinement. In Proceedings of the American Control Conference (ACC), Boston, MA, USA, 6–8 July 2016; pp. 6983–6988.

34. Goel, A.; Bernstein, D.S. Parameter estimation for nonlinearly parameterized gray-box models. In Proceedings of the Annual American Control Conference (ACC), Milwaukee, WI, USA, 27–29 June 2018; pp. 5280–5285.

35. Goel, A.; Bernstein, D.S. Data-Driven Parameter Estimation for Models with Nonlinear Parameter Dependence. In Proceedings of the 2018 IEEE Conference on Decision and Control (CDC), Miami Beach, FL, USA, 17–19 December 2018; pp. 1470–1475.

36. Hulburt, H.M.; Katz, S. Some problems in particle technology: A statistical mechanical formulation. *Chem. Eng. Sci.* **1964**, *19*, 555–574. [CrossRef]

37. Eberhart, R.; Kennedy, J. Particle Swarm Optimization. In Proceedings of the IEEE International Conference on Neural Networks, Perth, Australia, 27 November–1 December 1995; pp. 1942–1948.

38. Sobol, I.M. Sensitivity estimates for nonlinear mathematical models. *Math. Model. Comput. Exp.* **1993**, *1*, 407–414.

39. Cosenza, A.; Mannina, G.; Vanrolleghem, P.A.; Neumann, M.B. Variance-based sensitivity analysis for wastewater treatment plant modelling. *Sci. Total Environ.* **2014**, *470*, 1068–1077. [CrossRef]

40. Saltelli, A. Making best use of model evaluations to compute sensitivity indices. *Comput. Phys. Commun.* **2002**, *145*, 280–297. [CrossRef]

41. Saltelli, A.; Ratto, M.; Andres, T.; Campolongo, F.; Cariboni, J.; Gatelli, D.; Tarantola, S. *Global Sensitivity Analysis: The Primer*; John Wiley & Sons: Hoboken, NJ, USA, 2008.

42. Salas, S.D.; Geraili, A.; Romagnoli, J.A. Optimization of Renewable Energy Businesses under Operational Level Uncertainties through Extensive Sensitivity Analysis and Stochastic Global Optimization. *Ind. Eng. Chem. Res.* **2017**, *56*, 3360–3372. [CrossRef]

43. Shampine, L.F.; Reichelt, M.W. The matlab ode suite. *SIAM J. Sci. Comput.* **1997**, *18*, 1–22. [CrossRef]
44. Porru, M.; Ożkan, L. Monitoring of Batch Industrial Crystallization with Growth, Nucleation, and Agglomeration. Part 2: Structure Design for State Estimation with Secondary Measurements. *Ind. Eng. Chem. Res.* **2017**, *56*, 9578–9592. [CrossRef]

Article

Making the Most of Parameter Estimation: Terpolymerization Troubleshooting Tips

Alison J. Scott [1], Vida A. Gabriel [2], Marc A. Dubé [2] and Alexander Penlidis [1,*

[1] Department of Chemical Engineering, Institute for Polymer Research, University of Waterloo, 200 University Avenue West, Waterloo, ON N2L 3G1, Canada

[2] Department of Chemical and Biological Engineering, Centre for Catalysis Research and Innovation, University of Ottawa, Ottawa, ON K1N 6N5, Canada

* Correspondence: penlidis@uwaterloo.ca

Received: 28 May 2019; Accepted: 5 July 2019; Published: 12 July 2019

Abstract: Multi-component polymers can provide many advantages over their homopolymer counterparts. Terpolymers are formed from the combination of three unique monomers, thus creating a new material that will exhibit desirable properties based on all three of the original comonomers. To ensure that all three comonomers are incorporated (and to understand and/or predict the degree of incorporation of each comonomer), accurate reactivity ratios are vital. In this study, five terpolymerization studies from the literature are revisited and the 'ternary' reactivity ratios are re-estimated. Some recent studies have shown that binary reactivity ratios (that is, from the related copolymer systems) do not always apply to ternary systems. In other reports, binary reactivity ratios are in good agreement with terpolymer data. This investigation allows for the comparison between previously determined binary reactivity ratios and newly estimated 'ternary' reactivity ratios for several systems. In some of the case studies presented herein, reactivity ratio estimation directly from terpolymerization data is limited by composition restrictions or ill-conditioned systems. In other cases, we observe similar or improved prediction performance (for ternary systems) when 'ternary' reactivity ratios are estimated directly from terpolymerization data (compared to the traditionally used binary reactivity ratios). In order to demonstrate the advantages and challenges associated with 'ternary' reactivity ratio estimation, five case studies are presented (with examples and counter-examples) and troubleshooting suggestions are provided to inform future work.

Keywords: copolymerization; design of experiments; reactivity ratio estimation; terpolymerization

---

1. Introduction

Terpolymer systems are becoming increasingly popular in a variety of applications, as they allow for the combination of various desirable properties into a single material [1,2]. The diversity of monomers allows for the production of countless new multi-component materials, all with particular combinations of desirable properties for custom applications. In the past decade alone, terpolymers have been investigated for use as electronic materials [3–5], biomedical materials [6–8], pressure sensitive adhesives [9–12], coatings [13–15], and for many more applications. A good understanding of terpolymerization kinetics (for each unique terpolymer system) ensures that appropriate formulations are selected to achieve desirable properties in the product terpolymer. Namely, 'ternary' reactivity ratios (that is, reactivity ratios estimated directly from terpolymerization data using an appropriate model, as opposed to reactivity ratios estimated from the corresponding binary copolymerizations, alternatively referred to as 'ternary copolymerizations') provide information about the degree of incorporation of each comonomer into the product terpolymer (based on the initial pre-polymer formulation). Therefore, we can select appropriate feed compositions to achieve a desirable terpolymer composition and microstructure [16–19].

Recent studies have shown that binary (copolymerization) reactivity ratios (estimated from copolymerization data) do not always apply to terpolymerizations [18,20]. Thus, in general, 'ternary' reactivity ratio estimation (directly from terpolymerization data) is recommended. 'Ternary' reactivity ratio estimation (estimating all six related reactivity ratios simultaneously from terpolymerization data using the recast Alfrey–Goldfinger model [18]) ensures that any interactions and contributions from the third comonomer are considered, which improves the kinetic and statistical accuracy of parameter estimates. When terpolymerization experiments are selected with parameter estimation in mind (that is, using an error-in-variables model (EVM)-based design of experiments for ternary systems [21]), it can be straightforward to estimate 'ternary' reactivity ratios using the information-rich dataset. However, in using historical data (which may not be designed for parameter estimation), experimental terpolymerization data may not be sufficient for analysis. Common limitations in terpolymerization studies include composition restrictions or a lack of experimental information content (minimal replication, formulations selected targeting final properties rather than information collection, etc.). Since working with historical (previously collected) data has some challenges, the case studies presented herein address those challenges and aim to improve 'ternary' reactivity ratio prediction to the extent which is possible. For example, additional targeted experiments (i.e., the sequential selection of experiments based on an existing dataset), experimental replication, and full-conversion (cumulative) analysis can supplement a pre-existing terpolymerization dataset. While additional targeted experiments and replication may require revisiting the lab (or simulating additional data), full-conversion data are often already available from earlier runs. Cumulative analysis can provide greater information content from fewer experimental runs [22].

Despite the fact that binary reactivity ratios are only a numerical approximation for ternary systems, several cases have reported that model predictions using binary reactivity ratios seem to be in good agreement with terpolymerization data (most recently [12]). In these cases, although binary reactivity ratios provide good prediction performance, there are still benefits associated with using 'ternary' reactivity ratios directly [20]. First, this ensures that no unfounded assumptions are made about the nature of the terpolymerization, where the addition of the third comonomer (and any interactions it might have with the other two comonomers) is carefully evaluated. Second, it reduces the experimental load required to characterize the system (e.g., two designed formulations of three copolymers require six experiments for binary reactivity ratios, compared to three designed formulations per single terpolymer requiring three experiments for 'ternary' reactivity ratios). This is especially important if a new (unknown) combination of comonomers is being investigated. Third, and perhaps most importantly, the kinetic and statistical accuracy of using 'ternary' reactivity ratio estimation in terpolymerization studies ensures accurate model predictions for the system being studied.

In the present work, several case studies are revisited to explore the advantages (and challenges) of directly estimating 'ternary' reactivity ratios from terpolymerization data. In the first two case studies (Section 3.1), common challenges in parameter estimation are addressed, such as process constraints (experimental/composition limitations) and numerical estimation constraints (ill-conditioned systems). Additional case studies (Section 3.2) are then presented to provide a direct comparison between binary and 'ternary' reactivity ratio estimates. When the above-mentioned challenges are not a factor, the advantages of estimating 'ternary' reactivity ratios are evident.

2. Background

2.1. Free Radical Terpolymerization

Terpolymerization systems (and multicomponent polymerizations in general) are part of an interesting and growing area of research, since there are countless combinations of monomers to be discovered. However, because of the wide range of possibilities, these more complex terpolymerization systems have not been studied as thoroughly as copolymerization systems.

The kinetics of terpolymerization systems were first described by Alfrey and Goldfinger [23]. Given that there are three different possibilities for the terminal monomer (on the growing radical), and three options for the added monomer, nine different propagation steps are possible according to the terminal model:

$$\sim\sim M_1 + M_1 \xrightarrow{k_{11}} \sim\sim M_1 - M_1$$

$$\sim\sim M_1 + M_2 \xrightarrow{k_{12}} \sim\sim M_1 - M_2$$

$$\sim\sim M_1 + M_3 \xrightarrow{k_{13}} \sim\sim M_1 - M_3$$

$$\sim\sim M_2 + M_1 \xrightarrow{k_{21}} \sim\sim M_2 - M_1$$

$$\sim\sim M_2 + M_2 \xrightarrow{k_{22}} \sim\sim M_2 - M_2$$

$$\sim\sim M_2 + M_3 \xrightarrow{k_{23}} \sim\sim M_2 - M_3$$

$$\sim\sim M_3 + M_1 \xrightarrow{k_{31}} \sim\sim M_3 - M_1$$

$$\sim\sim M_3 + M_2 \xrightarrow{k_{32}} \sim\sim M_3 - M_2$$

$$\sim\sim M_3 + M_3 \xrightarrow{k_{33}} \sim\sim M_3 - M_3$$

The standard Alfrey–Goldfinger (AG) equations use ratios of instantaneous mole fractions in the terpolymer (i.e., F_i/F_j) as responses. However, measurements taken from experimental work are typically single mole fractions, not ratios, which means that evaluating (F_i/F_j) ratios results in lost information and a distorted error structure [18]. Thus, the AG model was re-derived (based on [18]) from basic terpolymerization kinetics for this study, so that each terpolymer mole fraction is presented as a single response (see Equations (1)–(3); details for this derivation are available upon request). This formulation is an improvement over the original AG model, as it promotes symmetrical estimation and error structures are not distorted. It also agrees with the recent work published by Kazemi et al. [18], in which the 'recast' version of the AG model was developed using the alternative ratio-based equations as a starting point.

$$F_1 - \frac{f_1\left(\frac{f_1}{r_{21}r_{31}}+\frac{f_2}{r_{21}r_{32}}+\frac{f_3}{r_{31}r_{23}}\right)\left(f_1+\frac{f_2}{r_{12}}+\frac{f_3}{r_{13}}\right)}{f_1\left(\frac{f_1}{r_{21}r_{31}}+\frac{f_2}{r_{21}r_{32}}+\frac{f_3}{r_{31}r_{23}}\right)\left(f_1+\frac{f_2}{r_{12}}+\frac{f_3}{r_{13}}\right)+f_2\left(\frac{f_1}{r_{12}r_{31}}+\frac{f_2}{r_{12}r_{32}}+\frac{f_3}{r_{13}r_{32}}\right)\left(f_2+\frac{f_1}{r_{21}}+\frac{f_3}{r_{23}}\right)+f_3\left(\frac{f_1}{r_{13}r_{21}}+\frac{f_2}{r_{23}r_{12}}+\frac{f_3}{r_{13}r_{23}}\right)\left(f_3+\frac{f_1}{r_{31}}+\frac{f_2}{r_{32}}\right)} = 0 \tag{1}$$

$$F_2 - \frac{f_2\left(\frac{f_1}{r_{12}r_{31}}+\frac{f_2}{r_{12}r_{32}}+\frac{f_3}{r_{13}r_{32}}\right)\left(f_2+\frac{f_1}{r_{21}}+\frac{f_3}{r_{23}}\right)}{f_1\left(\frac{f_1}{r_{21}r_{31}}+\frac{f_2}{r_{21}r_{32}}+\frac{f_3}{r_{31}r_{23}}\right)\left(f_1+\frac{f_2}{r_{12}}+\frac{f_3}{r_{13}}\right)+f_2\left(\frac{f_1}{r_{12}r_{31}}+\frac{f_2}{r_{12}r_{32}}+\frac{f_3}{r_{13}r_{32}}\right)\left(f_2+\frac{f_1}{r_{21}}+\frac{f_3}{r_{23}}\right)+f_3\left(\frac{f_1}{r_{13}r_{21}}+\frac{f_2}{r_{23}r_{12}}+\frac{f_3}{r_{13}r_{23}}\right)\left(f_3+\frac{f_1}{r_{31}}+\frac{f_2}{r_{32}}\right)} = 0 \tag{2}$$

$$F_3 - \frac{f_3\left(\frac{f_1}{r_{13}r_{21}}+\frac{f_2}{r_{23}r_{12}}+\frac{f_3}{r_{13}r_{23}}\right)\left(f_3+\frac{f_1}{r_{31}}+\frac{f_2}{r_{32}}\right)}{f_1\left(\frac{f_1}{r_{21}r_{31}}+\frac{f_2}{r_{21}r_{32}}+\frac{f_3}{r_{31}r_{23}}\right)\left(f_1+\frac{f_2}{r_{12}}+\frac{f_3}{r_{13}}\right)+f_2\left(\frac{f_1}{r_{12}r_{31}}+\frac{f_2}{r_{12}r_{32}}+\frac{f_3}{r_{13}r_{32}}\right)\left(f_2+\frac{f_1}{r_{21}}+\frac{f_3}{r_{23}}\right)+f_3\left(\frac{f_1}{r_{13}r_{21}}+\frac{f_2}{r_{23}r_{12}}+\frac{f_3}{r_{13}r_{23}}\right)\left(f_3+\frac{f_1}{r_{31}}+\frac{f_2}{r_{32}}\right)} = 0 \tag{3}$$

The newly derived equations are only valid for the instantaneous case. Using low conversion data makes it possible to assume that terpolymer composition drift is negligible (that is, at low conversions, the measurable cumulative copolymer composition is approximately equal to its instantaneous value). However, this restrictive assumption introduces additional sources of error, including significant experimental difficulties.

As an alternative, a cumulative ternary composition model can be employed to estimate 'ternary' reactivity ratios using the full conversion trajectory. The cumulative model shown in Equations (4)–(6) relates the cumulative terpolymer composition for each comonomer ($\overline{F}_i$) to the mole fraction of monomer in the initial feed ($f_{i,0}$), the mole fraction of unreacted monomer in the polymerizing mixture (f_i), and the overall molar conversion (X_n).

$$\overline{F}_1 = \frac{f_{1,0} - f_1(1 - X_n)}{X_n} \tag{4}$$

$$\overline{F}_2 = \frac{f_{2,0} - f_2(1 - X_n)}{X_n} \tag{5}$$

$$\overline{F}_3 = \frac{f_{3,0} - f_3(1 - X_n)}{X_n} \tag{6}$$

If we cannot assume constant composition (that is, if composition drift is no longer negligible), f_i must be evaluated over conversion X_n, as shown in Equations (7)–(9).

$$\frac{df_1}{dX_n} = \frac{f_1 - F_1}{1 - X_n} \tag{7}$$

$$\frac{df_2}{dX_n} = \frac{f_2 - F_2}{1 - X_n} \tag{8}$$

$$\frac{df_3}{dX_n} = \frac{f_3 - F_3}{1 - X_n} \tag{9}$$

The models used to describe terpolymerization processes are simply extensions of the copolymer case, though slightly more complex. It is important to realize that ternary systems are unique and should not simply be viewed as a consolidation of the three analogous copolymer systems. When a third monomer is added to a polymerization formulation it can alter the polymerization characteristics and affect the degree of incorporation of all three comonomers [20]. Thus, whenever possible, terpolymerization kinetics should be evaluated and investigated in their own right.

The reader should note here from the outset that the term "'ternary' reactivity ratio" (used throughout this paper) refers to the collection of the six reactivity ratios that exist in Equations (1)–(3) ($r_{12}, r_{13}, r_{21}, r_{23}, r_{31}, r_{32}$). Although each individual parameter represents a relationship between two of the three comonomers, all six are estimated simultaneously from terpolymerization data, and therefore have an indirect influence on each other. For example, even though $r_{12} = k_{11}/k_{12}$ (and therefore only represents the propagation relationships for comonomers 1 and 2), it may still be influenced by the presence of monomer 3 in the polymerizing mixture. One may argue that since only two monomers are involved in the definition of each reactivity ratio, they are still technically 'binary' in nature. However, the term "'ternary' reactivity ratio" has been employed herein to emphasize that these reactivity ratios are estimated directly from terpolymerization data using the recast Alfrey–Goldfinger model. This is discussed further in what follows. It is our hope that creating this 'binary'/'ternary' distinction (related to both the source data and the series of equations employed for estimation) has simplified the reading and understanding of this work.

2.2. Reactivity Ratio Estimation

For the case studies that follow, the error-in-variables-model (EVM) was employed for the estimation of reactivity ratios. Through the EVM and direct numerical integration, we are able to estimate reactivity ratios using cumulative composition data (where the experimental data are available) [22]. This analysis provides additional advantages, including eliminating unnecessary assumptions and avoiding the experimental challenges associated with collecting low-conversion data.

The EVM methodology is an established tool for both selecting optimal feed compositions for reactivity ratio estimation experiments and for estimating reactivity ratios using non-linear parameter estimation. Therefore, only a brief overview is presented in what follows.

2.2.1. Design of Experiments for 'Ternary' Reactivity Ratio Estimation

When a series of experiments is designed in an optimal way, it becomes possible to minimize the number of experiments while increasing the information content from those experiments. Optimally designed experiments typically have much smaller joint confidence regions (JCRs), which are indicative of higher precision reactivity ratio estimates [24].

The EVM considers error in all variables, thus, using an experiment design technique within the EVM context helps to account for the error in both the independent variables (feed compositions) and the dependent variables (terpolymer compositions). The EVM can also take into account any experimental limitations, which ensures that the mathematical results of the experimental design are physically viable [21]. Of particular note is a practical heuristic for designing experiments for 'ternary' reactivity ratio estimation, which suggests that (statistically speaking) the optimal feed compositions always fall into the same range. When the terminal model is valid (as per Section 2.1), a multiplicative error structure is assumed and there are no other process constraints. The optimal feed compositions are as shown (in the shaded areas) in Figure 1. Three terpolymerization formulations, each rich in one comonomer, are sufficient to estimate 'ternary' reactivity ratios [21]. In a recent application, this was further confirmed while investigating the terpolymerization kinetics of 2-acrylamido-2-methylpropane sulfonic acid, acrylamide, and acrylic acid [20].

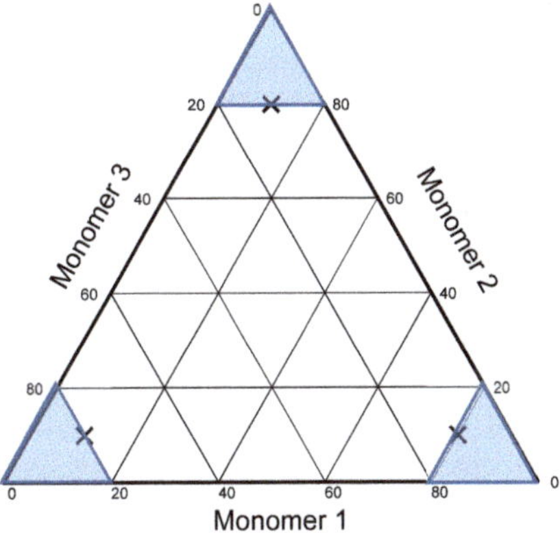

Figure 1. Optimally designed experiments for 'ternary' reactivity ratio estimation (as per [21]).

2.2.2. Reactivity Ratio Estimation Using EVM

The implementation of the EVM is based on Equations (10) and (11). Equation (10) relates the vector of known measurements ($\underline{x}_i$) to the vector of their unknown true values ($\underline{\xi}_i$) and an error term, $k\underline{\varepsilon}_i$ (where k is a constant that reflects the magnitude of measurement uncertainty, usually estimated from process information). Equation (11) is the model for the system, which shows the relationship between the true values of the variables ($\underline{\xi}_i$) and the true (but unknown) parameter values to be estimated ($\underline{\theta}$).

$$\underline{x}_i = \underline{\xi}_i + k\underline{\varepsilon}_i \text{ where } i = 1, 2 \ldots, n \tag{10}$$

$$\underline{g}\left(\underline{\xi}_i,\ \underline{\theta}\right) = 0 \text{ where } i = 1, 2 \ldots, n \tag{11}$$

In both Equations (10) and (11), i represents the trial number (out of n trials), and the underlined terms are either vectors or matrices.

The goal of the EVM is to minimize the sum of squares between the observed and predicted values, both in terms of the parameter estimates and the process variables [24]. The nested-iterative EVM algorithm [25] accomplishes this by using two nested loops, where the outer loop searches for parameter estimates while the inner loop identifies estimates of the true values of the variables involved. Mathematically, the following objective function (Equation (12)) is minimized:

$$\phi = \frac{1}{2}\sum_{i=1}^{n} r_i(\underline{\bar{x}}_i - \underline{\hat{\xi}}_i)'\underline{V}^{-1}\left(\underline{\bar{x}}_i - \underline{\hat{\xi}}_i\right) \tag{12}$$

where r_i is the number of replicates for the ith trial, $\overline{x_i}$ is the average of the r_i measurements (x_i), $\hat{\xi}_i$ is an estimate of the true values of the variables (ξ_i), and $\underline{V}$ is the variance-covariance matrix of the variables (which provides information about measurement error of the variables involved).

In addition to estimating model parameters and model variables, the EVM also evaluates the precision of the parameter estimates. Since several parameters are being estimated simultaneously, JCRs are plotted. JCRs are typically elliptical contours that quantify the level of uncertainty in the parameter estimates. Smaller JCRs indicate higher precision and therefore more confidence in the estimation results.

3. Results

In what follows, two main challenges associated with parameter estimation will be considered: Composition limitations (process constraints) and ill-conditioned systems (numerical estimation constraints). The discussion will be primarily informed by case studies from the literature. Then, we will revisit additional case studies that have shown good model agreement when using binary reactivity ratios to predict terpolymer properties. This provides a comparison of binary and ternary prediction performance for a variety of terpolymerization systems.

3.1. Addressing Composition Restrictions & Ill-Conditioned Systems

In some cases, system limitations and/or product requirements do not necessarily allow for statistically designed experiments, such as those described in Section 2 (each rich in a single comonomer). This paucity of data presents a challenge for researchers, especially when they hope to estimate reactivity ratios for subsequent microstructural predictions. The question is: What do we do when a terpolymer system has composition restrictions? How can we analyze a limited dataset in such a way that we are confident in our parameter estimates? In what follows, we take a closer look at the challenges associated with two such systems and provide some suggestions for overcoming the experimental limitations.

3.1.1. HOST/EAMA/PAG

In a recent study, Pujari et al. [5] synthesized a terpolymer of 4-hydroxystyrene (HOST; monomer 1), 2-ethyl-2-adamantyl methacrylate (EAMA; monomer 2), and a photoacid generator (PAG; monomer 3) to use in chemically amplified resists. Specifically, the PAG used for terpolymerization was triphenylsulfonium salt 4-(methacryloxy)-2,3,5,6-tetrafluorobenzenesulfonate (F4 PAG).

Pujari et al. [5] provided an interesting contrast between the copolymerization and terpolymerization behavior for the same comonomers. The authors reported that of the three comonomers, only HOST would homopolymerize. The HOST/EAMA copolymers were easily synthesized and HOST/PAG copolymers were achievable when the feed composition of PAG was below 10 mol%. However, their attempts to synthesize an EAMA/PAG copolymer were entirely unsuccessful (at a variety of distinct feed compositions). Finally, they synthesized HOST/EAMA/PAG terpolymers from several different feed compositions, in spite of their inability to synthesize some of the analogous copolymers. This is an example of a system where binary (copolymerization) data would not be suitable for predicting terpolymerization behavior.

Since Pujari et al. [5] had a specific application in mind, all terpolymer formulations had similar initial compositions. In the feed, mole fractions were selected within the following ranges: $0.25 \leq f_{1,0} \leq 0.40$, $0.50 \leq f_{2,0} \leq 0.75$, and $0.01 \leq f_{3,0} \leq 0.10$ (with $\sum f_i = 1.0$). Of particular interest for the current investigation is the very low mole fraction of PAG. At most, the PAG content in the feed was only 10 mol%. These experiments are not designed with reactivity ratio estimation in mind, rather, application requirements take precedence. However, these (low conversion) terpolymerization data were still used to estimate reactivity ratios and predict terpolymer properties.

Pujari et al. [5] were able to estimate 'ternary' reactivity ratios using Procop 2.3 [26]. However, because this dataset has minimal information content, the results exhibit a multiplicity of solutions. Table 1 compares the reactivity ratio estimates reported by Pujari et al. [5] with three successive estimations (labelled A through C) using the instantaneous model via the EVM (with the preliminary estimate that $r_{ij} = 0.500$ for all parameters).

Table 1. Reactivity ratio estimates for the terpolymerization of 4-hydroxystyrene (HOST)/2-ethyl-2-adamantyl methacrylate (EAMA)/photoacid generator (PAG). Data from [5].

	r_{12}	r_{21}	r_{13}	r_{31}	r_{23}	r_{32}
Reported by Pujari et al. [5]	0.05	0.12	0.81	0.12	0.05	0.79
Current Study (Error-in-Variables Model (EVM) Estimation)						
Inst. Estimation (A)	0.1063	0.2155	0.6754	0.0001	0.0877	0.0307
Inst. Estimation (B)	0.1767	0.2394	**99.9721**	0.0002	0.0897	0.7035
Inst. Estimation (C)	0.0003	0.1753	2.5329	0.0001	0.0728	**99.9997**

Clearly, the estimation stage and results using the instantaneous model with EVM are numerically unstable (and therefore unreliable). There are multiple solutions for this estimation. This is a limitation of using a dataset that only contains limited process information arising from PAG-poor formulations. Although common sense suggests that a reactivity ratio pair of $(r_{13}, r_{31}) = (99.9721, 0.0002)$ seems unlikely, as seen in estimation (B), it is numerically possible. This is a numerical artefact, as reactivity ratios reach the 'upper bound' (UB) of the parameter estimation program (UB = 100 for all parameter values during estimation).

All reactivity ratios presented in Table 1 give almost identical prediction performance when the PAG fraction is low. As shown in Figure 2, all model predictions 'fit' the experimental data (reported at one conversion level) equally well. The model predictions using the original reactivity ratio estimates reported by Pujari et al. [5] are not shown here, but they also 'fit' the experimental data well and fall within the ranges of Figure 2.

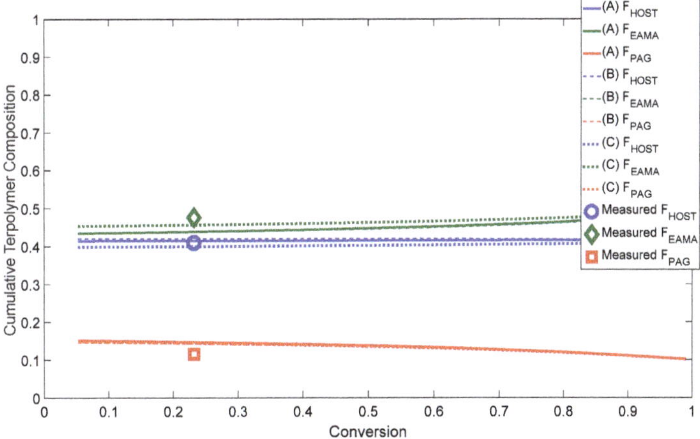

Figure 2. HOST/EAMA/PAG terpolymer composition prediction for $f_{1,0}/f_{2,0}/f_{3,0} = 0.4/0.5/0.1$ (experimentally measured composition for TPF10 from Pujari et al. [5]).

In contrast, if we predict terpolymerization behavior beyond the available experimental data (e.g., a PAG-rich formulation: $f_{1,0}/f_{2,0}/f_{3,0} = 0.1/0.1/0.8$), these considerably different reactivity ratio estimates give distinctly different results (see Figure 3). This is an extreme case, selected for demonstration purposes, and may not be achievable experimentally. However, this inconsistent prediction performance is observed for any PAG-rich recipe (with $f_{3,0}$ as low as 0.4).

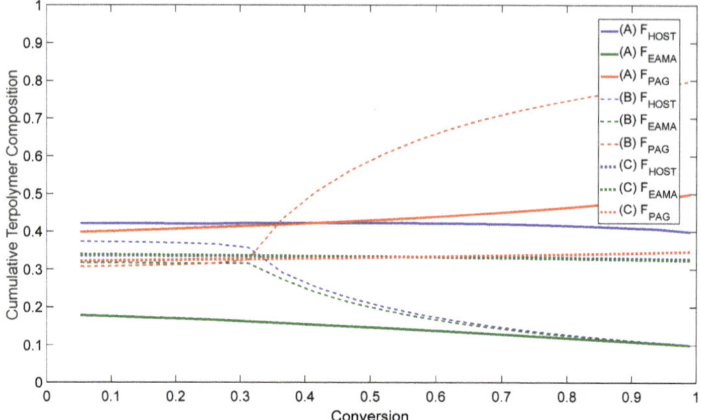

Figure 3. HOST/EAMA/PAG terpolymer composition prediction for $f_{1,0}/f_{2,0}/f_{3,0} = 0.1/0.1/0.8$.

The predictions shown in Figure 3 highlight the importance of statistically well-designed experiments. In spite of the experimental limitations in this case (limiting the PAG content to 10 mol%), introducing a run with as much PAG as possible would likely eliminate the numerical instabilities (and therefore improve the reliability of these estimation results).

3.1.2. BA/BMA/Limonene

Another terpolymerization case that is subject to composition restrictions (as well as ill-conditioning) is the terpolymerization of n-butyl acrylate (BA; monomer 1), butyl methacrylate (BMA; monomer 2), and D-limonene (lim; monomer 3), recently studied by Ren et al. [27]. D-limonene, a renewable monoterpene, is advantageous in terms of its polymer sustainability. When used as a comonomer it reduces the polymerization rate and molecular weight averages. Therefore, no more than 40 mol% lim was used in any feed composition for the study. The feed compositions used in the original investigation [27] are shown in Figure 4.

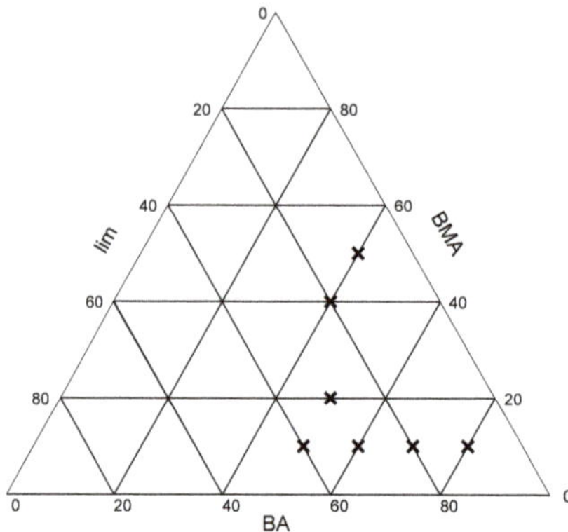

Figure 4. Feed compositions for the n-butyl acrylate (BA)/butyl methacrylate (BMA)/D-limonene (lim) terpolymer, as reported by Ren et al. [27].

Given that most experimental data are collected under lim-poor conditions, we might expect more error in the reactivity ratio estimates associated with the limonene comonomer. For both the instantaneous model (using low conversion data) and the cumulative model (using all available data), 'ternary' reactivity ratio estimation was performed three times (always with M_1 = BA, M_2 = BMA, and M_3 = lim). The results of each estimation are shown in Table 2. In this case, the binary reactivity ratio estimates (as reported by Ren et al. [27], collected from previous work by Dubé and colleagues [28–30]) were used as the preliminary estimates.

Table 2. Reactivity ratio estimates for the terpolymerization of BA/BMA/lim with experimental data from Ren et al. [27].

	r_{12}	r_{21}	r_{13}	r_{31}	r_{23}	r_{32}
Reported by Ren et al. [27]	0.46	2.008	6.08	0.007	6.096	0.046
Current Study (EVM Estimation with Low Conversion Data)						
Inst. Estimation (A)	0.3729	1.4350	5.0098	$<10^{-3}$	**35.5943**	**13.0409**
Inst. Estimation (B)	0.4986	1.4982	**5.2291**	**27.9826**	52.3811	$<10^{-3}$
Inst. Estimation (C)	0.3729	1.4350	5.0098	$<10^{-3}$	**35.5924**	**36.9594**
Current Study (EVM Estimation with Full Conversion Data)						
Cum. Estimation (A)	0.2787	0.9949	4.9888	0.0001	**29.6738**	**12.2200**
Cum. Estimation (B)	0.2733	0.9214	5.0104	0.0039	**27.1517**	**10.0072**
Cum. Estimation (C)	0.4081	0.9987	**5.4698**	**2.4651**	30.0914	0.0001

As we saw for the HOST/EAMA/PAG system, the estimation is numerically unstable. However, in general, the cumulative estimation results seem less ill-conditioned than in the instantaneous analysis. This is likely due to the increased information content provided when analyzing composition data over the full conversion range with a cumulative model [31].

Looking closer at the instantaneous estimation results, we see relatively good agreement between instantaneous estimations A and C. However, in both cases, the r_{23} and r_{32} estimates are both much greater than 1. Such a case has not been observed in free-radical copolymerization. There are some reports in the literature that have shown both estimates >1, but this is likely due to experimental error or a different copolymerization model being active. In this case, it is likely due to error, but the degradative chain transfer mechanism (due to the presence of limonene) may also contribute here. The uncertainty in this system is confirmed by comparing instantaneous estimations A and C to instantaneous estimation B. The fact that estimation results based on the same dataset have considerable convergence issues (likely due to local optima) suggests that there is not sufficient information for reactivity ratio estimation directly from the terpolymerization data.

Similar behavior is observed for the cumulative case. Here, cumulative estimations A and B are similar, whereas the third estimation, estimation (C), shows more variation. Again, the estimation results indicate that two reactivity ratios (for a given comonomer pair) are both greater than 1. The same comonomer pairs are of concern here: The BMA/lim comonomer pair (r_{23} and r_{32}) in cumulative estimations A and B, and the BA/lim pair (r_{13} and r_{31}) in cumulative estimation C. This behavior can be examined further by plotting the joint confidence regions (JCRs, or error ellipses) for each of the point estimates obtained using cumulative analysis (Figure 5).

As shown in Figure 5a, the JCRs associated with r_{23} and r_{32} are very large. The largest uncertainty was for estimation A, but subsequent estimations showed similar results. In comparison, the JCRs for the other parameters look like point estimates. This indicates substantial and disproportionate uncertainty in the estimates, especially for the BMA/lim comonomer pair.

As we focus in on the other parameter estimates (for the BA/BMA pair and the BA/lim pair), the JCRs become much smaller (note the change in scale between Figure 5a,b). Comparing BA/BMA to BA/lim, the most uncertainty is clearly for the BA/lim comonomer pair (especially for estimation (C)). In comparison, the JCRs for BA/BMA are very small, which gives us a much higher degree of certainty compared to the other estimates (see Figure 5c).

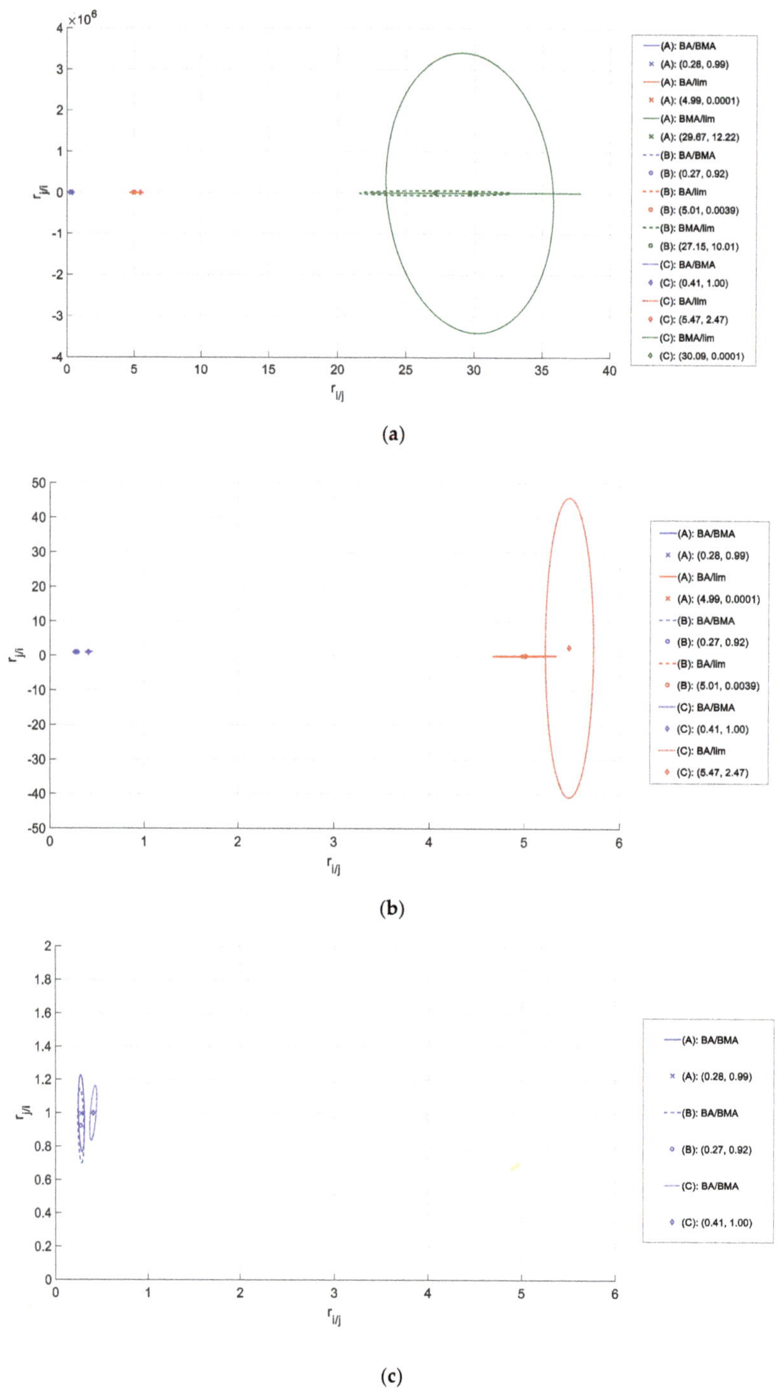

Figure 5. 'Ternary' reactivity ratio estimates for the terpolymerization of BA/BMA/lim, with data from Ren et al. [27].

The more precise estimation of r_{12} and r_{21} (that is, for the BA/BMA pair) is not coincidental. As mentioned earlier, the experimental data collected (Figure 4) only included formulations with low mole fractions of limonene. Therefore, we would suggest that a lack of lim-rich data has contributed to the poor estimation performance for the BA/BMA/lim terpolymer. This agrees with previous copolymerization observations reported by Scott et al. [31].

To demonstrate the importance of using well-designed data for 'ternary' reactivity ratio estimation, we have simulated supplemental data for the BA/BMA/lim system. Experimental data were simulated using the binary reactivity ratio estimates (which, as per the original investigation, give acceptable predictions of terpolymer behavior up to full conversion levels [27]). Two feed compositions, $f_{1,0}/f_{2,0}/f_{3,0}$ = 0.1/0.8/0.1 and $f_{1,0}/f_{2,0}/f_{3,0}$ = 0.1/0.1/0.8 (that is, BMA-rich and lim-rich formulations), were selected to supplement the original dataset. In both cases, the total conversion range was divided into 19 points (between 0 and 0.99, in steps of 0.052) and the corresponding monomer composition mole fractions were calculated via direct numerical integration. Then, the cumulative terpolymer compositions were calculated using the Skeist equation ([32], as per [22]). Random error was added to all data to mimic real experimental observations: A 1% error was added to the conversion and feed composition ($f_{i,0}$) data, while a 2% error was added to the cumulative terpolymer composition ($\bar{F}_i$) data. (Note that typically a 5% error is standard for $\bar{F}_i$ data, but using those levels for this system would occasionally make the simulated limonene content negative, given the low incorporation of limonene). The simulated data are provided in Table A1 of Appendix A.

For the instantaneous case, three low conversion data points from each (simulated) feed composition were added to the dataset. As shown in Table 3, the addition of these data 'stabilized' the estimation, and the reactivity ratio estimates obtained were now almost well-behaved. However, the newly estimated parameters still pose a concern: For the r_{23} and r_{32} pair (BMA/lim comonomers), both reactivity ratios were much greater than 1 for the low conversion case (instantaneous). In fact, r_{32} was estimated at 100.00 from three consecutive assessments. As explained previously (in Section 3.1.1), this is a numerical artefact, where the parameter has reached the 'upper bound' of the estimation program. However, the consistency of this result invites further investigation.

Table 3. Reactivity ratio estimates for terpolymerization of BA/BMA/lim with supplemental data (experimental data from [27] and simulated data from current work).

	r_{12}	r_{21}	r_{13}	r_{31}	r_{23}	r_{32}
Reported by Ren et al. [27]	0.46	2.008	6.08	0.007	6.096	0.046
Current Study (EVM Estimation with Experimental and Simulated Low Conversion Data)						
Inst. Estimation (A)	0.3663	1.4317	7.5324	0.1003	**10.1289**	**100.00**
Inst. Estimation (B)	0.3662	1.4317	7.5293	0.1004	**10.1307**	**100.00**
Inst. Estimation (C)	0.3663	1.4317	7.5324	0.1003	**10.1289**	**100.00**
Current Study (EVM Estimation with Experimental and Simulated Full Conversion Data)						
Cum. Estimation (A)	0.3009	2.4961	6.5236	0.0009	8.1873	0.0106
Cum. Estimation (B)	0.3167	1.9085	5.7962	0.0017	5.9647	0.0087
Cum. Estimation (C)	0.2876	1.5613	8.5751	0.0315	7.7202	0.1240

It is our understanding that the r_{32} = 100.00 result is due to the ill-conditioned nature of the terpolymer system. Physically, we can explain it as follows: The limonene incorporation is very low at low conversion levels, such that k_{32} (the rate constant for terminal limonene radicals adding BMA monomer units) is tending to 0, given the reactivity ratio definition, $r_{32} = k_{33}/k_{32}$, and as $k_{32} \to 0$, $r_{32} \to \infty$. With this logic, we can explain the observation that r_{32} is continually hitting the upper bound of the estimation program.

In contrast, the cumulative analysis (which uses the composition data, both experimental and simulated, over the full conversion range) gives more stable results. Specifically, the estimation results show that $r_{32} < 1$. Clearly, supplementing the terpolymerization dataset with optimal formulations (as per the EVM framework) significantly improves the stability and trustworthiness of ternary parameter estimates. Ideally, even more experimental data would be collected for the lim-rich system to offset the error associated with low limonene incorporation.

3.2. Improved Performance with Ternary Data

The above examples present a variety of challenges: Datasets with low information content or ill-conditioned systems (with reactivity ratios of different orders of magnitude) can make estimation difficult. If the estimation steps are unstable, then one does not have confidence in the final estimates. Unfortunately, even with many data points, we do not always have all the required information. Therefore, the design of experiments for reactivity ratio estimation is key.

In the cases that follow, we present some case studies that highlight the advantages of analyzing terpolymerization data directly using 'ternary' reactivity ratio estimation. All three case studies were originally modelled using analogous binary reactivity ratios, where the terpolymerization data have been revisited and reanalyzed.

3.2.1. BA/MMA/EHA

A recent study by Gabriel and Dubé [12] investigated the terpolymer of BA (monomer 1), methyl methacrylate (MMA; monomer 2), and 2-ethylhexyl acrylate (EHA; monomer 3), which is a material of interest for pressure sensitive adhesives. First, the authors determined the reactivity ratio pairs for two of the associated copolymers (BA/EHA and MMA/EHA), and subsequently used these binary reactivity ratios, along with literature values for the BA/MMA reactivity ratios, to predict the terpolymer composition. The terpolymer model prediction (using binary reactivity ratios) showed good agreement with the collected data, as described in the original work [12].

In spite of the good results achieved using binary reactivity ratios, 'ternary' reactivity ratio estimation directly from terpolymerization data presents some additional advantages. First, we can consider the experimental load: Rather than nine experimental runs, as described by Gabriel and Dubé [12] (and additional prior work for estimating the BA/MMA reactivity ratios [33]), only three different feed compositions are required. Since Gabriel and Dubé [12] selected ternary feed compositions according to the EVM 'rule-of-thumb' for 'ternary' reactivity ratio estimation [21], we can use their data to re-estimate reactivity ratios directly from the terpolymerization data.

First, only the low conversion data were used for an instantaneous analysis. Now, because these data points were collected for model validation (not necessarily parameter estimation), only 7 data points are available below 20% conversion. These data (shown in Table A2 of the Appendix A) were used for 'ternary' reactivity ratio estimation using the recast Alfrey–Goldfinger equation (recall Equations (1)–(3) and the EVM (Equations (10)–(12)). There are two observations of note here: (1) The estimation is stable, much more so than the case studies presented in Section 3.1, and (2) the estimation is symmetrical. That is, regardless of which monomer is defined as monomer 1, monomer 2, or monomer 3, the estimated parameters are the same. As an example, two variations are shown below and compared to the original (binary) estimation. Here, reactivity ratios are labelled according to the monomer name (rather than monomer number) for further clarity. Also, the colors shown in Table 4 are associated with the colors of the JCRs in Figure 6.

Table 4. Reactivity ratio estimates for terpolymerization of BA/MMA/2-ethylhexyl acrylate (EHA) from low conversion data (experimental data from [12]).

	$r_{BA/MMA}$	$r_{MMA/BA}$	$r_{BA/EHA}$	$r_{EHA/BA}$	$r_{MMA/EHA}$	$r_{EHA/MMA}$
Reported by Gabriel and Dubé [12] (binary reactivity ratio estimates)	0.34	2.02	0.99	1.62	1.50	0.32
Inst. Estimation ($M_1/M_2/M_3$ = BA/MMA/EHA)	0.41	1.49	1.21	8.52	0.81	0.36
Inst. Estimation ($M_1/M_2/M_3$ = MMA/EHA/BA)	0.41	1.49	1.20	8.45	0.81	0.36

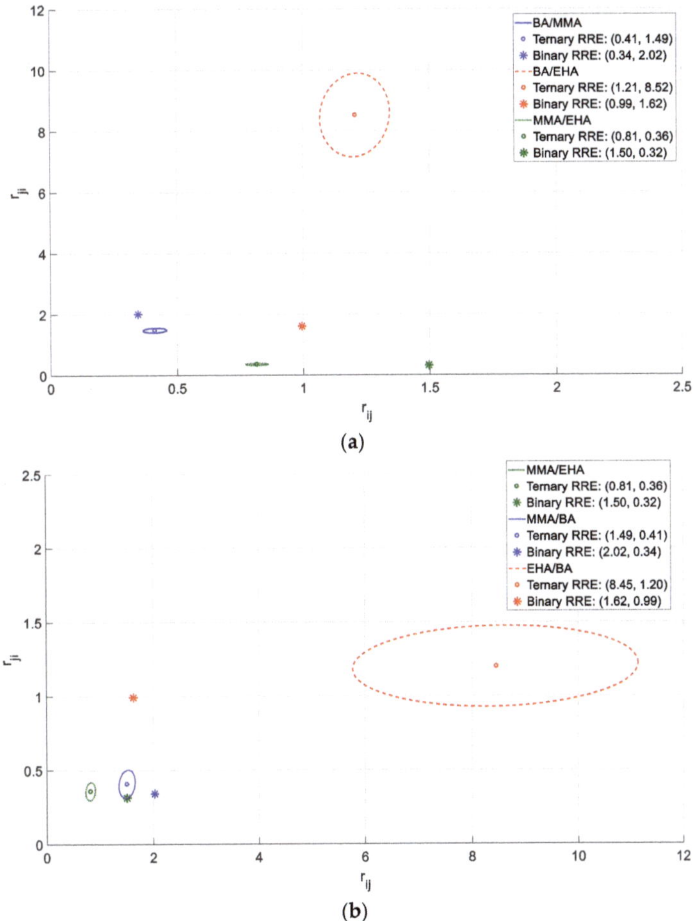

Figure 6. 'Ternary' reactivity ratio estimates for the terpolymerization of BA/MMA/EHA for (**a**) $M_1/M_2/M_3$ = BA/MMA/EHA and (**b**) $M_1/M_2/M_3$ = MMA/EHA/BA, with instantaneous data from [12].

In general, the 'ternary' reactivity ratios follow the same trends as the original (binary) reactivity ratio estimates (that is, if $r_{ij} > r_{ji}$ for the binary case, the same relationship holds for the ternary case). However, $r_{MMA/EHA}$ falls below 1.00 when estimated directly from the terpolymerization data. This suggests that $k_{MMA/MMA} > k_{MMA/EHA}$ in the binary case (homopropagation of MMA is preferable to the crosspropagation of MMA and EHA), but that $k_{MMA/MMA} < k_{MMA/EHA}$ in the ternary case (homopropagation becomes dominated by the crosspropagation of MMA/EHA).

Another notable difference is the significant error present in the BA/EHA system. This is shown in both plots (of Figure 6), as the JCR for BA/EHA is much larger than the other JCRs. This may be related to the absolute value of the parameter estimates. As shown in a recent study [31], uncertainty becomes much greater for larger parameter values. Since $r_{EHA/BA}$ is larger than the other reactivity ratio estimates (by as much as 20 times, in some cases), the same relative error (assumed to be 5% for this system) will have a much larger absolute value in $r_{EHA/BA}$ compared to the other parameter estimates. This behavior has been described for the copolymer case [31], but the difference in parameter estimates was observed within a single JCR (that is, the elliptical JCR was stretched in the direction of the larger parameter estimate). In this, a terpolymer case, the JCR associated with the comonomer pair containing larger parameter estimates is greater in both directions. The absolute value of the error seems magnified, likely because most other reactivity ratio estimates for the system are around or below 1.00. Another item of note is that both reactivity ratios for the BA/EHA pair are again >1. This may be for the same reasons discussed earlier for the BA/BMA/lim system. An additional reason may be related to the fact that the copolymerization of BA and EHA may lead to branched molecule formation and even microgel formation, which would complicate analysis further.

Next, we can look at the full conversion dataset, where 'ternary' reactivity ratios can be estimated using the EVM and the cumulative terpolymerization model. All terpolymerization data from the original study [12] were used herein, and the results are shown in Table 5 (and Figure 7). Again, the estimation is stable and symmetrical, which can be attributed to carefully designed data. As an aside, the estimation program also converged much more quickly, where parameters were estimated in under an hour (on an Intel(R) Core™ i7-860 processor) compared to (on average) 50 h of computation for the ill-conditioned system described earlier.

Table 5. Reactivity ratio estimates for the terpolymerization of BA/MMA/EHA from all terpolymerization data (experimental data from [12]).

	$r_{BA/MMA}$	$r_{MMA/BA}$	$r_{BA/EHA}$	$r_{EHA/BA}$	$r_{MMA/EHA}$	$r_{EHA/MMA}$
Reported by Gabriel and Dubé [12] (binary reactivity ratio estimates)	0.34	2.02	0.99	1.62	1.50	0.32
Cum. Estimation ($M_1/M_2/M_3$ = BA/MMA/EHA)	0.41	1.60	2.01	7.59	0.74	0.35
Cum. Estimation ($M_1/M_2/M_3$ = MMA/EHA/BA)	0.41	1.60	2.06	7.66	0.74	0.35

The values estimated using all terpolymerization data (full conversion) are similar to the results of the instantaneous parameter estimation (compare Table 4 to Table 5). Also, in comparing Figure 6 to Figure 7, the JCR areas are reduced when the full conversion dataset is used for analysis (note that the scales are the same for the easy comparison of Figure 6a to Figure 7a and of Figure 6b to Figure 7b). This is in agreement with previous studies within our group [18,22] and makes sense physically. Since more experimental data are available for analysis (18 data points over all conversion levels instead of 7 low conversion data points), the uncertainty associated with the parameter estimates is reduced. Also, since the instantaneous analysis used low conversion data up to 20%, the requisite assumption that no composition drift occurs may not be valid for all of the data [22]. Interestingly, a direct comparison of Figures 6 and 7 shows little or no JCR overlap (between the instantaneous and cumulative analysis), in spite of the fact that the trends remain consistent. Additional replication or sequential design of experiments could be used to further supplement this dataset, as has been described for the previous case studies (recall Section 3.1).

As for the instantaneous case, the most error associated with the cumulative analysis is present in the BA/EHA comonomer pair, and both reactivity ratios are greater than 1 (which is physically unlikely for any given comonomer pair). Again, this is likely due to the large values of the parameter estimates, which translate to a higher absolute value of the error (since we assume the same relative error for all experimental data and resulting parameter estimates). Also, as mentioned in the evaluation of the instantaneous results, other copolymerization mechanisms (branching, etc.) may be active specifically for the BA/EHA comonomer pair. However, without further analysis, no specific conclusions can be drawn about this system.

Finally, we can look at the prediction performance of these reactivity ratios (compared to the original binary reactivity ratio estimates). Since some of the more substantial differences in reactivity ratios were related to the EHA monomer (especially $r_{EHA/BA}$ and $r_{MMA/EHA}$), we can look at the model prediction for the EHA-rich terpolymer. The model predictions (using both 'binary' and 'ternary' reactivity ratios) and a comparison to the experimental data from Gabriel and Dubé [12] are shown in Figure 8. Only the prediction performance of the cumulative analysis (for 'ternary' reactivity ratio estimation) is provided in Figure 8, where, despite slight differences between the instantaneous and cumulative analysis results, the model prediction performance was very similar for both sets of reactivity ratio estimates.

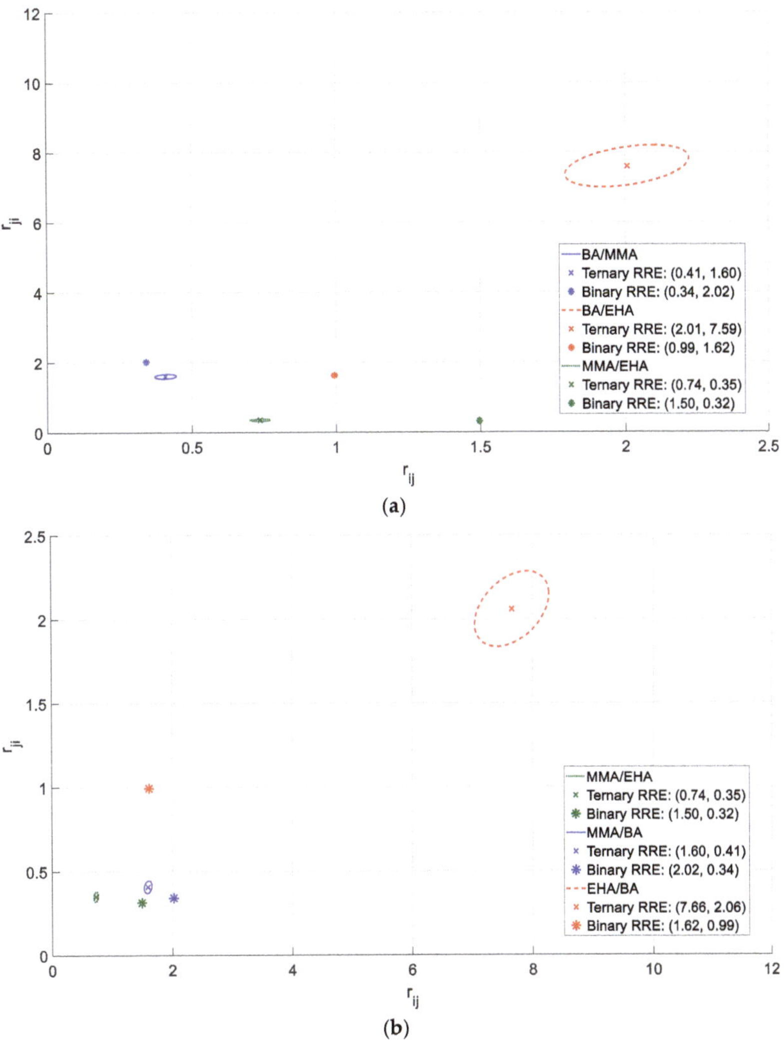

Figure 7. 'Ternary' reactivity ratio estimates for the terpolymerization of BA/MMA/EHA for (**a**) $M_1/M_2/M_3$ = BA/MMA/EHA and (**b**) $M_1/M_2/M_3$ = MMA/EHA/BA, with cumulative data from [12].

A direct comparison reveals that while the binary predictions are acceptable, the 'ternary' reactivity ratios further improve the prediction performance of the cumulative terpolymer composition model. In fact, a statistical comparison of the EHA-rich data (experimental data versus the two model predictions) shows that using the 'ternary' reactivity ratio estimates in the model leads to an 85% reduction in prediction error (total sum of square errors). Similar results were observed for the other terpolymer formulations but are not shown herein for the sake of brevity. These differences in prediction performance may further be accentuated if the estimated reactivity ratios are used in the sequence length part of the model.

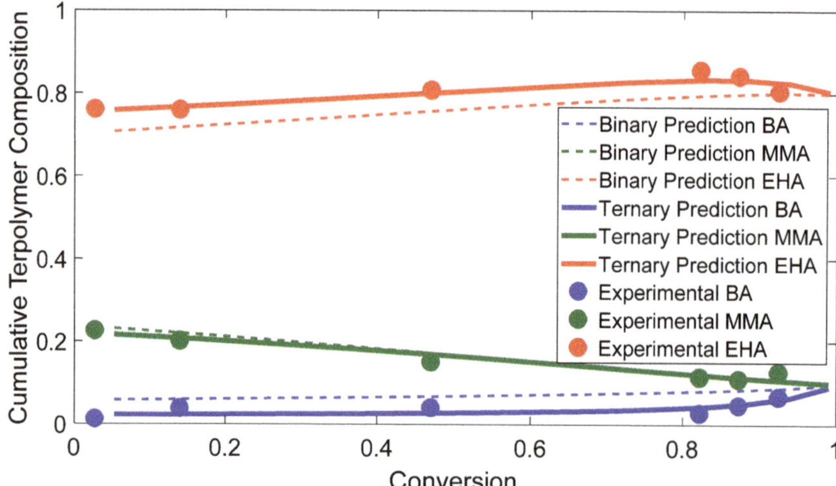

Figure 8. Prediction of cumulative terpolymer composition for BA/MMA/EHA ($f_{BA,0}/f_{MMA,0}/f_{EHA,0}$ = 0.1/0.1/0.8) (experimental data and binary predictions from [12]).

This case study has shown that when experiments are well-designed, 'ternary' reactivity ratio estimates can be obtained from small datasets. This allows for more resources to be directed towards careful replication and supplemental data collection. The results also suggest that binary and 'ternary' reactivity ratio estimates may be similar when the comonomers have similar structures and the polymerization is not affected by the solution properties. However, binary reactivity ratios are not always applicable to terpolymer systems (as has been shown recently [18,20]). In this case, the binary reactivity ratios gave reasonable prediction performance, but the 'ternary' reactivity ratios showed even better prediction performance based on fewer experimental data (and, hence, less effort overall).

3.2.2. Sty/MMA/MA

A study by Schoonbrood [34] looked at the terpolymerization kinetics for the styrene (Sty)/MMA/methyl acrylate (MA) (monomer 1/monomer 2/monomer 3 = Sty/MMA/MA) terpolymer. Again, according to standard practice, binary reactivity ratios (obtained from copolymerizations in the literature) were used to predict terpolymerization behavior. During this study, only low conversion data were reported. At the time (1994), this was 'best practice', where low conversion (instantaneous) data were typically used for parameter estimation. Parameter estimation from cumulative composition data was not part of typical practice, especially with the more complex system of equations representing terpolymerization kinetics. Low conversion data allow for a computationally simpler parameter estimation process but require some assumptions about a lack of composition drift in the system [22].

The experimentally determined (assumed as instantaneous) terpolymer compositions were compared to the model prediction. As reported in the original work, good agreement was observed between the predicted and measured values [34]. Given the available terpolymerization data, we can use the recast Alfrey–Goldfinger model (with the EVM) to re-estimate the terpolymer reactivity ratios directly from the terpolymerization data. The estimation is stable and symmetrical. A comparison of reactivity ratio estimates is presented in Table 6 and the prediction performance is evaluated in Table 7.

Table 6. Reactivity ratio estimates for the terpolymerization of styrene (Sty)/MMA/methyl acrylate (MA). Experimental data from [34].

	r_{12}	r_{21}	r_{13}	r_{31}	r_{23}	r_{32}
Reported by Schoonbrood [34] (binary reactivity ratio estimates)	0.48	0.42	0.73	0.19	2.49	0.29
Instantaneous Estimation (current work)	0.57	0.51	1.82	0.20	2.49	0.23

Table 7. Comparison of model predictions for the Sty/MMA/MA terpolymerization. Experimental data and original predictions from [34].

Feed Composition			Experimental Data			Original Predictions			Current (EVM) Predictions		
$f_{sty,0}$	$f_{MMA,0}$	$f_{MA,0}$	$\bar{F}_{sty}$	$\bar{F}_{MMA}$	$\bar{F}_{MA}$	$\bar{F}_{sty}$	$\bar{F}_{MMA}$	$\bar{F}_{MA}$	$\bar{F}_{sty}$	$\bar{F}_{MMA}$	$\bar{F}_{MA}$
0.10	0.10	0.80	0.27	0.28	0.45	0.26	0.19	0.55	0.26	0.23	0.51
0.10	0.20	0.70	0.24	0.38	0.38	0.24	0.33	0.44	0.23	0.37	0.40
0.10	0.30	0.60	0.21	0.45	0.33	0.23	0.43	0.34	0.21	0.48	0.31
0.20	0.10	0.70	0.44	0.18	0.39	0.37	0.16	0.47	0.40	0.20	0.41
0.20	0.20	0.60	0.37	0.39	0.24	0.35	0.28	0.37	0.36	0.33	0.31
0.20	0.30	0.50	0.33	0.39	0.28	0.34	0.38	0.28	0.34	0.43	0.23
0.20	0.50	0.30	0.29	0.60	0.11	0.32	0.53	0.15	0.31	0.57	0.12
0.30	0.20	0.50	0.42	0.27	0.31	0.43	0.26	0.31	0.46	0.31	0.24
0.30	0.30	0.40	0.40	0.39	0.20	0.42	0.35	0.23	0.43	0.40	0.17
0.30	0.40	0.30	0.41	0.44	0.15	0.40	0.43	0.16	0.41	0.47	0.12
0.30	0.50	0.20	0.39	0.54	0.07	0.39	0.50	0.10	0.39	0.54	0.07
0.40	0.30	0.30	0.48	0.34	0.18	0.48	0.34	0.18	0.50	0.38	0.12
0.40	0.40	0.20	0.50	0.36	0.14	0.47	0.42	0.11	0.47	0.45	0.08
0.50	0.20	0.30	0.56	0.24	0.21	0.55	0.25	0.21	0.59	0.28	0.13
0.50	0.40	0.10	0.50	0.49	0.01	0.52	0.42	0.06	0.53	0.43	0.04
0.50	0.30	0.20	0.52	0.43	0.06	0.53	0.34	0.13	0.56	0.36	0.08

Although the prediction performance looks similar, the current (EVM) prediction shows a decrease in the sum of square errors for all three comonomer compositions, especially $\bar{F}_{MMA}$ and $\bar{F}_{MA}$. In evaluating the total sum of square errors, the current work provides a 32% decrease in prediction error over the original analysis. To supplement this result, we can also examine the residuals for both the original and current predictions. As shown in Figure 9, the spread (that is, the vertical distance from 0) is reduced for the current predictions, where the residuals are smaller overall.

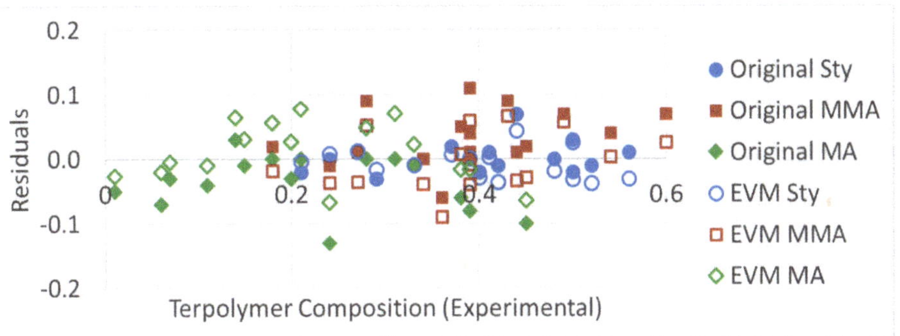

Figure 9. Comparison of residuals for Sty/MMA/MA terpolymer composition predictions.

Given these results (and those discussed previously), there is clearly an advantage for estimating 'ternary' reactivity ratios directly from terpolymerization data. If medium to high conversion data were available, they could have been used to supplement the dataset or to reduce the number of experiments required. However, even with this low conversion dataset, estimating 'ternary' reactivity ratios directly from terpolymerization data is feasible and preferable to using binary data.

3.2.3. AN/Sty/MMA

The terpolymerization of acrylonitrile (AN; monomer 1), Sty (monomer 2), and MMA (monomer 3) studied by Brar and Hekmatyar [35] provides us with some interesting experimental data. In addition to reporting the terpolymer composition data, they also reported microstructural (triad fraction) information. Thus, there is potential to re-estimate the 'ternary' reactivity ratios for AN/Sty/MMA and evaluate their ability to predict composition and sequence length distribution.

The original investigation used six experiments (no replication is mentioned) and the feed compositions selected provide a good amount of experimental information. As shown in the triangular diagram of Figure 10, there are three 'outer' formulations further along the outside of the triangle (designated with circles in Figure 10). Although (to the best of our knowledge) these were not statistically designed experiments, the fact that there is one formulation rich in each comonomer provides useful data for reactivity ratio estimation [21]. In fact, when the reactivity ratios are estimated using only these three trials, the parameter estimation results are as expected.

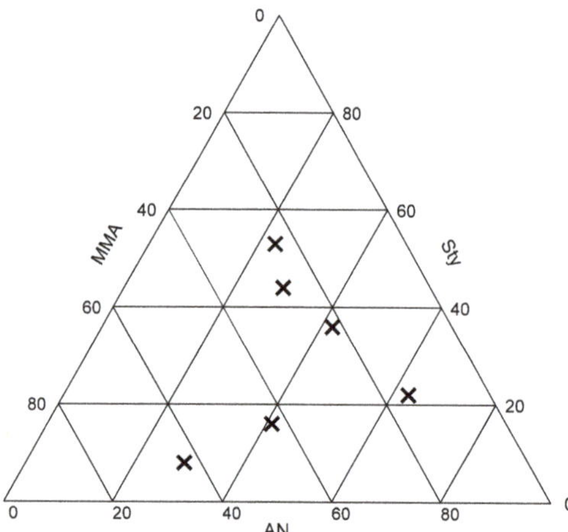

Figure 10. Terpolymerization feed compositions for AN/Sty/MMA. Data from [35].

Ideally, additional replication would be performed for all six formulations (perhaps even to higher conversion levels), but these results certainly represent carefully measured experimental (process) data (hence, good information content/lower experimental error), leading to a numerically stable estimation situation, thus ensuring that reactivity ratios can be successfully estimated even from a limited dataset.

Figure 11 shows a comparison of estimation results from these three (more optimal) points to the estimation results from the full (six) trial set. Clearly, the three 'internal' data points supplement the composition data, but do not significantly alter the reactivity ratio estimation results. Also, the 'ternary' reactivity ratio estimates are in good agreement with the previously reported binary reactivity ratios for the associated copolymers.

For all three comonomer pairs, the binary reactivity ratios are within the JCRs for the ternary estimates. Thus, the prediction performance (for both terpolymer composition and microstructure) will be similar, regardless of which parameters are used. For the purposes of demonstration, analysis of the composition and microstructure of one terpolymer sample (experimentally determined by Brar and Hekmatyar [35]) is summarized in Table 8. Note that triad fractions are defined only by the first letter of the monomer name, for example, triad fraction ASM represents the AN-Sty-MMA triad sequence.

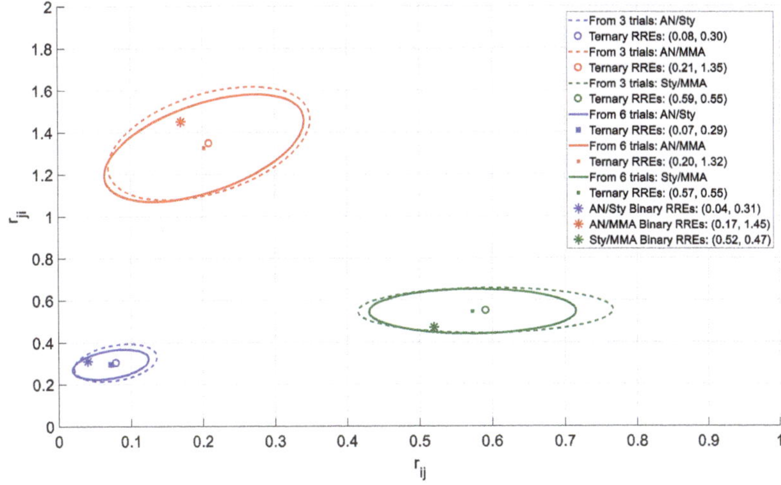

Figure 11. 'Ternary' reactivity ratio estimates (RREs) for the terpolymerization of AN/Sty/MMA with the data from Brar and Hekmatyar [35].

Table 8. Analysis of (a) composition and (b) microstructure for AN/Sty/MMA (experimental data from [35]). RRE = reactivity ratio estimates.

(a)			
Monomer	**Experimental [35]**	**Original prediction (from binary RREs) [35]**	**Current prediction (from ternary RREs)**
AN	0.30	0.28	0.30
Sty	0.48	0.48	0.48
MMA	0.22	0.24	0.22

(b)			
Triad	**Experimental [35]**	**Original prediction (from binary RREs) [35]**	**Current prediction (from ternary RREs)**
AAA	0.01	0.01	0.00
SAS	0.69	0.73	0.63
MAM	0.03	0.00	0.03
AAS	0.06	0.04	0.06
AAM	0.01	0.00	0.01
SAM	0.20	0.22	0.27
SSS	0.05	0.05	0.05
ASA	0.21	0.25	0.27
MSM	0.08	0.08	0.06
SSA	0.27	0.22	0.24
SSM	0.16	0.12	0.12
ASM	0.23	0.28	0.26
MMM	0.05	0.04	0.04
AMA	0.05	0.02	0.03
SMS	0.42	0.40	0.39
MMA	0.04	0.06	0.07
MMS	0.21	0.26	0.26
AMS	0.23	0.22	0.21

4. Conclusions

Through a series of case studies, we have demonstrated with examples and counter-examples both the challenges and advantages of estimating 'ternary' reactivity ratios directly from terpolymerization data. We highlighted some difficulties that may arise when studying multi-component polymerizations due to the nature of such systems and related experimental limitations. These limitations are usually translated to the paucity of experimental information content. The lack of sufficient information on polymer composition, combined with uncertain levels of experimental error (since independent replication is usually non-existent) usually leads to numerically ill-conditioned systems during the estimation steps. This in its turn results in a multiplicity of (and related confusion with) reactivity ratio values. Now, if the above already important limitations are superimposed to a lack of experimental design (i.e., the optimal selection of feed compositions), the problem is compounded with the extra dimension of added pitfalls with the use of undesigned (happenstance) data.

When experimental design is employed from the outset and is combined with appropriate parameter estimation techniques using carefully measured (and replicated) data, i.e., when every effort is made to make the terpolymerization system numerically 'well-behaved', then accurate and reliable estimates of 'ternary' reactivity ratios can be obtained. The case studies examined gave examples of both instantaneous (low conversion) and cumulative (medium-high conversion) data analysis and demonstrated the advantages of using the cumulative model. Some systems exhibited similarities between the 'binary' and 'ternary' reactivity ratio estimates, and the predictions related to composition and triad fractions were improved.

In some other terpolymerization systems, the reactivity ratios of the binary copolymerization pairs do not apply to ternary systems [20]. Trying to predict terpolymerization behavior from binary reactivity ratios can require making unfounded assumptions about system-specific polymerization kinetics. Researchers often assume that the presence of the third comonomer is exactly additive via simple superposition, hence the behavior of the three comonomers is independent of each other. This is the main assumption, akin to assuming that interaction terms are non-existent in a model. Evaluating subsets of the experimental data collected (that is, comonomer pairs) to represent ternary systems can ultimately result in the oversimplification of complex processes. Using appropriate kinetic models can result in better prediction performance and a higher degree of confidence in the resulting parameter estimates.

Overall, and as we have demonstrated in the current work, 'ternary' reactivity ratio estimation directly from terpolymerization data can provide improved understanding about a complex terpolymer system. Dealing for the first time with a completely unknown terpolymerization system, the current approach offers, if nothing else, a systematic and 'safe' approach to accumulating experimental evidence about the system in question in fewer experimental trials, and hence with less experimental effort, but with reliable parameter estimates which can be fine-tuned further later, as one becomes more familiar with the terpolymerization system in a sequential-iterative-optimal manner.

Author Contributions: A.J.S. analyzed the data and wrote the paper under the supervision of A.P.; V.A.G. and M.A.D. contributed experimental data for analysis and provided insights about the analysis of results.

Funding: This paper received no specific external funding, but all authors are grateful to the Natural Sciences and Engineering Research Council (NSERC) of Canada and the Canada Research Chair (CRC) program for financial support.

Conflicts of Interest: The authors declare no conflict of interest.

Appendix A Additional Data for Case Studies

Table A1. Simulated data for the BA/BMA/lim terpolymerization (supplementing experimental data from Ren et al. [27]; see Section 3.1.2).

X	$f_{BA,0}$	$f_{BMA,0}$	$f_{lim,0}$	$\bar{F}_{BA}$	$\bar{F}_{BMA}$	$\bar{F}_{lim}$
*0.041	0.098	0.791	0.111	0.053	0.931	0.016
*0.102	0.101	0.793	0.106	0.086	0.904	0.010
*0.153	0.100	0.802	0.098	0.044	0.928	0.028
0.221	0.111	0.800	0.089	0.038	0.916	0.046
0.269	0.085	0.804	0.112	0.072	0.895	0.033
0.313	0.101	0.801	0.098	0.055	0.936	0.010
0.361	0.103	0.795	0.102	0.062	0.897	0.041
0.420	0.091	0.808	0.102	0.076	0.890	0.034

Table A1. *Cont.*

X	$f_{BA,0}$	$f_{BMA,0}$	$f_{lim,0}$	$\bar{F}_{BA}$	$\bar{F}_{BMA}$	$\bar{F}_{lim}$
0.460	0.102	0.797	0.101	0.066	0.904	0.031
0.514	0.109	0.801	0.090	0.054	0.894	0.052
0.578	0.114	0.794	0.092	0.039	0.931	0.030
0.620	0.097	0.802	0.101	0.089	0.893	0.018
0.671	0.108	0.799	0.093	0.090	0.879	0.031
0.728	0.094	0.801	0.106	0.119	0.866	0.015
0.780	0.100	0.802	0.097	0.062	0.893	0.046
0.833	0.102	0.798	0.100	0.072	0.880	0.047
0.892	0.107	0.797	0.095	0.052	0.881	0.067
0.936	0.117	0.797	0.085	0.122	0.838	0.041
0.983	0.109	0.793	0.098	0.119	0.795	0.086
*0.055	0.088	0.107	0.805	0.376	0.333	0.292
*0.110	0.103	0.093	0.805	0.387	0.294	0.319
*0.160	0.098	0.101	0.801	0.367	0.317	0.316
0.209	0.088	0.109	0.803	0.366	0.298	0.337
0.260	0.119	0.096	0.785	0.363	0.286	0.351
0.308	0.091	0.103	0.806	0.322	0.269	0.409
0.359	0.099	0.105	0.796	0.249	0.275	0.476
0.419	0.101	0.099	0.801	0.225	0.246	0.528
0.469	0.098	0.101	0.800	0.215	0.211	0.574
0.520	0.106	0.090	0.804	0.159	0.207	0.634
0.576	0.102	0.098	0.801	0.168	0.183	0.648
0.624	0.090	0.100	0.809	0.138	0.180	0.683
0.672	0.116	0.091	0.793	0.152	0.162	0.686
0.727	0.103	0.100	0.797	0.129	0.149	0.722
0.781	0.112	0.090	0.798	0.112	0.129	0.759
0.834	0.097	0.101	0.802	0.135	0.113	0.752
0.890	0.105	0.097	0.798	0.111	0.120	0.770
0.935	0.098	0.102	0.800	0.112	0.120	0.769
0.993	0.087	0.110	0.803	0.102	0.092	0.807

Table A2. Low conversion data for the BA/MMA/EHA terpolymerization (experimental data from Gabriel and Dubé [12]; see Section 3.2.1).

X	$f_{BA,0}$	$f_{MMA,0}$	$f_{EHA,0}$	$\bar{F}_{BA}$	$\bar{F}_{MMA}$	$\bar{F}_{EHA}$
0.027	0.098	0.097	0.805	0.013	0.226	0.761
0.140	0.098	0.097	0.805	0.039	0.201	0.760
0.033	0.100	0.800	0.100	0.039	0.843	0.118
0.088	0.100	0.800	0.100	0.062	0.824	0.114
0.171	0.100	0.800	0.100	0.087	0.814	0.099
0.003	0.800	0.099	0.100	0.654	0.236	0.110
0.026	0.800	0.099	0.100	0.654	0.236	0.110

References

1. Rudin, A. Copolymerization. In *Elements of Polymer Science and Engineering: An Introductory Text and Reference for Engineers and Chemists*; Elsevier Science & Technology Books: Amsterdam, The Netherlands, 1998; Volume 7, pp. 241–276.
2. Scott, A.J.; Penlidis, A. Copolymerization. In *Reference Module*; Elsevier: Amsterdam, The Netherlands, 2017.
3. Smith, O.L.; Kim, Y.; Kathaperumal, M.; Gadinski, M.R.; Pan, M.; Wang, Q.; Perry, J.W. Enhanced permittivity and energy density in neat poly (vinylidene fluoride-trifluoroethylene-chlorotrifluoroethylene) terpolymer films through control of morphology. *ACS Appl. Mater. Interfaces* **2014**, *6*, 9584–9589. [CrossRef] [PubMed]
4. Li, Q.; Wang, Q. Ferroelectric polymers and their energy-related applications. *Macromol. Chem. Phys.* **2016**, *217*, 1228–1244. [CrossRef]

5. Pujari, N.S.; Wang, M.; Gonsalves, K. Co and terpolymer reactivity ratios of chemically amplified resists. *Polymer* **2017**, *118*, 201–214. [CrossRef]
6. Zhou, J.; Liu, J.; Cheng, C.J.; Patel, T.R.; Weller, C.E.; Piepmeier, J.M.; Jiang, Z.; Saltzman, W.M. Biodegradable poly (amine-co-ester) terpolymers for targeted gene delivery. *Nat. Mater.* **2012**, *11*, 82–90. [CrossRef] [PubMed]
7. Chung, H.; Grubbs, R.H. Rapidly cross-linkable DOPA containing terpolymer adhesives and PEG-based cross-linkers for biomedical applications. *Macromolecules* **2012**, *45*, 9666–9673. [CrossRef]
8. Chauhan, N.P.S. Preparation and characterization of bio-based terpolymer derived from vanillin oxime, formaldehyde, and p-hydroxyacetophenone. *Des. Monomers Polym.* **2014**, *17*, 176–185. [CrossRef]
9. Mecham, S.; Sentman, A.; Sambasivam, M. Amphiphilic silicone copolymers for pressure sensitive adhesive applications. *J. Appl. Polym. Sci.* **2010**, *116*, 3265–3270. [CrossRef]
10. Bowman, H.; Hudson, B.W. Terpolymers as Pressure Sensitive Adhesives. U.S. Patent 9598532B2, 21 March 2017.
11. Roberge, S.; Dubé, M.A. Emulsion-based pressure sensitive adhesives from conjugated linoleic acid/styrene/butyl acrylate terpolymers. *Int. J. Adhes. Adhes.* **2016**, *70*, 17–25. [CrossRef]
12. Gabriel, V.A.; Dubé, M.A. Bulk free-radical co- and terpolymerization of n-butyl acrylate/2-ethylhexyl acrylate/methyl methacrylate. *Macromol. React. Eng.* **2018**, *13*, 1800057. [CrossRef]
13. Hong, F.; Xie, L.; He, C.; Liu, J.; Zhang, G.; Wu, C. Effects of hydrolyzable comonomer and cross-linking on anti-biofouling terpolymer coatings. *Polymer* **2013**, *54*, 2966–2972. [CrossRef]
14. Pollack, K.A.; Imbesi, P.M.; Raymond, J.E.; Wooley, K.L. Hyperbranched fluoropolymer-polydimethylsiloxane-poly(ethylene glycol) cross-linked terpolymer networks designed for marine and biomedical applications: Heterogeneous nontoxic antibiofouling surfaces. *ACS Appl. Mater. Interfaces* **2014**, *6*, 19265–19274. [CrossRef] [PubMed]
15. Yousefi, F.K.; Jannesari, A.; Pazokifard, S.; Saeb, M.R.; Scott, A.J.; Penlidis, A. Terpolymerization of triisopropylsilyl acrylate, methyl methacrylate, and butyl acrylate: Reactivity ratio estimation. *Macromol. React. Eng.* **2019**, 1900014. [CrossRef]
16. Slocombe, R. Multicomponent polymers. I. Three-component systems. *J. Polym. Sci.* **1957**, *XXVI*, 9–22. [CrossRef]
17. Hagiopol, C. *Copolymerization: Toward a Systematic Approach*; Plenum Publishers: New York, NY, USA, 1999.
18. Kazemi, N.; Duever, T.A.; Penlidis, A. Demystifying the estimation of reactivity ratios for terpolymerization systems. *AIChE J.* **2014**, *60*, 1752–1766. [CrossRef]
19. Scott, A.J.; Duever, T.A.; Penlidis, A. The role of pH, ionic strength and monomer concentration on the terpolymerization of 2-acrylamido-2-methylpropane sulfonic acid, acrylamide and acrylic acid. *Polymer* **2019**, *177*, 214–230. [CrossRef]
20. Scott, A.J.; Penlidis, A. Binary vs. ternary reactivity ratios: Appropriate estimation procedures with terpolymerization data. *Eur. Polym. J.* **2018**, *105*, 442–450. [CrossRef]
21. Kazemi, N.; Duever, T.A.; Penlidis, A. Design of optimal experiments for terpolymerization reactivity ratio estimation. *Macromol. React. Eng.* **2015**, *9*, 228–244. [CrossRef]
22. Kazemi, N.; Duever, T.A.; Penlidis, A. Reactivity ratio estimation from cumulative copolymer composition data. *Macromol. React. Eng.* **2011**, *5*, 385–403. [CrossRef]
23. Alfrey, T.; Goldfinger, G. Copolymerization of systems of three and more components. *J. Chem. Phys.* **1944**, *12*, 322–323. [CrossRef]
24. Kazemi, N.; Duever, T.A.; Penlidis, A. A powerful estimation scheme with the error-in-variables model for nonlinear cases: Reactivity ratio estimation examples. *Comput. Chem. Eng.* **2013**, *48*, 200–208. [CrossRef]
25. Reilly, P.M.; Reilly, H.V.; Keeler, S.E. Parameter estimation in the error-in-variables model. *J. R. Stat. Soc. Ser. C (Appl. Stat.)* **1993**, *42*, 693–701.
26. Mao, R.; Huglin, M.B. A new linear method to calculate monomer reactivity ratios by using high conversion copolymerization data: Terminal model. *Polymer* **1993**, *34*, 1709–1715. [CrossRef]
27. Ren, S.; Zhang, L.; Dubé, M.A. Free-radical terpolymerization of n-butyl acrylate/butyl methacrylate/D-limonene. *J. Appl. Polym. Sci.* **2015**, *132*, 42821. [CrossRef]
28. Zhang, Y.; Dubé, M.A. Copolymerization of n-butyl methacrylate and D-limonene. *Macromol. React. Eng.* **2014**, *8*, 805–812. [CrossRef]

29. Ren, S.; Trevino, E.; Dubé, M.A. Copolymerization of limonene with n-butyl acrylate. *Macromol. React. Eng.* **2015**, *9*, 339–349. [CrossRef]

30. Ren, S.; Hinojosa-Castellanos, L.; Zhang, L.; Dubé, M.A. Bulk free-radical copolymerization of n-butyl acrylate and n-butyl methacrylate: Reactivity ratio estimation. *Macromol. React. Eng.* **2017**, *11*, 1600050. [CrossRef]

31. Scott, A.J.; Penlidis, A. Computational package for copolymerization reactivity ratio estimation: Improved access to the error-in-variables-model. *Processes* **2018**, *6*, 8. [CrossRef]

32. Skeist, I. Copolymerization: The composition distribution curve. *J. Am. Chem. Soc.* **1946**, *68*, 1781–1784. [CrossRef]

33. Dubé, M.A.; Penlidis, A. A systematic approach to the study of multicomponent polymerization kinetics—the butyl acrylate/methyl methacrylate/vinyl acetate example: 1. Bulk copolymerization. *Polymer* **1995**, *36*, 587–598. [CrossRef]

34. Schoonbrood, H.A.S. *Emulsion Co- and Terpolymerization: Monomer Partitioning, Kinetics and Control of Microstructure and Mechanical Properties*; Technische Universiteit Eindhoven: Eindhoven, The Netherlands, 1994.

35. Brar, A.S.; Hekmatyar, S.K. Microstructure determination of the acrylonitrile-styrene-methyl methacrylate terpolymers by NMR spectroscopy. *J. Appl. Polym. Sci.* **1999**, *74*, 3026–3032. [CrossRef]

Article

Handling Constraints and Raw Material Variability in Rotomolding through Data-Driven Model Predictive Control

Abhinav Garg, Hassan A. Abdulhussain, Prashant Mhaskar * and Michael R. Thompson

Department of Chemical Engineering, McMaster University, Hamilton, ON L8S 4L8, Canada
* Correspondence: mhaskar@mcmaster.ca; Tel.: +1-905-525-9140 (ext. 23273)

Received: 29 July 2019; Accepted: 5 September 2019; Published: 10 September 2019

Abstract: This work addresses the problems of uniquely specifying and robustly achieving user-specified product quality in a complex industrial batch process, which has been demonstrated using a lab-scale uni-axial rotational molding process. In particular, a data-driven modeling and control framework is developed that is able to reject raw material variation and achieve product quality which is specified through constraints on quality variables. To this end, a subspace state-space model of the rotational molding process is first identified from historical data generated in the lab. This dynamic model predicts the evolution of the internal mold temperature for a given set of input move trajectory (heater and compressed air profiles). Further, this dynamic model is augmented with a linear least-squares based quality model, which relates its terminal (states) prediction with key quality variables. For the lab-scale process, the chosen quality variables are sinkhole area, ultrasonic spectra amplitude, impact energy and shear viscosity. The complete model is then deployed within a model-based control scheme that facilitates specifying on-spec products via limits on the quality variables. Further, this framework is demonstrated to be capable of rejecting raw material variability to achieve the desired specifications. To replicate raw material variability observed in practice, in this work, the raw material is obtained by blending the matrix resin with a resin of slightly different viscosity at varying weight fractions. Results obtained from experimental studies demonstrate the capability of the proposed model predictive control (MPC) in meeting process specifications and rejecting raw material variability.

Keywords: subspace identification; polymer processing; model predictive control; rotational molding; batch process modeling and control

---

1. Introduction

Rotational Molding or rotomolding is a plastics processing batch process used for the manufacture of seamless hollow plastic products [1]. The process consists of distinct heating and cooling phases. The process constitutes filling a mold with powdered charge and rotating it slowly in a heated oven (using heaters) causing the resin to soften and stick to the walls. Continuously rotating the mold during the subsequent cooling phase produces an end product with even wall thickness. The products tend to be quite large including household water tanks and fuel tanks for agricultural equipment and marine vessels, which means that poor control over part quality results in substantial waste costs for a company. The key objectives in rotational molding are to obtain products, from one batch to another, with desired characteristics consistently while avoiding incomplete sintering or degradation due to extended overheating. Minimized wastage requires a quality control framework capable of compensating for lot-to-lot variance in the properties of the initial charge, and determining process 'recipes' meeting user-specified bounds on the quality variables, especially in for cases like the rotomolding example where the quality variables are not available for measurement during the process [2].

One approach to achieving the objectives is to develop a first-principles/physics-based model that will predict the evolution of the process variables based on candidate input variables. That model can be used to design a controller to achieve an on-spec product. Realistic incorporation of heat transfer, discrete particle dynamics and polymer rheology for this first-principles approach necessitate a very complex model structure with a high number of parameters. As is often the case, the development of a realistic mechanistic model is a difficult task, and even if developed, might be very challenging to maintain, or to use for optimization and control. Thus the first principles model-based model predictive control (MPC) [3–5] implementations remain elusive for rotomolding control.

In the absence of good first principles model, rotational molding processes have utilized recipe-based open loop policy. Thus, input trajectories that could yield a desired product are determined through a large number of experiments for a new product, yielding high waste rates. The assumption here is that by repeating previous successful input profiles, a desired product could be replicated. However, this approach is susceptible to disturbances, and, equally importantly, requires to be entirely redone for a different set of quality attribute requirements.

Another approach used to partially reject disturbances is trajectory tracking [6]. In this approach, a key measured process variable (such as the internal mold temperature), is tracked to a predefined set-point trajectory. However, even perfect trajectory tracking may not yield the desired product quality because of the possible change in the relationship between the measured/tracked variable and final quality variable due to varying process conditions across batches.

An abundance of historical data in most industrial processes has motivated the development of data-driven modeling and batch quality control approaches. Partial least squares (PLS) is one of the most popular and widely used batch process modeling methods. These models capture the essence of the process dynamics in a projected (lower) dimensional space known as latent space [7,8]. colorredThe PLS model structure resembles linear time-varying (LTV) models. These models predict deviations from mean past trajectories and relate to differences in final product quality. The PLS based modeling approach requires the batches to be of the same duration, which is seldom a case in practice. The remedy is to recognize an appropriate alignment variable for the training and validation batches. While this problem can generally be handled in the context of process monitoring, or control of fixed length batches, the use of these techniques for batches where the batch duration itself could be a decision variable, remains challenging.

More recently, data-driven model designs for continuous processes [9–11] were adapted and a subspace identification based batch modeling and control approach was proposed [12]. Here, the past batch database is used to determine a linear time-invariant (LTI) state-space model of the batch process. The approach, in order to capture nonlinear process dynamics, draws on utilizing a high number of 'states' as required. The merit of this approach lies in its ability to accommodate batches with variable duration without the need for batch alignment, as demonstrated through varied applications [2,12–15].

In a recent work [2], this approach has been used for rotomolding to achieve desired product quality specifications. The results [2] demonstrate the efficacy of the approach in terms of achieving improved product quality in validation batches, and the capability to achieve a specified product by expressing appropriate weights on the quality variables in the objective function of the resultant MPC optimization and control implementation. In practice, the rotomolding operation often has to grapple with minor yet significant raw material variability, manifested as varying flowability of the input charge from one molded part to another. The other challenge with the former approach [2] of indirectly enforcing quality requirements through the objective function is that a careful design of the objective function is required. The practitioner has to identify the desired batches from historical data and decide on the weights in the objective function such that minimization/maximization of quality variables will result. A much more intuitive, practical and desirable way for the practitioner to specify quality is through placing constraints on the quality variables.

Motivated by these considerations, this work presents a subspace identification based modeling and control approach for handling user-specified quality through constraints and demonstrates explicitly the ability to reject raw material variability in a rotational molding process. For the presented study, the raw material variability observed in practice is replicated in this case by blending the matrix polymer resin with a near-similar resin with slightly different flow properties. The rest of the paper is structured as follows: Section 2 describes the rotomolding process and reviews the batch subspace identification approach. The proposed MPC design is presented in Section 3 and closed-loop experimental results presented in Section 4. Finally, a few concluding remarks are made in Section 5.

2. Preliminaries

In this section, a brief overview of the rotational molding process used in this work and batch subspace identification based process modeling technique is given. This section forms the basis for the proposed novel MPC design and corresponding closed-loop results presented in Sections 3 and 4 respectively.

2.1. Rotational Molding

In this work, a high density powder (ExxonMobil^{TM} HD 8660.29, Imperial Oil, Sarnia, ON, Canada) and a linear low density polyethylene powder (ExxonMobil^{TM} LL 8460.29, Imperial Oil, Sarnia, ON, Canada) were utilized. Both resins were donated by Imperial Oil Ltd. The melt flow index (MFI) of HD 8660.29 is 2 g/10 min, while the MFI of LL 8460.29 is 3 g/10 min. Both MFI were tested with a standard weight of 2.16 kg according to ASTM Standard D1238-13. The melt temperature of HD 8660.29 and LL 8460.29 are 129 °C and 126 °C respectively, according to the vendor.

The samples were prepared using a laboratory-scale uni-axial rotational molding machine (see Figure 1, McMaster University, Hamilton, ON, Canada) with a custom LabVIEW program (National Instruments Corporation, Austin, TX, USA) for setting the inputs to the system. The input includes power to the left and right panels of the oven heater and the compressed air supply, with the measured variable being the internal air temperature profile inside the mold. The internal temperature was monitored using a K-type thermocouple mounted in the center of the mold, while heater temperature was monitored by similar thermocouples mounted against the respective panels. For every batch, a charge of 100 g of HD 8660.29 (matrix resin) or HD 8660.29 blended with a small fraction of LL 8460.29 was loaded into the mold. The mold was then closed and rotated at a constant speed of 4 RPM (rotations per minute). The oven was heated to 300 °C and lowered from above to surround the mold. A MATLAB (The Mathworks Inc., Natick, MA, USA) script was run to control the trajectory of the internal air temperature. After the mold reached the optimal number of control steps determined by the proposed approach, the oven was removed and a stream of forced air applied at a speed of 2.5 m/s was used to cool the mold to 80 °C (internal temperature). Thereafter, the sample is taken out of the mold for further characterization. The final dimensions of the mold were a $90 \times 90 \times 3$ mm cube though only the four peripheral walls are coated with polymer during processing. The measurement technique for the quality variables is described in the next section.

2.1.1. Surface Void Analysis

The quality of a molded part is determined based on the completeness by which the polymer particles sinter together during heating. Both the strength and appearance of the part are related to this factor of processing. One measurable quality parameter to assess the completeness of sintering is to visually inspect the molded wall for voids. To highlight the surface voids of an otherwise white-colored sample, a low viscosity lubricant containing a mixture of copper and graphite particles was rubbed onto one of the faces of a sample, with the excess lubricant removed using a paper towel. An image of the face is taken using a digital camera and image analysis software was utilized to estimate the void area for a 40×40 mm section. The sinkhole area was calculated by dividing the void area by the area of the section (1600 mm^2). Figure 2 shows the processed image of a product with a sinkhole area of

0.2%, 2%, and 3.95%. A good product is often expected to have the smallest void area which suggests completeness of sintering and thus better appearance and impact properties of the product.

Figure 1. Experimental setup for the rotational molding.

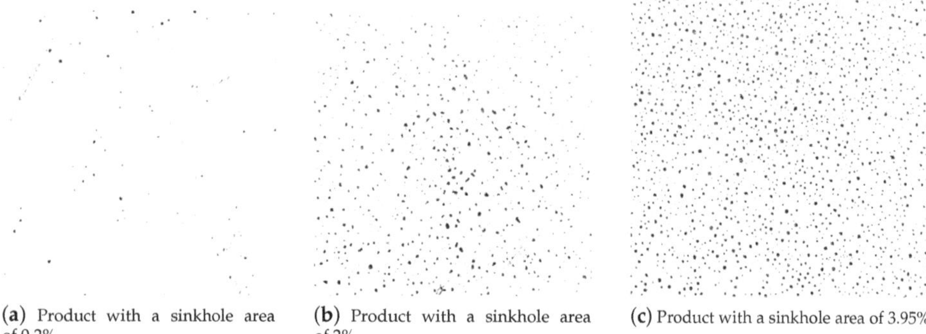

(**a**) Product with a sinkhole area of 0.2%

(**b**) Product with a sinkhole area of 2%

(**c**) Product with a sinkhole area of 3.95%

Figure 2. Surface void analysis.

2.1.2. Ultrasonic Spectroscopy

While mechanical testing can destructively test the quality of a part for completeness of sintering, new non-destructive methods based on acoustics are gaining acceptance and this study sought to maximize the possibility of data collection related to the rotomolding process. Ultrasonic testing for one of the faces of the uncut sample was carried out using the same setup as reported in previous studies [16]. The acoustic transducers (R15a, resonant and F30, broadband by Physical Acoustics, Physical Acoustics Corporation, Princeton Junction, NJ, USA) were attached to the face of the sample using high vacuum grease (Dow Corning), being kept 55 mm apart. A series of signals were emitted successively between 135 to 165 kHz in 1 kHz steps. Each signal propagated through the sample and was recorded at a rate of 4 MHz using a data acquisition system by National Instruments. The detected signals were converted to the frequency domain using a fast Fourier transform and combined into a single spectrum. This quality variable corresponded to the maximum amplitude (from all 31 signals emitted).

2.1.3. Impact Test

Samples were cut and frozen at $-40\,^{\circ}\mathrm{C}$ for 24 h in preparation for dart impact testing. The initial falling dart height was selected at 0.762 m (2.5 ft) and moved using the staircase method based on whether the sample failed from impact at its current height. A standard dart weight of 6.804 kg (15 lbs) was utilized for this test. If the sample failed, then the height of the dart is decreased by 0.1524 m (0.5 ft), whereas if the sample did not fail, then the height of dart should be increased by 0.1524 m. The impact energy was calculated based on the height where 50% of the samples failed (each face of the mold being used for analysis from each batch, thus resulting in four measurements, to ultimately yield an average value).

2.1.4. Rheology

As any polymer is heated, it experiences thermal damage. Initially, the harm is minor but as time progresses, the damage can significantly affect physical properties. During rotational molding, this degradation proceeds over the course of molding a part and at a certain period of time, the benefits to complete sintering are offset by the damage to the polymer. Degradation of a Sample was monitored in this study by measuring resin viscosity using a 25 mm parallel plate rheometer (DHR TA Instruments, TA Instruments, New Castle, DE, USA). A frequency sweep test was implemented at a strain of 0.15 covering shear rates between 0.1 to 200 s^{-1} for a temperature of 190 $^{\circ}\mathrm{C}$. Three tests are performed per sample to account for measurement variation. The data was fit using the Cross model as shown in Equation (1) using the TRIOS instruments software (TA Instruments, New Castle, DE, USA):

$$\eta = \eta_\infty + \frac{\eta_0 - \eta_\infty}{1 + (C\dot{\gamma})^m} \tag{1}$$

Here, η is shear viscosity, η_∞ is the infinite-shear viscosity parameter, η_0 is the zero-shear viscosity parameter, C is the Cross time constant, $\dot{\gamma}$ is the shear rate constant and m is the Cross rate constant. In this work, only the zero-shear viscosity was considered as the representative quality parameter for thermal degradation. Other parameters were not evaluated in the quality model.

2.2. Subspace Identification Approach for Batch Processes

To obtain a desired product quality consistently across batches, a model based control framework is necessary. This requires a good process model capable of predicting the process and quality variable evolution for a candidate input sequence. In our proposed approach, we achieve this by first identifying a dynamic model for predicting the internal mold temperature given a candidate heater power and compressed air flow rate profile. This model is then augmented with another model that captures the relationship between the final quality and the end point state prediction of the dynamic model.

In the present work, a deterministic subspace identification algorithm, adapted for batch processes, is used to identify the dynamic process model. The deterministic identification problem is that of determining the order n and system matrices of a state space model given input and output measurements. The identified model takes the following form:

$$x_{k+1}^d = \mathbf{A}x_k^d + \mathbf{B}u_k, \tag{2a}$$

$$y_k = \mathbf{C}x_k^d + \mathbf{D}u_k, \tag{2b}$$

where $\mathbf{A} \in \mathbb{R}^{n \times n}$, $\mathbf{B} \in \mathbb{R}^{n \times m}$, $\mathbf{C} \in \mathbb{R}^{l \times n}$, $\mathbf{D} \in \mathbb{R}^{l \times m}$ are the associated system matrices and are determined up to within a similarity transformation.

Subspace identification methods are non-iterative in nature and compute the unknown parameters using matrix algebra (see, e.g., [17,18]) which distinguishes them from classical system identification approach (see, e.g., [19]). A number of algorithms have been proposed in subspace identification among which some of the most prominent ones include canonical variate analysis (CVA) [20], numerical algorithms for subspace state space system identification (N4SID) [21] and multivariable output error state space algorithm (MOESP) [22]. These algorithms differ only in the weighting of the matrix used at the singular value decomposition step as shown in [23]. However, these algorithms were designed primarily for continuous processes (see, e.g., [11,24,25]). Thus, the training datasets are obtained through identification experiments carried around a desired steady-state condition. This facilitates straightforward construction of Hankel matrices within the subspace identification algorithm.

In contrast, in a batch process data is collected from multiple batches with the objective of identifying a model for the transient dynamics of the process. For instance, consider the output measurements of a batch process denoted as $\mathbf{y}^{(b)}[k]$, where k is the sampling instant since batch initiation, and b denotes the batch index. The output Hankel matrix for a batch b is given by:

$$\mathbf{Y}_{1|i}^{(b)} = \begin{bmatrix} \mathbf{y}^{(b)}[1] & \mathbf{y}^{(b)}[2] & \cdots & \mathbf{y}^{(b)}\left[j^{(b)}\right] \\ \vdots & \vdots & & \vdots \\ \mathbf{y}^{(b)}[i] & \mathbf{y}^{(b)}[i+1] & \cdots & \mathbf{y}^{(b)}\left[i+j^{(b)}-1\right] \end{bmatrix} \tag{3}$$

The key question then is how to appropriately utilize data from multiple batches. In essence, the significant difference in the structure of data collected in batch processes, in comparison to continuous processes, calls for specific adaptation of the batch database and subspace identification algorithms. A naive concatenation of the historical batches data into one 'continuous' data-set would result in the incorrect assumption that starting point of one batch is similar to the previous batch's end-point. The solution is to construct a single pseudo-Hankel matrix for both input and output data such that it recognizes the batch nature of the data. This is done by constructing pseudo-Hankel matrices of the following form [12]:

$$\mathbf{Y}_{1|i} = \begin{bmatrix} \mathbf{Y}_{1|i}^{(1)} & \mathbf{Y}_{1|i}^{(2)} & \cdots & \mathbf{Y}_{1|i}^{(nb)} \end{bmatrix} \tag{4}$$

where, nb is the number of batches used for training. Similarly, pseudo-Hankel matrices for input data are formed. This method does not require batches to be of same duration and thus mean-centring around nominal trajectory or calculation of an alignment variable are not required. This is in contrast to the time-dependent modeling approaches such as PLS.

Past the formation of the pseudo Hankel matrices, the deterministic algorithm presented in [26], and appropriately adapted in [12] is used for modeling of the rotational molding process. Note that in principle, the batch subspace identification algorithm discussed above can be used with appropriately adapting existing subspace identification algorithms. After obtaining the state trajectories, the system matrices are estimated using ordinary least squares (see [27] for detailed discussion of the algorithm).

2.3. Model Identification

The current lab scale rotational molding process consists of three inputs and one output variable. Control action for two heaters and a compressed air supply constitutes the process manipulated inputs while mold internal temperature is the measured output. For the model identification, a database consisting of 10 different batches is used. The database is segregated into training and validation batches, utilizing seven batches for training, while keeping the remaining three for validation purposes. As evidenced by the validation test, the training batches provide a sufficiently rich data set for model identification. The richness is due to the variation in the heating cycle time across batches thereby covering a large enough operating space for the process. A model only for the heating cycle is identified in this work. No control action was active during the cooling phase.

Subsequently, a state-space model of order two was identified using the proposed approach as discussed in Section 2.2. The model order was selected to ensure good prediction of the internal mold temperature in the validation batches. The open-loop model predictions for one of the validation batches was as shown in Figure 3. Model predictions from 70th, 75th and 85th sampling instants are shown. It can be observed from the figure that as the batch progresses, more data becomes available which results in improved model predictions.

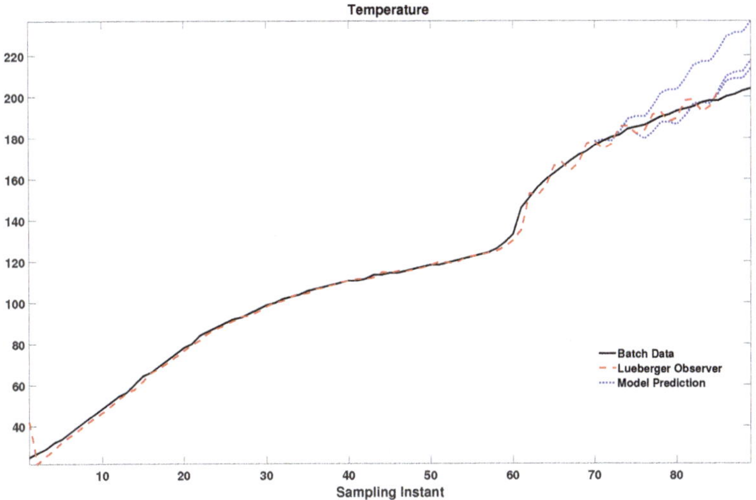

Figure 3. Model validation on a 'new' batch (heating phase only).

The identified state-space model alone is not sufficient to predict the quality of the finished sample at the completion of batch. For this, it is augmented with a least squares based linear quality model obtained by relating the terminal states of the state-space model to the quality measurements of the training batches (see Table 1) as follows.

$$Q_{t_f} = \hat{L}_m x[t_{f_{heat}}] + e, \tag{5}$$

where Q_{t_f} denote quality measurements at the termination of batch at time t_f, $\hat{L}_m$ is the matrix relating the terminal states to the terminal quality, $x[t_{f_{heat}}]$ are terminal states of the subspace model (i.e., at completion of the heating cycle at time $t_{f_{heat}}$) and e represents white noise. The predictions for the validation batch using the identified quality model are as listed in the Table 2. It can be observed that the predictions in general improve as batch progresses and are close to the actual values. In summary, the state space model together with this quality model completely describes the dynamics of the rotomolding process. Note that in this work internal mold temperature prediction alone formed the criterion for order selection of the state space model. An alternative is to incorporate the quality model predictions as well in the decision making.

Table 1. Quality measurements for training batches.

Batches	Q_1	Q_2	Q_3	Q_4
Batch 1	4.42	19.27	5.70	3596
Batch 2	5.46	5.43	5.18	4260
Batch 3	4.15	4.90	5.70	4276
Batch 4	5.41	4.59	5.70	4230
Batch 5	4.30	4.15	4.67	4441
Batch 6	3.88	4.31	4.15	4377
Batch 7	4.78	4.65	4.67	4387

Table 2. Quality measurements predictions for validation batch.

Batches	Q_1	Q_2	Q_3	Q_4
Actual	1.45	17.30	5.18	4681
Predicted ($t = 70$)	4.5	16.3	6.10	4740
Predicted ($t = 75$)	4.1	14.5	5.5	4296
Predicted ($t = 85$)	4.1	15.3	5.5	4292

3. Model Predictive Control Design for Rotational Molding Process

The state space model along with the quality model, as described in the previous section, results in a model capturing the process dynamics sufficiently well. The second step in data-driven modeling and control approach is then to incorporate this data-driven model with an MPC scheme to achieve desired product quality. A primary requirement on the model, for it to be suitable in MPC, is its ability to make reasonable prediction of the process evolution. In our proposed approach, good model predictions are direct function of the initial state estimates and the state-space (and quality) model matrices. As the initial state information for a new batch is not available upfront, the approach requires state estimation during initial operation of the batch. To achieve this, a PI controller is used till the internal mold temperature reached 130 °C, a predefined threshold in this work. During this phase, control actions for the two heaters and the compressed air supply are obtained by set-point tracking of the two oven temperatures (fixed at 300 °C). Subsequently, a state estimator (a Luenberger observer in the present manuscript, see Equation (6)), is run to obtain state estimates. During this phase of the controller design, essentially the mechanism obtains information regarding the phenomena of heating, adhesion and melting of the powder in that particular batch. Once a reasonable state estimate has been obtained (gauged by the accuracy of the estimated output), the controller is switched to MPC which uses the state information to appropriately control the rest of the sintering (heating) phase. A standard Luenberger observer takes the following form:

$$\hat{x}[k+1] = A\hat{x}[k] + Bu[k] + L(y[k] - \hat{y}[k]), \tag{6}$$

where $\mathbf{L}$ is the observer gain and is designed based on the user specified eigenvalues of $(\mathbf{A} - \mathbf{LC})$. These are chosen to ensure $(\mathbf{A} - \mathbf{LC})$ is stable (within the unit circle). In MATLAB, this can be achieved using the place command.

In batch operations, the control objective and the manner in which the control horizon evolves are very different in comparison to continuous operation. This requires appropriate controller design for batch processes. Further, in rotational molding control, the length of the heating cycle and thus the batch duration is also a decision variable in the MPC controller design.

The control action was computed and implemented every 10 s through the MATLAB-LabView interface. At a sampling instance l, the optimal input trajectory till the end of the batch was obtained through solution of the following optimization problem:

$$\min_{\mathbf{U}_{f},l_f} \beta \hat{Q}_{t_f}[l_f - l] \tag{7a}$$

$$\text{s.t.} \quad U_{j,min} \leq u_f[k] \leq U_{j,max}, \quad \forall\ 0 \leq k \leq l_f - l \tag{7b}$$

$$|u_f[0] - u[l-1]| \leq \delta, \tag{7c}$$

$$|u_f[k] - u[k-1]| \leq \delta, \quad \forall\ 1 \leq k \leq l_f - l \tag{7d}$$

$$\hat{x}[0] = \hat{x}[l] \tag{7e}$$

$$l_f \in \{t_{switch} + 300, t_{switch} + 350, t_{switch} + 400\} \tag{7f}$$

$$\Lambda \hat{Q}_{t_f} \leq \Gamma \tag{7g}$$

$$\hat{x}[k+1] = \mathbf{A}\hat{x}[k] + \mathbf{B}u_f[k] \tag{7h}$$

$$\hat{y}[k] = \mathbf{C}\hat{x}[k] + \mathbf{D}u_f[k] \quad \forall\ 0 \leq k \leq l_f - l \tag{7i}$$

$$\hat{Q}_{t_f} = L_m \hat{x}[l_f] \tag{7j}$$

with,

$$\hat{Q} = \begin{bmatrix} \hat{Q}_1 & \hat{Q}_2 & \hat{Q}_3 & \hat{Q}_4 \end{bmatrix}^T \tag{8a}$$

$$\beta = \begin{bmatrix} 1 & 0 & 0 & 1/1000 \end{bmatrix} \tag{8b}$$

$$\Lambda = \begin{bmatrix} 1 & 0 & 0 & 0 \\ 0 & 0 & 0 & 1 \end{bmatrix} \tag{8c}$$

$$\Gamma = \begin{bmatrix} 2 \\ 12000 \end{bmatrix} \tag{8d}$$

where, $\mathbf{U}_f = [\mathbf{u}_f[0], \mathbf{u}_f[1], ...\mathbf{u}_f[l_f - l]]$ is the decision variable consisting of two heaters and the air supply control action for the remainder of the batch, l_f denotes the heating cycle termination time. The set of possible termination times, l_f is specified in Equation (7f) based on experience. $\delta = \begin{bmatrix} 30 & 30 & 30 \end{bmatrix}^T$ is the permitted rate of input change specified in Equation (7c) and (7d), and $U_{j,min}$ and $U_{j,max}$ are the specified lower and upper bounds on the manipulated variable (Equation (7b)) with $U_{j,min}$ and $U_{j,max}$ being $\begin{bmatrix} 0 & 0 & 0 \end{bmatrix}^T$ and $\begin{bmatrix} 100 & 100 & 100 \end{bmatrix}^T$ respectively. In addition, t_{switch} denotes the time (in *sec*) at which the controller switches to MPC from PI controller. Equation (7e) denotes the Luenberger observer generated state estimate at the lth instant. Further, Equation (7g) represents the user-specified constraints on the quality variables. Finally, Equation (7j) specifies the quality model which predicts the terminal product quality using the terminal states (of the heating cycle) which in-turn are predicted by the dynamic model as specified in the Equation (7h) and (7i). In Equation (8a), $\hat{Q}_1$, $\hat{Q}_2$, $\hat{Q}_3$ and $\hat{Q}_4$ refers to the predicted values of the four quality variables namely, sinkhole area coverage (%), average ultrasonic spectra amplitude (dB), impact energy (Kg.m) and zero-shear viscosity (Pa–s) respectively.

The above optimization problem is essentially a mixed integer linear program (MILP) but is instead solved in a brute force fashion as three linear programs using linprog in MATLAB. Thus, exploiting the limited choices provided for the batch duration, the optimization over the time duration is simply carried out by comparing the optimal solution corresponding to each specific duration, and the best solution is implemented. That is, for each of the three candidate batch duration, the optimization problem is solved, and the objective function evaluated subject to constraints. Subsequently the best solution is chosen. In essence, for each possible batch termination time, a constrained quadratic program is solved. In the present application, only three different values of batch duration are evaluated, thus the computational complexity remains fairly low. Note that in principle, higher resolution of the batch end-times could be evaluated, resulting in increased computational complexity. Given the plant-model mismatch, the resultant benefit might not be significant, thus motivating the relatively modest exploration of the optimally over the batch duration. Thus, at each time step, the controller computes a trajectory for the best duration that would yield the on-spec product, and updates this value as more information is received from the process. Further, the optimization problem is solved in a hierarchical fashion to guarantee feasibility. First, the original optimization problem with all the constraints is solved. If the algorithm runs into feasibility issues, the optimization problem is relaxed by removing the constraint on the quality measurements in Equation (7g). This relaxed version is guaranteed to find a feasible solution due to the nature of the problem ensuring an implementable solution. In our experience, the feasibility is always recovered as the process moves to a new point, and more process output information becomes available. To facilitate the implementation of the proposed control approach on the lab scale experimental setup, MATLAB is interfaced with LabView which in-turn interfaces with the sensors and actuators and works as a data acquisition system.

4. Closed-Loop Experimental Results

The proposed rotomolding modeling and control approach is validated through implementation on the lab-scale experimental setup. Closed-loop implementations with two different objectives are carried out. We first illustrate the ability of the controller to deliver an on-spec product, and then evaluate the ability of the controller to handle raw material variability due to the blending of the matrix resin with a similar resin with slightly different viscosity (judged by its melt index).

In the first implementation, the efficacy of the controller was investigated on five new batches. l_f in all of these batches were selected as $t_{switch} + 30$ by the MPC in the last iteration; however, it selected other batch lengths as well during the course of control. The feedback control algorithm, proposed in this work, achieved excellent quality results while meeting the desired product constraints (see Table 3). The internal mold temperature and input profiles are shown in Figure 4. Note that the constraints imposed on the quality variables are significantly tighter compared to the training case (Q_1 constrained to be less than 2%—something that is not achieved in any of the training batches, the least Q_1 observed being 3.88%. On the other hand, the control design recognizes that the constraint on Q_4 is significantly larger than that observed in the training batches, and is able to appropriately push the value of Q_4 higher than that observed in the training, while still respecting the constraint, in turn allowing the tight Q_1 constraint to be met. This ability of the controller to meet target specifications is significant; and generally very difficult to achieve in practice (without the use of such a model-based control design). That is, determining the internal temperature profile (leave alone the heater power and compressed air flow) that would ensure that quality constraints are met, is a very challenging problem to address. In contrast, the present modeling and control approach utilizing a combination of a causal dynamic model and quality model is able to deliver on-spec products consistently.

Table 3. Zero-blend constrained case.

Batches	Q_1 $\leq 2\%$	Q_2	Q_3	Q_4 $\leq 12,000\ Pa{-}s$
MPC Batch 1	1.45	17.30	5.18	4681
MPC Batch 2	2.00	20.30	4.67	7730
MPC Batch 3	0.25	14.03	6.22	4722
MPC Batch 4	1.80	18.80	5.18	4726
MPC Batch 5	1.48	25.20	4.67	5262

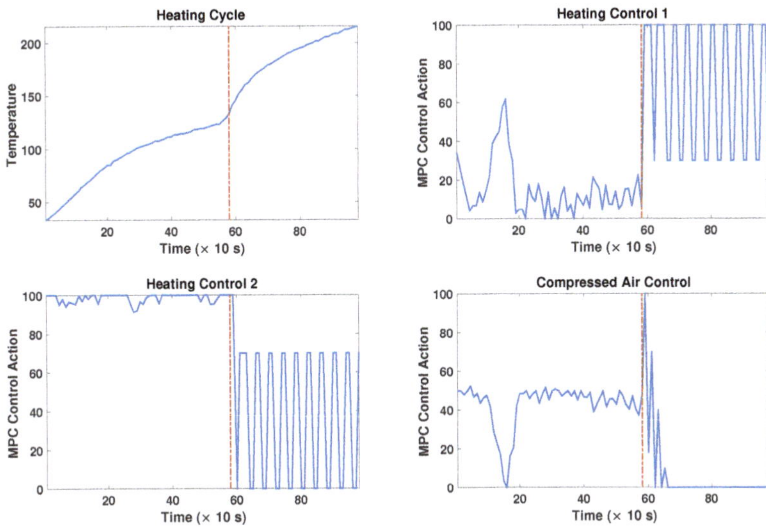

Figure 4. Zero-blend constrained case.

In the second set of batches, the efficacy of the proposed approach to meet user-specified constraints on product quality under raw material variability is evaluated. In this case the constraints were changed to another possible requirement by the practitioners defined by $\Lambda = \begin{bmatrix} 1 & 0 & 0 & 0 \\ 0 & 0 & 1 & 0 \end{bmatrix}$ and $\Gamma = \begin{bmatrix} 2 \\ 5 \end{bmatrix}$. The parameter β in this case was changed to $\beta = \begin{bmatrix} 1 & 0 & -1 & 0 \end{bmatrix}$ and $l_f \in \{t_{switch} + 400, t_{switch} + 450, t_{switch} + 500\}$. The rest of the MPC problem was identical to the previous case. The product qualities obtained for this case are listed in Table 4. The internal mold temperature and the input profiles for one of the blends are shown in Figure 5. Again, for all the blends, going up to 10% blending, the controller consistently achieves on-spec product. Note that the initial period of state estimation, where the control design 'learns', i.e., the estimation of the state of the system is influenced by the dynamics observed in the current batch. This, together with the feedback element, allows the controller to reject the induced raw material variability while achieving the on-spec product.

Table 4. Constrained case with varying blends.

Batches	Blending %	Q_1 $\leq 2\%$	Q_2	Q_3 $\geq 5\ \mathrm{Kg \cdot m}$
MPC Batch 1	2	0.15	27.63	6.22
MPC Batch 2	2	0.11	29.07	5.70
MPC Batch 3	4	0.11	25.90	5.70
MPC Batch 4	4	0.06	26.12	5.70
MPC Batch 5	6	0.15	27.89	5.70
MPC Batch 6	6	0.09	27.48	5.70
MPC Batch 7	8	0.11	25.93	5.70
MPC Batch 8	8	0.17	27.16	5.18
MPC Batch 9	10	0.12	27.56	5.18
MPC Batch 10	10	0.18	26.93	5.70

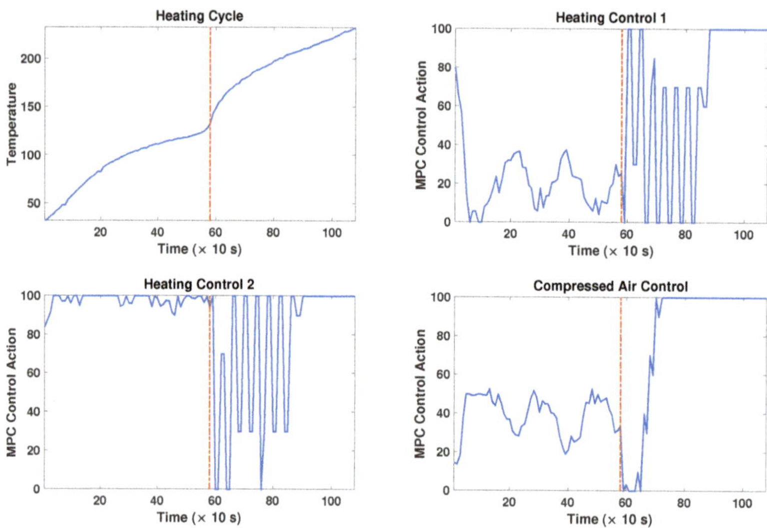

Figure 5. Varying blends constrained case.

Remark 1. *In the present application, the focus was on achieving on-spec product as specified by constraints on the quality variables and demonstrate effective rejection of raw material variability. Thus, other considerations, such as smoothness of input moves were not included in the control design. One could include these considerations explicitly in the control design, and implement these either via appropriate constraints on the rate of input change or putting penalties on the rate of change of input variables in the objective function.*

Remark 2. *Note that the present results demonstrate the ability to model the complex dynamics with the power to the two heaters and the cooling fan being treated as three separate inputs. In reality, all three inputs result in a unified heating/cooling effect. The success of this approach also suggests that alternatively, at the model identification step, a principle component analysis can be carried out to reduce the three inputs to one effective input (the principle component representing the heating/cooling). Then model identification step can be set up to use that principle component as part of the modeling process and in turn in the MPC controller. In yet another alternative, the two heaters could be reduced to one input, and the cooling input could be kept separate, and finally an MPC can be designed that focuses on achieving the desired product while minimizing resource usage. Thus, the cost of various inputs could be directly accounted for in the control calculation to design an economic MPC to enable production that directly maximizes profit. In this fashion, regional effects such as costs of electricity could be directly accounted for in the control calculations.*

Remark 3. *From an industry practitioner's point of view, it is important to meaningfully visualize the evolution of the process during the operation. It serves two purposes: (a) visualize the performance of the identified process model in the changing process conditions to assess any need for model update (for instance, see [28]), and (b) predict the evolution of product quality during the batch operation to gauge the possibility of achieving the desired product given a resin. If the monitoring approach determines that the given resin simply cannot yield the desired product, then the particular batch may be terminated early to prevent waste of additional resources.*

Remark 4. *The modeling approach proposed in this work results in a linear dynamic state space model coupled with a static quality model. The use of this model within an optimization framework results in a convex problem which is easy to solve. When implementing this approach in a commercial setting, the structure of the model will* *still be same. Therefore, the strategy would directly scale up for commercial use as the computations for control moves will still be tractable.*

Remark 5. *In another work [1,29], the authors explored the possibility of using acoustics data, collected by performing ultrasonic tests on the mold, as an alternative to destructive methods for quality assessment. Future work will also focus on integrating more detailed acoustics data with our modeling approach for modeling and control of the product quality.*

5. Conclusions

In this work, a data-driven modeling and control framework is developed capable of handling the problems of uniquely specifying and robustly achieving user-specified product quality in an experimental uni-axial rotomolding process. To this end, a subspace state-space model of the process is identified from historical data which predicts the evolution of the internal mold temperature from a given set of input profiles. This dynamic model is further augmented with a static quality model relating its terminal state predictions with key quality variables. The overall model is then deployed within an MPC scheme that enables achieving on-spec products based on the specified limits on the quality variables along with rejecting the raw material variability. The natural raw material variability observed in the industry is replicated in the pilot experimental setup by blending, in various percentages, the matrix resin with a slightly different one. The experimental results corroborate the ability of the proposed control framework to reject raw material variation and achieve desired product quality specified through explicit constraints on quality variables.

Author Contributions: Conceptualization, P.M. and M.R.T.; methodology, A.G. and P.M.; software, A.G.; experiments, H.A.A. and A.G.; writing—original draft preparation, A.G.; writing—review and editing, A.G., P.M., H.A.A. and M.R.T.; supervision, P.M. and M.R.T.

Funding: Financial support from the McMaster Advanced Control Consortium (MACC) and the Natural Sciences and Engineering Research Council of Canada (Collaborative Research and Development Grant) is gratefully acknowledged.

Acknowledgments: Authors thank Imperial Oil and Ron Cooke for the donation of polyethylene and technical advice on rotomolding process.

References

1. Gomes, F.P.; Garg, A.; Mhaskar, P.; Thompson, M.R. Quality Monitoring of Rotational Molded Parts Using a Nondestructive Technique. In Proceedings of the ANTEC 2018—Orlando, FL, USA, 7–10 May 2018; Society of Plastics Engineers: Brookfield, CT, USA, 2018; ISBN 123-0-1234567-8-9.
2. Garg, A.; Gomes, F.P.; Mhaskar, P.; Thompson, M.R. Model predictive control of uni-axial rotational molding process. *Comput. Chem. Eng.* **2019**, *121*, 306–316. [CrossRef]
3. Bonvin, D. Optimal operation of batch reactors—A personal view. *J. Process Control* **1998**, *8*, 355–368. [CrossRef]
4. Aumi, S.; Mhaskar, P. Integrating Data-Based Modeling and Nonlinear Control Tools For Batch Process Control. *AIChE J.* **2012**, *58*, 2105–2119. [CrossRef]
5. Aumi, S.; Corbett, B.; Mhaskar, P.; Clarke-Pringle, T. Data-based modeling and control of Nylon-6, 6 batch polymerization. *IEEE Trans. Control Syst. Technol.* **2013**, *21*, 94–106. [CrossRef]
6. Abu-Al-Nadi, D.I.; Abu-Fara, D.I.; Rawabdeh, I.; Crawford, R.J. Control of rotational molding using adaptive fuzzy systems. *Adv. Polym. Technol.* **2005**, *24*, 266–277. [CrossRef]
7. Flores-Cerrillo, J.; MacGregor, J.F. Control of particle size distributions in emulsion semibatch polymerization using mid-course correction policies. *Ind. Eng. Chem. Res.* **2002**, *41*, 1805–1814. [CrossRef]
8. Flores-Cerrillo, J.; MacGregor, J.F. Latent variable MPC for trajectory tracking in batch processes. *J. Process Control* **2005**, *15*, 651–663. [CrossRef]

9. Ljung, L. *System Identification: Theory for the User*; Pearson Education: London, UK, 1998.

10. Kadali, R.; Huang, B.; Rossiter, A. A data driven subspace approach to predictive controller design. *Control Eng. Pract.* **2003**, *11*, 261–278. [CrossRef]

11. Pour, N.D.; Huang, B.; Shah, S.L. Subspace Approach to Identification of Step-Response Model from Closed-Loop Data. *Ind. Eng. Chem. Res.* **2010**, *49*, 8558–8567. [CrossRef]

12. Corbett, B.; Mhaskar, P. Subspace identification for data-driven modeling and quality control of batch processes. *AIChE J.* **2016**, *62*, 1581–1601. [CrossRef]

13. Garg, A.; Corbett, B.; Mhaskar, P.; Hu, G.; Flores-Cerrillo, J. Subspace-based model identification of a hydrogen plant startup dynamics. *Comput. Chem. Eng.* **2017**, *106*, 183–190. [CrossRef]

14. Garg, A.; Mhaskar, P. Subspace Identification-Based Modeling and Control of Batch Particulate Processes. *Ind. Eng. Chem. Res.* **2017**, *56*, 7491–7502. [CrossRef]

15. Garg, A.; Mhaskar, P. Utilizing big data for batch process modeling and control. *Comput. Chem. Eng.* **2018**, *119*, 228–236. [CrossRef]

16. Gomes, F.P.; West, W.T.; Thompson, M.R. Effects of annealing and swelling to initial plastic deformation of polyethylene probed by nonlinear ultrasonic guided waves. *Polymer* **2017**, *131*, 160–168. [CrossRef]

17. Tangirala, A.K. *Principles of System Identification: Theory and Practice*; Taylor & Francis: Abingdon, UK, 2014.

18. Qin, S.J. An overview of subspace identification. *Comput. Chem. Eng.* **2006**, *30*, 1502–1513. [CrossRef]

19. Raghavan, H.; Tangirala, A.K.; Bhushan Gopaluni, R.; Shah, S.L. Identification of chemical processes with irregular output sampling. *Control Eng. Pract.* **2006**, *14*, 467–480. [CrossRef]

20. Larimore, W. Canonical variate analysis in identification, filtering, and adaptive control. In Proceedings of the 29th IEEE Conference on Decision and Control, Honolulu, HI, USA, 5–7 December 1990; Volume 2, pp. 596–604. [CrossRef]

21. Overschee, P.V.; Moor, B.D. Two subspace algorithms for the identification of combined deterministic-stochastic systems. In Proceedings of the 31st IEEE Conference on Decision and Control, Tucson, AZ, USA, 16–18 December 1992; Volume 311, pp. 75–93. [CrossRef]

22. Verhagen, M.; Dewilde, P. Subspace model identification Part 1. The output-error state-space model identification class of algorithms. *Int. J. Control* **1992**, *56*, 1187–1210. [CrossRef]

23. Van Overschee, P.; De Moor, B. A unifying theorem for three subspace system identification algorithms. *Automatica* **1995**, *31*, 1853–1864. [CrossRef]

24. Wang, J.; Qin, S.J. A new subspace identification approach based on principal component analysis. *J. Process Control* **2002**, *12*, 841–855. [CrossRef]

25. Qin, S.J.; Lin, W.; Ljung, L. A novel subspace identification approach with enforced causal models. *Automatica* **2005**, *41*, 2043–2053. [CrossRef]

26. Moonen, M.; De Moor, B.; Vandenberghe, L.; Vandewalle, J. On- And Off-Line Identification Of Linear State Space Models. *Int. J. Control* **1989**, *49*, 219–232. [CrossRef]

27. Mhaskar, P.; Garg, A.; Corbett, B. *Modeling and Control of Batch Processes*; Advances in Industrial Control; Springer International Publishing: Cham, Switzerland, 2019. [CrossRef]

28. Kheradmandi, M.; Mhaskar, P. Adaptive Model Predictive Batch Process Monitoring and Control. *Ind. Eng. Chem. Res.* **2018**, *57*, 14628–14636. [CrossRef]

29. Gomes, F.P.; Garg, A.; Mhaskar, P.; Thompson, M.R. Data-Driven Advances in Manufacturing for Batch Polymer Processing Using Multivariate Nondestructive Monitoring. *Ind. Eng. Chem. Res.* **2019**. [CrossRef]

 *processes*

Article

Categorization of Failures in Polymer Rapid Tools Used for Injection Molding

Anurag Bagalkot, Dirk Pons *, Digby Symons and Don Clucas

Department of Mechanical Engineering, University of Canterbury, 20 Kirkwood Ave, Christchurch 8041, New Zealand; anurag.bagalkot@pg.canterbury.ac.nz (A.B.); digby.symons@canterbury.ac.nz (D.S.); don.clucas@canterbury.ac.nz (D.C.)

* Correspondence: dirk.pons@canterbury.ac.nz; Tel.: +64-33-695-826

Received: 17 December 2018; Accepted: 25 December 2018; Published: 2 January 2019

Abstract: Background—Polymer rapid tooling (PRT) inserts for injection molding (IM) are a cost-effective method for prototyping and low-volume manufacturing. However, PRT inserts lack the robustness of steel inserts, leading to progressive deterioration and failure. This causes quality issues and reduced part numbers. Approach—Case studies were performed on PRT inserts, and different failures were observed over the life of the tool. Parts molded from the tool were examined to further understand the failures, and root causes were identified. Findings—Critical parameters affecting the tool life, and the effect of these parameters on different areas of tool are identified. A categorization of the different failure modes and the underlying mechanisms are presented. The main failure modes are: surface deterioration; surface scalding; avulsion; shear failure; bending failure; edge failure. The failure modes influence each other, and they may be connected in cascade sequences. Originality—The original contributions of this work are the identification of the failure modes and their relationships with the root causes. Suggestions are given for prolonging tool life via design practices and molding parameters.

Keywords: rapid tooling; additive manufacturing; failure modes; injection molding

---

1. Introduction

1.1. Background

New product development (NPD) is a key success factor for growing organizations, as they are constantly re-engineering and developing new products to stay competitive [1,2]. Transitioning new products from the research and development (R&D) stage to the prototyping stage, and finally to the manufacturing stage is a common problem for organizations of all sizes [3]. The stages of NPD from an engineering perspective are shown in Figure 1.

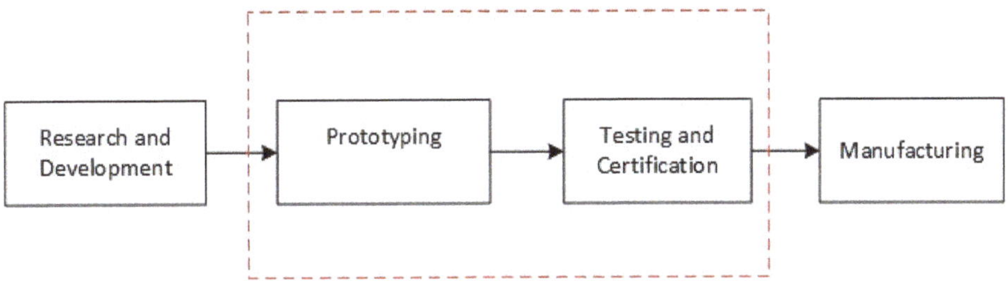

Figure 1. Stages of a new product development (NPD) process from an engineering perspective.

The middle phase of the NPD process highlighted in Figure 1 is often the most difficult phase, due to cost and time constraints; especially for innovative start-up firms and small organizations [4]. In the case of polymer products, injection molding (IM) is one of the most widely used polymer processing techniques. Conventionally, IM tools are machined out of blocks of steel, aluminium, or copper–beryllium alloys, and the choice of mold material is dependent on factors such as the molding material, the complexity of the part, the required life, and the available budget [5–7]. In cases of low-volume production, the capital cost of the injection molds is the largest cost component of an injection molded part, followed by the molding material and the processing costs [8]. Complex product geometry requires complex tooling, which increases the cost and lead times if manufactured by conventional methods, which does not work well for keeping the costs of prototype tooling low [6,9]. Hence, the NPD process for injection molded parts typically faces issues (delays) during the prototyping and testing phases. Since more than 35% by weight of all polymers is by IM, the highest demand for low cost tooling is from the IM industry [8]. To stay competitive, industries are looking to reduce the wastage of raw materials, shorten product lead times, and eliminate the need for expensive tooling [10]. In the above situations, industries increasingly turn to additive manufacturing (AM) to solve these challenges via direct part production [11]. However, there are disadvantages: lack of material availability; varied material properties; inaccurate representations of the final part; poor surface finish. For accurate evaluations of a prototype, industries prefer to manufacture prototypes by using the same materials and the same process as the final part. In such situations, industries may use rapid tooling (RT).

1.2. Polymer Rapid Tooling (PRT)

RT is a process that uses AM techniques to build tools at low cost. RT negates the need for complex conventional machining operations and direct labor, and instead uses an additive approach of building objects layer by layer [12,13]. The initial push for developing objects quickly without the need for complex tooling came from the American automotive industry [14]. RT involves both the use of polymer and metal AM systems. However, the cost of metal AM systems and operating consumables are significantly higher than polymer-based AM systems. Rapid Tools built using polymer AM systems are referred to as polymer rapid tooling (PRT). Fluid-based AM processes such as Stereolithography (SLA) and Material Jetting (MJ) are the most commonly used processes for manufacturing PRT inserts for injection molding. Selective laser sintering (SLS) and fused deposition modelling (FDM) have also been used, but not as commonly as the fluid-based AM processes [15].

1.3. Failure of Polymer Rapid Tools

This section deals with the various types of failures, and their causes as presented in the literature. Most of the failures were classified as mechanical failures that occurred due to high injection pressure and higher melt temperature. The second most common reason of failure was due to insufficient draft on the walls of the tool. Several suggestion have also been identified by authors to improve tool life and to avoid failure.

1.3.1. Failures Due to Mold and Melt Temperature

The first PRT inserts made using SLA technology were not robust, and they failed during the start-up of the injection molding process; they had a very low thermal conductivity, which caused thermal degradation, and resulted in mechanical failure [16]. The life of a PRT insert is largely dependent on the resin being molded, and the molding parameters in use [17]. The co-efficient of thermal expansion in polymers varies with the temperature; at temperatures above the glass transition temperature (T_g) of the material, the co-efficient of expansion increases significantly. This causes the mold to expand drastically and sometimes fail [18]. Adjusting the process parameters to keep the tool temperature below its glass transition temperature (T_g), to avoid softening of the tool, is one suggested solution to avoid such failure [19]. However, this solution is only possible if the molding polymer allows a lower mold temperature; for polycarbonates requiring higher mold temperatures, it is not applicable.

1.3.2. Failures Due to Injection Pressure

Failure typically occurs after several shots. Since the shear and bending forces induced by IM should not increase, it is therefore suggested that failure occurs due to a change in the mechanical properties of the tool [16]. PRT inserts were seen to be deforming and failing catastrophically, due to the pressure exerted by the molten polymer during the injection and packing stages of the molding process [20]. Inserts tend to fail if the stresses created by the flowing polymer are more than the yield strength of the tool at that temperature [21]. The injection of the polymer is also known to cause stresses on features, which may lead to crack propagation and the eventual failure of the insert [22]. Static friction between the polymer mold and part determine the ejection force, and the static friction increases with an increase in the mold surface area. Cooling cycles can determine the amount of shrinkage, and thereby the ejection forces as well; higher ejection forces will lead to tool wear and gradual failure [17]. The surface roughness of molds plays an important role in the ejection forces; small build-up layers and high gloss-finished molds experience smaller ejection forces [23]. There is also a possibility of surface smoothening over the life of molds which can be good for tool life, but they can lead to excessive flashing. 3D-printed tool designers have recommend having draft angles of about 5% in polymer molds for the ease of ejection of parts; higher draft angles will lead to easier part ejections [24].

1.4. Opportunities for Development

SLA-based materials such as Accura Bluestone® from 3D systems [25], and Somos® Perform from DSM [26], are more suited for produce PRT inserts than the MJ materials, such as Digital ABS and Visijet M3-X reported in this paper [27]. These materials contain reinforcement, and they have better thermal performance. However, the cost of the AM machines that are required to print these materials is high, and is generally not accessible to smaller industries. The journal literature for polymer-based RT is predominantly based on the SLA process, and it lacks data for other processes such as MJ and FDM.

The cost of the AM machines and material used in the MJ process are significantly lower compared to the SLA process. Hence, there is still a place for RT tools that are printed via the MJ process. This is especially for start-ups and small organizations, where keeping costs low is one of the biggest challenges. Currently there is ongoing work into the use of the MJ process to create low-cost PRT inserts, and methods for strengthening the PRT inserts. The application of metal layers (coatings) onto the PRT inserts has been suggested to improve their properties, such as wear resistance and hardness [28,29]. Compositing FDM parts with high-strength resins filled into the voids printed has also been known to improve the strength of the PRT inserts [30]. However, PRT inserts made from the MJ process tend to fail abruptly, and they have life issues. The life of these inserts is dependent on factors that are not very well documented. Understanding the failure modes and the possible causes of failure, and developing a process to prolong the tool life can be quite beneficial for creating low-cost tooling.

Therefore, this paper focuses on the categorization of failures that occur during the injection molding process, using PRT inserts manufactured by the material jetting technique.

1.5. Aim

PRT inserts lack the strength and robustness of conventional steel and aluminium inserts, and they typically only last for 50–100 shots before the surface starts deteriorating, and features on the insert begin to fail. The materials for PRT inserts made from the MJ process are cross-linked polymer systems (thermosets). These do not necessarily have a melting temperature, but instead have a glass transition temperature (T_g). Digital ABS®, the most commonly used material for PRT, has a T_g of 53 °C.

The current industrial trend is to use PRT inserts to mold polymers with relatively low processing temperatures of about 150–200 °C. These include commodity resins such as polypropylene (PP), acrylonitrile butadiene styrene (ABS), and polystyrene (PS). These resins have a good melt flow index, and they require mold temperatures in the range of 45–50 °C which is below the T_g of Digital ABS®. Lexan-943-A® is a resin that is commonly used by the aerospace industry for cabin interiors. However, the mold temperature that is required for molding Lexan-943-A is 80–95 °C, which is higher than the T_g of Digital ABS®, and the processing temperature is in the range of 285–315 °C. This presents a challenging opportunity to develop a process that would help in achieving low volume production (10–100 parts) of aerospace cabin interior parts (polymer) using PRT inserts.

The main aim of this study was to identify and categorize the different failures that occur in a PRT insert when it is used for injection molding resins with processing temperatures of above 275 °C. We are particularly interested in analyzing real-world components, and determining the best practices for users. An understanding of failure modes and the underlying mechanisms will help with predictions of tool failure, and assist with improvements in the operating life of PRT inserts.

2. Materials and Methods

We applied a case study methodology to real production parts. Three case studies were performed, of progressive complexity, both in terms of molding and part geometry. The first case study was a standard test specimen used for flame and toxicity testing in the aerospace industry; the geometry was flat and did not involve any complex features. The challenge in case study 1 was to completely fill the mold without damaging the PRT insert. The second case study was an electronic enclosure used in the navigation industry; the geometry was complex, as it had features such as thin walls, bosses, and ribs. The third case study was a finger guard used in the aerospace industry; the mold geometry did not have a flat parting line, and had walls with no draft on them. The complexity in case study 3 was to avoid scalding of the tool due to the lack of air vents. The details of the case studies are shown in Table 1.

Table 1. The type of part, and the reason for each of the case study.

Case Study	Type of Part	Reason for Study
Case Study 1	Standard Flame Test Specimen	Feasibility Testing
Case Study 2	Electronic Enclosure	Understanding Failures
Case Study 3	Aircraft Interior Part	Improving Tool Life

In all of the case studies, the molding was been performed with an aerospace resin: Lexan 943-A. The key issues are that the mold temperature required for processing Lexan 943-A is higher than the T_g of the PRT material. Hence the mold is operating under conditions of extreme thermal overload.

The case studies were devised based on a progressive regime of improving the tool life of PRT inserts. The findings from how the failures occurred in case study 1 were used to inform the design of case study 2, etc. For example, case study 1 gave important insights into the need to control the injection pressure to avoid failure at first shot, and case study 2 gave insights into the cooling time. The PRT insert from case study 1 that had failed on the fifth shot was examined; the runner wall had sheared, due to the incoming melt pressure. For case study 2, we used a low injection pressure, and built it up over each shot, until the pressure was sufficient to fill the cavity. The mold surface and features were examined after each shot, for any potential defects. If any defects such as chipping, erosion, or cracks were observed, the parameters for that shot were highlighted. Shot size, melt temperature, and mold temperature were all kept constant; the injection pressure was increased until it was sufficient to fill approximately 85% of the mold. Examinations were done on the failure regions to determine the type and cause of the failure.

The process for setting up each case study, and for extracting failure insights is shown in Figure 2.

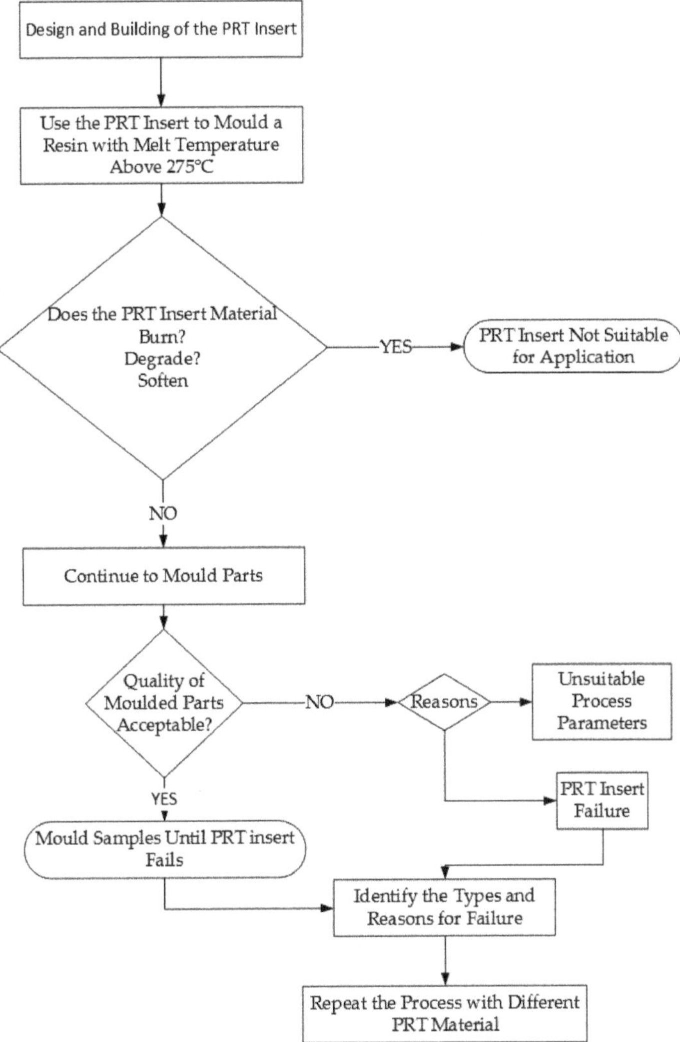

Figure 2. Process workflow.

The different failures observed in the PRT inserts occurred at different times during the molding cycle. We identified the shot number, during which we first observed the signs of each failure type. Each part was inspected to detect failures, and likewise the tool. We continued the molding process, while the tool damage was minor (surface deterioration, surface scalding, bending), and terminated the test when catastrophic tool failure occurred.

2.1. Case Study Setups

2.1.1. Case Study 1—FAR Test Specimen

The flame and toxicity testing for aerospace parts are performed, according to the standard tests specified in Federal Aviation Regulation (FAR) 25.853 and FAR 25.855. FAR 25.853 requires the standard test specimen to be manufactured via the same process as the final part. The part chosen for this case study was a standard test specimen used for the vertical Bunsen burner test for cabin and cargo compartment materials. It is a flat rectangular plate (304.80 × 50.80 × 2.56 mm), and the solid model that was used for printing the core insert is shown in Figure 3.

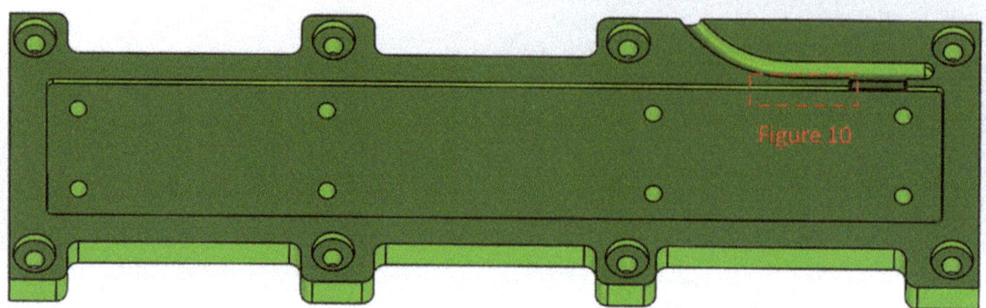

Figure 3. Solid model of the cavity insert used for printing.

SOLIDWORKS 2016® was used to design the core and cavity inserts. The inserts were printed on a Stratasys OBJET 350 Connex 3 Polyjet Machine. Digital ABS® was the material that was used for printing the inserts with a 30-micron layer height setting, and glossy print mode was used. The inserts were fitted into a standard master unit die (MUD) base, and a 230-ton TOYO IM machine was used. No post-processing was done, other than water jet cleaning of the inserts to remove the wax support material. The process parameters used for injection molding is shown in Table 2.

Table 2. Process parameters recommended vs used, for case study 1.

Parameter	Value
Resin Manufacturer	Sabic Innovative Plastics
Resin Name	Lexan 943-A
Resin Type	Polycarbonate/Amorphous
Mold Temperature	80 °C
Melt Temperature	310 °C
Maximum Injection Pressure Set	180 MPa
Maximum Pressure Used	146.6 MPa
Fill Time	3.2 s
Switchover Point	95% by volume
Highest Temperature of the Melt	289 °C
Cooling Time	55 s
Mold Open Time	Initially 20 s, was subsequently kept open until the mold temperature reduced to 80 °C

2.1.2. Case Study 2—Electronic Enclosure

Case study 2 represents a part with more complex geometric features, including a boss, thin walls, and thin core pins; see Figure 4. An aerospace resin Lexan 943-A from Sabic Australia PTY LTD Melbourne was injected. The part chosen was already in production using a steel insert, and was used as a reference. The inserts were printed on a Stratasys OBJET 350 Connex 3 Polyjet Machine from Stratasys, Minnesota, United States. Digital ABS® from Stratasys, Minnesota, United States was the material used for printing the inserts, with a 30 micron layer height setting and glossy print mode being used. The inserts were not hand-polished, as there was a risk of damaging the parting surface. A 450 ton injection molding machine with a maximum injection pressure of 170 MPa was used. The mold temperature was set at 45 °C to keep the tool below its T_g, which is about 53 °C. The recommended mold temperature from the resin supplier was 105 °C. Table 3 compares the process parameters for the meal molds recommended by the resin supplier, and the parameters used for the study.

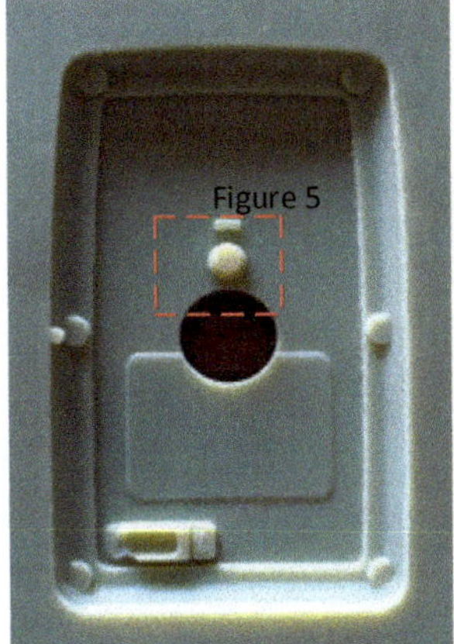

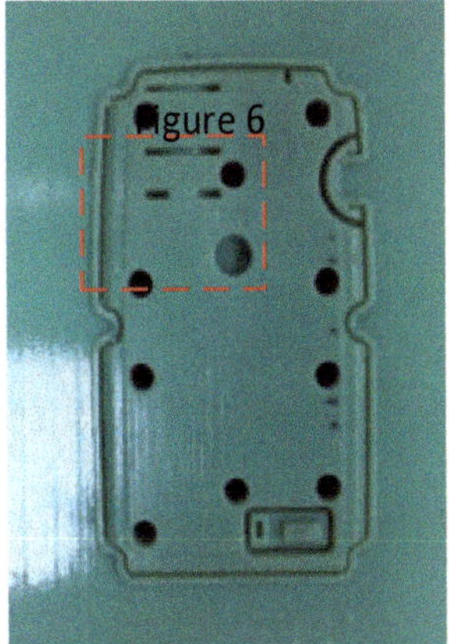

Figure 4. Core and cavity insert.

Table 3. Process parameters recommended vs used for case study 2.

Parameter	Recommended Value	Actual Value Used
Resin Manufacturer	Sabic Innovative Plastics	Sabic Innovative Plastics
Resin Name	Lexan 943-A	Lexan 943-A
Resin Type	Polycarbonate/Amorphous	Polycarbonate/Amorphous
Mold Temperature	105 °C	45 °C
Melt Temperature	285 °C	305 °C
Maximum Injection Pressure Set	180 MPa	50 MPa
Maximum Pressure Used	N/A	41 MPa
Fill Time	0.87 s	0.55 s
Switchover Point	95% by volume	90% by volume
Highest Temperature of the melt	289 °C	311 °C
Cooling Time	15 s	45 s
Mold Open Time	30 s	Mold kept open until mold reduced to the target temperature of 45 °C after every cycle

2.1.3. Case Study 3—Aerospace Part

For case study 3, we used a production part for the aerospace industry; see Figure 5, which shows the core and cavity insert. The inserts were printed onto a 3D systems ProJet MJP 3600 series printer from 3D systems, South Carolina, United States. The material used was Visijet M3-X® from 3D systems, South Carolina, United States. An ultra-high definition (750 × 750 dpi) print mode with a 29 μm layer thickness setting was used to print the inserts. For post processing, the inserts were placed in an oven at 100 °C for 1 h; they were then cooled and scrubbed with a hot detergent to remove all the wax (support material).

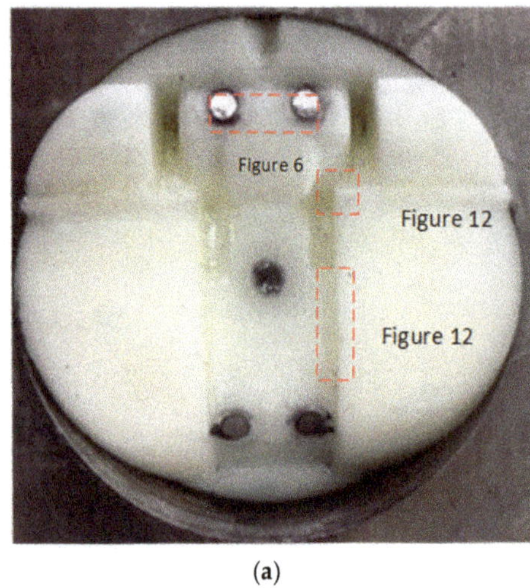

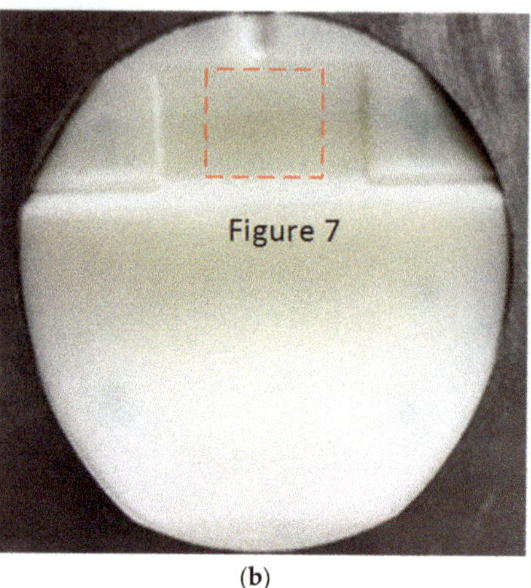

(a) (b)

Figure 5. (**a**) Core insert for case study 3, fitted inside a master unit die. (**b**) Cavity insert for case study 3, fitted inside a master unit die. The tool outer diameter is 55 mm.

A modified method of the process parameter settings was used for the initial start-up process. A Moldex3D flow simulation were used to determine the process parameters. The process parameters used for the study are shown in Table 4.

Table 4. Process parameters recommended vs used, for case study 3.

Parameter	Recommended Value	Actual Value Used
Resin Manufacturer	Sabic Innovative Plastics	Sabic Innovative Plastics
Resin Name	Lexan 943-A	Lexan 943-A
Resin Type	Polycarbonate/Amorphous	Polycarbonate/Amorphous
Mold Temperature	105 °C	45 °C
Melt Temperature	285 °C	285 °C
Maximum Pressure Set	180 MPa	70 MPa
Maximum Pressure Used	N/A	62 MPa
Fill Time	0.5 s	1.2 s
Switchover Point	95%	90%
Highest Temperature of the Melt	315 °C	300 °C
Cooling Time	10 s	25 s
Mold Open Time	5 s	Until the mold temperature after every shot is reduced to 45 °C.

3. Results

Various common failures were identified during the three case studies. We have categorized the failures into three failure modes: surface failures (crack formation); delamination (crack growth); feature failure (fracture). The individual failure mechanisms are described below.

3.1. Surface Failures (Stage 1—Crack Formation)

3.1.1. Surface Deterioration/Micro-Spallation

As the molten polymer enters the cavity, it begins to cool, while the PRT inserts begin to rise in temperature; at a certain stage of the molding cycle, the tool temperature is above the T_g of the tooling material, which causes the tool to soften. During the injection stage, as the polymer flows, there is a constant shear between the molten polymer and the top surface of the tool, resulting in erosion/spallation of the tool. The erosion tends to increase, with polymers having a low melt flow index, as they usually require a higher injection pressure. The erosion was microscopic, and could not be seen by the naked eye on the inserts in between shots, and the tool surface had failed completely by the end of the production cycle. The surface erosion was detected only when the parts were viewed by using a microscope. In Figure 6a, the brown protrusions are pieces of the tool material that have been eroded and stuck to the part, thus creating small voids over the tool surface. Since the tool surface is no longer flat, the next molded part will have protrusions that are similar to the eroded area. In Figure 6b, similar protrusions are evident as in Figure 6a, but now they are the same color as the part. The surface degradation worsened after each shot. A magnified image of the deteriorated surface can be seen in Figure 6c; the magnified image shows both the protrusions molded from the previous shot in white, and the brown protrusions are new fragments of the eroded surface from the current shot.

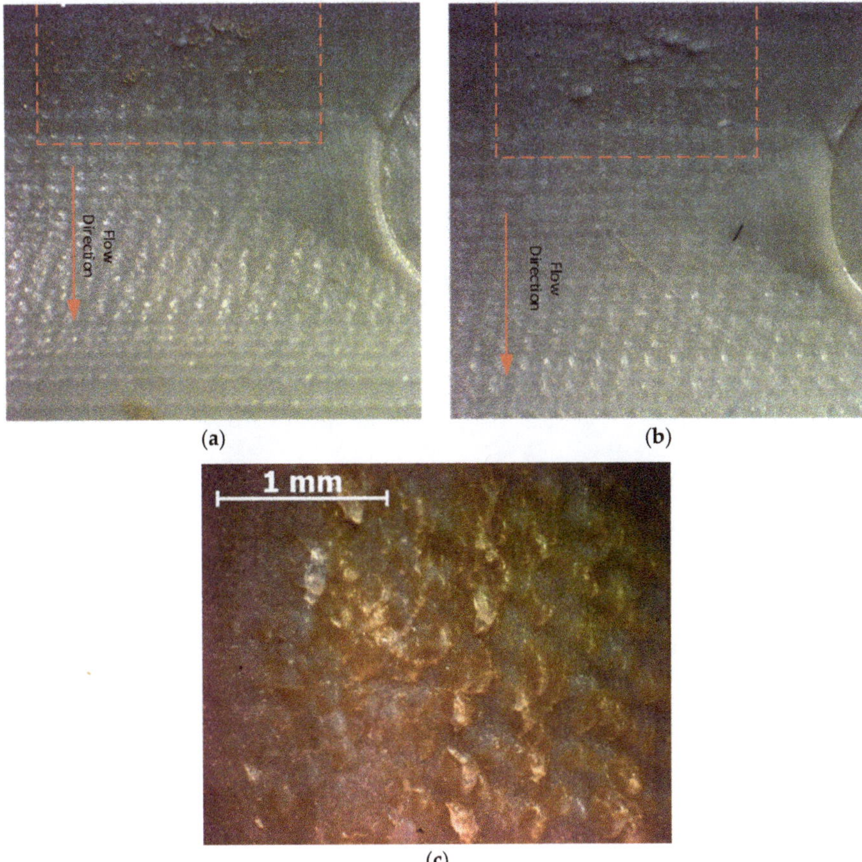

Figure 6. (a) Tool material eroded and deposited on the seventh shot from the tool. (b) Protrusions on the eighth part, due to surface deterioration of the tool in the same vicinity. (c) Magnified image of the area of the eroded tool surface.

After each molding cycle, the parts were inspected for quality, and once the part showed any signs of quality defect, we inspected the tool to assess tool damage. No quality defect was identified for the first six shots. A small area of deteriorated surface was seen on the seventh shot, on the tool and the part. The surface deterioration worsened after each subsequent shot.

3.1.2. Surface Scalding

Surface scalding refers to the burning of the tool surface, due to the incoming polymer melt or the hot air/gases developed during the molding process. This damage mode is characterized by browning of the tool in small patches, see Figure 7. Since MJ resins are thermosets, they do not melt; at higher temperatures, they usually degrade. Even though the tool temperature was kept below the T_g (53 °C) of the tooling material before every shot, the tool temperature would rise as the molten polymer entered the tool. The first two case studies had a planar parting surface, which made it easier for air vents to be printed, but for case study 3, the part had a curved parting surface, and the faces were intersecting with the MUD base, and hence, no air vents were printed on the tool. Surface scalding was not seen in the first two case studies, and was only seen in the third case and lack of air venting was suspected to be one of the reasons for the tool degradation. In addition, the periphery of the tool had discolored, and showed signs of scalding. This failure was evident at the 12th molding shots. This rapidly progressed to catastrophic failure at the 13th shot, as the tool surface material was removed.

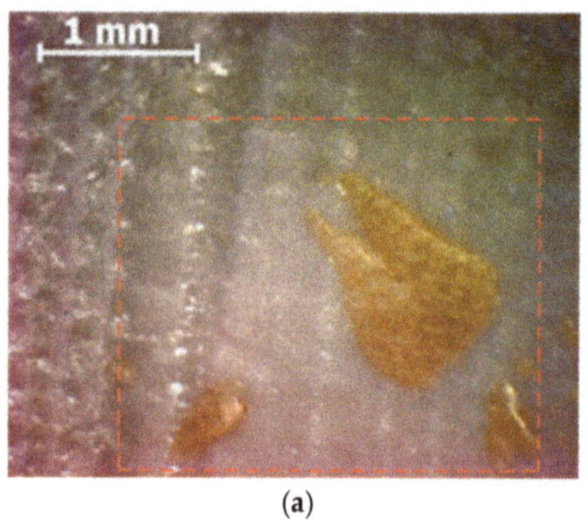

(a)

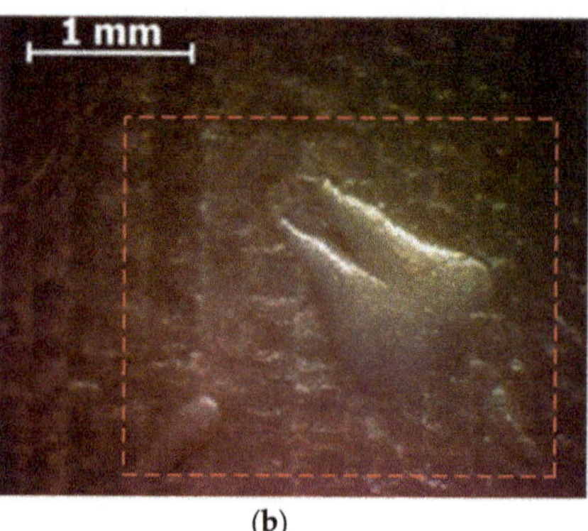

(b)

Figure 7. (**a**) Surface scalding patch attached to the part in shot 12. (**b**) Protrusion mold in a similar area, shape, and size because of the surface scalding, in shot 13. The scalded tool material has adhered to the part in (**a**), and hence, the tool has lost material.

3.2. Avulsion/Delamination (Stage 2—Crack Growth)

Once the mold was 95% filled, a second stage (hold) pressure was applied, to force excess melt into the cavity, to compensate for the shrinkage of the injected polymer. This second-stage pressure resulted in adhesion between the deteriorated surface of the PRT insert and the part. Avulsion failure was observed in case study 2 on the core insert; see Figures 8 and 9. The central hole on the core insert was initially observed to be deteriorating, and as the cracks extended, the part was seen to be avulsing chunks of material from the insert during each shot.

The insert at this stage was at a higher temperature than the T_g of the tool. We infer that this causes the tool to lose its strength, and also when the part is ejected; hence, avulsion of the mold surface may occur. The term avulsion refers to the ripping and peeling of the surface material of a body. Our proposed explanation is that during the first few shots, surface damage occurs, which leads to surface cracks (micro-spallation). During subsequent shots, molten polymer enters these micro-cracks and enlarges them; as the cracks grow, more molten polymer is forced into the cracks before solidifying. During ejection, this solidified polymer causes layers of the mold to be torn off. Avulsion failure is a progressive rather than an abrupt failure mode.

Figure 8. Avulsion failure region on the PRT insert from case study 2. This failure mode is characterized by progressive peeling of the substrate material. The image shows the catastrophic end-state of the process.

Figure 9. Molded part showing peeled off tool material from the top surface. This occurred during the ejection of the part. (Area highlighted is the tool material pulled out).

3.3. Feature Damage (Stage 3—Fracture of Features)

3.3.1. Shear Failure

During the injection stage, the polymer melt is forced into the insert under pressure. The polymer melt while filling the cavity exerts a force on the mold features, such as walls, bosses, and ribs. This force is responsible for a stress which, if exceeds the ultimate tensile strength of the material at that temperature, may result in the failure of the feature. The injection pressure is highest during the start of the injection cycle, and gradually reduces; this means that the features of the insert that are the closest to the injection point (gate) are more susceptible to failure. Shear failure was observed during case study 1 on the runner wall of the PRT insert; see Figure 10. The high melt pressure during the initial process parameter setup phase caused the wall to shear and cause a catastrophic failure of the mold. The wall of the runner sheared off on the first shot, but it did not completely break off from the mold surface; during the subsequent shots, the wall broke off the mold surface. Figure 3 shows the wall of the runner that sheared off. This was classified as shear failure, due to the low aspect ratio (height/thickness) of the wall. This was a critical failure, as it was a runner wall, and we could not produce any more parts from the insert—it would be unsafe to try. We infer that shear failure may be a particular risk in areas with low aspect ratio thin walls that are close to the gate (melt entrance).

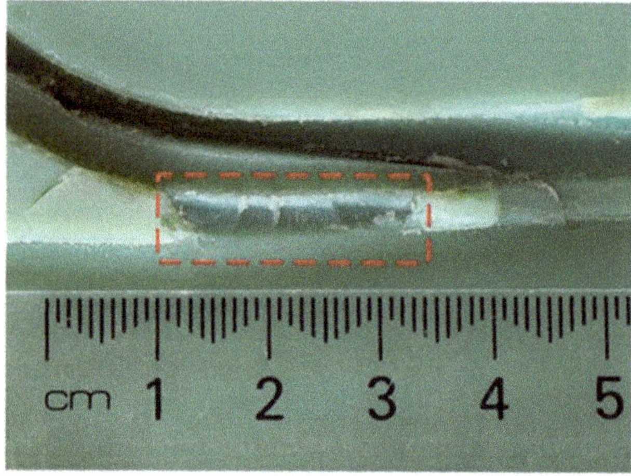

Figure 10. Shear failure of the runner wall (boxed area), due to high injection pressure. See Figure 3 for the location on the tool.

3.3.2. Bending Failure

A bending failure was observed in case 2; see Figure 11. This feature on the tool was supposed to create a hole on the part, but the boss feature was broken off the tool after five shots, and all the parts molded after it had a thick section instead of a hole. Although this made the part functionally defective, it was not considered a critical failure, as it was still safe to operate the tool. The melt pressure was causing the boss to bend during the initial filling stage. This eventually led to a crack at the bottom surface, and failure of the boss in subsequent shots. The boss was situated directly 5 mm in front of the injection point; this meant that the boss was experiencing the highest pressure of all of the features on the tool.

The incoming polymer melt during the injection stage exerts a pressure on the front face of a feature, leading to a deflection; as the melt front reaches the back face of the feature, there is a reduced pressure difference between the back and front faces, and the deflection of the feature reduces. This cycle is repeated during each molding cycle, and may be responsible for the development of a crack, which may eventually lead to a complete failure of the PRT insert. Features with high aspect ratios (height/thickness) should in principle, be the most vulnerable to bending failure.

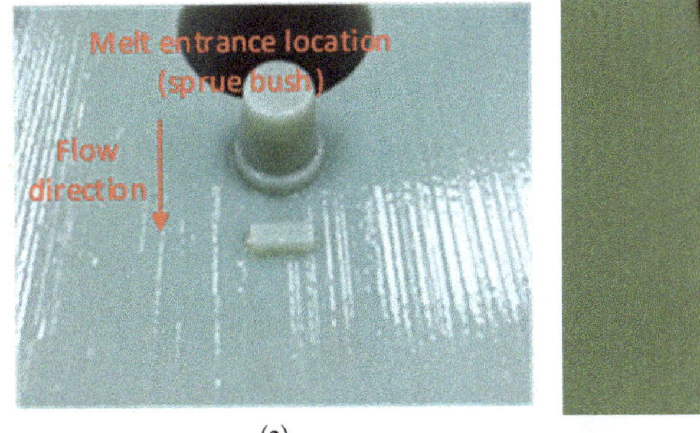

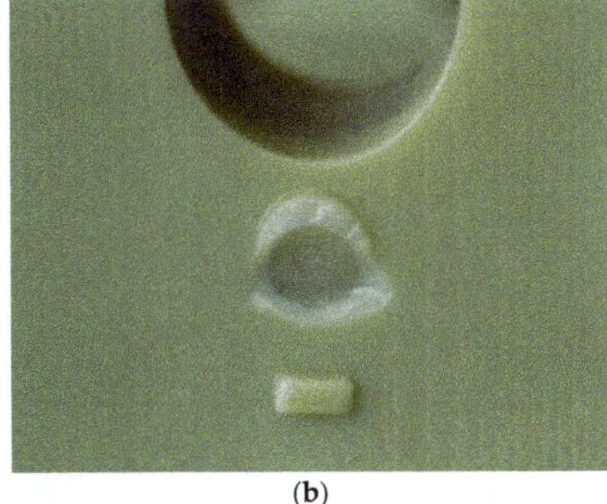

(a)   **(b)**

Figure 11. (**a**) Boss feature before molding. (**b**) Bending failure of the boss feature due to high injection pressure. See Figure 4 for the location on the tool.

3.3.3. Edge Fractures

The edges of the PRT insert deteriorated with each shot, and small chips of the material were seen to be eroding from the edges; see Figure 12. This was particularly worse near the areas of the runner (melt entrance) where the melt pressure was high; the edge deterioration was less on the edges at the end of the melt flow.

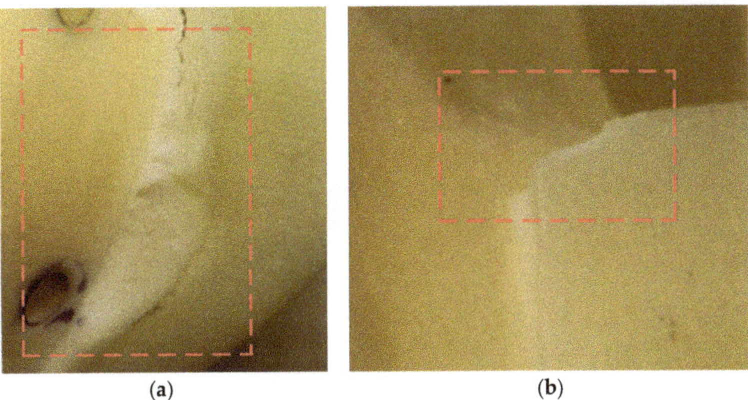

(a)   **(b)**

Figure 12. (**a**) Chipped edge of the tool caused during the ejection of the part. (**b**) Chipped corner of the tool due to injection pressure. See Figure 5 for the location on the tool.

The other reason for the edge failure is the draft on the walls. During ejection, there is constant friction between the part and the mold walls. The mold during the ejection stage is above its T_g and has low yield strength. The part while ejecting starts to degrade the surface (erosion) during the early shots, and results in chipping when there is significant deterioration. An advanced version of the edge failure was observed in case study 3. During case study 1, edge failure was not considered as a significant failure, because it did not pose any safety issues, and it was not flashing. However, the PRT insert in case study 2 was only used for five shots, after which it failed via the shear failure mode. In case study 3 the tool survived the initial process parameter setup phase, and the edges deteriorated progressively from the ninth to the 13th shot. During the ejection of the 14th shot, a large shard of material was pulled out. After this shot, the tool was flashing, due to the failure of the edge that was on the parting line of the tool. Surface erosion was seen on all of the edges along the flow path, and chipping was only seen on edges that had no draft. At the end of the 19th shot, the fracture was about 5 mm wide, and this was the final failure on the tool.

3.4. Categorization of Failures

Based on the above case studies, and after examining the molds and sectioning the inserts in the cracked region, we identified a list of common failure modes. The main categories of failure were identified as: shear failure; bending failure; avulsion; surface deterioration; edge failure; surface scalding. Table 5 summarizes the types of failure, the probable reasons and the common regions of occurrence for each type of failure. This table may be useful as a guide for part, tool, and process design.

Table 5. Categorization of failures.

Failure Modes			Type of Failure	Observed Regions	Possible Root Causes	Shot Number during Which Failure Was First Detected
1	Surface Failure (Crack Formation)	1.1	Surface Deterioration	All over the surface, higher at the melt entrance	Low melt temperature, high mold temperature	7
		1.2	Surface Scalding	Regions around the gate and periphery of the tool	Shear heating, high tool temperature, lack of air venting	13
2	Avulsion (Crack Growth)	2	Avulsion	Directly in front of the gate and close to features	Ejection force, long cooling cycle, draft angle	14
3	Feature Failure (Fracture)	3.1	Shear Failure	Thin walls and raised features (low aspect ratio)	High injection pressure, high tool temperature, gate location	3
		3.2	Bending Failure	Raised features, high aspect ratio	High injection pressure, high tool temperature, gate location	5
		3.3	Edge Failure	Edges closer to the gate	Low draft angle, stress concentration, gate location	13

4. Discussion

4.1. Towards a Theory of Failure

We propose that multiple failure mechanisms are operating: thermal degradation of the surface of the tool material (scalding); crack growth due to creep (surface deterioration); localized tensile failure of surface (avulsion); crack formation due to applied stresses on the tool features (bending and shear); crack growth due to the intrusion of pressurized molten material, with subsequent avulsion of tool material. The failure modes appear to influence each other, and be connected in cascade sequences.

Injection pressure, injection temperature, and mold temperature are identified as major factors that contribute to the failure mechanisms. The combined effect is a progressive failure that leads to catastrophic failure after a few molding cycles.

Three phases in the Failure Process.

The results have identified the types of locations where the failures initiate, and the appearance of the final failure. Generally, the progression shows three stages: crack initiation, crack growth, and structural failure.

First stage: Occurrence of surface damage (surface deterioration, scalding), which then results in a crack. The surface damage occurs due to the tool temperature, melt temperature, and shear heating. The latter is affected by the injection pressure.

Second stage: Crack growth (avulsion). Once a crack forms, its growth is rapid. A large extension of the crack was evident at each shot. In conventional applications of creep-fatigue, the crack is hidden inside the bulk of the material. However, in the case of IM, there is an additional mechanism that is highly deleterious. This is the combination of the intrusion of melt material into the crack, the solidification of that material, and the subsequent violent avulsion thereof at ejection. The ejection occurs at the opening of the tool, and is driven by ejector pins. Hence, there is another source of external loading on the features, other than injection pressure, that occurs at the end of each cycle. This process causes extensive damage to the cracked region, and this sets up the geometry for further damage at the next cycle.

Third Stage: Rapid destruction of the features within the tool, and a catastrophic deterioration of the integrity of the features, and sometimes of the tool (shear and bending failure). This process is similar to the generally accepted creep-fatigue failure process, but is accelerated by the intrusion effect. We propose the following conceptual map of the failure process; see Figure 13. The green color in the image represents the reason for failure and the red color represents the underlying failure mechanisms.

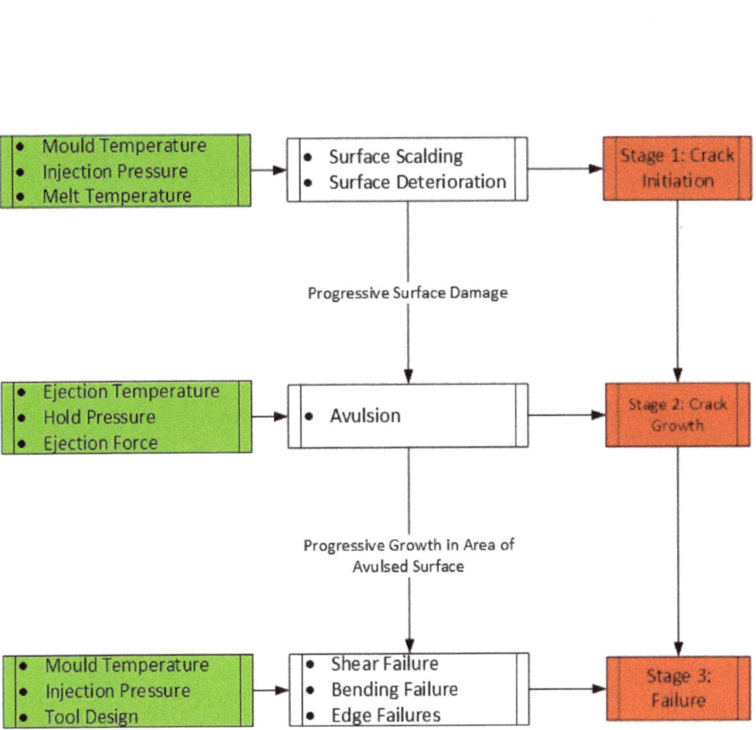

Figure 13. A conceptual map of the interaction between the root causes and the failure modes.

4.2. Implications for Practitioners

The main implication for tool designers is to avoid gating (melt entrance) near thin walls and raised features. This is because the injection pressure near the gate is the highest, and can cause shear and bending failure of these features. Tool designers could consider using flow simulation software to evaluate the pressure consequences of different gating locations. They could also identify the problematic areas so that molding technicians can monitor these locations after each shot.

Suggested Guidelines to Prolong Tool Life

The case studies have identified the stages of injection molding, and the tool locations where failure occurred. Based on this, guidelines for design and process-setting are proposed:

Tool Design Guidelines

I. Add thickness to walls nearer to the gate, to give them extra strength. The current industry practice is to reduce the amount of printed material, as it is expensive, but this practice leaves the tool vulnerable, and may not serve the end purpose of cost reduction.

II. Avoid choosing gate locations that are close to features with high aspect ratios, as otherwise these could be vulnerable to early failure. The effect of injection pressure on features is high at the start of the filling phase, and is much reduced at the end of flow.

III. Add flow leaders in the tool to help with flow in regions, rather than increasing the pressure during molding.

IV. Design tools to have sufficient air venting. Air venting reduces the chances of surface scalding.

V. Avoid designing ribs and features that are perpendicular to the flow path, to reduce the area of contact between polymer melt and the feature.

Process Setting Guidelines

I. The conventional method (used for steel and aluminium tools) for process parameter setting begins with using the highest injection pressure setting on the machine, and optimizing it on the subsequent shots. This may result in improper pressure setting, and can lead to failure at start up, as seen in case study 1. Instead, mold flow simulations could be used to determine the pressure that is required to fill the cavity. Consider starting molding with 50% of that pressure, and build up to the required pressure. This helps to avoid shear and bending failure at start up. (This is the first-stage injection pressure. The hold pressure should be 50% of the injection pressure at start, and then optimized too.)

II. Always maintain the tool temperature below the T_g of the tooling material, use air jet cooling, or increase the mold open time between cycles. The yield strength of the tool reduces as the temperature increases. The effect is pronounced at around the T_g of the material. The mold does not cool uniformly, and it is critical to check for hotspots on the tool, and cool them before the next shot.

III. Use mold release sprays after each shot, to help with ejection. Mold release reduces adhesion, and may reduce the risk of avulsion.

IV. A long part cooling time will result in the part shrinking and gripping onto mold surfaces, causing problems during ejection. Cooling times should be kept as low as possible to avoid this. However, sometimes this will result in a part not being fully formed, in such cases, increase the cooling time in intervals of 5 s.

4.3. Limitations of the Work

In this work we did not optimize gate location beforehand—instead, the location was determined primarily from the usual perspective of design convenience, rather than tool life. It is possible that a different gate location might have reduced some of the observed failures. In this study, we used only the recommended settings from the machine manufacturers (3D Systems and Stratasys). We did not experiment with changing the settings for layer thickness, nozzle speed, or infill density. It is possible that experimenting with different print settings and different post processing routes could alter the failure mechanisms. The failure analysis shown here is only relevant for 3D-printed tools. Such tools are increasingly being used in industry, but generally for molding benign materials (low melt temperatures) and not the more demanding materials that are shown here.

4.4. Implications for Future Research

The surface delamination (avulsion) is an interesting failure that starts below the surface of the PRT insert. Further research could be directed at better understanding this process, and how to avoid or suppress it for as long as possible. This is important because the inserts could still be safely used when small features were breaking off, but they became unusable only after avulsion occurred. Large-scale avulsion quickly degenerates into catastrophic failure of the tool. Hence, a better understanding the causes seems to be an important future direction of research. The effect is anticipated to be material-specific.

A possible future line of enquiry could be to use a design of experiments (DoE) approach to determine which variables contribute most to the failure process, and how they relate. This is likely to be an expensive process, given the destructive nature of the testing. In the current work, we have used real parts, but this would be disfavored for a DoE approach. Instead it may be preferable to use small parts with representative features.

Print settings and post-processing operations such as post-curing, polishing, and cleaning methods may also affect the failure processes, and hence, the life of the PRT inserts. Hence, a possible further line of enquiry could be to study the effect of these parameters.

Another possible research direction could be the microscopic analysis of fracture surfaces. This might help identify the relative contributions of creep vs fatigue vs thermal degradation effects at various stages of the failure. Our observations are that the white Visijet material is difficult to observe under optical microscopy, due to its poor contrast, but other colors or microscopy techniques may be more successful.

5. Conclusions

Case study investigations of PRT inserts used for IM suggest that the main failure modes are: shear failure; bending failure; avulsion; surface deterioration; edge failure; surface scalding. We have proposed a crack propagation theory, and we provide a conceptual map relating these failure modes to the root causes. The failure modes influence each other, and they may be connected in cascade sequences. Several suggestions are given for prolonging the tool life, via design practices and setting parameters. The latter includes the process of setting up the molding parameters. The original contributions of this work are the identification of failure modes, and the relationship with root causes and the proposed crack propagation theory.

This paper identifies the critical process parameters, and their effects on different areas of the PRT insert. The conceptual map for failures presented indicates a gradual progressive failure of the PRT insert, from initial surface quality defects, through to catastrophic feature failure. Mold temperature was a critical parameter, since it affected all areas of the tool, and contributed to multiple failure modes. Injection pressure was only significant in certain areas of the tool (injection pressure had little to no effect on failure of PRT inserts at the end of flow, and had the highest effect on features closest to the start of the flow path). Tool design was a factor in areas of the tool that had tall free-standing features.

The wider purpose of this work was to study the feasibility of: (a) using PRT inserts to mold resins with melt temperatures above 250 °C; (b) determining the size of the parts that can be successfully molded by using a PRT insert; (c) determining the variety of features that can be safely molded by using a PRT insert; (d) determining the number of parts that can be molded before failure, and (e) the quality control of the molded parts. In this study, we were able to answer point (a): it is feasible to mold resins with high melting temperatures of up to 350 °C, and we can expect about 14 parts from the mold before failure, for the cases under examination. Consequently, it is possible to obtain a limited run of injection molded parts from an inexpensive 3D-printed tool. This is important, because the mechanical behavior of an injected molded part can be very different to a 3D-printed part. Nonetheless further research is required to address the other variables (b) to (e), and to increase the part count.

Author Contributions: This work was conducted by A.B., and supervised by D.P., D.C., and D.S. The ideas for failure modes were developed by A.B. and D.P. All authors contributed to the writing of the paper.

Funding: This research was funded by Callaghan Innovation, grant number TALB 1501/PROP-47676-FELLOW TALBOT.

Acknowledgments: We thank Talbot Technologies Ltd. for the opportunity to pursue this research. Special thanks to Steve W, Steve O, Lance F, and Ben A at Talbot Technologies for continued support through the project.

Conflicts of Interest: The authors declare no financial conflict of interest in this project. The funding agency and industry partners did not control the research, nor influence the content of the paper.

References

1. Weber, A. 3D printing goes from prototyping to production. *Assembly* **2018**, *61*, 58–63.
2. Annacchino, M.A. *The Pursuit of New Product Development: The Business Development Process*; Butterworth-Heinemann: Amsterdam, The Netherland; Boston, MA, USA, 2007.
3. Brethauer, D.M. *New Product Development and Delivery: Ensuring Successful Products through Integrated Process Management*; AMACOM: New York, NY, USA, 2002.
4. Otto, K.N.; Wood, K.L. *Product Design: Techniques in Reverse Engineering and New Product Development*; Pearson Custom Pub.: Boston, MA, USA, 2006.
5. Rosato, D.V.; Rosato, D.V. *Injection Molding Handbook: The Complete Molding Operation Technology, Performance, Economics*, 2nd ed.; Chapman & Hall: New York, NY, USA, 1995.
6. Kazmer, D.O. Mold Cost Estimation. In *Injection Mold Design Engineering*; Hanser: Cincinnati, OH, USA, 2007; pp. 37–66.
7. Goodship, V. *Practical Guide to Injection Moulding*; Rapra Technology: Shawbury, UK, 2004.
8. Kazmer, D.O. Introduction. In *Injection Mold Design Engineering*; Hanser: Cincinnati, OH, USA, 2007; pp. 1–15. [CrossRef]
9. Yarlagadda, P.K.D.V.; Wee, L.K. Design, development and evaluation of 3D mold inserts using a rapid prototyping technique and powder-sintering process. *Int. J. Prod. Res.* **2006**, *44*, 919–938. [CrossRef]
10. Wagner, S.M.; Walton, R.O. Additive manufacturing's impact and future in the aviation industry. *Prod. Plan. Control* **2016**, *27*, 1124–1130. [CrossRef]
11. Ong, H.S.; Chua, C.K.; Cheah, C.M. Rapid Moulding Using Epoxy Tooling Resin. *Int. J. Adv. Manuf. Technol.* **2002**, *20*, 368–374. [CrossRef]
12. Gardan, J. Additive manufacturing technologies: State of the art and trends. *Int. J. Prod. Res.* **2016**, *54*, 3118–3132. [CrossRef]
13. Noble, J.; Walczak, K.; Dornfeld, D. Rapid Tooling Injection Molded Prototypes: A Case Study in Artificial Photosynthesis Technology. *Procedia CIRP* **2014**, *14*, 251–256. [CrossRef]
14. Upcraft, S.; Fletcher, R. The rapid prototyping technologies. *Assem. Autom.* **2003**, *23*, 318–330. [CrossRef]
15. Chua, C.K.; Leong, K.F.; Liu, Z.H. Rapid Tooling in Manufacturing. In *Handbook of Manufacturing Engineering and Technology*; Nee, A.Y.C., Ed.; Springer: London, UK, 2015; pp. 2525–2549.
16. Hopkinson, N.; Dickens, P. Predicting stereolithography injection mould tool behaviour using models to predict ejection force and tool strength. *Int. J. Prod. Res.* **2000**, *38*, 3747–3757. [CrossRef]
17. Sa Ribeiro, A., Jr.; Hopkinson, N.; Ahrens, C.H. Thermal effects on stereolithography tools during injection moulding. *Rapid Prototyp. J.* **2004**, *10*, 4. [CrossRef]
18. Fernandes, A.D.C.; De Souza, A.F.; Howarth, J.L.L. Mechanical and dimensional characterisation of polypropylene injection moulded parts in epoxy resin/aluminium inserts for rapid tooling. *Int. J. Mater. Prod. Technol.* **2016**, *52*, 15. [CrossRef]
19. Dos Santos, W.N.; De Sousa, J.A.; Gregorio, R., Jr. Thermal conductivity behaviour of polymers around glass transition and crystalline melting temperatures. *Polym. Test.* **2013**, *32*, 987–994. [CrossRef]
20. Rahmati, S.; Dickens, P. Stereolithography for injection mould tooling. *Rapid Prototyp. J.* **1997**, *3*, 53–60. [CrossRef]

21. Park, H.; Cha, B.; Cho, S.; Kim, D.; Choi, J.H.; Pyo, B.-G.; Rhee, B. A study on the estimation of plastic deformation of metal insert parts in multi-cavity injection molding by injection-structural coupled analysis. *Int. J. Adv. Manuf. Technol.* **2016**, *83*, 2057–2069. [CrossRef]
22. Rahmati, S.; Dickens, P. Rapid tooling analysis of Stereolithography injection mould tooling. *Int. J. Mach. Tools Manuf.* **2007**, *47*, 740–747. [CrossRef]
23. Harris, R.; Hopkinson, N.; Newlyn, H.; Hague, R.; Dickens, P. Layer thickness and draft angle selection for stereolithography injection mould tooling. *Int. J. Prod. Res.* **2002**, *40*, 719–729. [CrossRef]
24. Harris, R.A.; Newlyn, H.A.; Dickens, P.M. Selection of mould design variables in direct stereolithography injection mould tooling. *Proc. Inst. Mech. Eng.* **2002**, *216*, 499–505. [CrossRef]
25. Accura Bluestone. Available online: https://www.3dsystems.com/materials/accura-bluestone (accessed on 25 June 2018).
26. DSM Perform. Available online: https://www.dsm.com/solutions/additive-manufacturing/en_US/products/for-stereolithography/somos-perform.html (accessed on 25 June 2017).
27. Digital ABS Plus. Available online: http://www.stratasys.com/materials/search/digital-abs-plus (accessed on 25 June 2017).
28. Rajaguru, J.C.; Duke, M.; Au, C. Investigation of electroless nickel plating on rapid prototyping material of acrylic resin. *Rapid Prototyp. J.* **2016**, *22*, 162–169. [CrossRef]
29. Rajaguru, J.; Duke, M.; Au, C. Development of rapid tooling by rapid prototyping technology and electroless nickel plating for low-volume production of plastic parts. *Int. J. Adv. Manuf. Technol.* **2015**, *78*, 31–40. [CrossRef]
30. Belter, J.T.; Dollar, A.M. Strengthening of 3D Printed Fused Deposition Manufactured Parts Using the Fill Compositing Technique. *PLoS ONE* **2015**, *10*, e0122915. [CrossRef] [PubMed]

Article

Modeling, Simulation, and Operability Analysis of a Nonisothermal, Countercurrent, Polymer Membrane Reactor

Brent A. Bishop and Fernando V. Lima *

Department of Chemical and Biomedical Engineering, West Virginia University, Morgantown, WV 26506, USA; babishop@mix.wvu.edu
* Correspondence: Fernando.Lima@mail.wvu.edu; Tel.: +1-304-280-0903

Received: 16 October 2019; Accepted: 2 January 2020; Published: 7 January 2020

Abstract: As interest in the modularization and intensification of chemical processes continues to grow, more research must be directed towards the modeling and analysis of these units. Intensified process units such as polymer membrane reactors pose unique challenges pertaining to design and operation that have not been fully addressed. In this work, a novel approach for modeling membrane reactors is developed in AVEVA's Simcentral Simulation Platform. The produced model allows for the simulation of polymer membrane reactors under nonisothermal and countercurrent operation for the first time. This model is then applied to generate an operability mapping to study how operating points translate to overall unit performance. This work demonstrates how operability analyses can be used to identify areas of improvement in membrane reactor design, other than just using operability mapping studies to identify optimal input conditions. The performed analysis enables the quantification of the Pareto frontier that ultimately leads to design improvements that both increase overall performance and decreases the cost of the unit.

Keywords: process intensification; operability; modularity; process modeling and simulation

1. Introduction

Modular equipment is an emerging technology that offers many potential benefits in terms of increased safety, flexibility, and ease of modification [1]. However, to take advantage of or access the benefits of modular plants, traditional processes must be significantly reduced in size for easier transportation and assembly. This design philosophy goes against the long-standing tradition of building massive chemical plants to take advantage of economies of scale. These modular plants cannot simply be smaller versions of traditional processes, but they must also be more efficient than their traditional counterparts.

This concept of increased efficiency of process units gave rise to the technologies associated with process intensification. Process intensification (PI) is defined as "any chemical engineering development that leads to substantially smaller, cleaner, and more energy efficient technologies." [1] In many cases, this is achieved by combining multiple phenomena in a single process unit (i.e., reactive distillation, reactive crystallization, [2] and membrane reactors) rather than performing step-by-step unit operations. Although PI can greatly improve the efficiency of the unit, intensified processes pose challenges to modeling and in the control/operations of these units.

As combinations of phenomena are introduced in a unit, the complexity of the resulting problem grows. In the case of a membrane reactor, reaction kinetics, heat duty, and membrane permeation must all be considered at every point along the reactor. For some cases, the problem can be simplified by making isothermal assumptions to remove the heat duty or enthalpy from permeation, cocurrent operation to avoid boundary value problems, and utilization of membrane materials such as palladium where only one component is capable of permeating into or out of the process. These simplifications greatly reduce the time required to run online optimization or modeling studies, but at the cost of a loss in accuracy.

These aforementioned phenomena are so interdependent, that the removal of any one of them can result in very different results. These simplifying assumptions may lead to a gap in understanding of the behavior of membrane reactors.

The first objective in this research is to present a modeling methodology that addresses these issues in which the fundamental phenomena present are broken down into individual blocks that are assembled to create the complex behavior of a membrane reactor. This method provides a reliable way of simulating a countercurrent, nonisothermal, and bidirectional-permeation membrane reactor model.

Second, with the use of this model, operability analysis [3,4] is performed for a polymer membrane reactor for the first time. The operability concept is used to study the relationship between unit performance and operational decisions. This analysis helps with improving the understanding of the benefits and drawbacks of polymer membranes and provides a direction for future work in the design of membrane materials.

Third, the operability analyses are also employed to drive future membrane reactor design decisions. Others have used operability analysis previously to improve membrane reactor design [3–6], but this proposed approach differs in a key way. Prior work in operability either looked at how design changes could be used to improve a nominal operating point or what the optimal nominal point of a given design would be. In this work, operability is used to identify the Pareto frontier for optimization in order to drive design decisions for overall performance improvement.

These three contributions form the necessary foundation for future work in developing an effective mixed-integer algorithm for determining an optimum modular design that improves unit operability. The outline of the rest of this paper is as follows. First, the membrane reactor modeling, simulation, and operability analysis approaches are introduced. These approaches are then employed for simulating the polymer membrane reactor unit and performing the operability mapping for identifying the Pareto frontier and redesigning the membrane reactor. The paper is closed with conclusions and some directions for future work.

2. Modeling, Simulation, and Operability Approach

2.1. Membrane Reactor Modeling Approach

For this study, a water-gas shift membrane reactor (WGS-MR) system in a shell and tube reactor is considered as shown in Figure 1. The membrane reactor model is developed as a one-dimensional, nonisothermal, single tube, and countercurrent unit. The unit is assumed to be operated at steady-state using pressure-driven flow calculations in the AVEVA SimCentral Simulation Platform (AVEVA Group plc. Cambridge, UK) The process side, which is a packed bed, is modeled using the Ergun Equation (1), while the sweep gas side uses the Colbrooke Equation (2) for modeling the pressure drop.

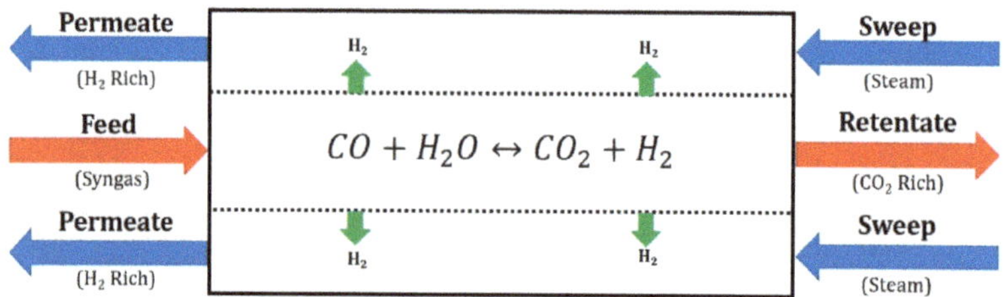

Figure 1. Schematic of the polymer membrane which consists of a tube packed with catalyst surrounded by a polymer membrane to allow simultaneous reaction and permeation.

Ergun equation, tube:

$$\frac{dp_t}{dz} = \frac{150\mu}{D_p{}^2}\frac{(1-\epsilon)^2}{\epsilon^3}v_s + \frac{1.75\rho}{D_p}\frac{(1-\epsilon)}{\epsilon^3}v_s|v_s| \tag{1}$$

where μ is the fluid's viscosity, D_p is the catalyst particle diameter, ε is the void fraction of the catalyst, v_s is the superficial velocity of the fluid, and ρ is the fluid density.

Colebrook equation, shell:

$$\frac{dp_s}{dz} = \frac{64f\mu v_s}{2D_h{}^2} \tag{2}$$

where D_h is the hydraulic diameter of the shell side and f is the Darcy friction factor that can be obtained by solving the implicit formula:

$$\frac{1}{\sqrt{f}} = -2\log\left(\frac{\epsilon}{3.7D_h} + \frac{2.51}{Re\sqrt{f}}\right) \tag{3}$$

On the tube side of the reactor, the water-gas shift reaction and membrane permeation take place, whereas only membrane permeation contributes to changes in the mole balance of the shell side. Therefore, the mole balance can be described by Equations (4) and (5) below, assuming plug-flow operation.

Mole balance, tube:

$$\frac{dF_{i,t}}{dz} = r_i A_t - J_i \pi d_t \tag{4}$$

where $F_{i,t}$ is the molar flow rate in the tube, r_i is the species reaction rate, A_t is the cross sectional area of a single tube, J_i is the molar flux across the membrane, and d_t is the diameter of a single tube. The subscript i denotes the species and z is the reactor axial coordinate. For the water-gas shift reaction, $r_i = r_{CO}$ for $i = CO$, H_2O, $r_i = -r_{CO}$ for $i = CO_2$, H_2, and $r_i = 0$ for N_2. The reaction rate, r_{CO}, is assumed to be for the low-temperature water-gas shift. ($Cu/ZnO/Al_2O_3$) catalyst and is modeled using the reaction provided in Reference [7].

Mole balance, shell:

$$-\frac{dF_{i,s}}{dz} = J_i \pi d_t \tag{5}$$

where $F_{i,s}$ is the flowrate in the shell. For this study, counter-current operation is used, and thus, a negative sign is required. The molar flux of a polymer membrane can be captured using a Fickian diffusion model with the partial pressure gradient of each component as the driving force:

$$J_i = \frac{Q_{i,o}}{\delta}\left(p_{i,t} - p_{i,s}\right) \tag{6}$$

in which $Q_{i,o}$ is the permeance, δ is the membrane thickness, and $p_{i,t}$ and $p_{i,s}$ are the partial pressures of species i on the tube and shell sides, respectively. For the case of the considered polybenzimidazole-based polymer membrane, the permeance values selected for this study come from References [8–11] and are shown in Table 1.

Table 1. Component permeance ($Q_{i,o}$) and permeability ($P_{i,o}$) values used reported in gas permeation units (GPU) and barrer, respectively, as well as the ideal selectivities ($\alpha_{H2/i}$) of hydrogen relative to other components in the syngas.

Component (i)	$Q_{i,o}$ (GPU)	$P_{i,o}$ (Barrer)	$\alpha_{H2/i}$
H_2	250.0	25.00	1.00
CO_2	8.9	0.89	28.1
H_2O	750.0	75.00	0.33
CO	2.5	0.25	100
N_2	2.5	0.25	100

Hence, in this case, all the species that are present are capable of crossing through the membrane. This is important to consider for later as the sweep gas, steam, is three times more permeable than the hydrogen through the polymer membrane whereas the other species permeabilities are very low, but not negligible.

Lastly, the energy balances for each side of the reactor are considered for the calculation of their temperatures. In a membrane reactor, there are three factors that contribute to changes in temperature: reactions, heat duty, and the Joules-Thomson effect as material permeates from one side of the membrane to the other. Considering a thin control volume along the length of the membrane reactor, these three effects can be modeled with the following energy balances:

Energy balance, tube:

$$0 = \frac{d(F_t H_t)}{dz} + \sum_i J_i \pi d_t \int_{p_t}^{p_s} \left(\frac{\partial H_i}{\partial p}\right)_{T_i} dp + U \pi d_t (T_t - T_s) \tag{7}$$

Energy balance, shell:

$$0 = \frac{d(F_s H_s)}{dz} + \sum_i J_i \pi d_t \int_{p_t}^{p_s} \left(\frac{\partial H_i}{\partial p}\right)_{T_i} dp + U \pi d_t (T_t - T_s) \tag{8}$$

in which the central term accounts for the Joule-Thomson effect caused by material leaving or entering each side through the membrane. p_t is the total tube pressure, p_s is the total shell pressure, and $\left(\frac{\partial H_i}{\partial p}\right)_{T_i}$ is the isothermal Joule-Thomson coefficient for species i at constant temperature T_i (where T_i is the temperature of the side species i originated). The last term in the balance accounts for the heat duty across the membrane where U is the overall heat transfer coefficient, which is assumed to have a value of 30 W/m^2-K for gas-to-gas convective heat transfer [12]. Normally, this value would change as a function of mass flowrate of the tube and shell sides; however, the dominating terms for heat duty on the lab scale reactor studied are the heat transfer area and the temperature gradient so potential U variations are expected to have a minimal effect when compared to the heats of reaction and Joule-Thomson effects. Finally, the temperature change due to the reaction is accounted for by the change in composition in the expanded form of the first term.

2.2. Block-Based Modeling and Simulation Approach

To simulate the polymer membrane reactor, a model was developed using AVEVA's equation-oriented SimCentral Simulation Platform [13] employing a block-based phenomena concept. This method allows for the simulation of complex membrane systems that may take too long in other platforms to calculate all the operating conditions required for a reasonable operability analysis to take place.

The model is constructed by discretizing along the length of the membrane reactor and modeling each of the produced sections using a model element referred to as "MemElem." MemElem is further broken down into submodels, where each of which captures the various phenomena that occur in a membrane reactor. A block diagram of how these submodels are connected to each other within the MemElem model is provided in Figure 2.

For a thin slice along the length of a membrane reactor, the tube and shell sides of the reactor can be assumed to operate as a continuous stirred-tank reactor (CSTR), where the outlets are equal to the current state of the reactor, depicted in Figure 2 as "TubeState" and "ShellState." These state submodels determine the thermodynamic state on each side of the reactor and send that information (in this case the molar composition (Z), pressure, and temperature) to the other submodels. The other submodels handle the calculations of the membrane permeation (Flux Model, PermStateS, and PermStateT) and reaction kinetics (Tube Rxn). The results from all the submodels are used by the MemElem model to calculate the mass and energy balances. Lastly, these thin sections of membrane can be combined in series to produce the full membrane reactor structure as seen in Figure 3.

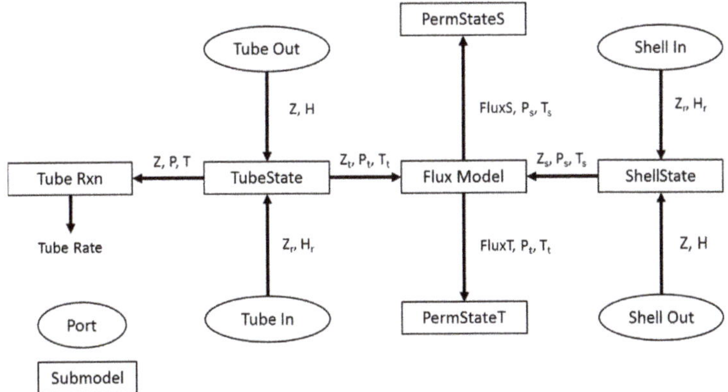

Figure 2. The submodel structure for the MemElem model.

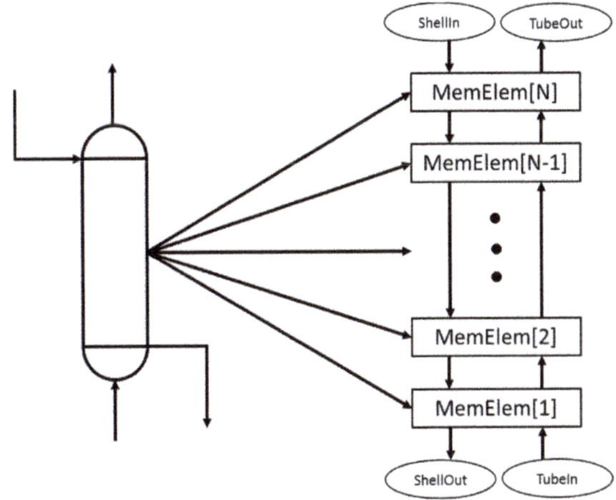

Figure 3. Each MemElem submodel $[1, 2, \ldots, N-1, N]$ is combined in series to complete the calculation of the full polymer membrane reactor unit.

This methodology allows for the simulation of countercurrent, nonisothermal, and bidirectional permeation of a polymer membrane system quickly enough that it can be solved in real-time in a dynamic simulation using an Intel® Core™ i7-4790 CPU @ 3.60GHz processor. Although for the purposes of this analysis, all simulations will be performed at steady-state conditions, the dynamic simulation and control of this reactor will be subject of subsequent studies.

By dividing the model into blocks that are dedicated to each of the phenomena that occur in the membrane reactor unit, additional features and unit customization are available to the user. By adding or removing blocks from the model, MemElem is thus capable of simulating a membrane reactor, membrane separator, packed-bed reactor, or heat exchanger. For example, if a user is interested in using MemElem to simulate a stand-alone packed-bed reactor with heat transfer, they would remove the Flux Model, PermStateS, and PermStateT blocks (see blocks in Figure 2). The potential benefits of being able to do this customization are covered in more depth in the Results section.

Although a block-based phenomena approach to modeling and simulating membrane reactor systems in an equation-oriented modeling environment provides the ability for conducting mixed-integer design optimizations, control studies, etc., the initialization process for this approach is challenging and the swapping in and out of the blocks may cause the model to become unsolved if the transition between solutions is discontinuous.

In equation-oriented modeling, it is necessary that the solver can find a unique solution at every step of the initialization process. This means two conditions should remain true throughout the initialization process: (1) the system of equations that describe the model must have a finite number of solutions; (2) the transition between intermediate solutions should be either continuous or smooth. If the system has infinitely many solutions, the solver will be unable to find a solution because the system is indeterminate. This can be a problem for the PermStateS and PermStateT blocks, which require N + 2 independent thermodynamic properties to be uniquely defined. During the initialization process, there is no material permeating across the membrane; therefore, there is no unique solution for what the N component mole fractions of the permeating material are and the PermState blocks can become unsolved. Arbitrary values could be sent to the PermState submodels so they remain solved, but this can lead to the second issue described above. If the provided arbitrary values are very different from the actual values in the final solution, the solver is unlikely to arrive at a solution, especially if the system is in countercurrent mode. Because the initialization approach is the most challenging barrier in the modeling of the countercurrent, nonisothermal, polymer membrane reactor, a comprehensive description of the developed approach is a necessary point of discussion.

This process begins by flowing the contents of the tube and shell side cocurrently with specified pressure drops for each side and no heat or mass transfer interactions between the two sides. This allows the TubeState and ShellState submodels in Figure 2 to solve for initial solutions for the thermodynamic states of the tube and shell side and gives MemElem the opportunity to establish an initial mass and energy balance for the system. Once the model is filled with initial values, the goal is then to introduce each element of the desired model in such a way that the two conditions mentioned above can be satisfied. The elements that need to be added to the initial model are: (1) pressure drop; (2) reaction kinetics; (3) countercurrent operation; (4) membrane permeation and heat transfer. This order of adding elements in steps 1–4 was determined to be the most effective in initializing the block-based modeling approach as each step provides the best possible guess for the next element to be added to the model.

Lastly, a series of "contact" variables, C, are implemented that facilitate the transition from step 3 to step 4. The standard formula used for these contact variables is of the form:

$$S = S_1 * (1 - C) + S_2 * C \tag{9}$$

where S is the solution used by the model, S_1 represents an initial solution, and S_2 represents a desired solution. By switching the value of the contact variable from 0 to 1, the value of S will switch from S_1 to S_2 continuously rather than discretely, making the model much easier for the solver to determine a solution. This same concept has also been applied at the block level of the model where contact variables can remove entire blocks from the model. This feature, combined with the fact that the block-based approach allows for simulating multiple unit operations by adding or removing blocks, creates a design space for a membrane reactor model using this MemElem concept that is homotopic. That is to say, every possible topological design for the unit is connected to every other design through a continuous deformation. This is especially appealing for a mixed-integer optimization problem, as the solution space can now be continuous and the transitions can take place without having to reinitialize the model.

This initialization procedure makes this model very useful for conducting operability analyses where many operating conditions must be simulated in quick succession. To demonstrate this, two performance metrics are selected for the membrane reactor unit. Reactor performance is assessed using the metrics of H_2 recovery (R_{H2}) and CO_2 capture (C_{CO2}) defined as follows:

$$R_{H_2} = \frac{H_2 \text{ in permeate}}{(H_2 + CO)\text{ in feed}} = \frac{F_{H_2,p}}{F_{H_2,f} + F_{CO,f}} \tag{10}$$

$$C_{CO_2} = \frac{\text{carbon in retentate}}{\text{carbon in feed}} = \frac{F_{CO,r} + F_{CO_2,r}}{F_{CO,f} + F_{CO_2,f}} \tag{11}$$

To achieve the different values for R_{H2} and C_{CO2}, the process flowrate and sweep gas flowrate are manipulated with the use of two flow control valves, FC1 and FC2, as shown in Figure 4.

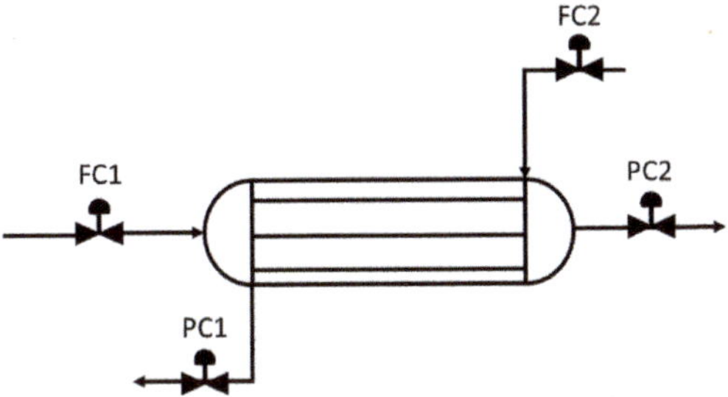

Figure 4. Depiction of the membrane reactor and control valves present in the simulation.

The valves are sized assuming they are 50% open at the nominal operating point. The manual positions of FC1 and FC2 are then varied between 10–100% open and the values of hydrogen recovery and carbon capture are recorded to produce the operability mapping in Section 3.2 below.

2.3. Operability Analysis

Operability ultimately serves as the interface between design and control [14]. Traditionally, when creating a plant, the unit will undergo a design phase for the pieces of equipment before it undergoes a design phase for the control system. However, as the interest in intensified processes grows, the literature has identified that "process design, operation, and control should be considered simultaneously, or in other terms, they should be fully integrated" [15] for intensified processes. Because operability can provide the interface between design and control, any intensified process (such as the polymer membrane reactor unit considered in this work) should undergo such an analysis to identify the control challenges upfront and to determine if any design changes can be implemented to mitigate these challenges. Set-point operability is used in this research to measure the performance of the membrane reactor design, as defined in the operability analysis for square systems [14]. The mapping of inputs ($u \in \mathbb{R}^m$) of a model (M) to its outputs ($y \in \mathbb{R}^p$) can be formulated the following way:

$$M = \begin{cases} \dot{x} = f(x, u) \\ y = g(x, u) \\ h_1(\dot{x}, x, y, \dot{u}, u) = 0 \\ h_2(\dot{x}, x, y, \dot{u}, u) \geq 0 \end{cases} \tag{12}$$

in which $x \in \mathbb{R}^n$ are the state variables and h_1 and h_2 are equality and inequality process constraints, respectively. Additionally, $\dot{x}$ and $\dot{u}$ represent time derivatives associated with x and u, respectively, and f and g are nonlinear process maps.

Using the operability mapping concept, there are two sets of inputs and outputs that are important in this analysis:

Available Input Set (AIS): The set of all operational inputs or manipulated variables that are available to produce change to the output of the process and is defined as:

$$AIS = \left\{ u \middle| u_i^{min} \leq u_i \leq u_i^{max}; 1 \leq i \leq m \right\} \tag{13}$$

For this study, the AIS for the polymer membrane reactor consists of the sweep gas flowrate and the syngas flow rate into the membrane unit.

Achievable Output Set (AOS): This set consists of all possible outputs that can be achieved, given the available input set and is mathematically defined as:

$$AOS_u = \{y | M(u); \forall u \in AIS\} \tag{14}$$

For the purposes of this study, the AOS consists of the achievable hydrogen recovery and carbon capture for the unit to analyze how the polymer membrane reactor performs under different operating conditions. In this study, the ideal performance for any membrane would be 100% recovery of hydrogen and 100% capture of CO_2, thus, high performance in these two outputs is desired. A schematic visual depiction of these spaces is provided in Figure 5.

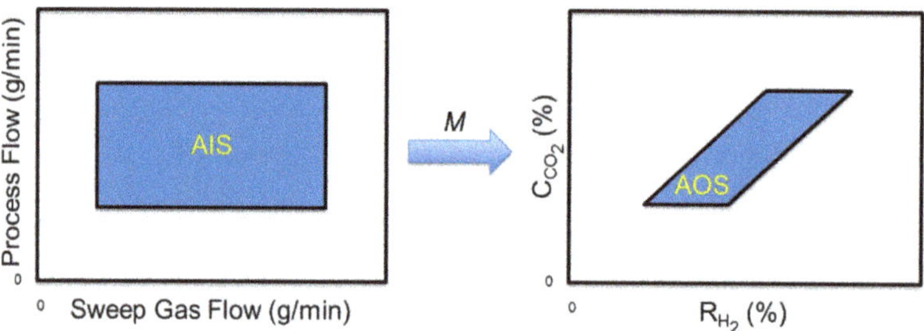

Figure 5. Example of an available input set (AIS) being mapped to an achievable output set (AOS) with the use of a process model (*M*).

Because the AOS corresponds to the set of all the outputs that can be achieved given the entire set of inputs, the ideal objective is to improve the design until the entire AOS meets some desired output specifications. In the ideal case, all the available inputs would map to a desired region of the output space. In the opposite case, if the unit is designed considering only the nominal operating point, a control system could face significant challenges, as the window for effective control shrinks significantly, making process disturbances much more difficult to reject. Therefore, the goal of the operability analyses in this paper is to determine design changes that can expand the AOS, and therefore, move the membrane reactor closer to the ideal case as described above.

3. Results

3.1. Simulation of the Polymer Membrane Reactor Model

The operating conditions for this study are taken from Reference [8]. The reactor utilizes as process stream a syngas feed that has undergone the necessary removal of impurities, while the chosen sweep gas is steam. The inlet composition of these streams can be found in Table 2.

Table 2. Molar composition of inlet streams, given in mole fraction.

Component	Syngas Feed	Steam Sweep
H_2	0.1933	0
CO_2	0.0568	0
H_2O	0.4886	1
CO	0.2443	0
N_2	0.017	0

The syngas feed enters the tube side of the reactor at 300 °C, 47.63 atm, and with a volumetric flow rate of 400 cm^3/min. The sweep gas enters the shell as pure, saturated steam at 25.86 atm and a volumetric flowrate of 400 cm^3/min. The reactor consists of a single tube (1.02 cm diameter) and shell (6.12 cm diameter) arrangement with a total reactor length of 300 cm. A membrane thickness of 100 nm is selected to reflect an industrially relevant thickness [9–11]. Under these conditions, a simulation was run using the block-based phenomena modeling approach and compared to the simulation data from Radcliffe et al. [8] for validation. The results of such comparison are shown in Table 3. In future work, the authors will pursue the additional validation of the model with experimental data.

Table 3. Comparison between the block-based proposed approach and literature [8].

Component	Proposed Approach (%)	Radcliffe et al. [8] (%)	Error (%)
X_{CO}	99.83	99.36	0.47
R_{H2}	98.12	98.38	−0.26
C_{CO2}	75.72	75.77	−0.07
$Purity_{CO2+H2O,r}$	95.55	96.05	−0.52
$Purity_{H2,p}$	42.00	42.05	−0.12

Using the nominal conditions described above, profiles for the permeation rates of hydrogen and steam are produced as shown in Figure 6.

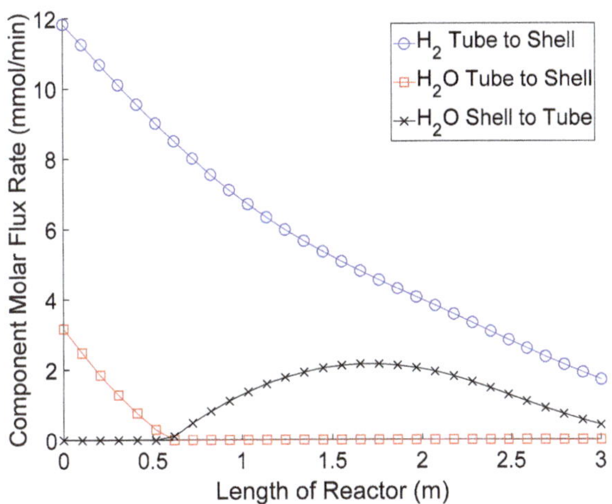

Figure 6. Profile of the permeation rates for hydrogen and steam down the length of the polymer membrane reactor.

Although the other components, CO_2, CO, and N_2, are also permeating through the membrane, they are permeating at a much lower rate than H_2 and H_2O, depicted in Figure 6. This result shows an interesting characteristic of the polymer membrane. For about the first sixth of the reactor, H_2O (steam) is being removed from the tube side along with the H_2 product. However, for the remainder of the reactor, steam is being injected into the tube side of the reactor. During this time, the sweep gas enters the catalytic tube side, reacts to become hydrogen, and then permeates back to the shell side as a product. This is a nice property for an equilibrium-limited reaction such as WGS, because not only is the membrane removing a main product, but it is also providing steam injection for enhanced conversion. This would be a useful property for cases where the CO/H_2O ratio of the feed becomes greater than one and there is insufficient steam to convert all the CO to CO_2.

Temperature is also an important factor to consider as WGS is known to produce a significant amount of heat and becomes equilibrium limited as the temperature increases. The temperature profiles for the tube and shell sides at the nominal operating condition and for a syngas feed at 200 °C can be seen in Figure 7.

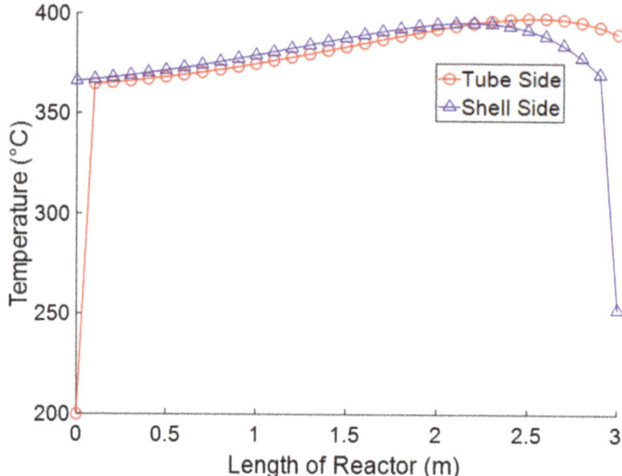

Figure 7. Temperature profiles of the tube and shell sides for nonisothermal operations.

It is important to note in Figure 7 the temperature spike at the beginning of the reactor. The conditions at the inlet of the reactor lie very far to the left of equilibrium. This is evidenced by the fact that steam is permeating out of the tube side at the beginning of the process. The reaction rates are much greater than in a normal WGS reactor because H_2 product is being removed, thus, also removing a major equilibrium limitation, but also introducing a large temperature spike. This problem of excess heat in polymer membrane reactors was reported by Singh et al. [11] who stated that "syngas operating temperatures in the vicinity of the water-gas shift (WGS) reactors ranges from 200 to 500 °C, depending on the WGS stage and catalyst used." This illustrates the importance of including the nonisothermal reactor operation in the modeling approach, as the reactor inlet temperature for our past study [8] leads to reactor temperatures that exceed the glass transition temperature (T_g = 450 °C) and would eventually cause degradation of the polymer membrane material.

These simulation results indicate many research pathways for the design of membrane reactors. Most membrane reactor models require simplifications to the problem in order to produce results in a tractable manner, especially when performing an operability analysis (or online optimization) that requires many simulations to be run to produce the output space for further analysis. However, these simplifications are harder to justify in the design and simulation of intensified processes. The purpose of combining phenomena such as mass transfer and kinetics is to see how the performance of the unit can be improved by having them work in tandem. However, as shown earlier with the temperature profile in Figure 7, there are also difficult challenges that arise when combining them. As more phenomena are combined in these process units, the more interdependent they become, and the more difficult it is to design the equipment to meet the desired output specifications [2,15,16].

Isothermal operation is often assumed for a few reasons. The reaction rate, equilibrium, and the membrane permeance are all affected by the temperature in the reactor. When this is coupled with the interdependence that the phenomena have with each other, it creates a very complex behavior and difficult calculation before even considering how the tube side of the membrane reactor might interact with the shell side for heat transfer calculations.

The difficulty in solving for the state of the tube or shell side of the membrane reactor is a reason the Joule-Thomson is normally not considered. In most cases, the Joule-Thomson effect has a minor impact on the accuracy of the temperature and requires accurate knowledge of the temperature at a given point in the reactor to calculate accurately. Many membranes, such as palladium-based membranes, are highly selective to only H_2, and therefore, do not need to consider this effect. Because of this, it is normally not worthwhile from a modeling perspective unless other nonisothermal factors are being considered in the model. However, in the case of polymer membranes with a high selectivity for H_2O permeation and nearly a 22 atm difference in pressure across the membrane wall, the Joules-Thomson effect may be more important to include.

The stated complexities are also the reason for avoiding countercurrent operation in membrane reactor simulation. The significant dependency on the states of each side of the reactor creates a very difficult boundary-value problem, where even slight changes to the conditions of one side of the reactor can lead to dramatically different solutions overall. In most traditional unit operations such as heat exchangers and reactors, the profiles will still have similar shapes, but will shift, compress, or contract as unit operations are varied. However, with the membrane reactor model, different operating points lead to dramatically different behaviors, and because of this, countercurrent operation is generally avoided.

Lastly, membranes such as the polybenzimidazole are able to have multiple components permeating in both directions simultaneously. This only becomes difficult when combined with countercurrent operation as this allows for a circulation effect to occur in the membrane reactor. For example, steam flows countercurrently to the tube side, permeates to the tube side, reacts to become hydrogen, flows cocurrently within the tube, and then permeates back to the shell side before leaving the reactor. Given all these aforementioned challenges, the proposed model of the comprehensive nonisothermal, countercurrent membrane reactor with bidirectional permeation is thus one of the significant contributions of this work.

3.2. Operability Analysis of Polymer Membrane Reactors

Once a reliable modeling approach is established, an operability analysis is conducted using the process model to see the behavior of the polymer membrane at various operating conditions. The results of this operability mapping can be found in Figure 8.

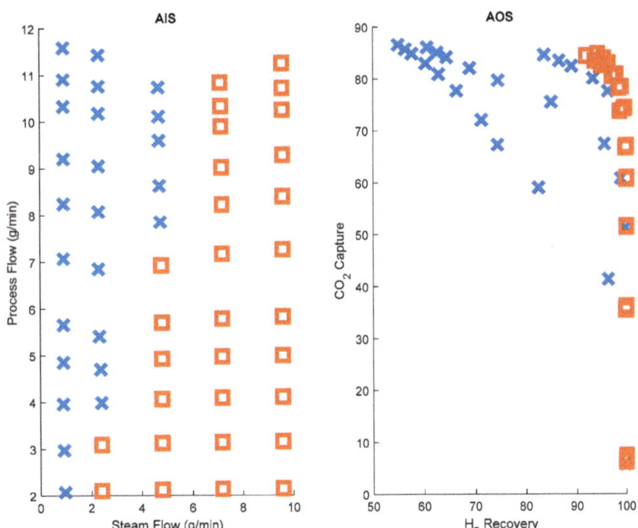

Figure 8. Input-output mapping of the polymer membrane reactor. For ease of understanding, parts of the mapping region (shown using the x and □ markings) are marked differently to see how the inputs are mapped to the output space.

This operability mapping in Figure 8 provides a good representation of the challenges caused by the nonlinearity of the system. It appears that if the flowrate of sweep gas increases to much more than that of the syngas (process flow), then the output becomes very constrained in terms of achievability (indicated by the square markers in Figure 8). This contrasts with the opposite mode of operation (sweep gas flow rate is much less than the syngas flowrate), where the two-dimensional input space is still mapped to another two-dimensional output space.

The achievable output set (AOS) also appears to form a Pareto front. The use of input-output mapping was also demonstrated by Messac and Mattson for generating Pareto points through physical programming [17]. Employing this concept in an operability framework proves to be a useful tool in identifying both operational shortcomings and potential design solutions for the polymer membrane reactor. To show this, an analysis is conducted to identify which points in the output space fit the definition of a Pareto optimal point. A point is considered to be Pareto optimal if one objective cannot be improved without degrading another objective. This analysis is performed using a MATLAB script that checks each point in the output set for Pareto optimality. Since the ideal membrane output would be 100% CO_2 capture with 100% H_2 recovery, a "best compromise" point is defined here as the Pareto optimal point with the shortest Euclidean distance between itself, and the ideal output point. This analysis is performed in MATLAB using the pdist function with the 'Euclidean' norm option. The result of this analysis is shown in Figure 9.

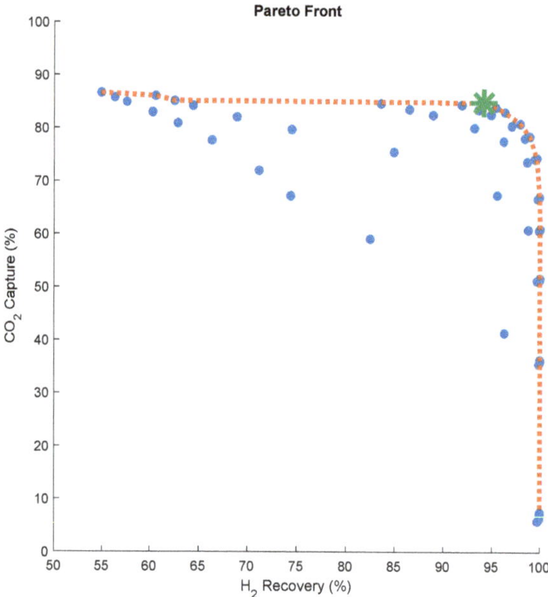

Figure 9. AOS points with the Pareto frontier included and the "best compromise" marked by a star.

This Pareto frontier in Figure 9, by definition, characterizes the best performance this polymer membrane reactor design can achieve, given the available inputs. To see how this operability analysis can help drive future design decisions, three operating points are selected from the Pareto front for further analysis based on where they are located in the output space; (i) high CO_2 capture, low H_2 recovery; (ii) low CO_2 capture, high H_2 recovery; (iii) a "best compromise" point (shown as a green star in Figure 9). The points at each end of the Pareto front were selected and for the purpose of easier comparison, the H_2 and H_2O fluxes associated with these points were calculated, shown in Figure 10.

The profiles in Figure 10 explain how the high CO_2 capture or high H_2 recovery is achieved. In Figure 10a, the process reaches an equilibrium within the first third of the reactor due to a relatively high flowrate of sweep gas. Slightly manipulating the sweep gas flowrate at this condition would reduce the permeation, but the membrane reactor would still reach this equilibrium. This results in very good H_2 recovery, but poor CO_2 capture as CO_2 continues to permeate out of the tube side despite the H_2 recovery process already being finished. Conversely, Figure 10b shows that steam injection is occurring at the end of the membrane reactor, leaving little room left in the reactor for the sweep gas to capture the newly produced H_2 product. This low sweep gas flowrate means that very little CO_2 will escape from the reactor, leading to optimum CO_2 capture, but at the cost of reducing H_2 recovery.

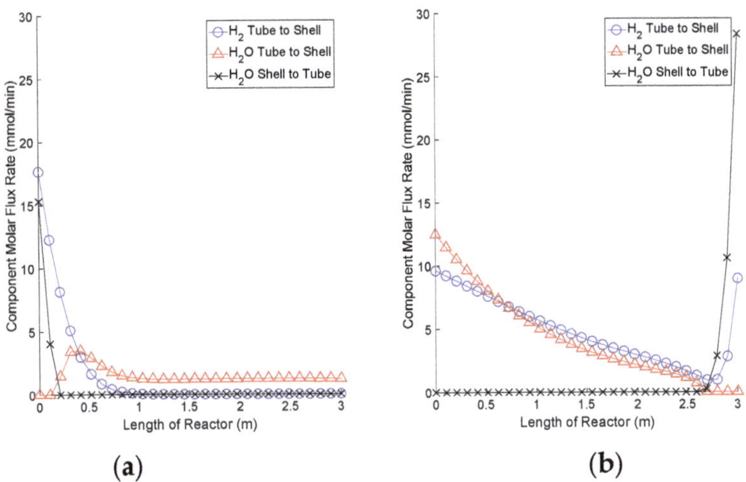

Figure 10. (**a**) Flux profile with the highest H_2 recovery. (**b**) Flux profile with the highest CO_2 capture.

The operating points selected for the simulations in Figure 10 show what is happening at the extreme ends of the Pareto frontier. For the next simulation, the best compromise point is selected from the elbow of the Pareto front to see how the polymer membrane reactor can achieve both high H_2 recovery and CO_2 capture. The results of this simulation are shown in Figure 11.

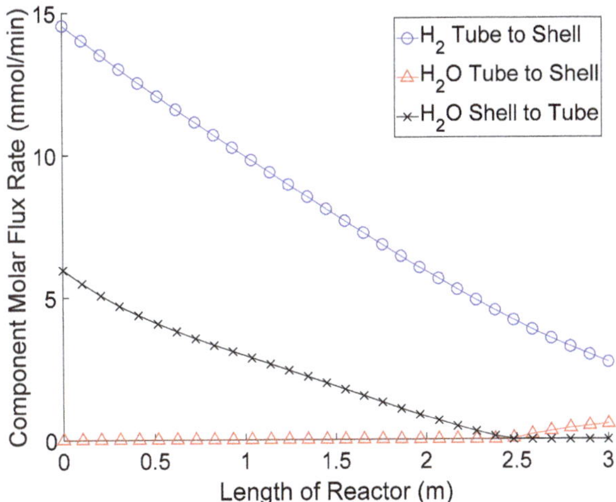

Figure 11. Flux profile for the "best compromise" operating condition (syngas-to-steam ratio = 0.92).

Unlike the simulation in Figure 10a, the process did not hit an equilibrium point where all potential H_2 recovery is completed. This means that although CO_2 is still being lost to the shell side, H_2 is also being recovered. Another improvement is that the steam injection process is completed before reaching the end of the membrane reactor. Taking what was learned from Figure 10b, this avoids hydrogen being formed at the end of the membrane reactor and not having space remaining to be captured. There are three lessons to be learned from this analysis; (i) steam injection appears to be beneficial at the beginning of the reactor; (ii) the end of the reactor should be primarily used for H_2 recovery; (iii) CO_2 continues to permeate out of the tube side at all points of the reactor (an undesired feature).

By further analyzing the Pareto frontier in Figure 9, the polymer membrane reactor does not struggle to recover H_2 but seems to be limited in its ability to capture CO_2. Therefore, the goal should be to push the Pareto front upward by improving the CO_2 capture of the unit. However, as previously mentioned, there is no operational change that can achieve this improvement. Therefore, a design change must be proposed to improve the process in this direction.

3.3. Redesign of the Polymer Membrane Reactor

The first two points learned from the operability analysis above focus on improving H_2 recovery and should remain features of any new design decision. The third point deals with CO_2 permeating out of the tube side, reducing the CO_2 capture of the unit and should be the focus of the design change. CO_2 permeation is increased by an increase in sweep gas flow in the presence of the membrane. Because sweep gas flow is needed for the H_2 recovery to occur, one obvious choice is to strategically remove the polymer membrane from certain areas of the reactor. Because membrane is needed at the beginning of the membrane reactor for the steam injection and at the end for H_2 recovery, the only place left is to remove it from the center of the reactor. The proposed redesign to avoid the issue of unnecessary CO_2 permeation is shown in Figure 12.

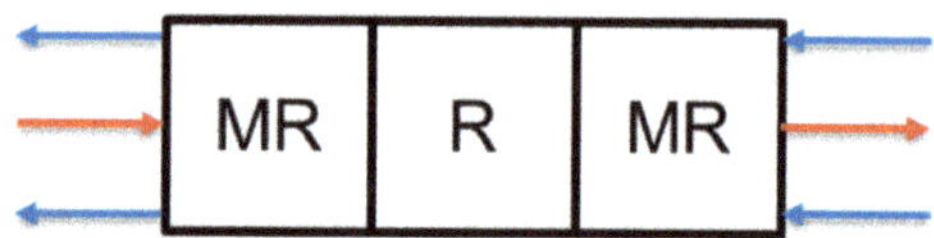

Figure 12. Proposed redesign of the polymer membrane reactor with membrane being removed from the center of the unit (MR = polymer membrane reactor, R = Reactor).

To simulate this redesign, the contact variables along with the block-based modeling approach described above can be used to allow for the removal of membrane without the need to reinitialize the model. Figure 13 shows the block-diagram representation for the packed-bed reactor section of the unit.

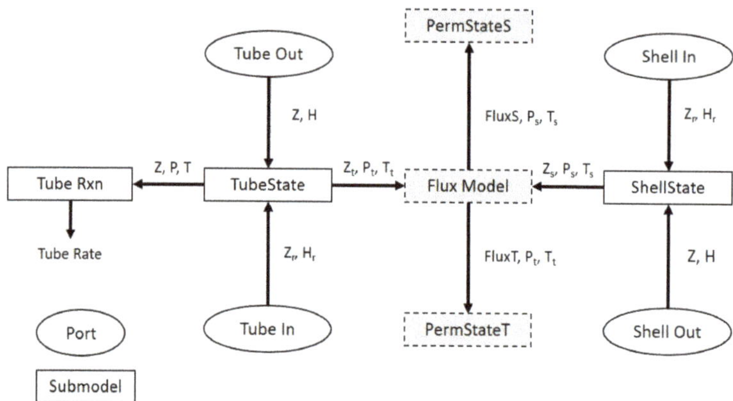

Figure 13. Diagram representing how removal of the Flux, PermStateT, and PermStateS submodels leads to the MemElem model of a packed-bed reactor.

The input space from the original operability analysis is applied to this new design modeled and simulated in the AVEVA SimCentral Simulation Platform. An example, using the best compromise operating point along with the new design, produces the flux profiles found in Figure 14.

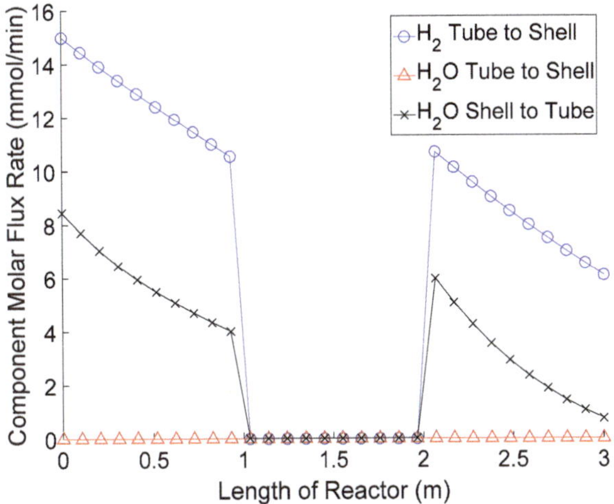

Figure 14. Flux profiles for the modified membrane reactor design with the membrane tube being replaced by a non-permeative tube.

For this new design, in Figure 14, steam injection is still taking place in the beginning of the membrane reactor and is almost complete by the end of the reactor while H_2 is being recovered at the end of the membrane reactor as desired. Additionally, it is important to compare the flux of CO_2 in both designs (in Figure 11 versus Figure 14) with the desire for reducing the CO_2 lost to the shell side. This comparison is shown in Figure 15.

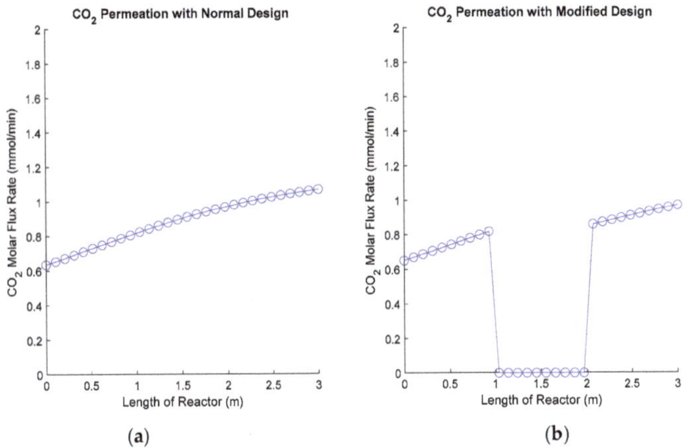

Figure 15. (**a**) Flux profile for the original design showing the continuous permeation of CO_2 out of the reactor. (**b**) Flux profile for the modified design where the polymer membrane was replaced with non-permeative tubes in the center.

By removing the polymer membrane from the center of the reactor, less CO_2 can permeate to the sweep gas, thus improving the capture percentage while maintaining a similar H_2 recovery. The Pareto analysis was conducted again on this new design and compared to the original design as shown in Figure 16.

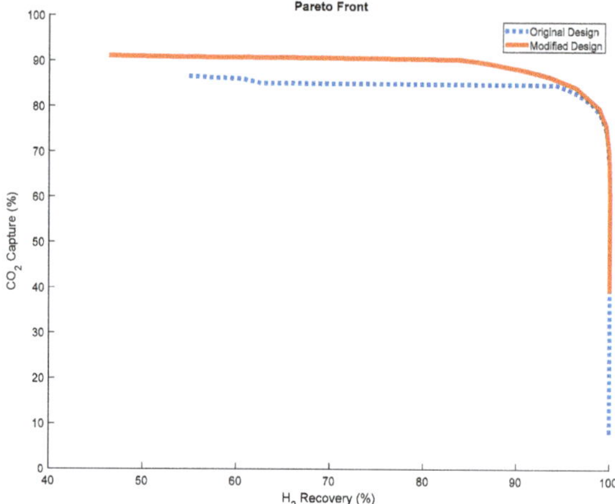

Figure 16. Comparison of the Pareto fronts for the original and modified designs.

Note that not only this new design change pushes the maximum CO_2 capture closer to 100%, it also increased the minimum possible CO_2 capture from 5.8% to 38.1%. Moreover, this design potentially brings economic benefits, as the amount of membrane material required to achieve the outputs is reduced. This study shows that operability analyses can provide insights for improving the design of membrane reactors in terms of performance targets and/or cost.

4. Conclusions and Future Work

This paper introduced a new modeling and operability analysis method for polymer membrane reactor studies. The modeling approach presented aimed to reduce the number of simplifying assumptions as much as possible to capture the complex behavior of polymer membranes. As a result, the following features were included in the membrane model for the study where normally some combination of them would be excluded: (i) nonisothermal operation—heat transfer across the tube wall, Joule-Thomson cooling and heating, heat of reaction, and thermal inertia for dynamic calculations; (ii) countercurrent operation; (iii) simultaneous, bidirectional permeation of multiple components.

Although this study is performed assuming steady-state operations, it should be noted that the polymer membrane reactor model also runs reliably in a dynamic simulation where additional features such as thermal inertia from the membrane materials can be considered. The hope is to continue to push these bounds in future work to further study the interdependence of phenomena and improve the overall operation of intensified equipment.

This modeling approach allowed for the first operability analysis of a polymer membrane reactor. In this work, an operability analysis showed a few important considerations when operating a polymer-based membrane reactor (1) nearly half of the available input space was mapped to a one-dimensional curve in the output space; and (2) although such reactors are able to recover H_2 effectively, it comes at the cost of lower CO_2 capture.

The first point is an undesirable feature of many membrane reactors and leads to problems for those who have worked in operability. If one is interested in which inputs are required to achieve a desired output, it is impossible to perform back calculations if more than one input can achieve that output. This fact would also be relevant for control studies associated with membrane. The second point highlights the tradeoff of switching to a polymer-based membrane rather than a more expensive palladium membrane, but also provides an objective for improving polymer membrane design.

This work also presented a new contribution for operability analyses. This type of analysis has always been useful for seeing how inputs translate to the output space. The use of operability to determine the Pareto front was employed in this study to motivate design changes to the WGS polymer membrane reactor unit. By applying this concept to the operability analysis, a best tradeoff point can be selected and compared to other operating points to determine possible design changes to improve the operability of the system. Although the process of running the analysis, diagnosing the problem areas, and using this info to develop a new design had a logical thought process applied to it, it is still an empirical approach. Future work will focus on the development of a mathematical optimization algorithm using the block-based modeling approach, and the operability and Pareto analyses shown in this work to design equipment with the objective of improving or expanding the entire output set given an available input set.

Author Contributions: This paper is a collaborative work among the authors. B.A.B. performed all simulations and wrote manuscript. F.V.L. helped with the paper writing and oversaw all technical aspects of the research work. All authors have read and agreed to the published version of the manuscript.

Funding: The authors gratefully acknowledge the financial support from AVEVA.

Acknowledgments: The authors thank Cal Depew, Ralph Cos, Richard Pelletier, and Larry Balcom of AVEVA for their technical guidance and advice on the membrane model.

Conflicts of Interest: The authors declare no conflict of interest.

References

1. Roy, S.; Eng, P. Consider Modular Plant Design. *AIChE CEP Magazine.* **2017**, *113*, 28–31.
2. Schembecker, G.; Tlatlik, S. Process synthesis for reactive separations. *Chem. Eng. Process. Process Intensif.* **2003**, *42*, 179–189. [CrossRef]
3. Carrasco, J.C.; Lima, F.V. Novel operability-based approach for process design and intensification: Application to a membrane reactor for direct methane aromatization. *AIChE J.* **2017**, *63*, 975–983. [CrossRef]
4. Carrasco, J.C.; Lima, F.V. Bilevel and parallel programing-based operability approaches for process intensification and modularity. *AIChE J.* **2018**, *64*, 3042–3054. [CrossRef]
5. Gazzaneo, V.; Lima, F.V. Multilayer Operability Framework for Process Design, Intensification, and Modularization of Nonlinear Energy Systems. *Ind. Eng. Chem. Res.* **2019**, *58*, 6069–6079. [CrossRef]
6. Fouty, N.J.; Carrasco, J.C.; Lima, F.V. Modeling and Design Optimization of Multifunctional Membrane Reactors for Direct Methane Aromatization. *Membranes* **2017**, *7*, 48. [CrossRef] [PubMed]
7. Choi, Y.; Stenger, H.G. Water gas shift reaction kinetics and reactor modeling for fuel cell grade hydrogen. *J. Power Sour.* **2003**, *124*, 432–439. [CrossRef]
8. Radcliffe, A.J.; Singh, R.P.; Berchtold, K.A.; Lima, F.V. Modeling and Optimization of High-Performance Polymer Membrane Reactor Systems for Water–Gas Shift Reaction Applications. *Processes* **2016**, *4*, 8. [CrossRef]
9. Li, X.; Singh, R.P.; Dudeck, K.W.; Berchtold, K.A.; Benicewicz, B.C. Influence of polybenzimidazole main chain structure on H_2/CO_2 separation at elevated temperatures. *J. Membr. Sci.* **2014**, *461*, 59–68. [CrossRef]
10. Berchtold, K.A.; Singh, R.P.; Young, J.S.; Dudeck, K.W. Polybenzimidazole composite membranes for high temperature synthesis gas separations. *J. Membr. Sci.* **2012**, *415–416*, 265–270. [CrossRef]
11. Singh, R.P.; Dahe, G.J.; Dudeck, K.W.; Welch, C.F.; Berchtold, K.A. High Temperature Polybenzimidazole Hollow Fiber Membranes for Hydrogen Separation and Carbon Dioxide Capture from Synthesis Gas. *Energy Procedia* **2014**, *63*, 153–159. [CrossRef]
12. Turton, R.; Bailie, R.; Whiting, W.; Shaeiwitz, J.; Bhattacharyya, D. *Analysis, Synthesis, and Design of Chemical Processes*, 4th ed.; Prentice Hall: Upper Saddle River, NJ, USA, 2012.
13. AVEVA. *Simcentral Simulation Platform: Process Simulation Reinvented*; AVEVA: Cambridge, UK, 2019.
14. Lima, F.V.; Jia, Z.; Ierapetritou, M.; Georgakis, C. Similarities and differences between the concepts of operability and flexibility: The steady-state case. *AIChE J.* **2010**, *56*, 702–716. [CrossRef]

15. Nikačević, N.M.; Huesman, A.E.M.; van den Hof, P.M.J.; Stankiewicz, A.I. Opportunities and challenges for process control in process intensification. *Chem. Eng. Process. Process Intensif.* **2012**, *52*, 1–15. [CrossRef]
16. Baldea, M. From Process Integration to Process Intensification. *Comput. Chem. Eng.* **2015**, *81*, 104–114. [CrossRef]
17. Messac, A.; Mattson, C.A. Generating Well-Distributed Sets of Pareto Points for Engineering Design Using Physical Programming. *Optim. Eng.* **2002**, *3*, 431–450. [CrossRef]

Article

Some Eccentricity-Based Topological Indices and Polynomials of Poly(EThyleneAmidoAmine) (PETAA) Dendrimers

Jialin Zheng [1], Zahid Iqbal [2], Asfand Fahad [3], Asim Zafar [3], Adnan Aslam [4], Muhammad Imran Qureshi [3,*] and Rida Irfan [5]

[1] School of Information Science and Engineering, Chengdu University, Chengdu 610106, China
[2] School of Natural Sciences, National University of Sciences and Technology, Sector H-12, Islamabad 44000, Pakistan
[3] Department of Mathematics, COMSATS University Islamabad, Vehari Campus, Vehari 61100, Pakistan
[4] Department of Natural Sciences and Humanities, University of Engineering and Technology, Lahore 54000, Pakistan
[5] Department of Mathematics, COMSATS University Islamabad, Sahiwal Campus, Sahiwal 57000, Pakistan
* Correspondence: imranqureshi@cuivehari.edu.pk; Tel.: +92-314-611-8139

Received: 30 May 2019; Accepted: 2 July 2019; Published: 9 July 2019

Abstract: Topological indices have been computed for various molecular structures over many years. These are numerical invariants associated with molecular structures and are helpful in featuring many properties. Among these molecular descriptors, the eccentricity connectivity index has a dynamic role due to its ability of estimating pharmaceutical properties. In this article, eccentric connectivity, total eccentricity connectivity, augmented eccentric connectivity, first Zagreb eccentricity, modified eccentric connectivity, second Zagreb eccentricity, and the edge version of eccentric connectivity indices, are computed for the molecular graph of a PolyEThyleneAmidoAmine (PETAA) dendrimer. Moreover, the explicit representations of the polynomials associated with some of these indices are also computed.

Keywords: PolyEThyleneAmidoAmine (PETAA) dendrimer; molecular topological indices; Eccentric connectivity index

---

1. Introduction

The rapid growth in the field of medicine is resulting in the production of unknown nanomaterials, crystalline materials and drugs. To investigate the chemical properties of these compounds, huge efforts of pharmaceutical researchers are required and are being made. One way to understand it is by using mathematics; in mathematical chemistry, many concepts of graph theory are being used to formulate the mathematical model for chemical phenomena. Molecules and molecular compounds can be considered as graphs, if we map atoms to vertices and chemical bonds to edges, respectively. Such graphs are called molecular graphs. A graph can be identified by a uniquely defined number (or finite numbers), a matrix or a polynomial that describe the graph. Topological index (TI) is a mathematical quantity which is assigned to a graph (or molecular graph structure). Topological indices (TIs) provide relationships between structure of a molecule and several physical properties, biological activity or chemical reactivity. A TI is an invariant, that is if $\mathrm{Top}(H_1)$ represents a TI of a graph H_1 and H_2 (another graph) with $H_1 \cong H_2$ then $\mathrm{Top}(H_1) = \mathrm{Top}(H_2)$.

The notions of TIs help pharmacists by providing several information based upon the structure of materials which reduce their workload. Computing TIs of a compound may help in approximating its medicinal behavior [1]. With the passage of time, the idea of understanding compounds through

TIs has gained significant importance in the field of medicine because it requires no chemical-related apparatus to study [2]. TIs are also used for studying quantitative structure–activity relationships (QSAR) and quantitative structure property relationships (QSPR) for predicting many properties of chemical compounds. QSPR and QSAR approaches are used to study the properties (chemical, physical and biological) of a chemical substance from its molecular structure. QSAR and QSPR are the significant descriptors of chemical compounds in mathematical formulation of various properties [3]. TIs, in recent years, have been used to study the molecular complexity, chemical documentation, chirality, isomer discrimination, QSAR/QSPR, similarity/dissimilarity, lead optimization, drug design and database selection, deriving multilinear regression models and rational combinatorial library design [1,4]. Absence of degeneracy and high discriminating power are two properties of an ideal topological index.

Dendrimers are macromolecular structures with a central core, an interior dendritic structure (also called branches), and an exterior functional surface group. Initially, these molecules were discovered (and studied) by E. Buhleier [5], D. Tomalia [6] and G.R. Newkome [7]. For a given size and structure, one may construct a dendrimer inductively by using a chemical synthesis approach with low polydispersity index. Mainly two approaches, divergent and convergent, are used to synthesize the dendrimers. There are many known dendrimers with biological properties such as chemical stability, solubility, polyvalency, electrostatic interactions, low cytotoxicity and self-assembling. Dendrimers have applications in blood substitution as they have fluorocarbon and hydrophillic moieties. These varying properties are useful in the field of medical science, such as diagnostic imaging and anticancer therapies. Other than that, the linear growth in the size of dendrimers makes them ideal delivery vehicle candidates for the study of effects of composition and size of polymers in biological properties such as blood plasma retention time, cytotoxicity, lipid bilayer interactions and filtration; see [8] and references therein.

The study of Polyamidoamine (PAMAM) dendrimers has remained the most important topic of research in this field due to properties. PAMAM Dendrimers have hydrophilic interiors and exteriors which play roles in its unimolecular micelle properties. Moreover, PAMAM-based carriers increase the possibility of bioavailability of problematic drugs. Hence, PAMAM nano carriers enhance the potential of the bioavailability of drugs which are not so soluble for efflux transporters; see [8] and references therein. PAMAM dendrimers are also used in gene delivery for encapsulation of biodegradable functional polymer films. Moreover, these dendrimers also have potential applications in biosensors; for example, Ferrocenyl dendrimer (Fc-D). PAMAMs also have a usage in the delivery of agrochemicals to make plants healthier and less susceptible to diseases, but the complex synthesis of PAMAM limits the clinical translation of PAMAM-based materials. Interestingly, PolyEThyleneAmidoAmine (PETAA) dendrimers with more complete and uniform structure than PAMAM possess several properties of PAMAM such as the number of chemical bonds between the dendrimer core and the surface, number of surface primary amino groups and the number of tertiary amino groups. Other than that, the unique synthesis process of the PETAA enhances its potential for large-scale production, which results more application in biomedical sciences [9]. Consequently, the study of PETAA becomes a very important topic of research.

The purpose of this paper is to compute several distance-based TIs. Moreover, we also formulate explicit polynomials corresponding to some of these indices.

2. Materials and Methods

In this article, we understand a molecular graph to be a simple graph, representing the carbon atom skeleton of an organic molecule (usually, of a hydrocarbon). Let $G(V, E)$ be a graph, where V and E are the sets of vertices and set of edges respectively. Vertices $u, v \in V$ are said to be adjacent if $\{u, v\} \in E$. The degree of a $v \in V$ is denoted by d_v and is the number of edges incident to the vertex v. A (u_1, u_n)-path with n vertices is defined as a graph with vertex set $\{u_i : 1 \leq i \leq n\}$ and edge set $\{u_i u_{i+1} : 1 \leq i \leq n - 1\}$. The distance between two vertices u and w is the length of the shortest path between u and w and is denoted by $d(u, w)$. For a given vertex $u \in V$, the eccentricity $\varepsilon(u)$ is defined as $\varepsilon(u) = \max\{d(w, u) | w \in V\}$.

Harold Wiener was the first who used the topological index in 1947. The path number of a graph, which is the sum of all distances between the carbon atoms of a molecule was introduced by H. Wiener, see [10]. Mathematically, the Wiener index ($W(G)$) is the count of all shortest distances in a graph G [11]. This molecular modeling is used to investigate the relationships between properties, structure and activity of chemical compounds. Another distance-based topological index is the eccentric-connectivity index $\xi(G)$ of the graph G which is defined as [12]

$$\xi(G) = \sum_{v \in V(G)} \varepsilon(v) d_v. \tag{1}$$

The comparisons of $\xi(G)$ and $W(G)$ in terms of estimation of biological activity have been done for several drugs, such as those used in Alzheimer's disease, see [13], hypertension, see [14], inflammation, see [15], HIV, see [16] and as diuretics, see [17]. In most of these cases, the prediction power of the $\xi(G)$ was much superior to those correspondingly obtained from the $W(G)$ which provide motivation to find the $\xi(G)$. Some properties and applications of this index are discussed in [18–21]. The eccentric-connectivity polynomial of G is defined as [22]

$$ECP(G, y) = \sum_{v \in V(G)} d_v y^{\varepsilon(v)}. \tag{2}$$

If the degrees of vertices are not used, then the total eccentricity index and total eccentric-connectivity polynomial are as follows [22]:

$$\varsigma(G) = \sum_{v \in V(G)} \varepsilon(v), \tag{3}$$

$$TECP(G, y) = \sum_{v \in V(G)} y^{\varepsilon(v)}. \tag{4}$$

For more information on different aspects of eccentric-connectivity and total eccentric-connectivity polynomials, one can see [23–26]. Zagreb indices of a graph G in terms of eccentricity were given in [27] as follows:

$$Z_1(G) = \sum_{v \in V(G)} (\varepsilon(v))^2, \tag{5}$$

$$Z_2(G) = \sum_{uv \in E(G)} \varepsilon(u)\varepsilon(v). \tag{6}$$

Some mathematical and computational properties of $Z_1(G)$ and $Z_2(G)$ have been established in [28–31]. The augmented eccentric-connectivity index of a graph G which is the generalization of $\xi(G)$, is defined as [32]

$$A_\varepsilon(G) = \sum_{v \in V(G)} \frac{M(v)}{\varepsilon(v)}, \tag{7}$$

here, $M(v) = \prod_{u \in N(v)} d_u$, $N(v)$ denoted the neighbors of v. It has been scrutinized in [33] that the $A_\varepsilon(G)$, is much better then the $W(G)$ to high discriminating power, higher predictive accuracy and practically no degeneracy. This index has been studied in [34,35]. The modified eccentric-connectivity index and polynomial are defined as

$$\Lambda(G) = \sum_{v \in V(G)} S_v \varepsilon(v), \tag{8}$$

$$MECP(G, y) = \sum_{v \in V(G)} S_v y^{\varepsilon(v)}, \tag{9}$$

where $S_v = \sum_{u \in N(v)}$, $N(v)$ denoted the neighbors of v. In [36,37] several chemical and mathematical aspects of $\Lambda(G)$ and $MECP(G, y)$ have been comprehensively studied. The edge version of the $\xi(G)$, denoted by $\varepsilon_e(G)$, was introduced by Xinli Xu and Yun Guo [38] as

$$\varepsilon_e(G) = \sum_{f \in E(G)} d(f)\varepsilon(f), \tag{10}$$

where $d(f)$ represents the degree of an edge and $\varepsilon(f)$ is defined as the largest distance of f with any other edge g in G. This index has been computed for different dendrimer structures in [39–42].

TIs are descriptors based upon the structure of a compound which brief facts about shape, branching, molecular size, presence of multiple bonds and heteroatoms in the numerical form [1]. Therefore, TIs may provide more insights into the interpretation of the molecular properties and/or is able to take part in a model like QSPR and QSAR for the forecast of different properties of the molecules. For instance, in [1,43–47] various properties of several chemical compounds were studied by using TIs. Based upon the computation of TIs and some known values of these properties, linear equations were developed in which TIs were the independent variable and a certain physical property was the dependent variable. Thereby, the computation of TIs is one of the flourishing lines of research. The significance of the above-mentioned indices in formulating mathematical models for various biological activities can be seen in [14,17,48–53]. In the next section, we compute these eccentricity-based indices for a PolyEThyleneAmidoAmine (PETAA) dendrimer.

3. Eccentricity-Based Indices and Polynomials for the PETAA Dendrimer

Let us denote the molecular graph of a PETAA dendrimer by $D(n)$, where n is representing the generation stage. Figure 1 depicts the molecular graph of a PETAA dendrimer for the growth stage $n = 5$. The cardinality of vertex set of $D(n)$ is $44 \times 2^n - 18$, and edge set is $44 \times 2^n - 19$. To compute the indices and polynomials described in previous section for $D(n)$, we determine the required values for the sets of representatives. Now, make two sets of representatives, say $A = \{\alpha_1, \alpha_2\}$ and $B = \{a_i, b_i, c_i, d_i, e_i, f_i, g_i, h_i\}$, where $0 \leq i \leq n$, as shown in Figure 2. The representatives, degree, S_v, $M(v)$, $\varepsilon(v)$ and frequency of each v in A and b are shown in Table 1. Let $\beta = 7(n + i + 1)$ throughout this section.

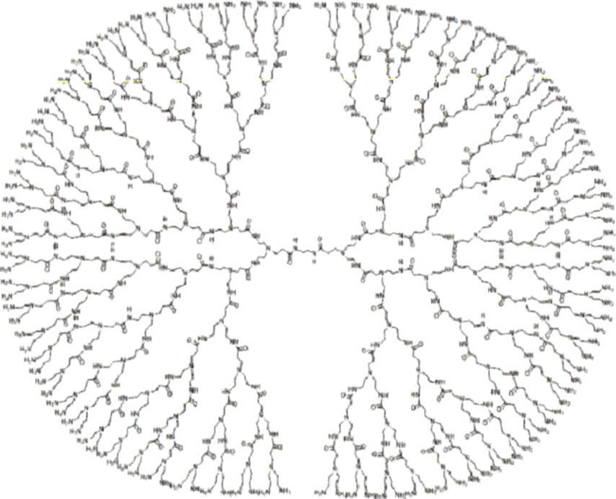

Figure 1. Chemical structure of PETAA dendrimer $D(5)$.

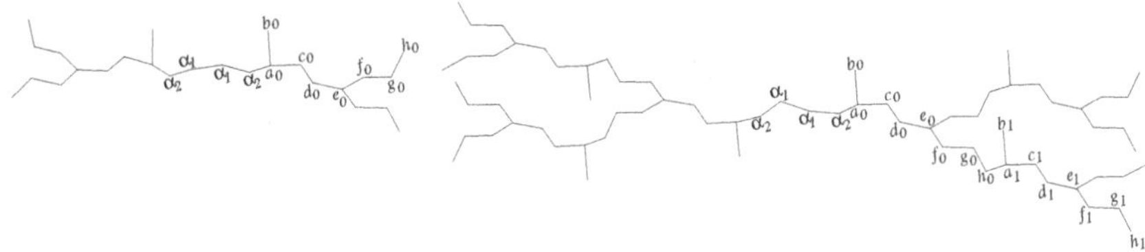

Figure 2. From left to right, $D(n)$ with $n = 0$ and $n = 1$.

Table 1. Sets A and B with degree of element v, S_v, $M(v)$, eccentricity, and frequency.

Representative	Degree	S_v	$M(v)$	Eccentricity	Frequency
α_1	2	4	4	$7n + 9$	2
α_2	2	5	6	$7n + 10$	2
a_i	3	5	4	$\beta + 4$	2^{i+1}
b_i	1	3	3	$\beta + 5$	2^{i+1}
c_i	2	5	6	$\beta + 5$	2^{i+1}
d_i	2	5	6	$\beta + 6$	2^{i+1}
e_i	3	6	8	$\beta + 7$	2^{i+1}
f_i	2	5	6	$\beta + 8$	2^{i+2}
g_i	2	4	4	$\beta + 9$	2^{i+2}
h_i when $i \neq n$	2	5	6	$\beta + 10$	2^{i+2}
h_n	1	2	2	$14n + 17$	2^{n+2}

Now with the help of Table 1, we are ready to compute different eccentricity-based indices and their corresponding polynomials for $D(n)$. We start with $\xi(G)$ for $D(n)$.

Theorem 1. *Let $D(n)$ be the molecular graph of PETAA Dendrimer. Then*

$$\xi(D(n)) = (1232n + 604)2^n - 266n + 62.$$

Proof. The values of $\varepsilon(v)$, d_v from Table 1 in Equation (1) yield

$$\xi(D(n)) = \xi(A) + \xi(B) = \sum_{v \in V(A)} \varepsilon(v)d_v + \sum_{v \in V(B)} \varepsilon(v)d_v$$

$$= (2 \times 2)(7n + 9) + (2 \times 2)(7n + 10) + \sum_{i=0}^{n} \bigg((3 \times 2^{i+1})(\beta + 4) + (1 \times 2^{i+1})(\beta + 5)$$

$$+ (2 \times 2^{i+1})(\beta + 5) + (2 \times 2^{i+1})(\beta + 6) + (3 \times 2^{i+1})(\beta + 7) + (2 \times 2^{i+2})(\beta + 8) \tag{11}$$

$$+ (2 \times 2^{i+2})(\beta + 9) \bigg) + 2^{n+2}(14n + 17) + \sum_{i=0}^{n-1} \bigg((2 \times 2^{i+2})(\beta + 10) \bigg).$$

Here, $\sum_{i=0}^{n} \bigg((3 \times 2^{i+1})(\beta + 4) + (1 \times 2^{i+1})(\beta + 5) + (2 \times 2^{i+1})(\beta + 5) + (2 \times 2^{i+1})(\beta + 6) + (3 \times$

$2^{i+1})(\beta + 7) + (2 \times 2^{i+2})(\beta + 8) + (2 \times 2^{i+2})(\beta + 9) \bigg) = 1064n \times 2^n + 512 \times 2^n - 266n + 10$,

and
$\sum_{i=0}^{n-1} \bigg((2 \times 2^{i+2})(\beta + 10) \bigg) = 112n \times 2^n + 24 \times 2^n - 24 - 56n.$

Using these values in Equation (11), we have

$$\xi(D(n)) = (1232n + 604)2^n - 266n + 62.$$

$\square$

If the degrees of vertices are not used, then Table 1 and Equation (3) produce the following result:

Corollary 1. *Let $D(n)$ be the molecular graph of PETAA Dendrimer. Then*

$$\varsigma(D(n)) = (616n + 324)2^n - 126n + 30.$$

In the next theorem, the $ECP(G, y)$ for $D(n)$ is derived.

Theorem 2. *Let $D(n)$ be the molecular graph of PETAA Dendrimer. Then*

$$ECP(D(n), y) = \frac{2(4y^5 + 4y^4 + 3y^3 + 3y^2 + 3)y^{7n+11}(2^{n+1}y^{7n+7} - 1)}{2y^7 - 1}$$
$$+ \frac{8y^{7n+17}(2^n y^{7n} - 1)}{2y^7 - 1} + 4y^{7n+9}(1 + y + 2^n y^{7n+8}).$$

Proof. The values of $\varepsilon(v)$, d_v from Table 1 in Equation (2) yield

$$ECP(D(n), y) = ECP(A, y) + ECP(B, y) = \sum_{v \in V(A)} d_v y^{\varepsilon(v)} + \sum_{v \in V(B)} d_v y^{\varepsilon(v)}$$

$$= (2 \times 2)y^{7n+9} + (2 \times 2)y^{7n+10} + \sum_{i=0}^{n} \Big((3 \times 2^{i+1})y^{\beta+4} + (1 \times 2^{i+1})y^{\beta+5}$$

$$+ (2 \times 2^{i+1})y^{\beta+5} + (2 \times 2^{i+1})y^{\beta+6} + (3 \times 2^{i+1})y^{\beta+7} + (2 \times 2^{i+2})y^{\beta+8} \qquad (12)$$

$$+ (2 \times 2^{i+2})y^{\beta+9} \Big) + 2^{n+2}y^{14n+17} + \sum_{i=0}^{n-1} \Big((2 \times 2^{i+2})y^{\beta+10} \Big).$$

Moreover, $\sum_{i=0}^{n} \Big((3 \times 2^{i+1})y^{\beta+4} + (1 \times 2^{i+1})y^{\beta+5} + (2 \times 2^{i+1})y^{\beta+5} +$

$(2 \times 2^{i+1})y^{\beta+6} + (3 \times 2^{i+1})y^{\beta+7} + (2 \times 2^{i+2})y^{\beta+8} + (2 \times 2^{i+2})y^{\beta+9} \Big) =$

$\dfrac{2(4y^5 + 4y^4 + 3y^3 + 2y^2 + 3y + 3)y^{7n+11}(2^{n+1}y^{7n+7} - 1)}{2y^7 - 1}$,

and

$\sum_{i=0}^{n-1} \Big((2 \times 2^{i+2})y^{\beta+10} \Big) = \dfrac{8y^{7n+17}(2^n y^{7n} - 1)}{2y^7 - 1}.$

Using these values in Equation (12), we get the desired result. $\square$

Similarly, the values from Table 1 in (4) produce the $TECP(G, y)$ for $D(n)$.

Corollary 2. *Let $D(n)$ be the molecular graph of PETAA Dendrimer. Then*

$$TECP(D(n), y) = \frac{2(2y^5 + 2y^4 + y^3 + y^2 + 2y + 1)y^{7n+11}(2^{n+1}y^{7n+7} - 1)}{2y^7 - 1}$$
$$+ \frac{4y^{7n+17}(2^n y^{7n} - 1)}{2y^7 - 1} + 2y^{7n+9}(1 + y + 2^{n+1}y^{7n+8}).$$

In the next theorem, we determine $Z_1(G)$ for $D(n)$.

Theorem 3. *Let $D(n)$ be the molecular graph of PETAA Dendrimer. Then*

$$Z_1(D(n)) = (8624n^2 + 9072n + 6876)2^n - 882n^2 + 420n - 1886.$$

Proof. By using the values of eccentricity from Table 1 in Equation (5), we get

$$Z_1(D(n)) = Z_1(A) + Z_1(B) = \sum_{v \in V(A)} [\varepsilon(v)]^2 + \sum_{v \in V(B)} [\varepsilon(v)]^2$$

$$= 2(7n+9)^2 + 2(7n+10)^2 + 2^{n+2}(14n+17)^2 + \sum_{i=0}^{n} \left(2^{i+1}(\beta+4)^2 + 2^{i+1}(\beta+5)^2 \right.$$

$$\left. + 2^{i+1}(\beta+5)^2 + 2^{i+1}(\beta+6)^2 + 2^{i+1}(\beta+7)^2 + 2^{i+2}(\beta+8)^2 + 2^{i+2}(\beta+9)^2 \right)$$

$$+ \sum_{i=0}^{n-1} \left(2^{i+2}(\beta+10)^2 \right). \tag{13}$$

Moreover, $\sum_{i=0}^{n} \left(2^{i+1}(\beta+4)^2 + 2^{i+1}(\beta+5)^2 + 2^{i+1}(\beta+5)^2 + 2^{i+1}(\beta+6)^2 + 2^{i+1}(\beta+7)^2 + \right.$

$\left. 2^{i+2}(\beta+8)^2 + 2^{i+2}(\beta+9)^2 \right) = 7056n^2 \times 2^n + 6832n \times 2^n - 882n^2 + 5292 \times 2^n + 56n - 1820,$

and

$\sum_{i=0}^{n-1} \left(2^{i+2}(\beta+10)^2 \right) = 784n^2 \times 2^n + 336n \times 2^n + 428 \times 2^n - 196n^2 - 168n - 428.$

By using these values in Equation (13), we get

$$Z_1(D(n)) = (8624n^2 + 9072n + 6876)2^n - 882n^2 + 420n - 1886,$$

which proves our theorem. $\square$

Now, we compute $A_\varepsilon(G)$ for $D(n)$ in the next theorem.

Theorem 4. *Let $D(n)$ be the molecular graph of PETAA Dendrimer. Then*

$$A_\varepsilon(D(n)) = \frac{8}{7n+9} + \frac{12}{7n+10} + \frac{2^{n+3}}{14n+17} + \left(\frac{8}{7n+11} + \cdots + \frac{2^{n+3}}{14n+11} \right)$$
$$+ \left(\frac{6}{7n+12} + \cdots + \frac{6 \times 2^n}{14n+12} \right) + \left(\frac{12}{7n+12} + \cdots + \frac{12 \times 2^n}{14n+12} \right)$$
$$+ \left(\frac{12}{7n+13} + \cdots + \frac{12 \times 2^n}{14n+13} \right) + \left(\frac{16}{7n+14} + \cdots + \frac{16 \times 2^n}{14n+14} \right)$$
$$+ \left(\frac{24}{7n+15} + \cdots + \frac{24 \times 2^n}{14n+15} \right) + \left(\frac{16}{7n+16} + \cdots + \frac{16 \times 2^n}{14n+16} \right)$$
$$+ \left(\frac{24}{7n+17} + \cdots + \frac{12 \times 2^n}{14n+10} \right).$$

Proof. By using the values of $M(v)$ and $\varepsilon(v)$ from Table 1 in (7), we get:

$$A_\varepsilon(D(n)) = A_\varepsilon(A) + A_\varepsilon(B) = \sum_{v \in V(A)} \frac{M(v)}{\varepsilon(v)} + \sum_{v \in V(B)} \frac{M(v)}{\varepsilon(v)}$$

$$= \frac{2 \times 4}{7n+9} + \frac{2 \times 6}{7n+10} + \frac{2 \times 2^{n+2}}{14n+17}$$

$$+ \sum_{i=0}^{n} \left(\frac{4 \times 2^{i+1}}{\beta+4} + \frac{3 \times 2^{i+1}}{\beta+5} + \frac{6 \times 2^{i+1}}{\beta+5} + \frac{6 \times 2^{i+1}}{\beta+6} \right.$$

$$\left. + \frac{8 \times 2^{i+1}}{\beta+7} + \frac{6 \times 2^{i+2}}{\beta+8} + \frac{4 \times 2^{i+2}}{\beta+9} \right) + \sum_{i=0}^{n-1} \left(\frac{6 \times 2^{i+2}}{\beta+10} \right). \tag{14}$$

Furthermore,
$$\sum_{i=0}^{n}\left(\frac{4\times 2^{i+1}}{\beta+4}+\frac{3\times 2^{i+1}}{\beta+5}+\frac{6\times 2^{i+1}}{\beta+5}+\frac{6\times 2^{i+1}}{\beta+6}+\frac{8\times 2^{i+1}}{\beta+7}+\right.$$
$$\left.\frac{6\times 2^{i+2}}{\beta+8}+\frac{4\times 2^{i+2}}{\beta+9}\right)=\left(\frac{8}{7n+11}+\cdots+\frac{2^{n+3}}{14n+11}\right)+\left(\frac{6}{7n+12}+\cdots+\frac{6\times 2^{n}}{14n+12}\right)+$$
$$\left(\frac{12}{7n+12}+\cdots+\frac{12\times 2^{n}}{14n+12}\right)+\left(\frac{12}{7n+13}+\cdots+\frac{12\times 2^{n}}{14n+13}\right)+\left(\frac{16}{7n+14}+\cdots+\frac{16\times 2^{n}}{14n+14}\right)+$$
$$\left(\frac{24}{7n+15}+\cdots+\frac{24\times 2^{n}}{14n+15}\right)+\left(\frac{16}{7n+16}+\cdots+\frac{16\times 2^{n}}{14n+16}\right),$$
and
$$\sum_{i=0}^{n-1}\left(\frac{6\times 2^{i+2}}{\beta+10}\right)=\left(\frac{24}{7n+17}+\cdots+\frac{12\times 2^{n}}{14n+10}\right).$$

The above two equations along with (14) yield the required result. $\square$

We now determine $\Lambda(D(n))$ for $D(n)$.

Theorem 5. *Let $D(n)$ be the molecular graph of PETAA Dendrimer. Then*

$$\Lambda(D(n))=(2744n+1332)2^{n}-602n+132.$$

Proof. By using the values S_v and $\varepsilon(v)$ from Table 1 in (8), we obtain

$$\Lambda(D(n))=\Lambda(A)+\Lambda(B)=\sum_{v\in V(A)}S_v\varepsilon(v)+\sum_{v\in V(B)}S_v\varepsilon(v)$$

$$=(2\times 4)(7n+9)+(2\times 5)(7n+10)+\sum_{i=0}^{n}\left((5\times 2^{i+1})(\beta+4)+(3\times 2^{i+1})(\beta+5)\right.$$

$$+(5\times 2^{i+1})(\beta+5)+(5\times 2^{i+1})(\beta+6)+(6\times 2^{i+1})(\beta+7)+(5\times 2^{i+2})(\beta+8)$$

$$\left.+(4\times 2^{i+2})(\beta+9)\right)+2\times 2^{n+2}(14n+17)+\sum_{i=0}^{n-1}\left((5\times 2^{i+2})(\beta+10)\right). \tag{15}$$

Here, $\sum_{i=0}^{n}\left((5\times 2^{i+1})(\beta+4)+(3\times 2^{i+1})(\beta+5)+(5\times 2^{i+1})(\beta+5)+(5\times 2^{i+1})(\beta+6)+(6\times$

$2^{i+1})(\beta+7)+(5\times 2^{i+2})(\beta+8)+(4\times 2^{i+2})(\beta+9)\right)=2352n\times 2^{n}+1136\times 2^{n}-588n+20,$

and

$\sum_{i=0}^{n-1}\left((5\times 2^{i+2})(\beta+10)\right)=280n\times 2^{n}+60\times 2^{n}-60-140n.$

Lastly by using these values in Equation (15), we have

$$\Lambda(D(n))=(2744n+1332)2^{n}-602n+132.$$

$\square$

In the following theorem, we determine the $MECP(G,y)$ for $D(n)$.

Theorem 6. *Let $D(n)$ be the molecular graph of PETAA Dendrimer. Then*

$$MECP(D(n),y)=\frac{2(8y^{5}+10y^{4}+6y^{3}+5y^{2}+8y+5)y^{7n+11}(2^{n+1}y^{7n+7}-1)}{2y^{7}-1}$$

$$+\frac{20y^{7n+17}(2^{n}y^{7n}-1)}{2y^{7}-1}+2y^{7n+9}(4+5y+2^{n+2}y^{7n+8}).$$

Proof. By using the values from Table 1 in Equation (9), we get:

$$MECP(D(n), y) = MECP(A, y) + MECP(B, y) = \sum_{v \in V(A)} S_v y^{\varepsilon(v)} + \sum_{v \in V(B)} S_v y^{\varepsilon(v)}$$

$$= (2 \times 4)y^{7n+9} + (2 \times 5)y^{7n+10} + \sum_{i=0}^{n} \left((5 \times 2^{i+1})y^{\beta+4} + (3 \times 2^{i+1})y^{\beta+5} \right.$$

$$+ (5 \times 2^{i+1})y^{\beta+5} + (5 \times 2^{i+1})y^{\beta+6} + (6 \times 2^{i+1})y^{\beta+7} + (5 \times 2^{i+2})y^{\beta+8}$$

$$\left. + (4 \times 2^{i+2})y^{\beta+9} \right) + 2 \times 2^{n+2}y^{14n+17} + \sum_{i=0}^{n-1} \left((5 \times 2^{i+2})y^{\beta+10} \right). \tag{16}$$

Moreover, $\sum_{i=0}^{n} \left((5 \times 2^{i+1})y^{\beta+4} + (3 \times 2^{i+1})y^{\beta+5} + (5 \times 2^{i+1})y^{\beta+5} + (5 \times 2^{i+1})y^{\beta+6} + (6 \times 2^{i+1})y^{\beta+7} + (5 \times 2^{i+2})y^{\beta+8} + (4 \times 2^{i+2})y^{\beta+9} \right) =$

$$\frac{2(8y^5 + 10y^4 + 6y^3 + 5y^2 + 8y + 5)y^{7n+11}(2^{n+1}y^{7n+7} - 1)}{2y^7 - 1},$$

and

$$\sum_{i=0}^{n-1} \left((5 \times 2^{i+2})y^{\beta+10} \right) = \frac{20y^{7n+17}(2^n y^{7n} - 1)}{2y^7 - 1}.$$

By using these values in Equation (16), we get the required result. $\square$

The edge partition of sets A and B with respect to the representatives of pairs of end vertices, eccentricity and degree of each edge of the corresponding representative, and their frequencies of occurrence are shown in Table 2. The eccentricities of vertices are taken from Table 1.

Table 2. The edge partition of set B with respect to the representatives of pairs of end vertices, eccentricity, and degree of each edge, and their frequencies.

Representative	Eccentricity	Frequency	Eccentricity of an Edge	Degree of an Edge
$[\alpha_1, \alpha_1]$	$[7n + 9, 7n + 9]$	1	$7n + 7$	2
$[\alpha_1, \alpha_2]$	$[7n + 9, 7n + 10]$	2	$7n + 8$	2
$[\alpha_2, a_0]$	$[7n + 10, 7n + 11]$	2	$7n + 9$	3
$[a_i, b_i]$	$[\beta + 4, \beta + 5]$	2^{i+1}	$\beta + 3$	2
$[a_i, c_i]$	$[\beta + 4, \beta + 5]$	2^{i+1}	$\beta + 3$	3
$[c_i, d_i]$	$[\beta + 5, \beta + 6]$	2^{i+1}	$\beta + 4$	2
$[d_i, e_i]$	$[\beta + 6, \beta + 7]$	2^{i+1}	$\beta + 5$	3
$[e_i, f_i]$	$[\beta + 7, \beta + 8]$	2^{i+2}	$\beta + 6$	3
$[f_i, g_i]$	$[\beta + 8, \beta + 9]$	2^{i+2}	$\beta + 7$	2
$[g_i, h_i]$ when $i \neq n$	$[\beta + 9, \beta + 10]$	2^{i+2}	$\beta + 8$	2
$[g_n, h_n]$	$[14n + 16, 14n + 17]$	2^{n+2}	$14n + 15$	1
$[h_i, a_{i+1}]$ when $i \neq n$	$[\beta + 10, \beta + 11]$	2^{i+2}	$\beta + 9$	3

In the next theorem, we determine $Z_2(G)$ for $D(n)$.

Theorem 7. *Let $D(n)$ be the molecular graph of PETAA Dendrimer. Then,*

$$Z_2(D(n)) = 2^n(8624n^2 + 8456n + 6552) - 931n^2 + 434n - 1979.$$

Proof. By using the values from Table 2 in Equation (6), we compute the second Zagreb eccentricity index of $D(n)$ as follows:

$$
\begin{aligned}
Z_2(D(n)) = Z_2(A) + Z_2(B) = &\sum_{uv \in E(A)} [\varepsilon(u)\varepsilon(v)] + \sum_{uv \in E(B)} [\varepsilon(u)\varepsilon(v)] \\
= &(7n+9)(7n+9) + 2(7n+9)(7n+10) + 2(7n+10)(7n+11) \\
&+ 2^{n+2}(14n+16)(14n+17) + \sum_{i=0}^{n} \Big(2^{i+1}(\beta+4)(\beta+5) + 2^{i+1}(\beta+4)(\beta+5) \\
&+ 2^{i+1}(\beta+5)(\beta+6) + 2^{i+1}(\beta+6)(\beta+7) + 2^{i+2}(\beta+7)(\beta+8) \\
&+ 2^{i+2}(\beta+8)(\beta+9) \Big) + \sum_{i=0}^{n-1} \Big(2^{i+2}(\beta+9)(\beta+10) + 2^{i+2}(\beta+10)(\beta+11) \Big).
\end{aligned}
\tag{17}
$$

Furthermore, $\sum_{i=0}^{n} \Big(2^{i+1}(\beta+4)(\beta+5) + 2^{i+1}(\beta+4)(\beta+5) + 2^{i+1}(\beta+5)(\beta+6) + 2^{i+1}(\beta+6)(\beta+7) + 2^{i+2}(\beta+7)(\beta+8) + 2^{i+2}(\beta+8)(\beta+9) \Big) = 6272n^2 \times 2^n + 5936n \times 2^n - 784n^2 + 4608 \times 2^n + 84n - 1604,$

and

$\sum_{i=0}^{n-1} \Big(2^{i+2}(\beta+9)(\beta+10) + 2^{i+2}(\beta+10)(\beta+11) \Big) = 1568n^2 \times 2^n + 672n \times 2^n - 392 \times n^2 + 856 \times 2^n - 336n - 856,$

which along with (17) give

$$
Z_2(D(n)) = 2^n(8624n^2 + 8456n + 6552) - 931n^2 + 434n - 1979,
$$

which was required. $\square$

Finally, we determine the $\varepsilon_e(D(n))$ for $D(n)$.

Theorem 8. *Let $D(n)$ be the molecular graph of PETAA Dendrimer. Then*

$$
\varepsilon_e(D(n)) = (1456n + 500)2^n - 336n + 144.
$$

Proof. By using the values from Table 2 in (10), we get

$$
\begin{aligned}
\varepsilon_e(D(n)) = \varepsilon_e(A) + \varepsilon_e(B) = &\sum_{f \in E(A)} d_f \varepsilon(f) + \sum_{f \in E(B)} d_f \varepsilon(f) \\
= &(2 \times 1)(7n+7) + (2 \times 2)(7n+8) + (2 \times 3)(7n+9) + \sum_{i=0}^{n} \Big((2 \times 2^{i+1})(\beta+3) \\
&+ (3 \times 2^{i+1})(\beta+3) + (2 \times 2^{i+1})(\beta+4) + (3 \times 2^{i+1})(\beta+5) + (3 \times 2^{i+2})(\beta+6) \\
&+ (2 \times 2^{i+2})(\beta+7) \Big) + 2^{n+2}(14n+15) + \sum_{i=0}^{n-1} \Big((2 \times 2^{i+2})(\beta+8) + (3 \times 2^{i+2})(\beta+9) \Big).
\end{aligned}
\tag{18}
$$

Here, $\sum_{i=0}^{n} \Big((2 \times 2^{i+1})(\beta+3) + (3 \times 2^{i+1})(\beta+3) + (2 \times 2^{i+1})(\beta+4) + (3 \times 2^{i+1})(\beta+5) + (3 \times 2^{i+2})(\beta+6) + (2 \times 2^{i+2})(\beta+7) \Big) = 1120n \times 2^n + 408 \times 2^n - 280n + 76,$

and

$\sum_{i=0}^{n-1} \Big((2 \times 2^{i+2})(\beta+8) + (3 \times 2^{i+2})(\beta+9) \Big) = 280n \times 2^n + 32 \times 2^n - 140n - 32.$

By using these values in Equation (18), we get

$$\varepsilon_e(D(n)) = (1456n + 500)2^n - 336n + 144.$$

This gives the required result. □

4. Discussion

In recent times, several techniques such as SARs model, QSAR model, QSPR model, least square regression analysis and other models are being used to predict chemical and biological behavior of compounds. By using graphs corresponding to compounds, non-experimental parameters are used to predict chemical behavior of these compounds. Among these parameters, several (distance based) indices have been formulated, such as the eccentric connectivity index (denoted by $\xi(G)$), total-eccentricity index ($\varsigma(G)$), augmented eccentric-connectivity index ($A_\varepsilon(G)$) and modified eccentric-connectivity index ($\Lambda(G)$). In order to demonstrate the importance of $\xi(G)$ in predicting biological activity, Gupta et. al [32] used nonpeptide N-benzylimidazole derivatives with respect to antihypertensive activity. After comparison between the results obtained by using $\xi(G)$ and the corresponding values obtained by using other TIs, they found the accuracy of prediction to be about 80%. In [12], Sharma et al. investigated the discrimination power of $\xi(G)$ with regard to biological/physical properties of molecules. By using $\xi(G)$ in several datasets, they obtained the correlation coefficients, ranging from 95% to 99%, with regard to physical properties of diverse nature. It was observed that the $\xi(G)$ shows excellent correlations with regard to analgesic activity of piperidinyl methyl ester and methylene methyl ester analogs. In the current work, we have extended the pool of information related to important PETTA dendrimers. This information can be used by the other researchers who are doing lab work. They may develop connections with the properties by using these values in different ways, for example, to decide the size of suitable dendrimers for a certain property. Furthermore, some physical and biological properties can be a topic of interest in the future.

5. Conclusions

The theoretical formulation studied in this article has received remarkable interest of researchers due to its extensive application in the fields of pharmacy, medical science, chemical engineering, and applied sciences. In this paper, we deal with the molecular structure of PETAA dendrimers. We compute some distance-based indices for PETAA dendrimers, along with the exact values of eccentric connectivity, total eccentricity, augmented eccentric connectivity, first Zagreb eccentricity, modified eccentric connectivity, second Zagreb eccentricity, and edge version of eccentric-connectivity indices for PETAA. The associated polynomials have also been computed for PETAA dendrimers. Since PETAA dendrimers have a wide variety of biomedical applications, therefore, these theoretical results may play a vital role in upcoming research in medical sciences.

Author Contributions: J.Z., Z.I., A.F., A.Z., A.A., M.I.Q., and R.I. contributed equally in the writing of this article.

Funding: This research is supported by Educational Reform Project of Chengdu University under grant number cdjgb2017070. Muhammad Imran Qureshi and Asfand Fahad gratefully acknowledge ORIC, COMSATS University Islamabad, Pakistan, for supporting their research under the grant of project number 16-52/CRGP/CIIT/VEH/17/1141.

Acknowledgments: The authors are also thankful to Dr. Zia Ul Haq Khan and Miss Maida Younis for their valuable input during the conduct of this research.

References

1.   Gozalbes, R.; Doucet, J.P.; Derouin, F. Application of topological descriptors in QSAR and drug design: history and new trends. *Curr. Drug Targets Infect. Disord.* **2002**, *2*, 93–102. [CrossRef] [PubMed]

2. Amic, D.; Beaa, D.; Lucic, B.; Nikolic, S.; Trinajstic, N. The vertexconnectivity index revisited. *J. Chem. Inf. Comput.* **1998**, *38*, 819–822. [CrossRef]

3. Ivanciuc, O. Topological Indices. In *Handbook of Chemoinformatics*; Gasteiger, J., Ed.; Wiley–VCH: Weinheim, Germany, 2003; Volume 3, pp. 981–1003, doi:10.1002/9783527618279.ch36.

4. Estrada, E.; Patlewicz, G.; Uriarte, E. From molecular graphs to drugs. A review on the use of topological indices in drug design and discovery. *Indian J. Chem. A* **2003**, *42*, 1315–1329.

5. Buhleier, E.; Wehner, W.; Vogtle, F. Cascade and Nonskid Chain like Syntheses of Molecular Cavity Topologies. *Synthesis* **1978**, *2*, 155–158. [CrossRef]

6. Tomalia, D.A.; Baker, H.; Dewald, J.R.; Hall, M.; Kallos, G.; Martin, S.; Roeck, J.; Ryder, J.; Smith, P. Dendritic macromolecules: Synthesis of starburst dendrimers. *Macromolecules* **1986**, *19*, 2466–2468. [CrossRef]

7. Newkome, G.R.; Yao, Z.; Baker, G.R.; Gupta, V.K. Micelles. Part 1. Cascade molecules: A new approach to micelles. A [27]-arborol. *J. Org. Chem.* **1985**, *50*, 2003–2004. [CrossRef]

8. Abbasi, F.; Aval, S.F.; Akbarzadeh, A.; Milani, M.; Nasrabadi, H.T.; Joo, S.W.; Hanifehour, Y.; Koshki, K.N.; Pashei-Asl, R. Dendrimers: Synthesis, applications, and properties. *Nano Scale Res. Lett.* **2014**, *247*, 1–10. [CrossRef] [PubMed]

9. Huang, B.; Tang, S.; Desai, A.; Lee, K.; Leroueil, P.R.; Baker, R., Jr. Noevl Poly EThyleneAmidoAmine PETAA dendrimers produced through a unique and highly efficient synthesis. *Polymer* **2011**, *52*, 5975–5984. [CrossRef]

10. Trinajstic, N. *Chemical Graph Theory*, 2nd ed.; CRC Press: Boca Raton, FL, USA, 1992.

11. Khalifeh, M.H.; Azari, H.Y.; Ashrafi, A.R. Vertex and edge PI indices of Cartesian product graphs. *Discret. Appl. Math.* **2008**, *156*, 1780–1789. [CrossRef]

12. Sharma, V.; Goswami, R.; Madan, A.K. Eccentric-connectivity index: A novel highly discriminating topological descriptor for structure property and structure activity studies. *J. Chem. Inf. Comput. Sci.* **1997**, *37*, 273–282. [CrossRef]

13. Kumar, V.; Madan, A.K. Application of graph theory: Prediction of glycogen synthase kinase-3 beta inhibitory activity of thiadiazolidinones as potential drugs for the treatment of Alzheimer's disease. *Eur. J. Pharm. Sci.* **2005**, *24*, 213–218. [CrossRef] [PubMed]

14. Sardana, S.; Madan, A.K. Application of graph theory: Relationship of molecular connectivity index, Wiener index and eccentric connectivity index with diuretic activity. *MATCH Commun. Math. Comput. Chem.* **2001**, *43*, 85–98.

15. Gupta, S.; Singh, M.; Madan, A.K. Application of graph theory: relationship of eccentric connectivity index and Wiener's index with anti-inflammatory activity. *J. Math. Anal. Appl.* **2002**, *266*, 259–268. [CrossRef]

16. Lather, V.; Madan, A.K. Topological models for the prediction of HIV-protease inhibitory activity of tetrahydropyrimidin-2-ones. *J. Mol. Graph. Model.* **2005**, *23*, 339–345. [CrossRef] [PubMed]

17. Sardana, S.; Madan, A.K. Topological models for prediction of antihypertensive activity of substituted benzylimidazoles. *J. Mol. Struct. (Theochem.)* **2003**, *638*, 41–49. [CrossRef]

18. Dureja, H.; Madan, A.K. Topochemical models for prediction of cyclin-dependent kinase 2 inhibitory activity of indole-2-ones. *J. Mol. Model.* **2005**, *11*, 525–531. [CrossRef]

19. Ilic, A.; Gutman, I. Eccentric-connectivity index of chemical trees. *MATCH Commun. Math. Comput. Chem.* **2011**, *65*, 731–744.

20. Kumar, V.; Madan, A.K. Application of graph theory: Prediction of cytosolic phospholipase A_2 inhibitory activity of propan-2-ones. *J. Math. Chem.* **2006**, *39*, 511–521. [CrossRef]

21. Zhou, B. On eccentric-connectivity index. *MATCH Commun. Math. Comput. Chem.* **2010**, *63*, 181–198.

22. Ashrafi, A.R.; Ghorbani, M.; Hossein-Zadeh, M.A. The eccentric-connectivity polynomial of some graph operations. *Serdica J. Comput.* **2011**, *5*, 101–116.

23. Hasni, R.; Arif, N.E.; Alikhani, S. Eccentric connectivity polynomials of some families of dendrimers. *J. Comput. Theor. Nanosci.* **2014**, *11*, 450–453. [CrossRef]

24. Ghorbani, M.; Iranmanesh, M. Computing eccentric connectivity polynomial of fullerenes. *Fuller. Nanotub. Carbon Nanostruct.* **2013**, *21*, 134–139. [CrossRef]

25. Ghorbani, M.; Hemmasi, M. Eccentric connectivity polynomial of C_{18n+10} fullerenes. *Bulg Chem. Commun.* **2013**, *45*, 5–8.

26. Fathalikhani, K.; Faramarzi, H.; Yousefi-Azari, H. Total eccentricity of some graph operations. *Electron. Notes Discret. Math.* **2014**, *45*, 125–131. [CrossRef]

27. Ghorbani, M.; Hosseinzadeh, M.A. A new version of Zagreb indices. *Filomat* **2012**, *26*, 93–100. [CrossRef]
28. Das, K.C.; Lee, D.W.; Graovac, A. Some properties of the Zagreb eccentricity indices. *ARS Math. Contemp.* **2013**, *6*, 117–125. [CrossRef]
29. Vukičević, D.; Graovac, A. Note on the comparison of the first and second normalized Zagreb eccentricity indices. *Acta Chim. Slov.* **2010**, *57*, 524–528.
30. Du, Z.; Zhou, B.; Trinajstić, N. Extremal properties of the Zagreb eccentricity indices. *Croat. Chem. Acta* **2012**, *85*, 359–362. [CrossRef]
31. Qi, X.; Du, Z. On Zagreb eccentricity indices of trees. *MATCH Commun. Math. Comput. Chem.* **2017**, *78*, 241–256.
32. Gupta, S.; Singh, M.; Madan, A.K. Connective eccentricity index: A novel topological descriptor for predicting biological activity. *J. Mol. Graph. Model.* **2000**, *18*, 18–25. [CrossRef]
33. Bajaj, S.; Sambi, S.S.; Gupta, S.; Madan, A.K. Model for prediction of anti-HIV activity of 2-pyridinone derivatives using novel topological descriptor. *QSAR Comb. Sci.* **2006**, *25*, 813–823. [CrossRef]
34. De, N. Relationship between augmented eccentric connectivity index and some other graph invariants. *Int. J. Adv. Math. Sci.* **2013**, *1*, 26–32. [CrossRef]
35. Došlić, T.; Saheli, M. Augmented eccentric-connectivity index. *Miskolc Math. Notes* **2011**, *12*, 149–157.
36. Alaeiyan, M.; Asadpour, J.; Mojarad, R. A numerical method for MEC polynomial and MEC index of one-pentagonal carbon nanocones, Fullerenes. *Nanotubes Carbon Nanostruct.* **2013**, *21*, 825–835. [CrossRef]
37. De, N.; Nayeem, S.M.A.; Pal, A. Modified eccentric-connectivity of Generalized Thorn Graphs. *Int. J. Comput. Math.* **2014**, *2014*, 436140. [CrossRef]
38. Xu, X.; Guo, Y. The edge version of eccentric connectivity index. *Int. Math. Forum* **2012**, *7*, 273–280.
39. Odaba, Z.N. The edge eccentric-connectivity index of dendrimers. *J. Comput. Theor. Nanosci.* **2013**, *10*, 783–784. [CrossRef]
40. Gao, W.; Iqbal, Z.; Ishaq, M.; Sarfraz, R.; Aamir, M.; Aslam, A. On Eccentricity-Based Topological Indices Study of a Class of Porphyrin-Cored Dendrimers. *Biomolecules* **2018**, *8*, 71–81. [CrossRef]
41. Iqbal, Z.; Ishaq, M.; Aslam, A.; Gao, W. On eccentricity-based topological descriptors of water-soluble dendrimers. *Z. Naturforsch.* **2018**, *74*, 25–33. [CrossRef]
42. Gao, W.; Iqbal, Z.; Ishaq, M.; Aslam, A.; Sarfraz, R. Topological aspects of dendrimers via distance-based descriptors. *IEEE Access* **2019**, *7*, 35619–35630. [CrossRef]
43. Devillers, J.; Balaban, A.T. *Topological Indices and Related Descriptors in QSAR and QSPR*; Gordon and Breach Science Publishers: Amsterdam, The Netherlands, 1999.
44. Randić, M. Quantitative structure-property relationship: Boiling points and planar benzenoids. *New J. Chem.* **1996**, *20*, 1001–1009.
45. Randić, M. Comparative regression analysis. Regressions based on a single descriptor. *Croatica Chem. Acta* **1993**, *66*, 289–312.
46. Randić, M.; Pompe, M. On characterization of the CC double bond in alkenes. *SAR QSAR Environ. Res.* **1999**, *10*, 451–471.
47. Karelson, M.; Lobanov, V.S. Quantum-Chemical descriptors in QSAR/QSPR studies. *Chem. Rev.* **1996**, *96*, 1027–1043. [CrossRef] [PubMed]
48. Dureja, H.; Gupta, S.; Madan, A.K. Predicting anti-HIV-1 activity of 6-arylbenzonitriles: Computational Approach using superaugmented eccentric connectivity topochemical indices. *J. Mol. Graph. Model.* **2008**, *26*, 1020–1029. [CrossRef] [PubMed]
49. Gupta, S.; Singh, M.; Madan, A.K. Eccentric distance sum: A novel graph invariant for predicting biological and physical properties. *J. Math. Anal. Appl.* **2002**, *275*, 386–401. [CrossRef]
50. Kumar, V.; Sardana, S.; Madan, A.K. Predicting anti-HIV activity of 2,3-diaryl-1,3 thiazolidin-4-ones: Computational approach using reformed eccentric connectivity index. *J. Mol. Model.* **2004**, *10*, 399–407. [CrossRef] [PubMed]
51. Iqbal, Z.; Ishaq, M.; Farooq, R. Computing different versions of atom-bond connectivity index of dendrimers. *J. Inform. Math. Sci.* **2017**, *9*, 217–229.

52. Kang, S.M.; Iqbal, Z.; Ishaq, M.; Sarfraz, R.; Aslam, A.; Nazeer, W. On Eccentricity-Based Topological Indices and Polynomials of Phosphorus-Containing Dendrimers. *Symmetry* **2018**, *10*, 237–246. [CrossRef]
53. Iqbal, Z.; Ishaq, M.; Aamir, M. On Eccentricity-Based Topological Descriptors of Dendrimers. *Iran. J. Sci. Technol. Trans. Sci.* **2018**. [CrossRef]

Article

Imbalance-Based Irregularity Molecular Descriptors of Nanostar Dendrimers

Zafar Hussain [1], Mobeen Munir [2], Shazia Rafique [3], Tayyab Hussnain [3], Haseeb Ahmad [2], Young Chel Kwun [4,*] and Shin Min Kang [5,6]

[1] Department of Mathematics and Statistics, The University of Lahore, Lahore 54500, Pakistan
[2] Department of Mathematics, Division of Science and Technology, University of Education, Lahore 54000, Pakistan
[3] Center for Excellence in Molecular Biology, Punjab University Lahore, Lahore 53700, Pakistan
[4] Department of Mathematics, Dong-A University, Busan 49315, Korea
[5] Department of Mathematics, Gyeongsang National University, Jinju 52828, Korea
[6] Center for General Education, China Medical University, Taichung 40402, Taiwan
* Correspondence: yckwun@dau.ac.kr

Received: 27 June 2019; Accepted: 29 July 2019; Published: 6 August 2019

Abstract: Dendrimers are branched organic macromolecules with successive layers of branch units surrounding a central core. The molecular topology and the irregularity of their structure plays a central role in determining structural properties like enthalpy and entropy. Irregularity indices which are based on the imbalance of edges are determined for the molecular graphs associated with some general classes of dendrimers. We also provide graphical analysis of these indices for the above said classes of dendrimers.

Keywords: nanostar dendrimer; irregularity measure; complexity of structure; $NS_1[p]$; $NS_2[p]$; $NS_3[p]$

1. Introduction

Algebra, topology, geometry and combinatorics are the main branches of mathematics which are employed to study the symmetries and irregularities of the structures of different substances. Dendrimers have consistently attracted the attention of both chemists as well as pure mathematicians because of the complexities of the underlying molecular graphs. Dendrimers are highly branched, star-shaped macromolecules with nanometer-scale dimensions. Dendrimers are constituted by main parts: A central core, an internal part called 'branch', and an exterior surface with functional surface groups. The varied combination of these components yields products of different shapes and sizes with shielded interior cores that are ideal candidates for applications in both biological and materials sciences. Some recent applications include drug delivery, gene transfection, catalysis, energy harvesting, photo activity, molecular weight and size determination, rheology modification, and nanoscale science and technology [1–3]

Graphs can be used to study theoretical and computational aspects of dendrimers. Recently this approach has proved remarkable in relating properties of substances with involved structural parameters [4–7]. Topological indices are used here as major ingredients [7–14]. Some nanotubes, modified electrodes, chemical sensors, micro- and macro-capsule, and colored glasses can be designed using nanostar dendrimers. The structure of polymer molecules in a plane depends on the adjacency of their units. For detailed insight, see [1–3,14–18] and references therein. Figure 1 shows the spatial arrangements of $NS_1[1]$, $NS_1[2]$ polypropylenimine octaamin dendrimers in plane. The recursive nature of these dendrimers is evident from this figure. Graph theoretic models of these dendrimers can potentially be used in fractals.

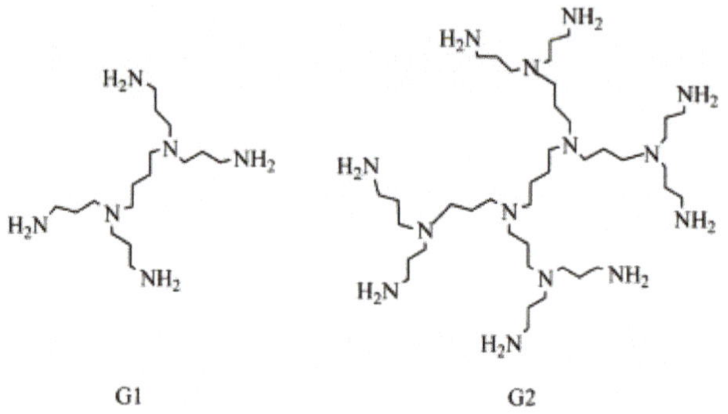

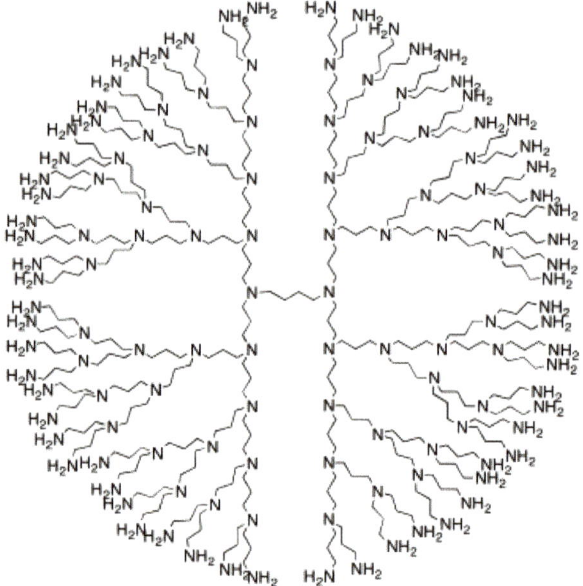

Figure 1. $NS_1[1]$ and $NS_1[2]$ polypropylenimine octaamin dendrimers.

In Figure 1, G_1 shows the structure of polypropylenimine octaamin dendrimers $NS_1[p]$ when $p = 1$, and G_2 represents the structure of $NS_1[p]$ when $p = 2$.

The next object will be polypropylenimine octaamin dendrimer $NS_2[p]$. Figure 2 is a graph theoretical representation for this dendrimer.

Figure 2. $NS_2[p]$ polypropylenimine octaamin dendrimers.

The third object of interest is the $NS_3[p]$, also known as polymer dendrimer. Figure 3 shows the molecular structure of this dendrimer.

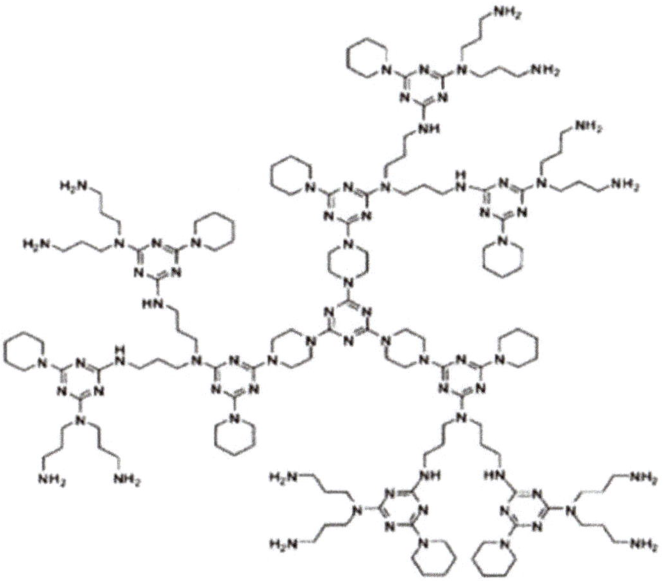

Figure 3. $NS_3[p]$ polymer de ndrimer.

The above three families have been used a lot recently for their theoretical properties. De et al. in [14], computed the F-index of nanostar dendrimers, Siddiqui et al. computed the Zagreb indices for different nanostar dendrimers in [15], and Madanshekaf computed the Randi index for some different classes of nanostar dendrimers in [16,17]. Munir, et al. computed M-polynomial and related indices of these nanostar dendrimers in [18], titania nanotubes in [19], polyhez nanotubes in [20] and circulant graphs in [21].

In the current article, we are interested in imbalance-based irregularity indices of the above discussed families of three dendrimers. We use techniques from combinatorics and graph theory to avoid the use of quantum mechanics, as has been done recently in most of the cases, see [7–14]. Important tools which are used for this purpose are structural and functional polynomials. These tools use structural parameters as inputs and the outputs are the key information that is used to determine properties of the material under discussion. Certain properties of matters like standard enthalpy, toxicity, entropy as well as reactivity and biological mechanics are theoretically based on these tools [4–6]. Estrada related the atom bond connectivity index with energies of the branched alkanes in [9]. Some applications of indices in pharmacy are given in [10] and in Quantitative structure activity analysis in [11].

2. Preliminaries and Notations

In this part we lay out some basic material and notations which will be used throughout the article. All graphs will be connected. We fix the symbol G for a simple connected graph, V for the set of vertices of G, E for the set of edges, du and dv are the degrees of vertices u and v, respectively. Topological index is an invariant of the graph that preserves the structural aspects of the graph. A degree based topological index is based on the end degrees of edges. A graph is said to be regular if every vertex of the graph has the same degree. A topological invariant is called irregularity index if the index vanishes for a regular graph and is non-zero for a non-regular graph. Regular graphs have been investigated a lot, particularly in mathematics. Their applications in chemical graph theory came to be known after the discovery of nanotubes and fullerenes. Paul Erdos emphasized the study of irregular graphs for the first time in history in [22]. At the Second Krakow Conference on Graph Theory (1994), Erdos officially posed an open problem about determination of the extreme size of highly irregular graphs of given order [23]. Since then, the irregular graphs and the degree of irregularity have become the basic open problem of graph theory. A graph in which each vertex has a different degree than other vertices is known as a perfect graph. Authors in [24], proved that no graph is perfect. The graphs lying in between are called quasi-perfect graphs in which each, except two vertices, have different degrees [25]. A simplified way of expressing the irregularity is the irregularity index. Irregularity indices have been studied recently in a novel way [26,27]. The first such irregularity index was introduced in [28]. Most of these indices used the concept of imbalance of an edge defined as $imball_{uv} = |du - dv|$, [25–27]. The Albertson index, AL (G), was defined by Alberston in [29] as $AL(G) = \sum_{UV\in E}|d_u - d_v|$. In this index, the imbalance of edges is computed. The irregularity index IRL(G) and IRLU(G) is introduced by Vukicevic and Gasparov, [30] as $IRL(G) = \sum_{UV\in E}|lnd_u - lnd_v|$, and $IRLU(G) = \sum_{UV\in E}\frac{|d_u-d_v|}{min(d_u,d_v)}$. Recently, Abdoo et al. introduced the new term "total irregularity measure of a graph G", which is defined as [31–33] $IRR_t(G) = \frac{1}{2}\sum_{UV\in E}|d_u - d_v|$. Recently, Gutman et al. introduced the σ (G) irregularity index of the graph G, which is described as $\sigma(G) = \sum_{UV\in E}(d_u - d_v)^2$ in [34]. The Randic index itself is directly related to an irregularity measure, which is described as $IRA(G) = \sum_{UV\in E}\left(d_u^{\frac{-1}{2}} - d_v^{\frac{-1}{2}}\right)^2$ in [35]. Further irregularity indices of similar nature can be traced in [34] in detail. These indices are given as $IRDIF(G) = \sum_{UV\in E}\left|\frac{d_u}{d_v} - \frac{d_v}{d_v}\right|$, $IRLF(G) = \sum_{UV\in E}\frac{|d_u-d_v|}{\sqrt{(d_ud_v)}}$, $LA(G) = 2\sum_{UV\in E}\frac{|d_u-d_v|}{(d_u+d_v)}$,

$IRD1 = \sum_{UV\in E}ln\{1 + |d_v - d_v|\}$, $IRGA(G) = \sum_{UV\in E}ln\frac{d_u+d_v}{2\sqrt{(d_ud_v)}}$, and $IRB(G) = \sum_{UV\in E}\left(d_u^{\frac{1}{2}} - d_v^{\frac{1}{2}}\right)^2$.

Futher details are given in [28–46]. Recently authors computed irregularity indices of a nanotube [47]. Recently Gao et al. computed irregularity measures of some dendrimer structures in [48] and molecular structures in [49]. Actually, the authors computed only four irregularity measures for some classes of dendrimers in [48]. These structures are used as long infinite chain macromolecules in chemistry and related areas. Hussain et al. computed these irregularity measures for some classes of benzenoid systems in [50].

3. Main Results

In this part, we give our main theoretical results.

Theorem 1. *Let $NS_1[p]$ be the polypropylenimine octaamin dendrimers, then the irregularity indices of $NS_1[p]$ are:*

1. $IRDIF(NS_1[p]) = 1.5\left(2^{p+1}\right) + 22.3372^p - 22.334$
2. $IRR(NS_1[p]) = 2^{p+1} + 22(2^p) - 22$
3. $IRL(NS_1[p]) = 0.69314718\left(2^{p+1}\right) + 10.070961\,(2^p) - 10.070961$
4. $IRLU(NS_1[p]) = 2^{p+1} + 15(2^p) - 15$
5. $IRLF(NS_1[p]) = 0.7071068(2^{p+1}) + 10.334278(2^p) - 10.334278$
6. $\sigma(NS_1[p]) = 2^{p+1} + 30(2^p) - 30$
7. $IRLA(NS_1[p]) = 0.6667(2^p) + 9.6(2^p) - 9.6$
8. $IRD1 = 0.69314\left(2^{p+1}\right) + 14.098(2^p) - 14.098$
9. $IRA(NS_1[p]) = 0.085786432(2^{p+1}) + 0.950245633(2^p) - 0.9502456337$
10. $IRGA(NS_1[p]) = 0.06036(2^{p+1}) + 0.8610954(2^p) - 0.8610954$
11. $IRB(NS_1[p]) = 0.171528753(2^{p+1}) + 3.557593539(2^p) - 3.557593539$
12. $IRR_t(NS_1[p]) = 12(2^p) - 11$

Proof. In order to prove the above theorem we have to consider Figure 1.

We can see that the edges $NS_1[p]$ admit the following partition in Table 1. □

Table 1. Edge partition of $NS_1[p]$ polypropylenimine octaamin dendrimers.

Edge Type (d_u, d_v)	Number of Edges
(1,2)	2^{p+1}
(1,3)	$4\,(2^p - 1)$
(2,2)	$\{12 \times 2^p - 11\}$
(2,3)	$14\,(2^p - 1)$

Now, using above Table 1 and definitions we have:

1. $IRDIF(G) = \sum_{UV \in E} \left| \frac{d_u}{d_v} - \frac{d_v}{d_v} \right|$

$$
\begin{aligned}
IRDIF(NS_1[p], x, y) &= 2^{p+1} \left| \frac{2}{1} - \frac{1}{2} \right| + 4(2^p - 1) \left| \frac{3}{1} - \frac{1}{3} \right| + \{12 \times 2^p - 11\} \left| \frac{2}{2} - \frac{2}{2} \right| + 14(2^p - 1) \left| \frac{3}{2} - \frac{2}{3} \right| \\
&= 2^{p+1} \left| \frac{2}{1} - \frac{1}{2} \right| + 4(2^p - 1) \left| \frac{3}{1} - \frac{1}{3} \right| + 14(2^p - 1) \left| \frac{3}{2} - \frac{2}{3} \right|.
\end{aligned}
$$

2. $IRR(G) = \sum_{UV \in E} imb(e)$

$$
\begin{aligned}
IRR(NS_1[p], x, y) &= 2^{p+1} |2 - 1| + 4(2^p - 1)|3 - 1| + \{12 \times 2^p - 11\}|2 - 2| + 14(2^p - 1) \\
&= 2^{p+1} + 4(2^p - 1)|2| + 14(2^p - 1).
\end{aligned}
$$

3. $IRL(G) = \sum_{UV \in E} |ln d_u - ln d_v|$

$$
\begin{aligned}
IRL(NS_1[p], x, y) &= 2^{p+1} |ln2 - ln1| + 4(2^p - 1)|ln3 - ln1| + \{12 \times 2^p - 11\}|ln2 - ln2| + 14(2^p - 1)|ln3 - ln2| \\
&= 2^{p+1} |ln2| + 4(2^p - 1)|ln3| + 14(2^p - 1)\left|ln\frac{3}{2}\right|.
\end{aligned}
$$

4. $IRLU(G) = \sum_{UV \in E} \frac{|d_u - d_v|}{min(d_u d_v)}$

$$
\begin{aligned}
IRLU(NS_1[p], x, y) &= 2^{p+1} \frac{|2-1|}{1} + 4(2^p - 1) \frac{|3-1|}{1} + \{12 \times 2^p - 11\} \frac{|2-2|}{2} + 14(2^p - 1) \frac{|3-2|}{2} \\
&= 2^{p+1} + 8(2^p - 1) + 7(2^p - 1).
\end{aligned}
$$

5. $IRLF(G) = \sum_{UV \in E} \frac{|d_u - d_v|}{\sqrt{(d_u d_v)}}$

$$
\begin{aligned}
IRLF(NS_1[p], x, y) &= 2^{p+1} \frac{|2-1|}{\sqrt{2}} + 4(2^p - 1) \frac{|3-1|}{\sqrt{3}} + \{12 \times 2^p - 11\} \frac{|2-2|}{\sqrt{4}} + 14(2^p - 1) \frac{|3-2|}{\sqrt{6}} \\
&= \frac{2^{p+1}}{\sqrt{2}} + \frac{8(2^p - 1)}{\sqrt{3}} + \frac{14(2^p - 1)}{\sqrt{6}}.
\end{aligned}
$$

6. $\sigma(G) = \sum_{UV \in E} (d_u - d_v)^2$

$$
\begin{aligned}
\sigma(NS_1[p], x, y) &= 2^{p+1} (2 - 1)^2 + 4(2^p - 1)(3 - 1)^2 + \{12 \times 2^p - 11\}(2 - 2)^2 + 14(2^p - 1)(3 - 2)^2 \\
&= 2^{p+1} + 16(2^p - 1) + 14(2^p - 1).
\end{aligned}
$$

7. $IRLA(G) = 2 \sum_{UV \in E} \frac{|d_u - d_v|}{(d_u + d_v)}$

$$
\begin{aligned}
IRLA(NS_1[p], x, y) &= 2 \left[2^{p+1} \frac{|2-1|}{(2+1)} + 4(2^p - 1) \frac{|3-1|}{(3+1)} + \{12 \times 2^p - 11\} \frac{|2-2|}{(2+2)} + 14(2^p - 1) \frac{|3-2|}{(3+2)} \right] \\
&= 2 \left[2^{p+1} \frac{|2-1|}{(2+1)} + 4(2^p - 1) \frac{|3-1|}{(3+1)} + 14(2^p - 1) \frac{|3-2|}{(3+2)} \right].
\end{aligned}
$$

8. $IRD1 = \sum_{UV \in E} ln\{1 + |d_v - d_v|\}$

$$
\begin{aligned}
IRD1(NS_1[n], x, y) &= 2^{p+1} ln\{1 + |2 - 1|\} + 4(2^p - 1)ln\{1 + |3 - 1|\} + \{12 \times 2^p - 11\}ln\{1 + |2 - 2|\} + 14(2^p - 1)ln\{1 + |3 - 2|\} \\
&= 2^{p+1} ln\{1 + |2 - 1|\} + 4(2^p - 1)ln\{1 + |3 - 1|\} + 14(2^p - 1)ln\{1 + |3 - 2|\}.
\end{aligned}
$$

9. $IRA(G) = \sum_{UV \in E} \left(d_u^{\frac{-1}{2}} - d_v^{\frac{-1}{2}} \right)^2$

$$IRA(NS_1[p], x, y) = 2^{p+1}\left(\tfrac{1}{\sqrt{2}} - \tfrac{1}{\sqrt{1}}\right)^2 + 4(2^p - 1)\left(\tfrac{1}{\sqrt{3}} - \tfrac{1}{\sqrt{1}}\right)^2 + \{12 \times 2^p - 11\}\left(\tfrac{1}{\sqrt{2}} - \tfrac{1}{\sqrt{2}}\right)^2 + 14(2^p - 1)\left(\tfrac{1}{\sqrt{3}} - \tfrac{1}{\sqrt{2}}\right)^2$$
$$= 2^{p+1}\left(\tfrac{1}{\sqrt{2}} - \tfrac{1}{\sqrt{1}}\right)^2 + 4(2^p - 1)\left(\tfrac{1}{\sqrt{3}} - \tfrac{1}{\sqrt{1}}\right)^2 + 14(2^p - 1)\left(\tfrac{1}{\sqrt{3}} - \tfrac{1}{\sqrt{2}}\right)^2.$$

10. We have $IRGA(G) = \sum_{UV \in E} \ln \dfrac{d_u + d_v}{2\sqrt{(d_u d_v)}}$

$$IRGA(NS_1[p], x, y) = 2^{p+1}\ln \tfrac{|2+1|}{2\sqrt{2}} + 4(2^p - 1)\ln \tfrac{|3+1|}{2\sqrt{3}} + \{12 \times 2^p - 11\}\ln \tfrac{|2+2|}{2\sqrt{4}} + 14(2^p - 1)\ln \tfrac{|3+2|}{2\sqrt{6}}$$
$$= 2^{p+1}\ln \tfrac{|2+1|}{2\sqrt{2}} + 4(2^p - 1)\ln \tfrac{|3+1|}{2\sqrt{3}} + 14(2^p - 1)\ln \tfrac{|3+2|}{2\sqrt{6}}.$$

11. $IRB(G) = \sum_{UV \in E} \left(d_u^{\frac{1}{2}} - d_v^{\frac{1}{2}} \right)^2$

$$IRB(NS_1[p], x, y) = 2^{p+1}\left(\sqrt{2} - \sqrt{1}\right)^2 + 4(2^p - 1)\left(\sqrt{3} - \sqrt{1}\right)^2 + \{12 \times 2^p - 11\}\left(\sqrt{2} - \sqrt{2}\right)^2 + 14(2^p - 1)\left(\sqrt{3} - \sqrt{2}\right)^2$$
$$= 2^{p+1}\left(\sqrt{2} - \sqrt{1}\right)^2 + 4(2^p - 1)\left(\sqrt{3} - \sqrt{1}\right)^2 + 14(2^p - 1)\left(\sqrt{3} - \sqrt{2}\right)^2.$$

12. $IRR_t(G) = \tfrac{1}{2}\sum_{UV \in E} |d_u - d_v|$

$$IRR_t(NS_1[p], x, y) = \tfrac{1}{2}\left[2^{p+1}|2 - 1| + 4(2^p - 1)|3 - 1| + \{12 \times 2^p - 11\}|2 - 2| + 14(2^p - 1)|3 - 2| \right]$$
$$= \tfrac{1}{2}\left[2^{p+1} + 4(2^p - 1)|2| + 14(2^p - 1) \right].$$

The following Table 2 shows the values of these irregularity indices for some test values of parameter p.

Table 2. Irregularity indices for $NS_1[p]$ polypropylenimine octaamin dendrimers.

Irregularity Indices	$p = 1$	$p = 2$	$p = 3$	$p = 4$	$p = 5$
$IRDIF(G) = \sum_{UV \in E}\left\|\frac{d_u}{d_v} - \frac{d_v}{d_u}\right\|$	28.340	79.014	180.335	383.005	788.344
$AL(G) = \sum_{UV \in E}\|d_u - d_v\|$	26	74	170	362	746
$IRLU(G) = \sum_{UV \in E}\frac{\|d_u - d_v\|}{min(d_u, d_v)}$	19	53	121	257	529
$IRLU(G) = \sum_{UV \in E}\frac{\|d_u - d_v\|}{\sqrt{(d_u d_v)}}$	13.1627	36.6596	83.6536	177.6415	365.6174
$IRF(G) = \sum_{UV \in E}(d_u - d_v)^2$	34	98	226	482	994
$IRLA(G) = 2\sum_{UV \in E}\frac{\|d_u - d_v\|}{(d_u + d_v)}$	12.268	34.136	77.872	165.344	340.288
$IRD1 = \sum_{UV \in E}\ln\{1 + \|d_v - d_v\|\}$	16.870	47.839	109.776	233.650	481.398
$IRA(G) = \sum_{UV \in E}\left(d_u^{\frac{-1}{2}} - d_v^{\frac{-1}{2}}\right)^2$	1.2934	3.5370	8.0243	16.9988	34.9479
$IRGA(G) = \sum_{UV \in E}\ln\frac{d_u + d_v}{2\sqrt{(d_u d_v)}}$	1.1025	3.0661	6.9934	14.8480	30.5571
$IRB(G) = \sum_{UV \in E}\left(d_u^{\frac{1}{2}} - d_v^{\frac{1}{2}}\right)^2$	4.2437	12.0450	27.6476	58.8528	121.2632
$IRR_t(G) = \tfrac{1}{2}\sum_{UV \in E}\|d_u - d_v\|$	13	37	85	181	373

Now we proceed to irregularity indices of $NS_2[p]$.

Theorem 2. *Let $NS_2[p]$ be the nanostar polypropylenimine octaamin dendrimers, then the irregularity indices of $NS_2[p]$ are:*

1. $IRDIF(NS_2[p]) = 1.5\left(2^{p+1}\right) + 5(2^p) - 5$
2. $IRR(NS_2[p]) = 2^{p+1} + 6(2^p) - 6$
3. $IRL(NS_2[p]) = 0.6931471\left(2^{p+1}\right) + 2.4328(2^p) - 2.4328$

4. $IRLU(NS_2[p]) = 2^{p+1} + 3(2^p) - 3$
5. $IRLF(NS_2[p]) = 0.707106(2^{p+1}) + 2.4494897(2^p) - 2.4494897$
6. $\sigma(NS_2[p]) = 2^{p+1} + 6(2^p) - 6$
7. $IRLA(NS_2[p]) = 0.6667(2^{p+1}) + 2.4(2^p) - 2.4$
8. $IRD1 = 0.6931471806(2^{p+1}) + 4.15888302(2^p) - 4.15888302$
9. $IRA(NS_2[p]) = 0.08578644(2^{p+1}) + 0.1010205144(2^p) - 0.1010205144$
10. $IRGA(NS_2[p]) = 0.0588915178(2^{p+1}) + 0.1224659836(2^p) - 0.1224659836$
11. $IRB(NS_2[p]) = 0.171578(2^{p+1}) + 0.606123086(2^p) - 0.606123086$
12. $IRR_t(NS_2[p]) = 4(2^p) - 3$

Proof. In order to prove the above theorem, we have to consider Figure 2. We can see that the edges of $NS_2[p]$ admit the following partition in Table 3. □

Table 3. Edge partition of nanostar polypropylenimine octaamin dendrimers.

Edge Type (d_u, d_v)	Number of Edges
(1,2)	2^{p+1}
(2,2)	$\{8 \times 2^p - 5\}$
(2,3)	$6(2^p - 1)$

We can see that the edges of $NS_2[p]$ admit the following partition in Table 3.
Now using above Table 3 and the above definitions, we have:

1. $IRDIF(G) = \sum_{UV \in E} \left| \frac{d_u}{d_v} - \frac{d_v}{d_v} \right|$

$$IRDIF(NS_2[p], x, y) = 2^{p+1} \left| \frac{2}{1} - \frac{1}{2} \right| + |\{8 \times 2^p - 5\}| \left| \frac{2}{2} - \frac{2}{2} \right| + 6(2^p - 1) \left| \frac{3}{2} - \frac{2}{3} \right|$$
$$= 2^{p+1} \left| \frac{2}{1} - \frac{1}{2} \right| + 6(2^p - 1) \left| \frac{3}{2} - \frac{2}{3} \right|.$$

2. $IRR(G) = \sum_{UV \in E} imb(e) \therefore imb(e) = |d_u - d_v|$

$$IRR(NS_2[p], x, y) = 2^{p+1}|2 - 1| + +\{8 \times 2^p - 5\}|2 - 2| + 6(2^p - 1)|3 - 2|$$
$$= 2^{p+1} + 6(2^p - 1).$$

3. $IRL(G) = \sum_{UV \in E} |lnd_u - lnd_v|$

$$IRL(NS_2[p], x, y) = 2^{p+1}|ln2 - ln1| + \{8 \times 2^p - 5\}|ln2 - ln2| + 6(2^p - 1)|ln3 - ln2|$$
$$= 2^{p+1}|ln2| + 6(2^p - 1)\left|ln\frac{3}{2}\right|.$$

4. $IRLU(G) = \sum_{UV \in E} \frac{|d_u - d_v|}{min(d_u d_v)}$

$$IRLU(NS_2[p], x, y) = 2^{p+1}\frac{|2-1|}{1} + \{8 \times 2^p - 5\}\frac{|2-2|}{2} + 6(2^p - 1)\frac{|3-2|}{2}$$
$$= 2^{p+1} + 3(2^p - 1).$$

5. $IRLF(G) = \sum_{UV \in E} \frac{|d_u - d_v|}{\sqrt{(d_u d_v)}}$

$$IRLU(NS_2[p], x, y) = 2^{p+1}\frac{|2-1|}{\sqrt{2}} + \{8 \times 2^p - 5\}\frac{|2-2|}{\sqrt{4}} + 6(2^p - 1)\frac{|3-2|}{\sqrt{6}}$$
$$= \frac{2^{p+1}}{\sqrt{2}} + \frac{6(2^p - 1)}{\sqrt{6}}.$$

6. $\sigma(G) = \sum_{UV \in E}(d_u - d_v)^2$

$$\begin{aligned}\sigma(NS_2[p], x, y) &= 2^{p+1}(2-1)^2 + \{8 \times 2^p - 5\}(2-2)^2 + 6(2^p-1)(3-2)^2 \\ &= 2^{p+1}(2-1)^2 + 6(2^p-1)(3-2)^2.\end{aligned}$$

7. $IRLA(G) = 2\sum_{UV \in E}\frac{|d_u - d_v|}{(d_u + d_v)}$

$$\begin{aligned}IRLA(NS_2[p], x, y) &= 2\left[2^{p+1}\frac{|2-1|}{(2+1)} + \{8 \times 2^p - 5\}\frac{|2-2|}{(2+2)} + 6(2^p-1)\frac{|3-2|}{(3+2)}\right] \\ &= 2\left[\frac{2^{p+1}}{3} + \frac{6(2^p-1)}{5}\right].\end{aligned}$$

8. $IRD1 = \sum_{UV \in E} ln\{1 + |d_v - d_v|\}$

$$\begin{aligned}IRD1(NS_2[p], x, y) &= 2^{p+1}ln\{1 + |2-1|\} + \{8 \times 2^p - 5\}ln\{1 + |2-2|\} + 6(2^p-1)ln\{1 + |3-2|\} \\ &= 2^{p+1}ln2 + 6(2^p-1)ln2 \ .\end{aligned}$$

9. $IRA(G) = \sum_{UV \in E}\left(d_u^{\frac{-1}{2}} - d_v^{\frac{-1}{2}}\right)^2$

$$\begin{aligned}IRA(NS_2[p], x, y) &= 2^{p+1}\left(\frac{1}{\sqrt{2}} - \frac{1}{\sqrt{1}}\right)^2 + \{8 \times 2^p - 5\}\left(\frac{1}{\sqrt{2}} - \frac{1}{\sqrt{2}}\right)^2 + 6(2^p-1)\left(\frac{1}{\sqrt{3}} - \frac{1}{\sqrt{2}}\right)^2 \\ &= 2^{p+1}\left(\frac{1}{\sqrt{2}} - \frac{1}{\sqrt{1}}\right)^2 + 6(2^p-1)\left(\frac{1}{\sqrt{3}} - \frac{1}{\sqrt{2}}\right)^2.\end{aligned}$$

10. $IRGA(G) = \sum_{UV \in E} ln\frac{d_u + d_v}{2\sqrt{(d_u d_v)}}$

$$\begin{aligned}IRGA(NS_2[p], x, y) &= 2^{p+1}ln\frac{|2+1|}{2\sqrt{2}} + \{8 \times 2^p - 5\}ln\frac{|2+2|}{2\sqrt{4}} + 6(2^p-1)ln\frac{|3+2|}{2\sqrt{6}} \\ &= 2^{p+1}ln\frac{3}{2\sqrt{2}} + 6(2^p-1)ln\frac{5}{2\sqrt{6}} \ .\end{aligned}$$

11. $IRB(G) = \sum_{UV \in E}\left(d_u^{\frac{1}{2}} - d_v^{\frac{1}{2}}\right)^2$

$$\begin{aligned}IRB(NS_2[p], x, y) &= 2^{p+1}\left(\sqrt{2} - \sqrt{1}\right)^2 + \{8 \times 2^p - 5\}\left(\sqrt{2} - \sqrt{2}\right)^2 + 6(2^p-1)\left(\sqrt{3} - \sqrt{2}\right)^2 \\ &= 2^{p+1}\left(\sqrt{2} - \sqrt{1}\right)^2 + 14(2^p-1)\left(\sqrt{3} - \sqrt{2}\right)^2.\end{aligned}$$

12. $IRR_t(G) = \frac{1}{2}\sum_{UV \in E}|d_u - d_v|$

$$\begin{aligned}IRR_t(NS_2[p], x, y) &= \frac{1}{2}\left[2^{p+1}|2-1| + + \{8 \times 2^p - 5\}|2-2| + 6(2^p-1)|3-2|\right] \\ &= \frac{1}{2}\left[2^{p+1} + 6(2^p-1)\right].\end{aligned}$$

Table 4 represents some calculated values of irregularity indices of $NS_2[p]$ for some test values of p.

Table 4. Irregularity indices for $NS_2[p]$ polypropylenimine octaamin dendrimers.

Irregularity indices	$p = 1$	$p = 2$	$p = 3$	$p = 4$	$p = 5$
$IRDIF(G) = \sum_{UV \in E}\left\lvert\frac{d_u}{d_v} - \frac{d_v}{d_v}\right\rvert$	11	27	59	123	251
$AL(G) = \sum_{UV \in E}\lvert d_u - d_v\rvert$	10	26	58	122	250
$IRL(G) = \sum_{UV \in E}\lvert ln d_u - ln d_v\rvert$	5.2054	12.844	28.1199	58.6727	119.7782
$IRLU(G) = \sum_{UV \in E}\frac{\lvert d_u - d_v\rvert}{min(d_u,d_v)}$	7	17	37	77	157
$IRLU(G) = \sum_{UV \in E}\frac{\lvert d_u - d_v\rvert}{\sqrt{(d_u d_v)}}$	5.2779	13.0053	28.4601	59.3697	121.1889
$IRF(G) = \sum_{UV \in E}(d_u - d_v)^2$	10	26	58	122	250
$IRLA(G) = 2\sum_{UV \in E}\frac{\lvert d_u - d_v\rvert}{(d_u + d_v)}$	5.0668	12.5336	27.4672	57.3344	117.0688
$IRD1 = \sum_{UV \in E}ln\{1 + \lvert d_v - d_v\rvert\}$	6.9315	18.0218	40.2025	84.5639	173.2867
$IRA(G) = \sum_{UV \in E}\left(d_u^{\frac{-1}{2}} - d_v^{\frac{-1}{2}}\right)^2$	0.4442	0.9894	2.0797	4.2605	8.62197
$IRGA(G) = \sum_{UV \in E}ln\frac{d_u + d_v}{2\sqrt{(d_u d_v)}}$	0.3580	0.8385	1.7995	3.7215	7.5655
$IRB(G) = \sum_{UV \in E}\left(d_u^{\frac{1}{2}} - d_v^{\frac{1}{2}}\right)^2$	1.2924	3.1909	6.9881	14.5823	29.7708
$IRR_t(G) = \frac{1}{2}\sum_{UV \in E}\lvert d_u - d_v\rvert$	5	13	29	61	125

Our next aim is to determine the of irregularity indices of polymer dendrimers.

Theorem 3. *Let $NS_3[p]$ be polymer dendrimer then the irregularity indices of $NS_3[p]$ are:*

1. $IRDIF(NS_3[p]) = 4.5(2^p) + 54.999978(2^{p-1}) - 14.999994$
2. $IRR(NS_3[p]) = 3(2^p) + 66(2^{p-1}) - 18$
3. $IRL(NS_3[p]) = 2.079441(2^p) + 26.76069(2^{p-1}) - 7.29837$
4. $IRLU(NS_3[p]) = 3(2^p) + 33(2^{p-1}) - 9$
5. $IRLF(NS_3[p]) = 2.12132(2^p) + 26.944384(2^{p-1}) - 7.348468$
6. $\sigma(NS_3[p]) = 3(2^p) + 66(2^{p-1}) - 18$
7. $IRLA(NS_3[p]) = 2(2^p) + 26.4(2^{p-1}) - 7.2$
8. $IRD1 = 2.079441(2^p) + 45.747702(2^{p-1}) - 12.476646$
9. $IRA(NS_3[p]) = 1.5(2^p) + 1.111242(2^{p-1}) - 0.303066$
10. $IRGA(NS_3[p]) = 0.176673(2^p) + 1.347126(2^{p-1}) - 0.367398$
11. $IRB(NS_3[p]) = 0.514719(2^p) + 6.667386(2^{p-1}) - 1.8378$
12. $IRR_t(NS_3[p]) = 1.5(2^p) + 33(2^{p-1}) - 9$

Proof. In order to prove the above theorem we have to consider Figure 3. □

We can see that the edges of $NS_3[p]$ admit the following partition in Table 5.

Table 5. Edge partition of polymer dendrimer.

Edges Type (d_u,d_v)	Number of Edges
(1,2)	3.2^p
(2,2)	$54(2^{p-1}) - 24$
(2,3)	$66(2^{p-1} - 1) + 48$
(3,3)	3.2^{p+1}

Now using above Table 5 and the above definitions we have:

1. $IRDIF(G) = \sum_{UV \in E} \left| \frac{d_u}{d_v} - \frac{d_v}{d_v} \right|$

$$\begin{aligned} IRDIF(NS_3[p], x, y) &= 3.2^p \left| \frac{2}{1} - \frac{1}{2} \right| + 54(2^{p-1}) - 24 \left| \frac{2}{2} - \frac{2}{2} \right| + 66(2^{p-1} - 1) + 48 \left| \frac{3}{2} - \frac{2}{3} \right| + 3.2^{p+1} \left| \frac{3}{3} - \frac{3}{3} \right|. \\ &= 3.2^p \left| \frac{2}{1} - \frac{1}{2} \right| + 66(2^{p-1} - 1) + 48 \left| \frac{3}{2} - \frac{2}{3} \right|. \end{aligned}$$

2. $IRR(G) = \sum_{UV \in E} imb(e) \therefore imb(e) = |d_u - d_v|$

$$\begin{aligned} IRR(NS_3[p], x, y) &= 3.2^p |2 - 1| + 54(2^{p-1}) - 24|2 - 2| + 66(2^{p-1} - 1) + 48|3 - 2| + 3.2^{p+1}|3 - 3| \\ &= 3.2^p + 66(2^{p-1} - 1) + 48. \end{aligned}$$

3. $IRL(G) = \sum_{UV \in E} |lnd_u - lnd_v|$

$$\begin{aligned} IRL(NS_3[p], x, y) &= 3.2^p |ln2 - ln1| + 54(2^{p-1}) - 24|ln2 - ln2| + 66(2^{p-1} - 1) + 48|ln3 - ln2| + 3.2^{p+1}|ln3 - ln3| \\ &= 3.2^p |ln2 - ln1| + 66(2^{p-1} - 1) + 48|ln3 - ln2|. \end{aligned}$$

4. $IRLU(G) = \sum_{UV \in E} \frac{|d_u - d_v|}{min(d_u d_v)}$

$$\begin{aligned} IRLU(NS_3[p], x, y) &= 3.2^p \frac{|2-1|}{1} + 54(2^{p-1}) - 24 \frac{|2-2|}{2} + 66(2^{p-1} - 1) + 48 \frac{|3-2|}{2} + 3.2^{p+1} \frac{|3-3|}{3} \\ &= 3.2^p + \frac{66(2^{p-1}-1)+48}{2}. \end{aligned}$$

5. $IRLF(G) = \sum_{UV \in E} \frac{|d_u - d_v|}{\sqrt{(d_u d_v)}}$

$$\begin{aligned} IRLF(NS_3[p], x, y) &= 3.2^p \frac{|2-1|}{\sqrt{2}} + 54(2^{p-1}) - 24 \frac{|2-2|}{\sqrt{4}} + 66(2^{p-1} - 1) + 48 \frac{|3-2|}{\sqrt{6}} + 3.2^{p+1} \frac{|3-3|}{\sqrt{9}} \\ &= \frac{3.2^p}{\sqrt{2}} + \frac{66(2^{p-1}-1)+48}{\sqrt{6}}. \end{aligned}$$

6. $\sigma(G) = \sum_{UV \in E} (d_u - d_v)^2$

$$\begin{aligned} \sigma(NS_3[p], x, y) &= 3.2^p (2-1)^2 + 54(2^{p-1}) - 24(2-2)^2 + 66(2^{p-1} - 1) + 48(3-2)^2 + 3.2^{p+1}(3-3)^2 \\ &= 3.2^p + 66(2^{p-1} - 1) + 48. \end{aligned}$$

7. $IRLA(G) = 2 \sum_{UV \in E} \frac{|d_u - d_v|}{(d_u + d_v)}$

$$\begin{aligned} IRLA(NS_3[p], x, y) &= 2 \left[3.2^p \frac{|2-1|}{(2+1)} + 54(2^{p-1}) - 24 \frac{|2-2|}{(2+2)} + 66(2^{p-1} - 1) + 48 \frac{|3-2|}{(2+2)} + 3.2^{p+1} \frac{|3-3|}{(3+3)} \right] \\ &= 2 \left[\frac{3.2^p}{(3)} + 66(2^{p-1} - 1) + \frac{48}{(5)} \right]. \end{aligned}$$

8. $IRD1 = \sum_{UV \in E} ln\{1 + |d_v - d_v|\}$

$$\begin{aligned} IRD1(NS_3[n], x, y) &= 3.2^p ln\{1 + |2 - 1|\} + 54(2^{p-1}) - 24ln\{1 + |2 - 2|\} + 66(2^{p-1} - 1) + 48ln\{1 + |3 - 2|\} + 3.2^{p+1} ln\{1 + |3 - 3|\} \\ &= 3.2^p ln2 + (66(2^{p-1} - 1) + 48)ln2. \end{aligned}$$

9. $IRA(G) = \sum_{UV \in E} \left(d_u^{\frac{-1}{2}} - d_v^{\frac{-1}{2}} \right)^2$

$$\begin{aligned} IRA(NS_3[p], x, y) &= 3.2^p \left(\frac{1}{\sqrt{2}} - \frac{1}{\sqrt{1}} \right)^2 + 54(2^{p-1}) - 24 \left(\frac{1}{\sqrt{2}} - \frac{1}{\sqrt{2}} \right)^2 + 66(2^{p-1} - 1) + 48 \left(\frac{1}{\sqrt{3}} - \frac{1}{\sqrt{2}} \right)^2 + 3.2^{p+1} \left(\frac{1}{\sqrt{3}} - \frac{1}{\sqrt{3}} \right)^2 \\ &= 3.2^p \left(\frac{1}{\sqrt{2}} \right)^2 + 66(2^{p-1} - 1) + 48 \left(\frac{1}{\sqrt{3}} - \frac{1}{\sqrt{2}} \right)^2. \end{aligned}$$

10. $IRGA(G) = \sum_{UV \in E} ln \dfrac{d_u + d_v}{2\sqrt{(d_u d_v)}}$

$$IRGA(NS_3[p], x, y) = 3.2^p ln \frac{|2+1|}{2\sqrt{2}} + 54(2^{p-1}) - 24ln \frac{|2+2|}{2\sqrt{3}} + 66(2^{p-1} - 1) + 48ln \frac{|3+2|}{2\sqrt{4}} + 3.2^{p+1} ln \frac{|3+3|}{2\sqrt{6}}$$
$$= 3.2^p ln \frac{|3|}{2\sqrt{2}} + (66(2^{p-1} - 1) + 48) ln \frac{|5|}{2\sqrt{6}}.$$

11. $IRB(G) = \sum_{UV \in E} \left(d_u^{\frac{1}{2}} - d_v^{\frac{1}{2}}\right)^2$

$$IRB(NS_3[p], x, y) = 3.2^p \left(\sqrt{2} - \sqrt{1}\right)^2 + 54(2^{p-1}) - 24\left(\sqrt{2} - \sqrt{2}\right)^2 + 66(2^{p-1} - 1) + 48\left(\sqrt{3} - \sqrt{2}\right)^2 + 3.2^{p+1}\left(\sqrt{3} - \sqrt{3}\right)^2$$
$$= 3.2^p \left(\sqrt{2} - \sqrt{1}\right)^2 + (66(2^{p-1} - 1) + 48)\left(\sqrt{3} - \sqrt{2}\right)^2.$$

12. $IRR_t(G) = \frac{1}{2} \sum_{UV \in E} |d_u - d_v|$

$$IRR_t(NS_3[p], x, y) = \frac{1}{2}\left[3.2^p |2 - 1| + 54(2^{p-1}) - 24|2 - 2| + 66(2^{p-1} - 1) + 48|3 - 2| + 3.2^{p+1}|3 - 3|\right]$$
$$= \frac{1}{2}\left[3.2^p + (66(2^{p-1} - 1) + 48)\right].$$

The following Table 6 represents some calculated values of irregularity indices of $NS_3[p]$ for some test values of p.

Table 6. Irregularity indices for $NS_3[p]$ polymer dendrimers.

Irregularity Indices	$p = 1$	$p = 2$	$p = 3$	$p = 4$	$p = 5$
$IRDIF(G) = \sum_{UV \in E}\left|\frac{d_u}{d_v} - \frac{d_v}{d_v}\right|$	48.9999	112.9999	240.9999	496.9998	1008.99
$AL(G) = \sum_{UV \in E}|d_u - d_v|$	54	126	270	558	1134
$IRL(G) = \sum_{UV \in E}|lnd_u - lnd_v|$	23.6212	54.541	116.379	240.058	487.414
$IRLU(G) = \sum_{UV \in E} \frac{|d_u - d_v|}{min(d_u, d_v)}$	30	69	147	303	615
$IRLU(G) = \sum_{UV \in E} \frac{|d_u - d_v|}{\sqrt{(d_u d_v)}}$	23.8385	55.025	117.399	242.147	491.644
$IRF(G) = \sum_{UV \in E}(d_u - d_v)^2$	54	126	270	558	1134
$IRLA(G) = 2\sum_{UV \in E} \frac{|d_u - d_v|}{(d_u + d_v)}$	23.2	53.6	114.4	236	479.2
$IRD1 = \sum_{UV \in E} ln\{1 + |d_v - d_v|\}$	37.4299	87.3365	187.1496	386.7760	786.0286
$IRA(G) = \sum_{UV \in E}\left(d_u^{\frac{-1}{2}} - d_v^{\frac{-1}{2}}\right)^2$	3.8082	7.9194	16.1419	32.5868	65.4768
$IRGA(G) = \sum_{UV \in E} ln \frac{d_u + d_v}{2\sqrt{(d_u d_v)}}$	1.3331	3.0335	6.4345	13.236	26.84015
$IRB(G) = \sum_{UV \in E}\left(d_u^{\frac{1}{2}} - d_v^{\frac{1}{2}}\right)^2$	5.8590	13.5558	28.9495	59.7368	121.3113
$IRR_t(G) = \frac{1}{2}\sum_{UV \in E}|d_u - d_v|$	27	63	135	279	567

4. Graphical Analysis, Discussions and Conclusions

In this part we give our comparative analysis of some of the irregularity indices of the above discussed dendrimers and the dependences of the irregularity indices on the parameter of the structures. Figures 4–7 contain three graphs of irregularity indices. The horizontal axis is used for step size p and the vertical axis shows the value of irregularity index. In the graphs, the red color shows the irregularity of $NS_1[p]$, the blue color shows the irregularity of $NS_2[p]$ and the green color shows the irregularity of the $NS_3[p]$. In each graph, three different colored curves are depicted which shows the behavior of the irregularity indices with increase in p.

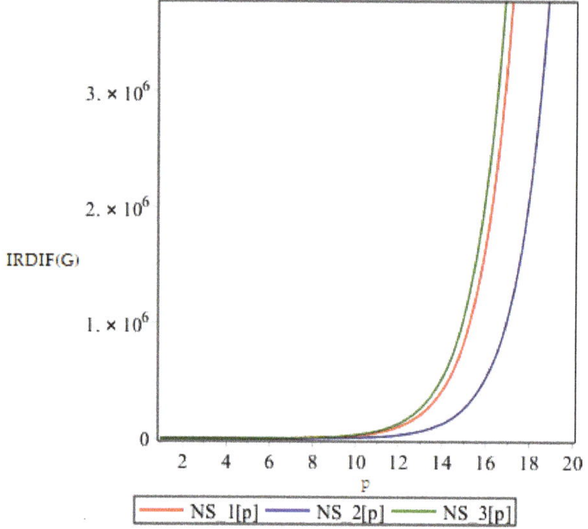

Figure 4. Graph of *IRDIF*(*G*).

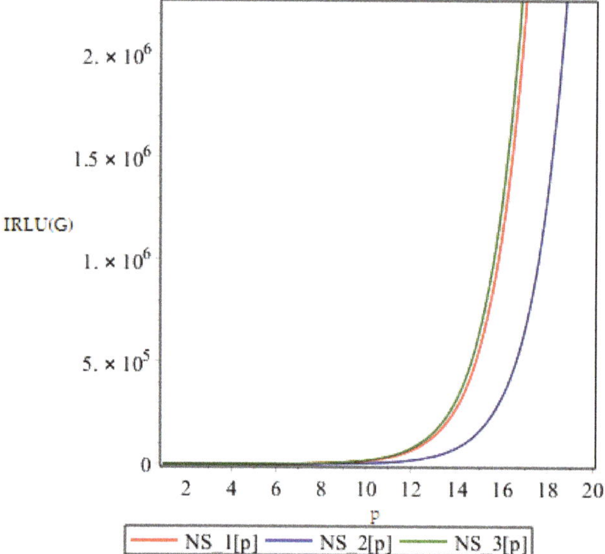

Figure 5. Graph of *IRLU*(*G*).

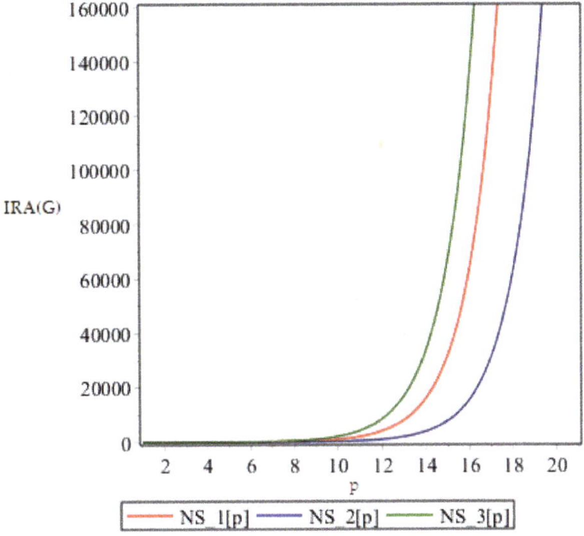

Figure 6. Graph of $IRA(G)$.

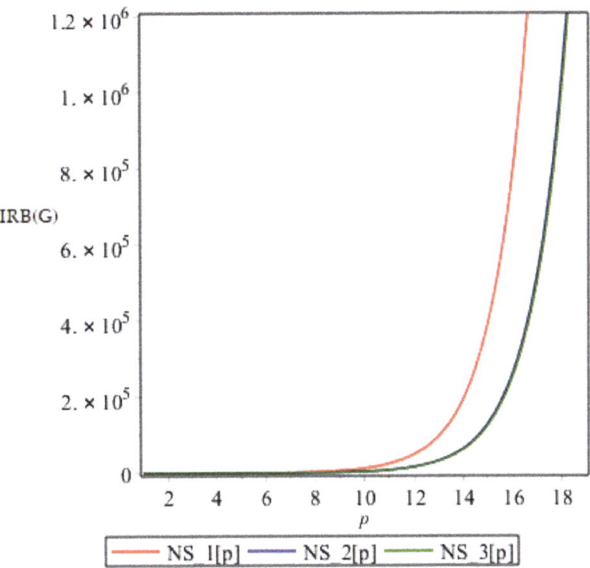

Figure 7. Graph of $IRB(G)$.

From above graph it seems obvious that irregularities have a slight increase with an increase in the step size p for the range $p \leq 12$. But after $p \geq 14$, these irregularity indices drastically increase with increase in p. So $NS_3[p]$ is the most irregular and asymmetric structure as far as most of the irregularity indices are concerned. Nanostar dendrimers are relatively less irregular, and $NS_1[p]$ are the most regular dendrimers. This trend is not restricted to only irregularity index $IRDIF$. Most of the irregularity indices behave pretty similarly as shown in the following figures. All other figures show the trends which can easily be understood in the Figures 5–7.

All above graphs (Figures 4–7) indicate that $NS_3[p]$ is highly asymmetric as far as all irregularity indices are concerned, whereas $NS_1[p]$ is less asymmetric, and $NS_2[p]$ is the most regular structure with respect to all indices. In *IRB*, $NS_1[p]$ and $NS_3[p]$ show the same irregularity behavior. These facts typically relate geometry and topology of the structure of these dendrimers and can be used in modelling purposes.

We foresee that our results could play an important role in determining properties of these dendrimers such as enthalpy, toxicity, resistance and entropy. Similar research has been done in [29], where authors discussed some properties of alkane isomers.

Author Contributions: Z.H. and M.M. gave the idea; H.A., T.H. and S.R. wrote the article; S.M.K. and Y.C.K. edited and verified the results.

Funding: This work was supported by the Dong-A University research fund.

Acknowledgments: We are greatly thankful to reviewers for their comments which have helped bring this article to its current form.

Conflicts of Interest: Authors declare no conflict of interests.

References

1. Hodge, P. Polymer science branches out. *Nature* **1993**, *362*, 18–19. [CrossRef]
2. Fischer, M.; Vögtle, F. Dendrimers: From design to applications—A progress report. *Angew. Chem. Int. Ed.* **1999**, *38*, 884–905. [CrossRef]
3. Wiener, E.C.; Auteri, F.P.; Chen, J.W.; Brechbiel, M.W.; Gansow, O.A.; Schneider, D.S.; Belford, R.L.; Clarkson, R.B.; Lauterbur, P.C. Molecular dynamics of ion-chelate complexes attached to dendrimers. *J. Am. Chem. Soc.* **1996**, *118*, 7774–7782. [CrossRef]
4. Rucker, G.; Rucker, C. On topological indices, boiling points, and cycloalkanes. *J. Chem. Inf. Comput. Sci.* **1999**, *39*, 788–802. [CrossRef]
5. Klavžar, S.; Gutman, I. A Comparison of the Schultz molecular topological index with the Wiener index. *J. Chem. Inf. Comput. Sci.* **1996**, *36*, 1001–1003. [CrossRef]
6. Gutman, I.; Polansky, O.E. *Mathematical Concepts in Organic Chemistry*; Springer: New York, NY, USA, 1986.
7. Randic, M. On the characterization of molecular branching. *J. Am. Chem. Soc.* **1975**, *97*, 6609–6615. [CrossRef]
8. Wiener, H. Structural determination of paraffin boiling points. *J. Am. Chem. Soc.* **1947**, *69*, 17–20. [CrossRef] [PubMed]
9. Estrada, E. Atomic bond connectivity and the energetic of branched alkanes. *Chem. Phys. Lett.* **2008**, *463*, 422–425. [CrossRef]
10. Kier, L.; Hall, L. *Molecular Connectivity in Chemistry and Drug Research*; Academic Press: New York, NY, USA, 1976.
11. Kier, L.B.; Hall, L.H. *Molecular Connectivity in Structure Activity Analysis*; Wiley: New York, NY, USA, 1986.
12. Deng, H.; Yang, J.; Xia, F. A general modeling of some vertex-degree based topological indices in benzenoid systems and phenylenes. *Comput. Math. Appl.* **2011**, *61*, 3017–3023. [CrossRef]
13. Zhang, H.; Zhang, F. The Clar covering polynomial of hexagonal systems. *Discret. Appl. Math.* **1996**, *69*, 147–167. [CrossRef]
14. De, N.; Nayeem, S.M.A. Computing the F-index of nanostar dendrimers. *Pac. Sci. A Nat. Sci. Eng.* **2016**, *18*, 14–21. [CrossRef]
15. Siddiqui, M.K.; Imran, M.; Ahmad, A. On Zagreb indices, Zagreb polynomials of some nanostar dendrimers. *Appl. Math. Comput.* **2016**, *280*, 132–139. [CrossRef]
16. Madanshekaf, A.; Ghaneei, M. Computing two topological indices of nanostars dendrimer, Optoelectron. *Adv. Mater. Rapid Commun.* **2010**, *4*, 1849–1851.
17. Madanshekaf, A.; Moradi, M. The first geometric-arithmetic index of some nanostar dendrimers. *Iran. J. Math. Chem.* **2014**, *5*, 1–6.
18. Munir, M.; Nazeer, W.; Rafique, S.; Kang, S.M. M-polynomial and related topological indices of Nanostar dendrimers. *Symmetry* **2016**, *8*, 97. [CrossRef]
19. Munir, M.; Nazeer, W.; Rafique, S.; Nizami, A.R.; Kang, S.M. M-polynomial and degree-based topological indices of Titania nanotubes. *Symmetry* **2016**, *8*, 117. [CrossRef]

20. Munir, M.; Nazeer, W.; Rafique, S.; Kang, S.M. M-Polynomial and Degree-Based Topological Indices of Polyhex Nanotubes. *Symmetry* **2016**, *8*, 149. [CrossRef]

21. Munir, M.; Nazeer, W.; Shahzadi, S.; Kang, S.M. Some invariants of circulant graphs. *Symmetry* **2016**, *8*, 134. [CrossRef]

22. Chartrand, G.; Erdos, P.; Oellermann, O. How to define an irregular graph. *Coll. Math. J.* **1988**, *19*, 36–42. [CrossRef]

23. Majcher, Z.; Michael, J. Highly irregular graphs with extreme numbers of edges. *Discret. Math.* **1997**, *164*, 237–242. [CrossRef]

24. Behzad, M.; Chartrand, G. No graph is perfect. *Am. Math. Mon.* **1967**, *74*, 962–963. [CrossRef]

25. Horoldagva, B.; Buyantogtokh, L.; Dorjsembe, S.; Gutman, I. Maximum size of maximally irregular graphs. *MATCH Commun. Math. Comput. Chem.* **2016**, *76*, 81–98.

26. Liu, F.; Zhang, Z.; Meng, J. The size of maximally irregular graphs and maximally irregular triangle–free graphs. *Graphs Comb.* **2014**, *30*, 699–705. [CrossRef]

27. Collatz, L.; Sinogowitz, U. Spektren endlicher Graphen. In *Abhandlungen aus dem Mathematischen Seminar der Universität Hamburg*; Springer: Berlin, Germany, 1957; Volume 21, pp. 63–77.

28. Bell, F.K. A note on the irregularity of graphs. *Linear Algebra Appl.* **1992**, *161*, 45–54. [CrossRef]

29. Albertson, M. The irregularity of a graph. *Ars Comb.* **1997**, *46*, 219–225.

30. Vukičević, D.; Graovac, A. Valence connectivities versus Randić, Zagreb and modified Zagreb index: A linear algorithm to check discriminative properties of indices in acyclic molecular graphs. *Croat. Chem. Acta* **2004**, *77*, 501–508.

31. Abdo, H.; Brandt, S.; Dimitrov, D. The total irregularity of a graph. *Discret. Math. Theor. Comput. Sci.* **2014**, *16*, 201–206.

32. Abdo, H.; Dimitrov, D. The total irregularity of graphs under graph operations. *Miskolc Math. Notes* **2014**, *15*, 3–17. [CrossRef]

33. Abdo, H.; Dimitrov, D. The irregularity of graphs under graph operations. *Discuss. Math. Graph Theory* **2014**, *34*, 263–278. [CrossRef]

34. Gutman, I. Topological Indices and Irregularity Measures. *J. Bull.* **2018**, *8*, 469–475.

35. Reti, T.; Sharfdini, R.; Dregelyi-Kiss, A.; Hagobin, H. Graph irregularity indices used as molecular dicriptors in QSPR studies. *MATCH Commun. Math. Comput. Chem.* **2018**, *79*, 509–524.

36. Hu, Y.; Li, X.; Shi, Y.; Xu, T.; Gutman, I. On molecular graphs with smallest and greatest zeroth-Corder general randic index. *MATCH Commun. Math. Comput. Chem.* **2005**, *54*, 425–434.

37. Caporossi, G.; Gutman, I.; Hansen, P.; Pavlovic, L. Graphs with maximum connectivity index. *Comput. Biol. Chem.* **2003**, *27*, 85–90. [CrossRef]

38. Li, X.; Gutman, I. Mathematical aspects of Randic, type molecular structure descriptors. In *Mathematical Chemistry Monographs*; University of Kragujevac and Faculty of Science: Kragujevac, Serbia, 2006.

39. Das, K.; Gutman, I. Some properties of the second Zagreb Index. *MATCH Commun. Math. Comput. Chem.* **2004**, *52*, 103–112.

40. Trinajstic, N.N.; Ikolic, S.; Milicevic, A.; Gutman, I. On Zagreb indices. *Kemija Industriji* **2010**, *59*, 577–589.

41. Milicevic, A.; Nikolic, S.; Trinajstic, N. On reformulated Zagreb indices. *Mol. Divers.* **2004**, *8*, 393–399. [CrossRef] [PubMed]

42. Gupta, C.K.; Lokesha, V.; Shwetha, S.B.; Ranjini, P.S. On the symmetric division DEG index of graph. *Southeast Asian Bull. Math.* **2016**, *40*, 59–80.

43. Balaban, A.T. Highly discriminating distance based numerical descriptor. *Chem. Phys. Lett.* **1982**, *89*, 399–404. [CrossRef]

44. Furtula, B.; Graovac, A.; Vukičević, D. Augmented Zagreb index. *J. Math. Chem.* **2010**, *48*, 370–380. [CrossRef]

45. Das, K.C. Atom-bond connectivity index of graphs. *Discret. Appl. Math.* **2010**, *158*, 1181–1188. [CrossRef]

46. Estrada, E.; Torres, L.; Rodríguez, L.; Gutman, I. An atom-bond connectivity index: Modeling the enthalpy of formation of alkanes. *Indian J. Chem.* **1998**, *37A*, 849–855.

47. Iqbal, Z.; Aslam, A.; Ishaq, M.; Aamir, M. Characteristic study of irregularity measures of some Nanotubes. *Can. J. Phys.* **2019**. [CrossRef]

48. Gao, W.; Aamir, M.; Iqbal, Z.; Ishaq, M.; Aslam, A. On Irregularity Measures of Some Dendrimers Structures. *Mathematics* **2019**, *7*, 271. [CrossRef]

49. Gao, W.; Abdo, H.; Dimitrov, D. On the irregularity of some molecular structures. *Can. J. Chem.* **2017**, *95*, 174–183.

50. Hussain, Z.; Rafique, S.; Munir, M.; Athar, M.; Chaudhary, M.; Ahmad, H.; Min Kang, S. Irregularity Molecular Descriptors of Hourglass, Jagged-Rectangle, and Triangular Benzenoid Systems. *Processes* **2019**, *7*, 413. [CrossRef]

Article

Comparison of Irregularity Indices of Several Dendrimers Structures

Dongming Zhao [1], Zahid Iqbal [2], Rida Irfan [3], Muhammad Anwar Chaudhry [4], Muhammad Ishaq [2], Muhammad Kamran Jamil [5] and Asfand Fahad [6,*]

[1] School of Automation, Wuhan University of Technology, Wuhan 430070, China; dmzhao@whut.edu.cn
[2] School of Natural Sciences, National University of Sciences and Technology, Sector H-12, Islamabad 44000, Pakistan; 786zahidwarraich@gmail.com (Z.I.); ishaq_maths@yahoo.com (M.I.)
[3] Department of Mathematics, COMSATS University Islamabad, Sahiwal Campus, Sahiwal 57000, Pakistan; ridairfan_88@yahoo.com
[4] Department of Mathematics and Statistics, Institute of Southern Punjab, Multan 66000, Pakistan; chaudhry@bzu.edu.pk
[5] Department of Mathematics, RIPHAH International University, Lahore 54000, Pakistan; m.kamran.sms@gmail.com
[6] Department of Mathematics, COMSATS University Islamabad, Vehari Campus, Vehari 61100, Pakistan
* Correspondence: asfandfahad1@yahoo.com; Tel.: +92-332-607-6976

Received: 21 August 2019; Accepted: 20 September 2019; Published: 27 September 2019

Abstract: Irregularity indices are usually used for quantitative characterization of the topological structures of non-regular graphs. In numerous problems and applications, especially in the fields of chemistry and material engineering, it is useful to be aware of the irregularity of a molecular structure. Furthermore, the evaluation of the irregularity of graphs is valuable not only for quantitative structure-property relationship (QSPR) and quantitative structure-activity relationship (QSAR) studies but also for various physical and chemical properties, including entropy, enthalpy of vaporization, melting and boiling points, resistance, and toxicity. In this paper, we will restrict our attention to the computation and comparison of the irregularity measures of different classes of dendrimers. The four irregularity indices which we are going to investigate are σ irregularity index, the irregularity index by Albertson, the variance of vertex degrees, and the total irregularity index.

Keywords: molecular graph; irregularity indices; dendrimers

---

1. Introduction

The rapid growth in the field of medicine has resulted in the production of unknown nanomaterials, crystalline materials, and drugs. To investigate the chemical properties of these compounds, huge efforts of the pharmaceutical researchers are required and are being made. One way to understand it is by using mathematics, and in mathematical chemistry many concepts of graph theory are being used to formulate the mathematical models for chemical phenomena. Molecules and molecular compounds can be considered as graphs if we correspond atoms to vertices and chemical bonds to edges respectively. Such graphs are called molecular graphs. The notion of topological indices (TIs) helps the pharmacists by providing some information based upon the structures of materials, which reduce their workload. Computing the TIs of a compound may help in approximating its medicinal behaviour [1]. With the passage of time, the idea of understanding compounds through TIs gained significant importance in the field of medicine because it does not require chemical-related apparatus to study [2]. TIs are being intensively studied for different graphs, especially for chemical graphs, for example, see [3–6]. TIs can be separated into various classes, specifically distance-based indices, degree-based indices, eigenvalue-based indices, and mixed indices. An important subclass of degree-based indices is the class of irregularity indices that measure the irregularity of the given graph. A topological invariant $TI(G)$ of a graph G is known as the irregularity index if $TI(G) \geq 0$ and $TI(G) = 0$, if and only if, it is a regular graph. Before the article [7], it was considered that the irregularity indices do not play a significant role to predict physico-chemical properties of organic molecules. In [7] authors performed a regression analysis to check and evaluate the applications of different graph irregularity indices for the estimation of physico-chemical properties of octane isomers. They showed that there exist many irregularity indices by which four octane isomer properties such as standard enthalpy of vaporization (DHVAP), Acentric factor (AcenFac), Entropy, and Enthalpy of vaporization (HVAP) can be estimated with a correlation coefficient greater than 0.9. Before proceeding further with the details related to irregularity indices, we include some important definitions.

Throughout the article, we denote vertex set of a graph G by $V(G)$ and edge set by $E(G)$. A regular graph is a graph whose all vertices have the same degree, otherwise it is called the irregular graph. A sequence $c_1, \ldots, c_{n'}$, where $c_i \in \mathbb{Z}^+$ for all $i = 1, \ldots, n'$, is called a degree sequence of a graph G, if a graph G exists with the property that $V(G) = \{v_1, \ldots, v_{n'}\}$ and $d_G(v_j) = c_j$. Let n_j denotes the number of vertices of degree j, where $j = 1, 2, \ldots, n - 1$. Let $e = uv \in E(G)$, the imbalance of e is defined as $imb(e) := |d_G(u) - d_G(v)|$. In 1997, the term "irregularity of a graph G" was introduced by Albertson [8]. It is denoted by $irr(G)$ and defined as follows:

$$irr(G) = \sum_{e \in E(G)} imb(e) \tag{1}$$

This invariant is also known as the third Zagreb index. It follows immediately that a graph has zero irregularity if and only if it is a regular graph. Albertson [8] showed that the irregularity of any graph is an even number. Furthermore, he also proposed upper bounds for irregularity of triangle-free graphs, bipartite graphs, and for trees. The relationships between the matching number and irregularity of unicyclic graphs and trees were examined in [9]. Hansen et al. [10] characterized the graphs with maximal irregularity. Abdo and Dimitrov [11] worked out for the irregularity of graph operations. In 2014, Abdo et al. [12] defined the total irregularity measure of a graph G, which was denoted and detailed as follows:

$$irr_t(G) = \frac{1}{2} \sum_{u,v \in V(G)} |d_G(u) - d_G(v)|. \tag{2}$$

The relationship between irregularity measures, characterization of graphs with extremal irregularity and the smallest graph with the same irregularity indices are explored in [13]. Fath-Tabar [14] set up some new bounds on the Zagreb indices using the irregularity of graphs. For the detail discussions about these graph invariants, we refer [15,16]. Very recently, Gutman et al. [17] introduced the σ irregularity index of a graph G, which is described as:

$$\sigma(G) = \sum_{uv \in E(G)} (d_G(u) - d_G(v))^2. \tag{3}$$

Some properties of this index have been presented in [18,19]. If the order and size of G is n and m, then the variance of G is defined as [20]:

$$Var(G) := \frac{1}{n} \sum_{j=1}^{n-1} n_j \left(j - \frac{2m}{n}\right)^2 = \frac{1}{n} \sum_{j=1}^{n} d_j^2 - \frac{1}{n^2} \left(\sum_{j=1}^{n} d_j\right)^2. \tag{4}$$

As for prerequisite, the reader is expected to be familiar with dendrimers. Recently, it has been noticed that the highly branched macromolecules have exceptionally different properties from the traditional polymers. Their structural properties have a huge impact on their applications. These hyperbranched molecules are called dendrimers. Moreover, the linear growth in the size of dendrimers makes them the ideal delivery vehicle candidates for the study of effects of composition, and size of polymer in the biological properties like cytotoxicity, blood plasma retention time, lipid bilayer interactions, and filtration, see [21] and references therein. These molecules were first discovered (and studied) by E.Buhleier [22], D. Tomalia [23], and G. R. Newkome [24]. There are many known dendrimers with biological properties such as chemical stability, solubility, polyvalency, electrostatic interactions, low cytotoxicity and self-assembling.

Although plenty of work has been executed on the distance and degree based indices of molecular graphs, the analyses of irregularity measures for chemical structures still need attention. In [25–28], the irregularity measures of various chemical structures were investigated. In this work, we are interested in the irregularity indices of the molecular graphs of different types of dendrimers. For some topological aspects of different complex dendrimers structures, we refer the interested reader to [29–37]. We will restrict our attention to three dendrimers Polyamidoamine (PAMAM), Poly(EThyleneAmidoAmine) (PETAA), and poly(PropylETherIMine)(PETIM) dendrimers.

2. Main Results

In this section, we will compute the irregularity indices of several classes of dendrimers. Firstly, we describe the significance of PAMAM dendrimers. Furthermore, we will also compute four irregularity indices for PAMAM dendrimers.

2.1. Irregularity Measures of PAMAM Dendrimers

The PAMAM dendrimers are a family of dendrimers which is made of repetitively branched subunits of amide and amine functionality. PAMAM dendrimers have hydrophilic interiors and exteriors, which play a role for its unimolecular micelle properties. Moreover, PAMAM-based carriers increase the possibility of bioavailability of problematic drugs. Hence, PAMAM nanocarriers enhance the potential of the bioavailability of drugs, which are not so soluble for efflux transporters, see [38,39]. Due to these properties, PAMAM dendrimers were extensively studied since their synthesis in 1985, see [40,41]. Donald A. Tomalia brought in the PAMAM (polyamidoamine) as a novel class of polymers, called starburst polymers [40]. From then on, the study of PAMAM dendrimers has remained the most prominent topic of research due to its various applications in different fields, including biomedical applications. For further detail and biomedical application, we refer [42].

Let $D_1(n)$ be the molecular graph of this dendrimers, where n represents the generation stage of $D_1(n)$. The molecular graph of PAMAM dendrimer has four branches, and the central core has four vertices. The total number of vertices in each branch is $13 \times 2^0 + 13 \times 2^1 + \cdots + 13 \times 2^{n-1} + 3 \times 2^n + 5 = 16 \times 2^n - 8$. Hence, the total number of vertices in this molecular graph is $4(16 \times 2^n - 8) + 4 = 64 \times 2^n - 28$. In [43], it is shown that for a given tree graph G (Tree is a connected graph which has no cycle), it follows that $|E(G)| = |V(G)| - 1$. Since PAMAM dendrimer is a tree graph, the total number of edges is $64 \times 2^n - 29$. The chemical structure of this dendrimer is shown in Figure 1.

Theorem 1. *Let $D_1(n)$ be the molecular graph of PAMAM dendrimer, where $n \geq 0$ is the generation. Then the irregularity indices are given by*

$$
\begin{aligned}
irr(D_1(n)) &= 48 \times 2^n - 22, \\
irr_t(D_1(n)) &= 552 \times 4^n - 476 \times 2^n + 102, \\
\sigma(D_1(n)) &= 64 \times 2^n - 30, \\
Var(D_1(n)) &= \frac{384 \times 4^n - 328 \times 2^n + 69)}{4(16 \times 2^n - 7)^2}.
\end{aligned}
$$

Proof. The order and size of $D_1(n)$ are $64 \times 2^n - 28$ and $64 \times 2^n - 29$, respectively. Let $V_i^3(n)$ be the set of vertices of degree i in $D_1(n)$. We can classify the vertices of $V(D_1(n))$ into three partite sets, the orders of these sets are $|V_1^3(n)| = 12 \times 2^n - 4$, $|V_2^3(n)| = 40 \times 2^n - 18$ and $|V_3^3(n)| = 12 \times 2^n - 6$.

Now, let $E_{jk}^3(n) \subset E(D_1(n))$ be the set of edges that have end vertices of degrees j and k. We make partition of the edge set $E(D_1(n))$ on the basis of degrees of end vertices of each edge which yields four subsets. The cardinalities of these partite subsets are $|E_{12}^3(n)| = 4 \times 2^n$, $|E_{13}^3(n)| = 8 \times 2^n - 4$, $|E_{22}^3(n)| = 24 \times 2^n - 11$, and $|E_{23}^3(n)| = 28 \times 2^n - 14$.

Followed by the above information and the expressions of irregularity indices manifested in Equations (1)–(4), the explicit formulas of these indices can be obtained in the following way:

$$
\begin{aligned}
irr(D_1(n)) &= \sum_{uv \in E(D_1(n))} |d_{D_1(n)}(u) - d_{D_1(n)}(v)| \\
&= \left(\sum_{uv \in E_{12}^3(n)} + \sum_{uv \in E_{13}^3(n)} + \sum_{uv \in E_{23}^3(n)} \right) |d_{D_1(n)}(u) - d_{D_1(n)}(v)| \\
&= 48 \times 2^n - 22.
\end{aligned}
$$

$$
\begin{aligned}
irr_t(D_1(n)) &= \frac{1}{2} \sum_{u,v \in V(D_1(n))} |d_{D_1(n)}(u) - d_{D_1(n)}(v)| \\
&= \frac{1}{2}\Big((12 \times 2^n - 4)(40 \times 2^n - 18) + (12 \times 2^n - 4)(12 \times 2^n - 6) \\
&\quad + (12 \times 2^n - 6)(40 \times 2^n - 18) \Big) \\
&= 552 \times 4^n - 476 \times 2^n + 102.
\end{aligned}
$$

$$
\begin{aligned}
\sigma(D_1(n)) &= \sum_{uv \in E(D_1(n))} (d_{D_1(n)}(u) - d_{D_1(n)}(v))^2 \\
&= \left(\sum_{uv \in E_{12}^3(n)} + \sum_{uv \in E_{13}^3(n)} + \sum_{uv \in E_{23}^3(n)} \right) (d_{D_1(n)}(u) - d_{D_1(n)}(v))^2 \\
&= 64 \times 2^n - 30.
\end{aligned}
$$

$$
\begin{aligned}
Var(D_1(n)) &= \frac{1}{n'} \sum_{v \in V(D_1(n))} d^2_{D_1(n)}(v) - \frac{1}{n'^2} \left(\sum_{v \in V(D_1(n))} d_{D_1(n)}(v) \right)^2 \\
&= \frac{1}{n'} \left(\sum_{v \in V_1^3} d^2_{D_1(n)}(v) + \sum_{v \in V_2^3} d^2_{D_1(n)}(v) + \sum_{v \in V_3^3} d^2_{D_1(n)}(v) \right) \\
&\quad - \frac{1}{n'^2} \left(\sum_{v \in V_1^3} d_{D_1(n)}(v) + \sum_{v \in V_2^3} d_{D_1(n)}(v) + \sum_{v \in V_3^3} d_{D_1(n)}(v) \right)^2 \\
&= \frac{1}{64 \times 2^n - 28}\big((1)^2(12 \times 2^n - 4) + (2)^2(40 \times 2^n - 18) \\
&\quad + (3)^2(12 \times 2^n - 6) \big) - \frac{1}{(64 \times 2^n - 28)^2}\big(((1)(12 \times 2^n - 4)) \\
&\quad + (2)(40 \times 2^n - 18) + (3)(12 \times 2^n - 6) \big)^2.
\end{aligned}
$$

By means of simple calculations, we derive that

$$Var(D_1(n)) = \frac{384 \times 4^n - 328 \times 2^n + 69)}{4(16 \times 2^n - 7)^2}.$$

□

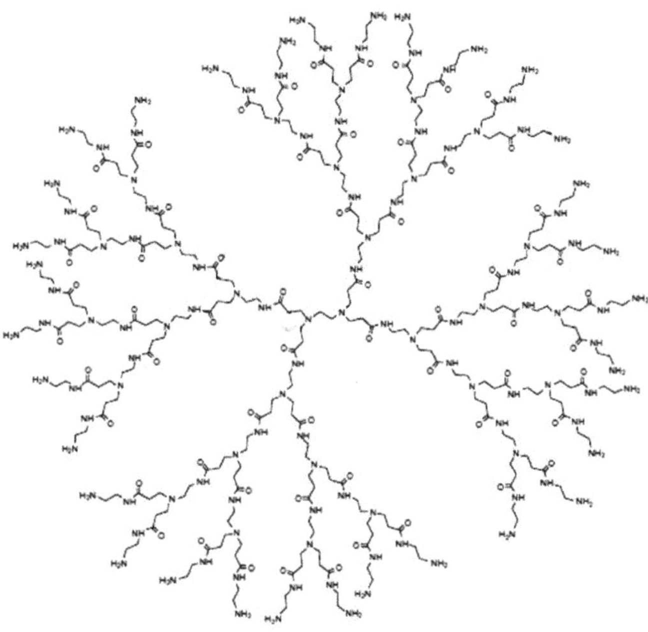

Figure 1. $D_1(n)$ with $n = 3$.

The values of the computed irregularity indices against the different generation stages of PAMAM dendrimers are shown in the Table 1.

Table 1. The values of irregularity indices of structure of Polyamidoamine (PAMAM) dendrimers against different generation stages.

Growth Stage	$irr(D_1(n))$	$irr_t(D_1(n))$	$\sigma(D_1(n))$	$Var(D_1(n))$
$D_1(1)$	74	1358	98	949/2500
$D_1(2)$	170	7030	226	4901/12,996
$D_1(3)$	362	31,622	482	22,021/58,564
$D_1(4)$	746	133,798	994	93,125/248,004
$D_1(5)$	1514	550,118	2018	382,789/1,020,100

The complex synthesis of PAMAM limits the clinical translation of the materials based on PAMAM. Interestingly, PolyEThyleneAmidoAmine (PETAA) dendrimers with more uniform and complete structure then PAMAM possesses several properties of PAMAM. In the next subsection, we will investigate four irregularity indices of PETAA dendrimers.

2.2. Irregularity Measures of PolyEThyleneAmidoAmine (PETAA) Dendrimers

PolyEThyleneAmidoAmine (PETAA) dendrimers have various properties of PAMAM dendrimers such as the number of bonds between the surface, and the dendrimer core is the same. Moreover, the number of surface primary amino groups and tertiary amino groups in PETAA dendrimers are also the same as in PAMAM dendrimers. Other than that, the unique synthesis process of the PETAA enhances its potential for the large-scale production, which results in more application in biomedical sciences [44]. Consequently, the study of PETAA becomes a very respected topic of research. The molecular graph of this dendrimers is denoted by $D_2(n)$, where n represents the generation of $D_2(n)$. The number of vertices in $D_2(n)$ is $44 \times 2^n - 19$ and the number of edges is $44 \times 2^n - 18$. The molecular graph $D_2(n)$ for $n = 5$ is shown in Figure 2.

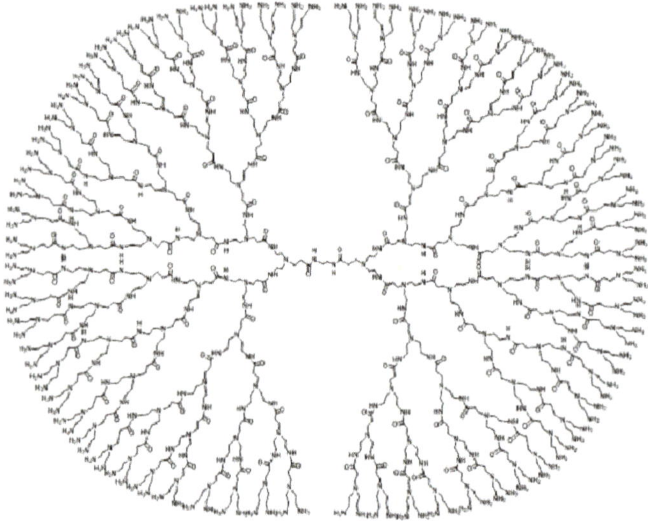

Figure 2. Chemical structure of Poly(EThyleneAmidoAmine) (PETAA) dendrimer $D(5)$.

Theorem 2. *For the molecular graph of PETAA dendrimers $D_2(n)$, the irregularity indices are the following:*

$$
\begin{aligned}
irr(D_2(n)) &= 32 \times 2^n - 13, \\
irr_t(D_2(n)) &= 256 \times 4^n - 204 \times 2^n + 40, \\
\sigma(D_2(n)) &= 40 \times 2^n - 17, \\
Var(D_2(n)) &= \frac{2(88 \times 4^n - 69 \times 2^n + 13)}{(22 \times 2^n - 9)^2}.
\end{aligned}
$$

Proof. The order and the size of $D_2(n)$ are $44 \times 2^n - 18$ and $44 \times 2^n - 19$ respectively. Let $V_i^2(n)$ be the set of vertices of degree i in $D_2(n)$. We can classify the vertices of $V(D_2(n))$ into three partite sets; the orders of these sets are $|V_1^2(n)| = 8 \times 2^n - 2$, $|V_2^2(n)| = 28 \times 2^n - 12$, and $|V_3^2(n)| = 8 \times 2^n - 4$.

Now, let $E_{jk}^2(n) \subset E(D_2(n))$ be the set of edges that have end vertices of degrees j and k. There are four types of edges in $E(D_2(n))$ based on the degrees of end vertices of each edge. The cardinalities of these partite sets are $|E_{12}^2(n)| = 4 \times 2^n$, $|E_{13}^2(n)| = 4 \times 2^n - 2$, $|E_{22}^2(n)| = 16 \times 2^n - 8$, and $|E_{23}^2(n)| = 20 \times 2^n - 9$.

Now, with the help of vertex and edge partitions and Equations (1)–(4), the irregularity indices can be computed in the following manner:

$$
\begin{aligned}
irr(D_2(n)) &= \sum_{uv \in E(D_2(n))} |d_{D_2(n)}(u) - d_{D_2(n)}(v)| \\
&= \left(\sum_{uv \in E^2_{12}(n)} + \sum_{uv \in E^2_{13}(n)} + \sum_{uv \in E^2_{23}(n)} \right) |d_{D_2(n)}(u) - d_{D_2(n)}(v)| \\
&= 32 \times 2^n - 13.
\end{aligned}
$$

$$
\begin{aligned}
irr_t(D_2(n)) &= \frac{1}{2} \sum_{u,v \in V(D_2(n))} |d_{D_2(n)}(u) - d_{D_2(n)}(v)| \\
&= \frac{1}{2} \Big((8 \times 2^n - 2)(28 \times 2^n - 12) + (8 \times 2^n - 2)(8 \times 2^n - 4) \\
&\quad + (8 \times 2^n - 4)(28 \times 2^n - 12) \Big) \\
&= 256 \times 4^n - 204 \times 2^n + 40.
\end{aligned}
$$

$$
\begin{aligned}
\sigma(D_2(n)) &= \sum_{uv \in E(D_2(n))} (d_{D_2(n)}(u) - d_{D_2(n)}(v))^2 \\
&= \left(\sum_{uv \in E^2_{12}(n)} + \sum_{uv \in E^2_{13}(n)} + \sum_{uv \in E^2_{23}(n)} \right) (d_{D_2(n)}(u) - d_{D_2(n)}(v))^2 \\
&= 40 \times 2^n - 17.
\end{aligned}
$$

$$
\begin{aligned}
Var(D_2(n)) &= \frac{1}{n'} \sum_{v \in V(D_2(n))} d^2_{D_2(n)}(v) - \frac{1}{n'^2} \left(\sum_{v \in V(D_2(n))} d_{D_2(n)}(v) \right)^2 \\
&= \frac{1}{n'} \left(\sum_{v \in V^2_1} d^2_{D_2(n)}(v) + \sum_{v \in V^2_2} d^2_{D_2(n)}(v) + \sum_{v \in V^2_3} d^2_{D_2(n)}(v) \right) \\
&\quad - \frac{1}{n'^2} \left(\sum_{v \in V^2_1} d_{D_2(n)}(v) + \sum_{v \in V^2_2} d_{D_2(n)}(v) + \sum_{v \in V^2_3} d_{D_2(n)}(v) \right)^2 \\
&= \frac{1}{44 \times 2^n - 18} \big((1)^2(8 \times 2^n - 2) + (2)^2(28 \times 2^n - 12) \\
&\quad + (3)^2(8 \times 2^n - 4) \big) - \frac{1}{(44 \times 2^n - 18)^2} \big(((1)(8 \times 2^n - 2)) \\
&\quad + (2)(28 \times 2^n - 12) + (3)(8 \times 2^n - 4) \big)^2.
\end{aligned}
$$

After simplification, we get

$$
Var(D_2(n)) = \frac{2(88 \times 4^n - 69 \times 2^n + 13)}{(22 \times 2^n - 9)^2}.
$$

$\square$

The Table 2 describes the values of the computed irregularity indices against the different generation stages of PETAA dendrimers.

Table 2. The values of irregularity indices of structure of PETAA dendrimers against different generation stages.

Growth Stage	$irr(D_2(n))$	$irr_t(D_2(n))$	$\sigma(D_2(n))$	$Var(D_2(n))$
$D_2(1)$	51	656	63	454/1225
$D_2(2)$	115	3320	143	2290/6241
$D_2(3)$	243	14,792	303	10,186/27,889
$D_2(4)$	499	62,312	623	42,874/117,649
$D_2(5)$	1011	255,656	1263	175,834/483,025

In 2003, another dendrimer poly (propyl ether imine) (PETIM) was synthesized [45] and reported as a carrier for the sustained delivery of the drug ketoprofen [46]. In the last subsection, we are going to highlight the irregularity-based topological indices of this dendrimers.

2.3. Irregularity Measures of Poly (propyl ether imine) (PETIM) Dendrimers

Zidovudine (AZT) is the first antiretroviral drug approved by the Food and Drug Administration (FDA) for the treatment of Acquired Immune Deficiency Syndrome (AIDS) [47]. In a recent investigation, AZT-loaded PETIM dendrimer for its sustained drug delivery was studied. The findings of the present study revealed that PETIM dendrimer is a better alternative for sustained drug delivery of zidovudine in comparison to present conventional therapy [48]. Let $D_3(n)$ be the molecular graph of this dendrimers, where n represents the generation stage of $D_3(n)$. The molecular graph $D_3(n)$ for $n = 5$ is shown in Figure 3.

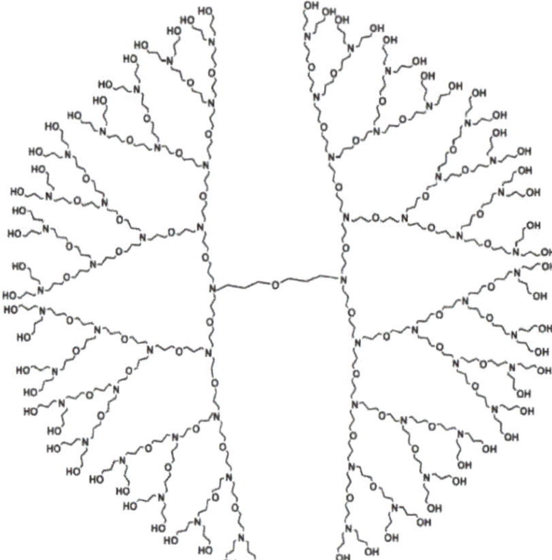

Figure 3. $D_3(n)$ with $n = 5$.

Theorem 3. *For the molecular graph* $D_3(n)$, *the irregularity indices can be represented by the following formulas:*

$$
\begin{aligned}
irr(D_3(n)) &= 8 \times 2^n - 6 = \sigma(D_3(n)), \\
irr_t(D_3(n)) &= 40 \times 4^n - 62 \times 2^n + 21, \\
\sigma(D_3(n)) &= 8 \times 2^n - 6, \\
Var(D_3(n)) &= \frac{96 \times 4^n - 140 \times 2^n + 42}{(24 \times 2^n - 23)^2}.
\end{aligned}
$$

Proof. The order and the size of $D_3(n)$ are $3 \times 2^{n+3} - 23$ and $24(2^n - 1)$ respectively. Let $V_i^1(n)$ be the set of vertices of degree i in $D_3(n)$. We can classify the vertices of $V(D_3(n))$ into three partite sets, the orders of these sets are $|V_1^1(n)| = 2^{n+1}$, $|V_2^1(n)| = 5 \times 2^{n+2} - 21$, and $|V_3^1(n)| = 2^{n+1} - 2$.

Now, let $E_{jk}^1(n) \subset E(D_3(n))$ be the set of edges that have end vertices of degrees j and k. There are three types of edges in $E(D_3(n))$ based on the degrees of end vertices of each edge. The cardinalities of these partite sets are $|E_{12}^1(n)| = 2^{n+1}$, $|E_{22}^1(n)| = 2(2 \times 2^{n+3} - 9)$, and $|E_{23}^1(n)| = 6(2^n - 1)$.

Followed by the above information, and the expressions of irregularity indices manifested in Equations (1)–(4), the explicit formulae of these indices can be formulated as follows:

$$
\begin{aligned}
irr(D_3(n)) &= \sum_{uv \in E(D_3(n))} |d_{D_3(n)}(u) - d_{D_3(n)}(v)| \\
&= \left(\sum_{uv \in E_{12}^1(n)} + \sum_{uv \in E_{23}^1(n)} \right) |d_{D_3(n)}(u) - d_{D_3(n)}(v)| \\
&= 8 \times 2^n - 6.
\end{aligned}
$$

$$
\begin{aligned}
irr_t(D_3(n)) &= \frac{1}{2} \sum_{u,v \in V(D_3(n))} |d_{D_3(n)}(u) - d_{D_3(n)}(v)| \\
&= \frac{1}{2} \left((2^{n+1})(5 \times 2^{n+2} - 21) + (5 \times 2^{n+2} - 21)(2^{n+1} - 2) \right) \\
&= 40 \times 4^n - 62 \times 2^n + 21.
\end{aligned}
$$

$$
\begin{aligned}
\sigma(D_3(n)) &= \sum_{uv \in E(D_3(n))} (d_{D_3(n)}(u) - d_{D_3(n)}(v))^2 \\
&= \left(\sum_{uv \in E_{12}^1(n)} + \sum_{uv \in E_{23}^1(n)} \right) (d_{D_3(n)}(u) - d_{D_3(n)}(v))^2 \\
&= 8 \times 2^n - 6.
\end{aligned}
$$

$$
\begin{aligned}
Var(D_3(n)) &= \frac{1}{n'}\sum_{v\in V(D_3(n))} d^2_{D_3(n)}(v) - \frac{1}{n'^2}\left(\sum_{v\in V(D_3(n))} d_{D_3(n)}(v)\right)^2 \\
&= \frac{1}{n'}\left(\sum_{v\in V_1^1} d^2_{D_3(n)}(v) + \sum_{v\in V_2^1} d^2_{D_3(n)}(v) + \sum_{v\in V_3^1} d^2_{D_3(n)}(v)\right) \\
&\quad - \frac{1}{n'^2}\left(\sum_{v\in V_1^1} d_{D_3(n)}(v) + \sum_{v\in V_2^1} d_{D_3(n)}(v) + \sum_{v\in V_3^1} d_{D_3(n)}(v)\right)^2 \\
&= \frac{1}{3\times 2^{n+3}-23}\left((1)^2(2^{n+1}) + (2)^2(5\times 2^{n+2} - 21)\right. \\
&\quad + (3)^2(2^{n+1}-2)) - \frac{1}{(3\times 2^{n+3}-23)^2}(((1)(2^{n+1})) \\
&\quad + (2)(5\times 2^{n+2}-21) + (3)(2^{n+1}-2))^2.
\end{aligned}
$$

Further simplifications yield

$$
Var(D_3(n)) = \frac{96\times 4^n - 140\times 2^n + 42}{(24\times 2^n - 23)^2}.
$$

$\square$

The Table 3 presents the values of the computed irregularity indices against the different generation stages of PETIM dendrimers. In the next section, we present some graphical analysis of irregularity measures (of the structures under discussion) based upon the results obtained in this section.

Table 3. The values of irregularity indices of structure of poly (propyl ether imine) (PETIM) dendrimers against different generation stages.

Growth Stage	$irr(D_3(n))$	$irr_t(D_3(n))$	$\sigma(D_3(n))$	$Var(D_3(n))$
$D_3(1)$	10	57	10	146/625
$D_3(2)$	26	413	26	1018/5329
$D_3(3)$	58	2085	58	5066/28,561
$D_3(4)$	122	9269	122	22,378/130,321
$D_3(5)$	250	38,997	250	93,866/55,5025

3. Graphical Analysis and Discussions

In this section, we present our theoretical outcomes and deduce which of the dendrimer structures depicted above is more irregular than the remaining ones with respect to a specific irregularity index. Each dendrimer structure relies on a single variable n. We plot the graph with irregularity index of $D_1(n)$, $D_2(n)$, and $D_3(n)$ (on a single graph) as a dependent variable and n as an independent variable. We observe that all the irregularity indices behave in a particular way with the increase in the value of n. We provided the graphical appearances of these indices with respect to the change in the growth stage of the dendrimer structure. In this graphical examination, the blue colour shows the graphical functioning of the PETIM dendrimer structure, the red colour shows the graphical behaviour of the PETAA dendrimer structure, and the green colour shows the graphical response of the PAMAM dendrimer structure. We present the values of various irregularity indices for the dendrimer structures in Tables 1–3. With the help of these tables, the comparative behavior of these irregularity indices for dendrimer structures has been expressed in Figures 4–7. These figures show that the molecular structure of PAMAM dendrimer is highly irregular as compared to the other dendrimer structures and the molecular structure of PETAA dendrimer is more irregular than the molecular structure of PETIM dendrimer.

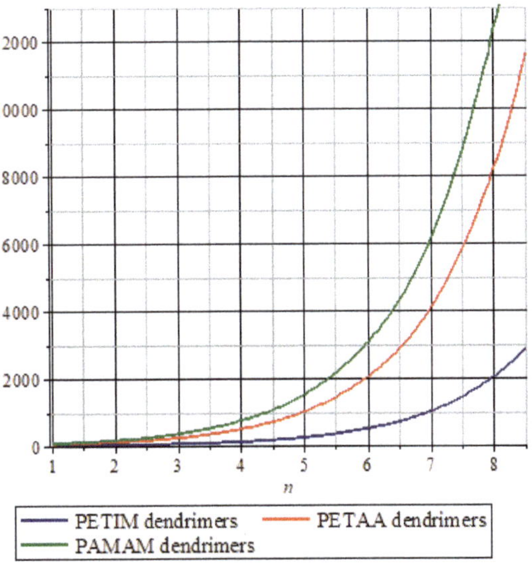

Figure 4. Albertson irregularity index.

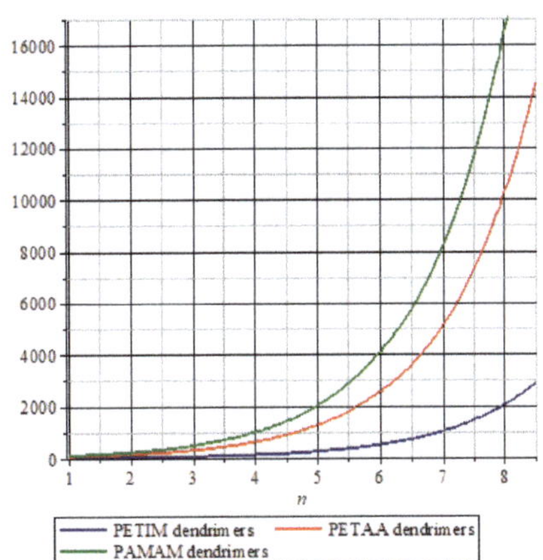

Figure 5. σ irregularity index.

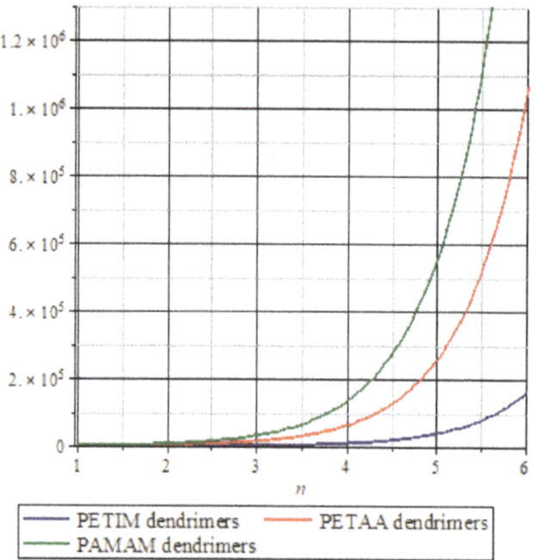

Figure 6. Total irregularity index.

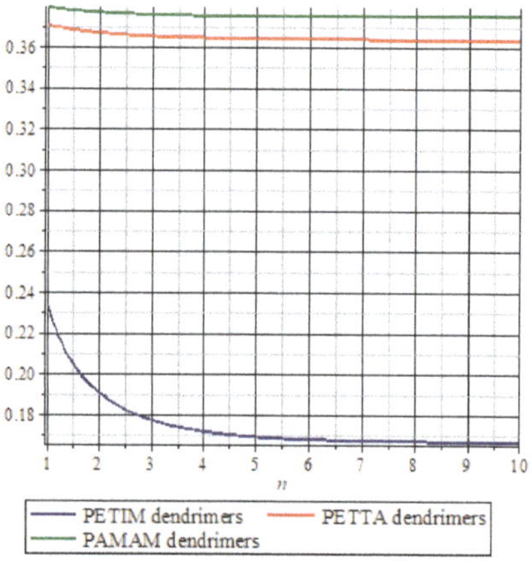

Figure 7. Variance index.

4. Conclusions

A modern trend in QSAR/QSPR studies is the use of properties which can be secure from the molecular structure without any other input data. The basic reason for this is to forecast the chemical properties of such a huge collection of compounds and drugs takes a large number of chemical inspections, thereby these tasks increase the burden of work of the chemical and pharmaceutical researchers. Thus, the strategy of estimating the topological indices has offered the explanations of such medicinal behaviour of various compounds and drugs. Hence, the irregularity indices for the molecular graphs of dendrimers are demonstrated by a mathematical derivation method and we presented the comparison of the molecular structures by using the graph structures. Our outcomes could perform a significant role in estimating and comparing the properties of these molecular structures.

Author Contributions: The contribution of the authors is as follows: Conceptualization, Z.I. and M.K.J.; methodology, Z.I.; validation, R.I., A.F. and M.A.C.; investigation, Z.I.; M.K.J.; writing–original draft preparation, Z.I., A.F., R.I.; writing—review and editing, D.Z., M.I.

Funding: This research received no external funding.

Acknowledgments: All the authors are thankful to their respective institutes for providing the research facilities.

Conflicts of Interest: The authors declare no conflict of interest.

References

1. Gozalbes, R.; Doucet, J.P.; Derouin, F. Application of topological descriptors in QSAR and drug design: History and new trends. *Curr. Drug Targets Infect. Disord.* **2002**, *2*, 93–102. [CrossRef] [PubMed]
2. Amic, D.; Beaa, D.; Lucic, B.; Nikolic, S.; Trinajstic, N. The vertex connectivity index revisited. *J. Chem. Inf. Comput. Sci.* **1998**, *38*, 819–822. [CrossRef]
3. Gao, Y.; Zhu, E.; Shao, Z.; Gutman, I.; Klobucar, A. Total domination and open packing in some chemical graphs. *J. Math. Chem.* **2018**, *5*, 1481–1492. [CrossRef]
4. Nasiri, R.; Fath-Tabar, G.H. The second minimum of the irregularity of graphs. *Electron. Notes Discret. Math.* **2014**, *45*, 133–140. [CrossRef]
5. Shao, Z.; Wu, P.; Zhang, X.; Dimitrov, D.; Liu, J.B. On the maximum ABC index of graphs with prescribed size and without pendent vertices. *IEEE Access* **2018**, *6*, 27604–27616. [CrossRef]
6. Ye, A.; Qureshi, M.I.; Fahad, A.; Aslam, A.; Jamil, M.K.; Zafar, A.; Irfan, R. Zagreb connection number index of nanotubes and regular hexagonal lattice. *Open Chem.* **2019**, *17*, 75–80. [CrossRef]
7. Réti, T.; Sharafdini, R.; Dregelyi-Kiss, A.; Haghbin, H. Graph irregularity indices used as molecular descriptor in QSPR studies. *Match Commun. Math. Comput. Chem.* **2018**, *79*, 509–524.
8. Albertson, M.O. The irregularity of a graph. *Ars Combin.* **1997**, *46*, 219–225.
9. Luo, W.; Zhou, B. On the irregularity of trees, unicyclic graphs with given matching number. *Util. Math.* **2010**, *83*, 141–147.
10. Hansen, P.; Mélot, H. Variable neighborhood search for extremal graphs. 9. bounding the irregularity of a graph. *Discret. Math. Theor. Comput. Sci.* **2005**, *69*, 253–264.
11. Abdo, H.; Dimitrov, D. The irregularity of graphs under graph operations. *Discuss. Math. Graph Theory* **2014**, *34*, 263–278. [CrossRef]
12. Abdo, H.; Brandt, S.; Dimitrov, D. The total irregularity of a graph. *Discret. Math. Theor. Comput. Sci.* **2014**, *16*, 201–206.
13. Dimitrov, D.; Škrekovski, R. Comparing the irregularity, the total irregularity of graphs. *Ars Math. Contemp.* **2015**, *9*, 45–50. [CrossRef]
14. Fath-Tabar, G.H. Old, new Zagreb indices of graphs. *Match Commun. Math. Comput. Chem.* **2011**, *65*, 79–84.
15. Abdo, H.; Cohen, N.; Dimitrov, D. Graphs with maximal irregularity. *Filomat* **2014**, *28*, 1315–1322. [CrossRef]
16. Nasiri, R.; Ellahi, H.R.; Gholami, A.; Fath-Tabar, G.H. The irregularity, total irregularity of Eulerian graphs. *Iran. J. Math. Chem.* **2018**, *9*, 101–111.
17. Gutman, I.; Togan, M.; Yurttas, A.; Cevik, A.S.; Cangul, I.N. Inverse problem for sigma index. *Match Commun. Math. Comput. Chem.* **2018**, *79*, 491–508.
18. Abdo, H.; Dimitrov, D.; Gutman, I. Graphs with maximal σ irregularity. *Discret. Appl. Math.* **2018**, *250*, 57–64. [CrossRef]
19. Gutman, I. Stepwise irregular graphs. *Appl. Math. Comput.* **2018**, *325*, 234–238. [CrossRef]
20. Bell, F.K. A note on the irregularity of graphs. *Linear Algebra Appl.* **1992**, *161*, 45–54. [CrossRef]
21. Abbasi, F.; Aval, S.F.; Akbarzadeh, A.; Milani, M.; Nasrabadi, H.T.; Joo, S.W.; Hanifehour, Y.; Koshki, K.N.; Pashei-Asl, R. Dendrimers: Synthesis, applications, and properties. *Nano Scale Res. Lett.* **2014**, *247*, 1–10. [CrossRef]
22. Buhleier, E.; Wehner, W.; Vogtle, F. Cascade and nonskid chain like syntheses of molecular cavity topologies. *Synthesis* **1978**, *2*, 155–158. [CrossRef]
23. Tomalia, D.A.; Baker, H.; Dewald, J.R.; Hall, M.; Kallos, G.; Martin, S.; Roeck, J.; Ryder, J.; Smith, P. Dendritic macromolecules: Synthesis of starburst dendrimers. *Macromolecules* **1986**, *19*, 2466–2468. [CrossRef]
24. Newkome, G.R.; Yao, Z.; Baker, G.R.; Gupta, V.K. Micelles. Part 1. Cascade molecules: A new approach to micelles. A [27]-arborol. *J. Org. Chem.* **1985**, *50*, 2003–2004. [CrossRef]
25. Abdo, H.; Dimitrov, D.; Gao, W. On the irregularity of some molecular structures. *Can. J. Chem.* **2017**, *95*, 174–183. [CrossRef]
26. Gao, W.; Aamir, M.; Iqbal, Z.; Ishaq, M.; Aslam, A. On irregularity measures of some dendrimers structures. *Mathematics* **2019**, *7*, 271. [CrossRef]
27. Gutman, I.; Hansen, P.; Mélot, H. Variable neighborhood search for extremal graphs 10. Comparison of irregularity indices for chemical trees. *J. Chem. Inf. Model.* **2005**, *45*, 222–230. [CrossRef]

28. Iqbal, Z.; Aslam, A.; Ishaq, M.; Aamir, M. Characteristic study of irregularity measures of some nanotubes. *Can. J. Phys.* **2019**. [CrossRef]

29. Gao, W.; Iqbal, Z.; Ishaq, M.; Sarfraz, R.; Aamir, M.; Aslam, A. On eccentricity-based topological indices study of a class of porphyrin-cored dendrimers. *Biomolecules* **2018**, *8*, 71. [CrossRef]

30. Gao, W.; Iqbal, Z.; Ishaq, M.; Aslam, A.; Aamir, M.; Binyamin, M.A. Bounds on topological descriptors of the corona product of *F*-sum of connected graphs. *IEEE Access* **2019**, *7*, 26788–26796. [CrossRef]

31. Gao, W.; Iqbal, Z.; Ishaq, M.; Aslam, A.; Sarfraz, R. Topological aspects of dendrimers via distance-based descriptors. *IEEE Access* **2019**, *7*, 35619–35630. [CrossRef]

32. Hussain, Z.; Munir, M.; Rafique, S.; Hussnain, T.; Ahmad, H.; Kwun, Y.C.; Kang, S.M. Imbalance-based irregularity molecular descriptors of nanostar dendrimers. *Processes* **2019**, *7*, 517. [CrossRef]

33. Iqbal, Z.; Aamir, M.; Ishaq, M.; Shabri, A. On theoretical study of Zagreb indices and Zagreb polynomials of water-soluble perylenediimide-cored dendrimers. *J. Inf. Math. Sci.* **2018**, *10*, 647–657.

34. Iqbal, Z.; Ishaq, M.; Aamir, M. On eccentricity-based topological descriptors of dendrimers. *Iran. J. Sci. Technol. Trans. Sci.* **2019**, *43*, 1523–1533. [CrossRef]

35. Iqbal, Z.; Ishaq, M.; Aslam, A.; Gao, W. On eccentricity-based topological descriptors of water-soluble dendrimers. *Z. Nat.* **2018**, *74*, 25–33. [CrossRef]

36. Kang, S.M.; Iqbal, Z.; Ishaq, M.; Sarfraz, R.; Aslam, A.; Nazeer, W. On eccentricity-based topological indices and polynomials of phosphorus-containing dendrimers. *Symmetry* **2018**, *10*, 237. [CrossRef]

37. Zheng, J.; Iqbal, Z.; Fahad, A.; Zafar, A.; Aslam, A.; Qureshi, M.I.; Irfan, R. Some eccentricity-based topological indices and polynomials of poly(EThyleneAmidoAmine) (PETAA) dendrimers. *Processes* **2019**, *7*, 433. [CrossRef]

38. Tomalia, D.A.; Baker, H.; Dewald, J.R.; Hall, M.; Kallos, G.; Martin, S.; Roeck, J.; Ryder, J.; Smith, P. Dendrimers II: Architecture, nanostructure and supramolecular chemistry. *Macromolecules* **1986**, *19*, 2466.

39. Froehling, P.E. Dendrimers and dyes a review. *Dyes Pigment.* **2001**, *48*, 187–195. [CrossRef]

40. Tomalia, D.A.; Baker, H.; Dewald, J.R.; Hall, M.; Kallos, G.; Martin, S.; Roeck, J.; Ryder, J.; Smith, P. A new class of polymers: Starburst-dendritic macromolecules. *Polym. J.* **1985**, *17*, 117–132. [CrossRef]

41. Alper, J. Rising chemical stars could play many roles. *Science* **1991**, *251*, 1562–1564. [CrossRef]

42. Araújo, R.V.; Santos, S.S.; Ferreira, E.I.; Giarolla, J. New advances in general biomedical applications of PAMAM dendrimers. *Molecules* **2018**, *23*, 2849.

43. Bondy, J.A.; Murty, U.S.R. *Graph Theory*; Springer: London, UK, 2008.

44. Huang, B.; Tang, S.; Desai, A.; Lee, K.H.; Leroueil, P.R.; Baker, J.R. Novel Poly(EthyleneAmidoAmine) (PETAA) dendrimers produced through a unique and highly efficient synthesis. *Polymer* **2011**, *52*, 5975–5984. [CrossRef]

45. Ramakrishna, T.; Jayaraman, N. Synthesis of poly(propyl ether imine) dendrimers and evaluation of their cytotoxic properties. *J. Org. Chem.* **2003**, *69*, 9694–9704.

46. Jain, S.; Kaur, A.; Puri, R.; Utreja, P.; Jain, A.; Bhide, M.; Ratnam, R.; Singh, V.; Patil, A.S.; Jayaraman, N. Poly propyl ether imine (PETIM) dendrimer, A novel non-toxic dendrimer for sustained drug delivery. *Eur. J. Med. Chem.* **2010**, *45*, 4997–5005. [CrossRef]

47. Liang, Z.; Gong, T.; Sun, X.; Tang, J.Z.; Zhang, Z. Chitosan oligomers as drug carriers for renal delivery of zidovudine. *Carbohydr. Polym.* **2012**, *87*, 2284–2290. [CrossRef]

48. Jain, S.K.; Sharma, A.; Mahajan, M.; Sankar, R. In-vitro and in-vivo evaluation of poly (propyl ether imine) (PETIM) dendrimer for sustained delivery of zidovudine. *J. Antivir. Antiretrovir.* **2013**, *5*. [CrossRef]

Article

Numerical Simulations of Polymer Solution Droplet Impact on Surfaces of Different Wettabilities[†]

Moussa Tembely [1,*], Damien Vadillo [2], Arthur Soucemarianadin [3] and Ali Dolatabadi [1,*]

[1]   Department of Mechanical, Industrial, and Aerospace Engineering, Concordia University, Montreal, QC H3G 1M8, Canada

[2]   AkzoNobel, Research, Development and Innovation, Gateshead NE10 0EX, UK; damien.vadillo@akzonobel.com

[3]   Laboratory for Geophysical and Industrial Flow(LEGI), Grenoble University, Grenoble 38000, France; souce@ujf-grenoble.fr

*   Correspondence: moussa.tembely@concordia.ca (M.T.); ali.dolatabadi@concordia.ca (A.D.)

†   This paper is an extended version of a paper published at the 4th Micro and Nano Flows Conference, London, UK, 7–10 September 2014.

Received: 30 August 2019; Accepted: 24 October 2019; Published: 2 November 2019

Abstract: This paper presents a physically based numerical model to simulate droplet impact, spreading, and eventually rebound of a viscoelastic droplet. The simulations were based on the volume of fluid (VOF) method in conjunction with a dynamic contact model accounting for the hysteresis between droplet and substrate. The non-Newtonian nature of the fluid was handled using FENE-CR constitutive equations which model a polymeric fluid based on its rheological properties. A comparative simulation was carried out between a Newtonian solvent and a viscoelastic dilute polymer solution droplet. Droplet impact analysis was performed on hydrophilic and superhydrophobic substrates, both exhibiting contact angle hysteresis. The effect of substrates' wettability on droplet impact dynamics was determined the evolution of the spreading diameter. While the kinematic phase of droplet spreading seemed to be independent of both the substrate and fluid rheology, the recoiling phase seemed highly influenced by those operating parameters. Furthermore, our results implied a critical polymer concentration in solution, between 0.25 and 2.5% of polystyrene (PS), above which droplet rebound from a superhydrophobic substrate could be curbed. The present model could be of particular interest for optimized 2D/3D printing of complex fluids.

Keywords: droplet impact; viscoelasticity; volume of fluid method

---

1. Introduction

The dynamics of the impact and spreading of liquid drops onto a solid substrate is a highly active subject of research for both academic and industrial purposes. Indeed, these phenomena are widely encountered in everyday life, such as raindrops impacting on a surface. From an industrial perspective, there is much interest in the physics of non-Newtonian drop–surface interactions because of their wide range of applications. As such, a detailed knowledge of droplet impingement onto solid materials is required for the overall process development and improvement of many engineering operations. The impact of fluid drops on solid surfaces leads to different outcomes, such as partial or total spreading, recoil, or splashing [1]. One essential condition for the accurate placement of drops—a stringent requirement for technological applications—is the ability to control the maximum spreading diameter through different means. Although the problem has been considered from several angles for over 50 years, it is only during the past 10 years that repeatable and consistent results have been reported both on well-characterized homogenous substrates and on purpose-designed media with either chemical or topological heterogeneities [2,3]. Very recently, Antonini et al. [4] performed a comprehensive study of the impact of drops on hydrophobic and superhydrophobic surfaces so as to help select the appropriate substrate for a given application. Complex fluid droplet impact has gained attention recently due to the key role it plays in applications ranging from 3D printing, polymer light-emitting diode technology, and lab-on-chip technology, to biotechnology for protein engineering [5–10]. The liquids involved in these processes are likely to exhibit non-Newtonian properties, such as viscoelasticity, which results from adding flexible polymers to solvent liquids. In order to address these challenges, numerical tools are essential for the accurate modeling of droplet dynamics on surfaces exhibiting different wettabilities.

Numerical models to handle droplet impact have been investigated since the initial work by Fukai et al. [11] based on the volume of fluid (VOF) method first introduced by Hirt and Nichols [12]. Since then, a wide range of numerical techniques have been developed for computing multiphase flow with moving interfaces, including, for example, the level set method [13], the front tracking method [14], and the lattice–Boltzmann method [15,16]. Among those cited, the VOF method has enjoyed a rather special place in the simulation of drop spreading for many reasons. These include its inherent mass conservation property, its suitability for problems where large surface topology changes occur, the ease of its implementation, and its reduced computational costs. However, it may be less accurate in interface calculations than, for example, the level set method, which is particularly adapted for resolving intricate topological changes of interfaces. In spite of this limitation, it is still the most preferred method for computations of drop impact and spreading where quite strong interfacial effects occur both at the substrate level and at the free surface of the drop [17,18]. To be more complete on the issue of numerical simulations of drop impact, it should be mentioned that the lattice–Boltzmann and phase field methods have recently been used for characterizing drop impact behavior. Additionally, another approach using the coupled level set and volume of fluid (CLSVOF) formulation has been implemented, with the VOF method dealing with the interface motion and the level set technique handling surface tension effects. This approach also remedied the problem of mass conservation typically encountered with level set methods.

A fundamental understanding of the droplet dynamics of complex fluids is a key element for future breakthroughs in the growing domain of micro-fabrication and micro-fluidics [7,9]. Regarding droplet impact, the question of whether a droplet will be deposited on the substrate or will eventually rebound is of particular interest, and a full understanding of the entire droplet impact dynamic is required.

Although Newtonian droplet impact has been extensively studied in the literature, the non-Newtonian viscoelastic droplet impact—despite its plethora of useful applications—has been only loosely documented both experimentally and numerically, even though recent works, showing an increasing interest for these fluids, make them an active research area. A finite element method was pioneered in Reference [19] to model the die swell of an Oldroyd-B fluid through an ad-hoc iterative technique for the free surface, based on the kinematic condition. The split Lagrangian–Eulerian method was extended by [20] to study viscoelastic jet breakup. Recently, Reference [21] simulated free surface flow using both marker-and-cell (MAC) and finite difference (FD) techniques. In addition, a mesh-free method using the smoothed particle hydrodynamics (SPH) technique has been applied to model an Oldroyd-B drop impact in 2D while neglecting the surface tension [22]. These different techniques neglect both the surface tension and the dynamic contact angle during spreading, along with the presence of air. Additionally, it is worth noting the interface-capturing techniques used to model non-Newtonian free surface flow such as the level set method [23] or the phase field method [24]. Finally, a one-dimensional approach [25] to model free surface flow of non-Newtonian fluid using an arbitrary Lagrangian–Eulerian (ALE) method should be mentioned, as it has shown good results in modeling filament thinning using the FENE-CR model.

Although the low viscosity of dilute polymers are of particular interest in inkjet fluids [26,27], it is only recently that they have been experimentally investigated using a controlled stretching rheometer [28]. The present work benefitted from the rheological characterization of these fluids to optimize the simulation.

Most numerical studies have considered polymeric internal flow, but few have considered droplet interaction with a solid substrate. Within the VOF framework, the present model (accounting for the substrate hysteresis) investigated viscoelastic droplets impact using a single mode FENE-CR model in an axisymmetric configuration. The numerical model enabled the effect of monodisperse polymer concentration in solution and the resulting liquid droplet dynamics on both hydrophilic and superhydrophobic surfaces to be investigated.

The paper is organized as follows. Section 2 details the equations governing the numerical model and the polymers solutions' rheological properties. After validation of the model in the Newtonian case, the main results regarding the polymer droplet impact on hydrophilic and superhydrophobic flat surfaces are discussed in Section 3. Section 4 concludes the paper.

2. Methodology

2.1. Numerial Methods

The equations governing the impact of a non-Newtonian droplet are based on the discretization of the mass and momentum equations within the VOF framework. In the present work, the focus was be on modeling the impact and spreading of a droplet of dilute polymer solution which displays a viscoelastic behavior. Therefore, the momentum conservation equation incorporated this effect through an additional stress tensor (τ_p), and the governing equations can be written as follows:

$$\nabla.\boldsymbol{V} = 0 \tag{1}$$

$$\frac{\partial(\rho\boldsymbol{V})}{\partial t} + \nabla.(\rho\boldsymbol{V}\boldsymbol{V}) = -\nabla p + \rho\boldsymbol{g} + \nabla.(\eta_s\nabla\boldsymbol{V}) + \nabla.\boldsymbol{\tau}_p + \gamma\kappa\nabla\alpha \tag{2}$$

where κ is the curvature of the free surface, γ is the surface tension, α is the phase fraction of the liquid phase, η_s is the viscosity of the solvent, g is the gravity acceleration, p is the pressure, and V the velocity.

In Equation (2), the continuum surface force (CSF) method of Brackbill et al. [29] was used to model the surface tension as a body force acting only on interfacial numerical cells, and the mean curvature at the interface is given by

$$\kappa = -\nabla.\left(\frac{\nabla\alpha}{|\nabla\alpha|}\right) \tag{3}$$

The outcome of an impacting drop is affected by various factors including droplet properties and kinematic and surface characteristics. In the present paper, the impact of a droplet of diameter D_0 impacting at an initial velocity of V_0 was simulated. To analyze droplet impact dynamics, we used the spreading factor, defined as the ratio between the spreading diameter and droplet initial diameter (D_0). In addition, the time was made dimensionless using the kinematic time scale $t_c = D_0/V_0$.

In order to model the polymer solution rheology, we used the dumbbell approximation based on a finitely extensible nonlinear elastic fluid in Reference [30], known as FENE-CR. This choice of FENE-CR was motivated by its superior performance for modeling viscoelastic behavior, as we have previously shown in the context of filament thinning and stretching [25,31,32]. The equation relating the stress tensor to conformation tensor A was expressed as follows:

$$\kappa = -\nabla.\boldsymbol{\tau}_p = Gf(\,R\,)\,(\boldsymbol{A} - \boldsymbol{I}) \tag{4}$$

where the elastic modulus is denoted by G and $f(R)$ the finite extensibility factor is given by

$$f(R) = \frac{1}{1 - R/L^2} \tag{5}$$

which relates the finite extensibility parameter L, which corresponds to the maximum possible extension of the dumbbell, to the parameter $R = \mathrm{Tr}(A)$.

The polymer impact on the flow behaviour can be expressed through the constitutive equations based the evolution of the conformation tensor:

$$\frac{dA}{dt} - \nabla V^T.A - A.\nabla v = -\frac{f(R)}{\lambda}(A - I) \tag{6}$$

where λ denotes the relaxation time and I the identity tensor.

Finally, the transport of the phase fraction equation was simulated through an interface compression method:

$$\frac{\partial \alpha}{\partial t} + V\nabla\alpha + \nabla.[V_c\alpha(1 - \alpha)] = 0 \tag{7}$$

The interface compression speed V_c, describing the relative velocity at the free surface between the fluids, followed the equation below [33]:

$$V_c = n_f\,min\left[C_\alpha\frac{\phi_f}{S_f}, max\left(\frac{\phi_f}{S_f}\right)\right] \tag{8}$$

where S_f and ϕ_f represent cell surface area and mass flux, respectively, while the coefficient C_α, set here to 1, defines the degree of compression at the interface. It is worth noting that the adoption of an interface compression scheme avoids the tedious geometrical reconstruction of the interface habitually done in implementation of the VOF method. In addition, the algebraic method used here can readily be extended to unstructured meshes. Further details of the numerical discretization schemes and techniques can be found in Reference [34].

Within the VOF framework, the different physical parameters in each cell of the domain were expressed through the liquid fraction as follows:

$$\xi = \alpha\xi_{liquid} + (1 - \alpha)\xi_{gas} \tag{9}$$

where ξ represents any physical properties, such as the density, velocity, or viscosity, for both the liquid and gas phases.

To handle the liquid–substrate interaction, a more physically-based dynamic contact angle model was used. Droplet impact dynamics are highly controlled by the manner in which its dynamic contact angle is modeled. In the present work, Kistler's dynamic contact angle model [34,35] was implemented:

$$\theta_d = f_H[Ca + f_H^{-1}(\theta_E)] \tag{10}$$

with the Hoffman function expressed as

$$f_H(s) = arccos\left\{1 - 2\tan h\left[5.16\left[\frac{s}{1 + 1.31s^{0.99}}\right]^{0.706}\right]\right\} \tag{11}$$

where the capillary number $Ca = \mu U_{cl}/\gamma$, in which U_{cl} corresponds to the contact line velocity, was numerically approximated by taking the velocity within the first cell above the substrate. The effect of the surface hysteresis was accounted for by replacing the equilibrium contact angle θ_E in Equation (9) by either the receding contact angle θ_R or the advancing contact angle θ_A according to the direction of the triple line velocity. By adopting this approach, our model became sensitive to the substrate hysteresis, which plays a significant role in simulating droplet impacts on surfaces with varied wettabilities. It is worth noting that unlike in Reference [36], the presented model did not rely heavily on experimental measurements of the dynamic contact angle evolution in order to match a given experiment.

Finally, the governing Equations (1)–(11) were implemented in OpenFOAM/C++ using second order linear upwind-biased schemes. The simulations (in axisymmetric geometry) were performed in parallel using the domain decomposition method. The pressure implicit with splitting of operators (PISO) algorithm was used to calculate the pressure, while the evolution of the conformation tensor was solved using a preconditioned bi-conjugate gradient technique; further details on implementation and discretization can be found in our previous work in References [10,34,37]. The convergence criteria set for the pressure, velocity, and conformation tensor fields were of the order of 10^{-6}.

2.2. Polymer Solution and Substrates Properties

In order to evaluate our numerical model, a viscoelastic liquid and its properties were measured based on a"Cambridge Trimaster" filament stretching and thinning experimental set-up [28]. This extensional rheometer proceeded by performing filament stretching at a constant velocity for a liquid initially placed between two pistons of initial diameter 1.2 mm. The two pistons, which operate on opposite sides of a belt, move away from each other at a prescribed distance, letting the mid-filament remain in a central position. The ensuing measurements of the filament mid-diameter allows accurate estimation the relaxation time of the liquid, which has been shown previously to be crucial in numerical modeling of filament thinning dynamics [25].

The numerical simulation investigated a Newtonian solvent—diethyl phthalate (DEP)—in addition to a polymer solution consisting of polystyrene (PS), 110,000 g/mol, dissolved into the DEP solvent. Table 1 presents the interfacial surface tension and viscosity values of the two liquids.

Table 1. Liquid physical properties. DEP: diethyl phthalate, PS: polystyrene.

Liquids	Interfacial Surface Tension (mN/m)	Viscosity (mPa.s)
DEP	37	14
DEP + 2.5% PS	37	31

Finally, the relaxation time measured from the polymer solution, DEP + 2.5 wt % PS, was $\lambda = 1.19$ ms, while the extensibility parameter L, accounting for the liquid physical properties, was taken as $L = 15$ [25].

Regarding fluid–substrate interaction, we considered two types of substrate in the numerical experiment, hydrophilic and superhydrophobic, the properties of which were similar to the aluminum and WX2100 properties for a water droplet, respectively [38]. Experimentally, it may be challenging to find a substrate exhibiting those chemical properties with respect the DEP and DEP + 2.5 wt. % PS, though the aim of this study was to highlight the relevant features in viscoelastic droplet dynamics, which have been poorly numerically documented in the literature, especially with the added effect of substrate hysteresis. The contact angles (CAs) consisting of the equilibrium contact angles, in addition to the hysteresis with the advancing and receding contact angles of these two substrates, are given in Table 2.

Table 2. Substrate contact angles (CAs).

Substrate	Equilibrium CA (°)	Advancing CA (°)
Hydrophilic (H)	74	90
Superhydrophobic (SH)	154	162

3. Results

3.1. Validation of Newtonian Droplet Impact Dynamics

We first tested the model in a Newtonian case for validation. For this purpose, we have provided a comparison of the VOF method using our modified Kistler's model to tackle the challenging case of droplet impact on a surface with contact angle hysteresis (CAH). The comparison was performed with an experiment performed in Antonini et al. [39] for the impact of a 2.5 mm diameter droplet impacting at 1 m/s with receding and advancing contact angles of 108°/169°. The results are shown in Figures 1 and 2 for the transient profile and spreading diameter evolution, respectively. We observed a good agreement between the experiment and simulation of the droplet upon impact on a surface exhibiting a significant contact angle hysteresis, CAH = 61°, highlighting the robustness and capability of our model and paving the way for an accurate assessment of the impact of non-Newtonian fluids.

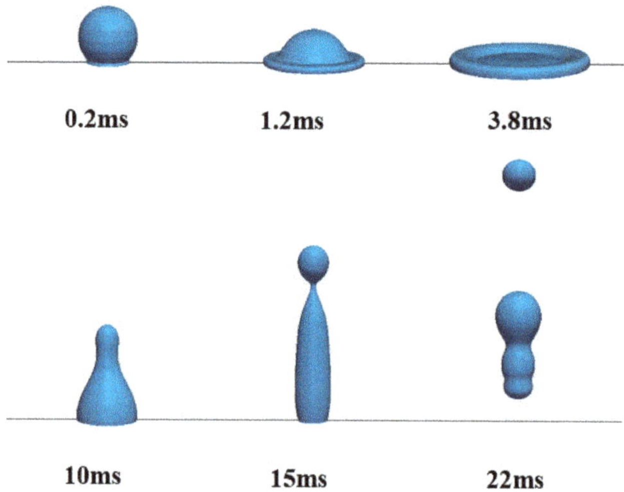

Figure 1. Transient profiles of a Newtonian droplet impacting a superhydrophobic surface.

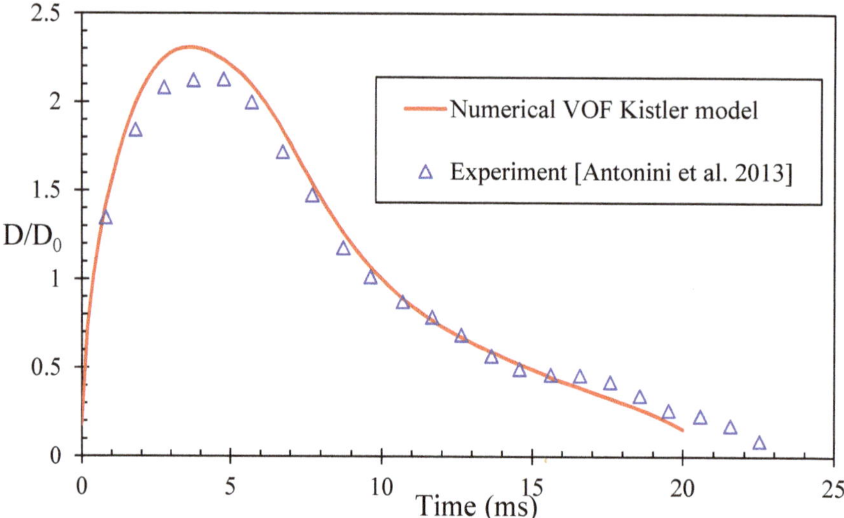

Figure 2. Comparison between simulation and experiment of the dimensionless spreading diameter evolution of a droplet upon impact on a superhydrophobic substrate with hysteresis. VOF: volume of fluid.

3.2. Impact of Polymer Solution Droplet

3.2.1. Hydrophilic Surface

We performed a comparative simulation of a Newtonian and a polymer solution droplet impact on a hydrophilic surface (Table 2). The transient evolution of 1 mm diameter drop at a velocity of 1 m/s is shown in Figure 3. The left-hand side corresponds to the Newtonian liquid, while the polymer droplet is shown on the right hand-side. We plotted the dimensionless spreading diameter (D/D_0) function of the dimensionless time ($t_c = t\,V_0/D_0$). We observed that during the kinematic phase ($t_c < 1$), droplet spreading seemed to be independent of the fluid model (Figure 4). However, at subsequent phases of the two droplets' spreading, we observed a much more marked difference between the two cases. Additionally, the viscoelastic liquid droplet displayed little oscillation while having a greater maximum spreading diameter, due to the dominance of the elastic normal stress over of the surface tension, which favored retraction. Interestingly, the liquid flowing out of the droplet center towards its periphery seemed to behave similarly to the filament thinning and stretching situation we reported in Reference [25].

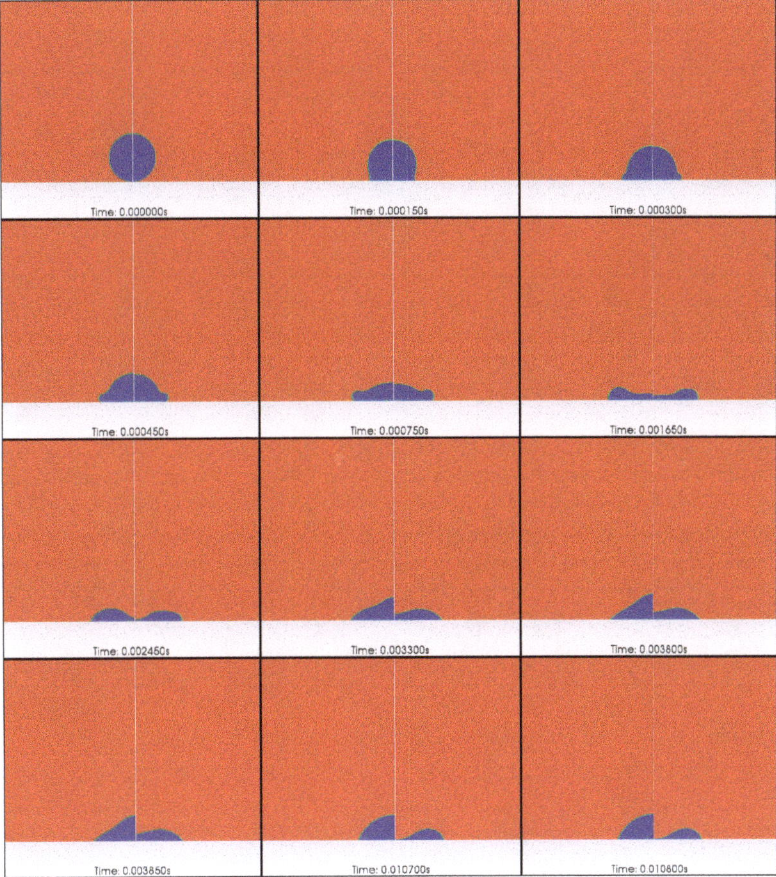

Figure 3. A comparative simulation of a 1 mm diameter droplet impacting at 1 m/s on a hydrophilic substrate between (**right**) a Newtonian solvent DEP and (**left**) polymer solution, DEP + 2.5 wt. % PS.

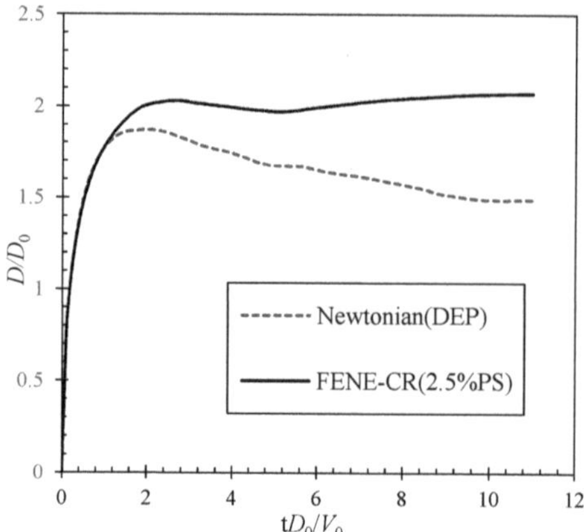

Figure 4. Evolution of the dimensionless spreading diameter on the hydrophilic surface.

3.2.2. Superhydrophobic Surface

We simulated the impact of a 1 mm droplet at 1 m/s on a superhydrophobic (SH) substrate comparing a Newtonian and polymer solution, which was modeled by a FENE-CR fluid. Comparative transient profiles between the two cases are depicted in Figure 5. The left-hand side corresponds to the Newtonian DEP solvent, while the right-hand side represents the polymer solution, which consisted of a dilution of polymer (PS) into the solvent. The addition of a minute quantity of polymer to a Newtonian liquid has been found to affect droplet dynamics, mainly during the recoiling stage for the impact conditions that were investigated in the present work. The kinetic energy of a impacting droplet is converted to the elastic and surface energy; that stored energy contributes to droplet retraction and rebound on superhydrophobic surface, after partial dissipation by viscous effect. As expected, the Newtonian (DEP) droplet rebounded and detached after impact on the superhydrophobic substrate; however, the rebound was suppressed with the dilute polymer solution (DEP + 2.5 wt. % PS), where the high elongational viscosity dissipated much of the drop kinetic energy during the spreading phase. As the Newtonian DEP droplet rebounded from the superhydrophobic surface, the dilute polymer settled at its equilibrium position during the wetting phase, as depicted in Figure 6.

Finally, the contrast between the Newtonian and the polymer solution behavior is quantified in Figure 7 by the dimensionless spreading diameter evolution. For superhydrophobic substrate, it is worth noting that the maximum spreading diameter in contact with substrate should not be confused with the maximum (deformation) diameter. The latter was higher for the viscoelastic fluid than the Newtonian one. In addition, we found that with a much lower concentration of polymer (DEP + 0.25 wt. % PS), the droplet bounced back again on the superhydrophobic substrate, which exhibited per se a higher retraction energy (Figure 7). Therefore, this implies the existence of a critical polymer concentration between 0.25 wt. % PS and 2.5 wt. % PS, above which droplet rebound can be suppressed on a superhydrophobic surface. The capability of our numerical model to retrieve such a feature is of particular interest for various processes, such as nutrients deposited as spray on plants and 2D/3D drop-on-demand printing.

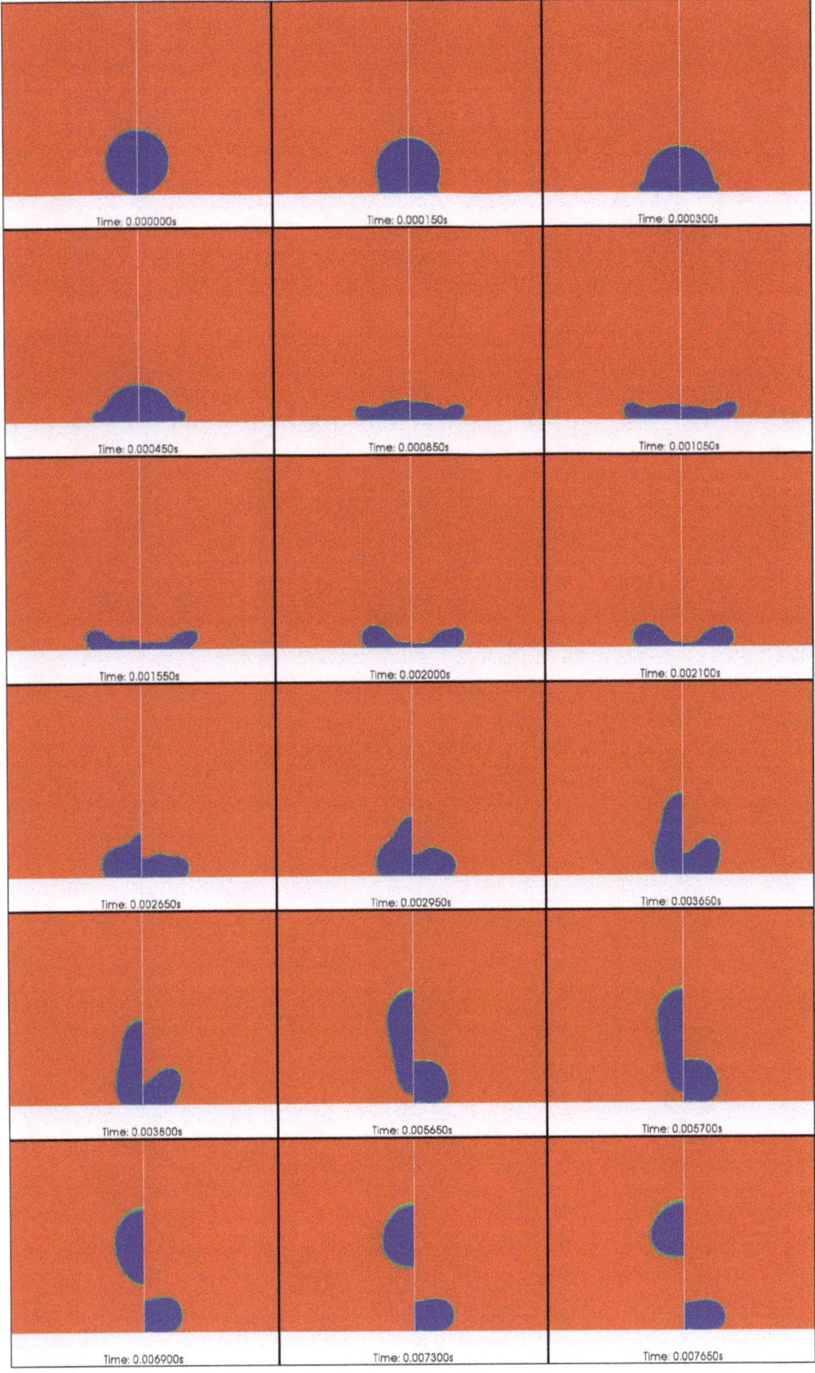

Figure 5. Simulated comparison between (**right**) the Newtonian solvent DEP and (**left**) viscoelastic polymer solution, DEP + 2.5 wt. % PS, for the impact of a 1 mm diameter droplet at 1 m/s on the superhydrophobic substrate.

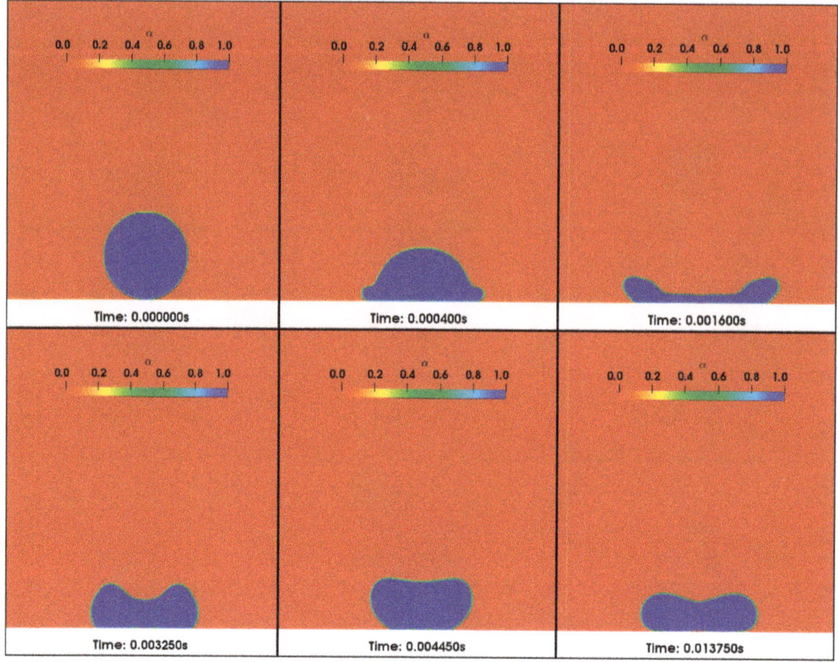

Figure 6. Impact of a 1 mm droplet of polymer solution, DEP + 2.5 wt. % PS, at 1 m/s on the superhydrophobic substrate.

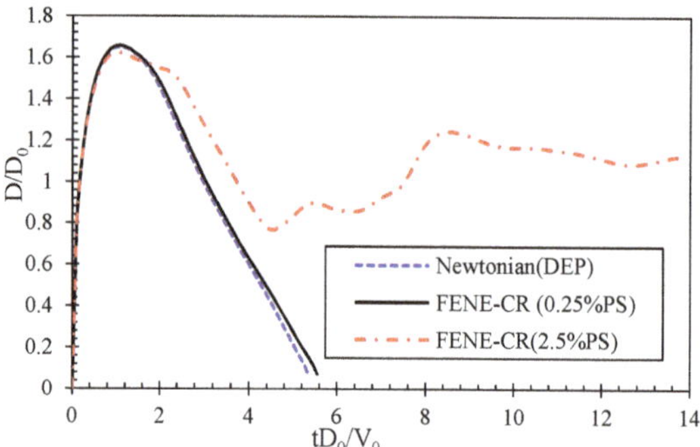

Figure 7. Simulated spreading diameter of 1 mm droplets of a Newtonian (DEP) fluid and two polymer solutions, DEP + 0.25 wt. % and DEP + 2.5 wt. % PS, impacting at 1 m/s.

4. Conclusions

Numerical simulations of droplet impact on different substrates based the VOF method were performed for both Newtonian and polymer solutions. The liquid–substrate interaction was also accounted for with a more realistic dynamic contact angle, as opposed to solely relying on the equilibrium contact angle. The constitutive equations considered for the viscoelastic fluids came from the FENE-CR model. While no noticeable difference was found in the early stages of droplet impact (beyond the kinematic phase), we observed that the substrate and fluid viscoelasticity influence became much more dominant on droplet dynamics during the recoiling phase particularly. The results of the simulation indicated that the dilute polymer solution droplet had a higher spreading diameter compared to a Newtonian solvent on a hydrophilic substrate. This can be explained by the dominance of the elastic normal stress over both the kinetic energy dissipation and surface tension force, which tends to favor droplet retraction. In addition, the existence of a critical polymer concentration—at which a droplet may no longer detach, even on superhydrophobic substrates—was inferred. The present model, based on a single mode FENE-CR fluid, could be extended to multimode constitutive models, along with tailored experiments for a comprehensive description of polymer droplet impact.

Author Contributions: M.T., D.V., A.S. and A.D. designed the study. M.T. performed the study and wrote the paper. All the authors contributed to the paper.

Funding: M.T. and A.D. gratefully acknowledge the financial support from the Natural Sciences and Engineering Research Council of Canada (NSERC).

Conflicts of Interest: The authors declare no conflict of interest.

References

1. Rioboo, R.; Marengo, M.; Tropea, C. Time evolution of liquid drop impact onto solid, dry surfaces. *Exp. Fluids* **2002**, *33*, 112–124. [CrossRef]
2. Mock, U.; Michel, T.; Tropea, C.; Roisman, I.; Rühe, J. Drop impact on chemically structured arrays. *J. Phys. Condens. Matter* **2005**, *17*, S595–S605. [CrossRef]
3. Vadillo, D.C.; Soucemarianadin, A.; Delattre, C.; Roux, D.C.D. Dynamic contact angle effects onto the maximum drop impact spreading on solid surfaces. *Phys. Fluids* **2009**, *21*, 122002. [CrossRef]
4. Antonini, C.; Amirfazli, A.; Marengo, M. Drop impact and wettability: From hydrophilic to superhydrophobic surfaces. *Phys. Fluids* **2012**, *24*, 102104. [CrossRef]
5. Tembely, M.; Vadillo, D.; Mackley, M.R.; Soucemarianadin, A. Towards an optimization of DOD printing of complex fluids. *Int. Conf. Digit. Print. Technol.* **2011**, *2011*, 86–92.
6. Clemens, W.; Fix, W.; Ficker, J.; Knobloch, A.; Ullmann, A. From polymer transistors toward printed electronics. *J. Mater. Res.* **2004**, *19*, 1963–1973. [CrossRef]
7. Xu, J.; Attinger, D. Drop on demand in a microfluidic chip. *J. Micromech. Microeng.* **2008**, *18*, 065020. [CrossRef]
8. Ottnad, T.; Kagerer, M.; Irlinger, F.; Lueth, T.C. Modification and further development of a drop on demand printhead for wax enabling future 3D-printing and rapid prototyping. In Proceedings of the 2012 IEEE/ASME International Conference on Advanced Intelligent Mechatronics (AIM), Kaohsiung, Taiwan, 11–14 July 2012; pp. 117–122.
9. Joensson, H.N.; Andersson-Svahn, H. Droplet microfluidics—A tool for protein engineering and analysis. *Lab Chip* **2011**, *11*, 4144. [CrossRef]
10. Tembely, M.; AlSumaiti, A.M.; Jouini, M.S.; Rahimov, K. The effect of heat transfer and polymer concentration on non-Newtonian fluid from pore-scale simulation of rock X-ray micro-CT. *Polymers* **2017**, *9*, 509. [CrossRef]
11. Fukai, J.; Zhao, Z.; Poulikakos, D.; Megaridis, C.M.; Miyatake, O. Modeling of the deformation of a liquid droplet impinging upon a flat surface. *Phys. Fluids A Fluid Dyn.* **1993**, *5*, 2588–2599. [CrossRef]
12. Hirt, C.; Nichols, B. Volume of fluid (VOF) method for the dynamics of free boundaries. *J. Comput. Phys.* **1981**, *39*, 201–225. [CrossRef]
13. Sussman, M.; Smereka, P.; Osher, S. A Level Set Approach for Computing Solutions to Incompressible Two-Phase Flow. *J. Comput. Phys.* **1994**, *114*, 146–159. [CrossRef]
14. Tryggvason, G.; Bunner, B.; Esmaeeli, A.; Juric, D.; Al-Rawahi, N.; Tauber, W.; Han, J.; Nas, S.; Jan, Y.J. A Front-Tracking Method for the Computations of Multiphase Flow. *J. Comput. Phys.* **2001**, *169*, 708–759. [CrossRef]
15. Shan, X.; Doolen, G. Multicomponent lattice-Boltzmann model with interparticle interaction. *J. Stat. Phys.* **1995**, *81*, 379–393. [CrossRef]
16. Dalgamoni H., N. and Yong X. Axisymmetric lattice Boltzmann simulation of droplet impact on solid surfaces. *Phys. Rev. E.* **2018**, *98*, 013102. [CrossRef] [PubMed]
17. Roisman, I.V.; Opfer, L.; Tropea, C.; Raessi, M.; Mostaghimi, J.; Chandra, S. Drop impact onto a dry surface: Role of the dynamic contact angle. *Colloids Surf. A Physicochem. Eng. Asp.* **2008**, *322*, 183–191. [CrossRef]

18. Šikalo, Š.; Wilhelm, H.-D.; Roisman, I.V.; Jakirlić, S.; Tropea, C. Dynamic contact angle of spreading droplets: Experiments and simulations. *Phys. Fluids* **2005**, *17*, 062103. [CrossRef]

19. Crochet, M.J.; Keunings, R. Finite element analysis of die swell of a highly elastic fluid. *J. Nonnewton. Fluid Mech.* **1982**, *10*, 339–356. [CrossRef]

20. Morrison, N.F.; Harlen, O.G. Viscoelasticity in inkjet printing. *Rheol. Acta* **2010**, *49*, 619–632. [CrossRef]

21. Tomé, M.F.; Paulo, G.S.; Pinho, F.T.; Alves, M.A. Numerical solution of the PTT constitutive equation for unsteady three-dimensional free surface flows. *J. Nonnewton. Fluid Mech.* **2010**, *165*, 247–262. [CrossRef]

22. Fang, J.; Owens, R.G.; Tacher, L.; Parriaux, A. A numerical study of the SPH method for simulating transient viscoelastic free surface flows. *J. Nonnewton. Fluid Mech.* **2006**, *139*, 68–84. [CrossRef]

23. Yu, J.-D.; Sakai, S.; Sethian, J.A. Two-phase viscoelastic jetting. *J. Comput. Phys.* **2007**, *220*, 568–585. [CrossRef]

24. Yue, P.; Feng, J.J. Phase-field simulations of dynamic wetting of viscoelastic fluids. *J. Nonnewton. Fluid Mech.* **2012**, *189*, 8–13. [CrossRef]

25. Tembely, M.; Vadillo, D.; MacKley, M.R.; Soucemarianadin, A. The matching of a "one-dimensional" numerical simulation and experiment results for low viscosity Newtonian and non-Newtonian fluids during fast filament stretching and subsequent break-up. *J. Rheol.* **2012**, *56*, 159–183. [CrossRef]

26. Dong, H.; Carr, W.W.; Morris, J.F. An experimental study of drop-on-demand drop formation. *Phys. Fluids* **2006**, *18*, 072102. [CrossRef]

27. Hoath, S.D.; Vadillo, D.C.; Harlen, O.G.; McIlroy, C.; Morrison, N.F.; Hsiao, W.-K.; Tuladhar, T.R.; Jung, S.; Martin, G.D.; Hutchings, I.M. Inkjet printing of weakly elastic polymer solutions. *J. Nonnewton. Fluid Mech.* **2014**, *205*, 1–10. [CrossRef]

28. Vadillo, D.C.; Tuladhar, T.R.; Mulji, A.C.; Jung, S.; Hoath, S.D.; Mackley, M.R. Evaluation of the inkjet fluid's performance using the "Cambridge Trimaster" filament stretch and break-up device. *J. Rheol.* **2010**, *54*, 261–282. [CrossRef]

29. Brackbill, J.; Kothe, D.; Zemach, C. A continuum method for modeling surface tension. *J. Comput. Phys.* **1992**, *100*, 335–354. [CrossRef]

30. Chilcott, M.D.; Rallison, J.M. Creeping flow of dilute polymer solutions past cylinders and spheres. *J. Nonnewton. Fluid Mech.* **1988**, *29*, 381–432. [CrossRef]

31. Vadillo, D.C.; Tembely, M.; Morrison, N.F.; Harlen, O.G.; MacKley, M.R.; Soucemarianadin, A. The matching of polymer solution fast filament stretching, relaxation, and break up experimental results with 1D and 2D numerical viscoelastic simulation. *J. Rheol.* **2012**, *56*, 1491–1516. [CrossRef]

32. Mackley, M.R.; Butler, S.A.; Huxley, S.; Reis, N.M.; Barbosa, A.I.; Tembely, M. The observation and evaluation of extensional filament deformation and breakup profiles for Non Newtonian fluids using a high strain rate double piston apparatus. *J. Nonnewton. Fluid Mech.* **2017**, *239*, 13–27. [CrossRef]

33. Rusche, H. Computational Fluid Dynamics of Dispersed Two-Phase Flows at High Phase Fractions. Ph.D. Thesis, Imperial College of Science, Technology and Medicine, London, UK, 2002.

34. Tembely, M.; Attarzadeh, R.; Dolatabadi, A. On the numerical modeling of supercooled micro-droplet impact and freezing on superhydrophobic surfaces. *Int. J. Heat Mass Transf.* **2018**, *127*, 193–202. [CrossRef]

35. Kistler, S.F. Hydrodynamics of Wetting. In *Wettability*; Berg, J.C., Ed.; CRC Press: Boca Raton, FL USA, 1993.

36. Yokoi, K.; Vadillo, D.; Hinch, J.; Hutchings, I. Numerical studies of the influence of the dynamic contact angle on a droplet impacting on a dry surface. *Phys. Fluids* **2009**, *21*, 072102. [CrossRef]

37. Jadidi, M.; Tembely, M.; Moghtadernejad, S.; Dolatabadi, A. A coupled level set and volume of fluid method in openfoam with application to compressible two-phase flow. In Proceedings of the 22nd Annual Conference of the CFD Society of Canada, Toronto, ON, Canada, 1–4 June 2014.

38. Graham, P.J.; Farhangi, M.M.; Dolatabadi, A. Dynamics of droplet coalescence in response to increasing hydrophobicity. *Phys. Fluids* **2012**, *24*, 112105. [CrossRef]

39. Antonini, C.; Villa, F.; Bernagozzi, I.; Amirfazli, A.; Marengo, M. Drop Rebound after Impact: The Role of the Receding Contact Angle. *Langmuir* **2013**, *29*, 16045–16050. [CrossRef]

Article

On the Recovery of PLP-Molar Mass Distribution at High Laser Frequencies: A Simulation Study

Shaghayegh Hamzehlou [1], M. Ali Aboudzadeh [2] and Yuri Reyes [3,*]

[1] POLYMAT, Kimika Aplikatu saila, Kimika Fakultatea, University of the Basque Country UPV/EHU, Avda Tolosa 72, 20018 Donostia-San Sebastián, Spain

[2] Departamento de Química Inorgánica, Universidad del País Vasco UPV/EHU, Barrio Sarriena, 48970 Leioa, Spain

[3] Departamento de Recursos de la Tierra, Universidad Autónoma Metropolitana Unidad Lerma (UAM-L), Av. Hidalgo 46, Col. La Estación, 52006 Lerma de Villada, Mexico

* Correspondence: yuri.reyes.mx@gmail.com

Received: 13 May 2019; Accepted: 23 July 2019; Published: 2 August 2019

Abstract: Due to the inherent difficulties in determination of the degree of branching for polymers produced in pulsed laser polymerization (PLP) experiments, the behavior of the degree of branching and backbiting reaction in high laser frequency and relatively high reaction temperatures have not been well-established. Herein, through a simulation study, the validity of different explanations on the recovery of PLP-molar mass distribution at high laser frequencies is discussed. It is shown that the reduction of the backbiting reaction rate at high laser frequency, and consequent decrease in the degree of branching, is not a necessary condition for recovering the PLP-molar mass distribution. The findings of this work provide simulation support to a previous explanation about the possibility of using high laser frequency for reliable determination of the propagation rate coefficient for acrylic monomers.

Keywords: PLP-SEC; n-butyl acrylate; degree of branching

---

1. Introduction

Having accurate rate coefficients of polymerization reactions is crucial to obtaining basic knowledge about complex polymerization reactions, modeling and better controlling the industrial processes. Pulsed laser polymerization (PLP) has been used to measure the rate coefficients of polymerization reactions of different monomers [1–5]. In particular, PLP together with size exclusion chromatography (PLP-SEC), has been recommended by International Union of Pure and Applied Chemistry (IUPAC) and successfully applied to calculate the propagation rate coefficient of several academically and industrially relevant monomers. A PLP-SEC experiment can be carried out as a function of temperature, and hence it is possible to calculate Arrhenius' parameters, pre-exponential factor, and activation energy [6]. In a PLP-SEC experiment, a reacting mixture containing monomer (optionally a solvent) and a photoinitiator is initially irradiated with a laser pulse at t_0, generating radicals, [R_0], in the order of 10^{-5} moL L^{-1} [7]. These radicals initiate the polymer chains that grow altogether during the so called dark period, i.e., when the mixture is not irradiated by the laser and therefore, there is not any new radical. At a later time, t_1, another laser pulse hits the reacting mixture and the new radicals terminate a large fraction of the growing chains (those who did not undergo termination during the dark period) or initiate new chains. In a posterior multiple of $t_1 - t_0$, i.e., t_d, another laser pulse hits the sample and this continues; t_d is controlled by the laser frequency or laser repetition rate. Obviously, some chains grow during more than one dark period. Once the polymerization finishes, either at a fixed time or after a number of laser pulses, the molar mass distribution (MMD) is measured by size exclusion chromatography (SEC), and the MMD or the chain length distribution (CLD) should show distinctive peaks at defined chain lengths according to the following equation:

$$L_n = nk_p[M]t_d \ n \geq 1 \tag{1}$$

where L_n is the chain length at which the CLD shows a critical point (calculated by the maximum of the first derivative), n is an integer number related to the number of dark periods that the chains were able to grow (i.e., $L_1 = 1L_0$, $L_2 = 2L_0$ and so forth); $[M]$ is the monomer concentration; and k_p is the propagation rate coefficient, as the unknown parameter. To provide reliable results in a PLP-SEC experiment, several consistency criteria must be fulfilled. First, at least two peaks should be present in the CLD, which are multiples of L_0; second, it is desirable to reach conditions for the independence of k_p on initiator concentration, laser power, laser frequency, and monomer concentration [1,8]. The PLP-SEC method has provided good results within a broad temperature range for some widely used monomers, such as styrene [2], methyl, and alkyl methacrylates [1,3], among others. Laser frequencies up to 100 Hz were used in most of the above-mentioned PLP-SEC experiments. Nevertheless, accurate estimation of k_p from a PLP experiment is still a challenge [9].

However, the PLP-shape of the MMD (distributions with clearly defined critical points) was not obtained for acrylic monomers, specifically for n-butyl acrylate at temperatures ≥ 30 °C. Although it was found that by carrying out the PLP reaction at sufficient low temperatures the PLP-MMD was recovered, the origin of this phenomena was not known initially [10,11]. It is well known now that in addition to the common reactions of a free radical polymerization (initiation, propagation, and termination), acrylic monomers undergo significant intramolecular chain transfer to polymer or backbiting. Intermolecular chain transfer reaction is present as well, which is important at high polymer concentration. However, this is not the case in a standard PLP-SEC experiment due to the low monomer-to-polymer conversion. In the backbiting reaction, a growing chain-end radical wraps around itself to abstract a hydrogen atom from a monomer unit on its own backbone via the formation of a six-membered ring [4,12]. The stability of the resulting mid-chain radical (MCR) has a significant effect on the polymerization kinetic and the microstructure of the polymer chains, leading to a featureless MMD, which in turn invalidates the k_p determination by PLP-SEC above c.a. 30 °C [13,14]. As the activation energy of backbiting is higher than the activation energy of the propagation of the chain-end radicals, the k_p of n-butyl acrylate (BA) could be determined accurately between -65 °C and 20 °C using frequencies up to 100 Hz [4].

It is worth noting that simulation studies dealing with the polymerization of acrylic monomers have been very important for better understanding of this phenomenon and better estimation of the propagation rate coefficient [15], and even later used to estimate the rate coefficients of the secondary reactions in acrylic radical polymerization [16,17]. Using a simulation approach, it was demonstrated that the increment of the laser frequency in a PLP-SEC experiment could lead to a better determination of the k_p of the chain-end radicals. For BA, at frequency around 300–500 Hz and temperature of 20 °C, the PLP-MMD is recovered and makes the average propagation coefficient (considering both MCR and chain-end radicals) independent from the laser frequency [14]. With the development of the laser technology, this simulation finding was experimentally confirmed. With a laser frequency of 500 Hz, it was possible to calculate the k_p of BA up to 70 °C. This temperature was 50 °C higher than the previously reported experiments, meeting the consistency criteria, showing up to three inflection points in the derivative of the CLD, providing a very good agreement of Arrhenius' parameters between the two data sets [18]. It was logical to carry out the PLP-SEC experiment at such frequency and higher temperature than in previous experiments for other acrylic monomers: up to 80 °C for isobornyl acrylate, tert-butyl acrylate and 1-ethoxyethyl acrylate [19], and for methyl acrylate and 2-ehtylhexyl acrylate [20,21]. Nevertheless, as an MCR can undergo β-scission at high temperatures, the MMD can be even more complex to analyze.

As in a PLP-SEC experiment, the conversion of monomer to polymer remains rather low, and it is difficult to measure the degree of branching directly (defined as the ratio of the quaternary carbons to the total polymerization events). Generally, the cumulative determination of the degree of branching is carried out by ^{13}C-NMR spectroscopy [22–25], which requires some milligrams of polymer, which is not easy to obtain in a standard PLP-SEC experiment; otherwise, the error in the ^{13}C-NMR determination invalidates such a measurement. Plessis et al. were able to measure experimentally the branching density in a PLP experiment using a reactor with larger volumes of reactants than in conventional PLP set-ups [26]. They studied the effect of temperature and monomer concentration on branching density in a PLP experiment. To the best of our knowledge, no experimental study has been done to investigate the effect of frequency on the branching level.

Now the question is why do reaction temperatures close to 70 °C, using a laser frequency of 500 Hz, lead to a PLP-MMD, allowing an adequate k_p determination for *n*-butyl acrylate? It is reported in the literature: "Diminishing the effects from the MCRs is hence a prerequisite for a successful PLP-SEC experiment" [18,27]. Nikitin et al. [7] stated that the fraction of radicals that are not subjected to backbiting is proportional to $\exp(-k_{bb}t)$, where k_{bb} is the backbiting rate coefficient and t is the time after generation of these radicals by a laser pulse. Therefore, as the laser frequency increases, the fraction of growing chains that undergoes backbiting decreases. Such explanation has been used to explain the PLP-MMD-recovering of other acrylic monomers at higher temperatures than previous experiments [21]. On the other hand, in a PLP Monte Carlo simulation, the concept of competitive processes was used to explain the recovery of PLP-MMD at high laser frequencies [28]. In this simulation, the events that each of the radicals undergo were selected based on the probability of the corresponding reaction mechanism, with the condition that the time needed to undergo the chosen reaction could not exceed the life expectancy of the radical. This assumption led to a decrease in the degree of branching and recovery of PLP-MMD distributions at high temperatures and laser frequencies [28]. Additionally, in a deeper study of reversible addition-fragmentation chain transfer polymerization (RAFT), the concept of competitive processes was used to explain the decrease in the degree of branching in such a controlled radical polymerization. In this process, the deactivation time of growing radicals (adjusted by changing the amount of the RAFT agent) is typically faster than the characteristic time for backbiting. Following the same idea, the authors explained that the reduction of the radical lifetime to chain termination induced by fast laser pulse can result in a decrease in backbiting rate and, hence, the branching level in the PLP process, allowing the recovery of the PLP-MMD [29]. In summary, there are two possible explanations for the recovering of the PLP-MMD as the laser frequency increases. According to the latest explanation, reduction in the backbiting rate and, hence, the branching level is a crucial requisite for recovery of the PLP-MMD. However, in the first explanation, only decreasing the fraction of the radicals that undergo backbiting while maintaining the same backbiting rate leads to recovery of PLP-MMD.

In this study, PLP-SEC experiments are simulated using the polymerization kinetics of BA (the most studied acrylic monomer); the behavior of backbiting and the degree of branching are analyzed during PLP-SEC experiments carried out at different temperatures, laser frequencies, and concentration of radicals generated in the laser pulse. The degree of branching of the polymer obtained in a PLP is studied and discussed under different experimental conditions; and finally, the simulation results are used to study the validity of different explanations for the recovery of PLP-MMD at high laser frequency. It is shown whether reduction of the backbiting rate/branching level is an essential requisite for recovery of the PLP-MMD.

2. Simulation Details

BA PLP experiments at different reaction conditions were simulated in the present study. The detailed kinetic scheme used in these simulations, and relevant kinetic parameters which were taken from the literature, are presented in Supporting Information. It includes all relevant reactions in homopolymerization of acrylates, such as inter/intra molecular transfer to polymer. β-scission was not included in the kinetic scheme due to its negligible effect at the simulated temperatures (maximum 60 °C) [30]. The simulated PLP experiments were implemented in Predici commercial software package [31]. To simulate the PLP polymerization by using an isothermal process at a given time (that is fixed by the laser frequency), the concentration of new radicals generated by the laser pulse ($[R_0]$) sharply increased to 10^{-5} moL L^{-1}. The total time of the simulation was 1 s in these simulations. Three different temperatures (20, 40, and 60 °C) and four laser frequencies (50, 100, 250, and 500 Hz) were employed. The monomer concentration was 7 moL/L. Also, in order to analyze the effect of radical concentration generated in the laser pulse on the CLD and on the degree of branching, PLP-SEC experiments at 20 °C with frequencies of 100 and 500 Hz and at two additional $[R_0]$, 10^{-4} moL L^{-1} and 10^{-6} moL L^{-1}, were simulated.

Instead of using the MMD of the polymer chains and polymeric radicals, the analysis was done using the CLD calculated by Predici in the Gel Permeation Chromatography (GPC) mode, size exclusion chromatography-chain length distribution (SEC-CLD). This simplified the analysis of the results, and the transformation process to the MMD was straightforward. Moreover, it should be pointed out that measuring the absolute molar mass of branched polymer chains by SEC is not trivial [20], and the calibration might introduce some error in the experimental measurements; such analysis is beyond the scope of the present study.

In addition to SEC-CLD of the inactive polymer chains, the simulation allows the calculation of the SEC-CLD of the chain-end radicals and of the MCR, the degree of branching, the fraction of MCRs that terminates with radicals generated by the laser pulse, and the ratio of the formed MCRs to the total propagation steps, among others.

3. Results and Discussion

In Figure 1 (left side), the final SEC-CLD of the polymer chains is presented for the simulations with $[R_0] = 10^{-5}$ moL L^{-1}. Each row of Figure 1 shows a different temperature. In the same figure (right side), the first derivative of each SEC-CLD is presented; the derivate was calculated by averaging the slopes of two adjacent points for each data point using the raw data, i.e., no smoothing was applied. At 20 °C and 50 Hz, it can be seen that there is a single peak in the CLD. The presence of this peak was confirmed in its corresponding derivative, which showed a sharp peak that resembles a PLP polymerization; however, a second peak was not observed. By increasing the frequency to 100 Hz, the peak at short chain-length became more visible and it shifted to nearly half of the value of chain-length of the previous condition; in the corresponding derivative, one can see that there is another small peak at larger chain-length, and this last peak is at the chain-length of the previous experiment at 50 Hz. In the SEC-CLD of the simulation at 250 Hz, two defined peaks can be seen in the CLD that were also present in the corresponding derivative. At the highest frequency, 500 Hz, two peaks at shorter chain-length were clearly observed, and in the derivative there were up to three peaks. For this temperature, the PLP-shape of the CLD was obtained at frequencies higher than, or equal to, 100 Hz. The higher the frequency of the laser pulse, the shorter the chains (shorter dark period) and the more defined the peaks in the SEC-CLD and in the correspondent derivative. From an experimental point of view, the acceptable k_p calculations were carried out up to this temperature at 100 Hz due to the presence of more than one peak in the derivative of the SEC-CLD [4], which was observed in the present results as well.

Next, we analyzed the results at 40 °C; it should be recalled here that at such temperature, it was not possible to obtain reliable experimental PLP-MMDs to properly calculate the k_p using frequencies up to 100 Hz. At 50 Hz, there were no defined peaks in the SEC-CLD or in its derivative, and a broad, featureless distribution was obtained. At 100 Hz, although the SEC-CLD was broad, there was a rather small peak in the SEC-CLD that was also observable in its derivative. The peak in the derivative resembles a PLP polymerization. In principle, this peak could be used to determine the k_p [7], but for the reliability of the results, the presence of at least a second peak at larger chain lengths is desired. Such second peak in the derivative of the SEC-CLD appeared when the frequency was 250 Hz, along with the formation of a broad peak at even larger chain lengths. For the simulation with 500 Hz, a PLP-shape of the CLD was totally recovered, and in the derivative it was possible to observe the presence of three peaks. Therefore, the information of the experiment with 500 Hz could be used to calculate k_p.

The highest temperature considered in the simulations was 60 °C (third row in Figure 1). For this temperature, the SEC-CLD at 50 and 100 Hz did not have the shape of a PLP experiment, i.e., there were no peaks in the SEC-CLD nor in the corresponding derivatives. At 250 Hz, only one peak in both graphs was formed, therefore, the use of such information to calculate the k_p would not be recommended. However, at 500 Hz, the SEC-CLD recovered the PLP-shape and in the derivative two peaks were present. Therefore, it was possible to use this information to calculate the k_p at such a temperature.

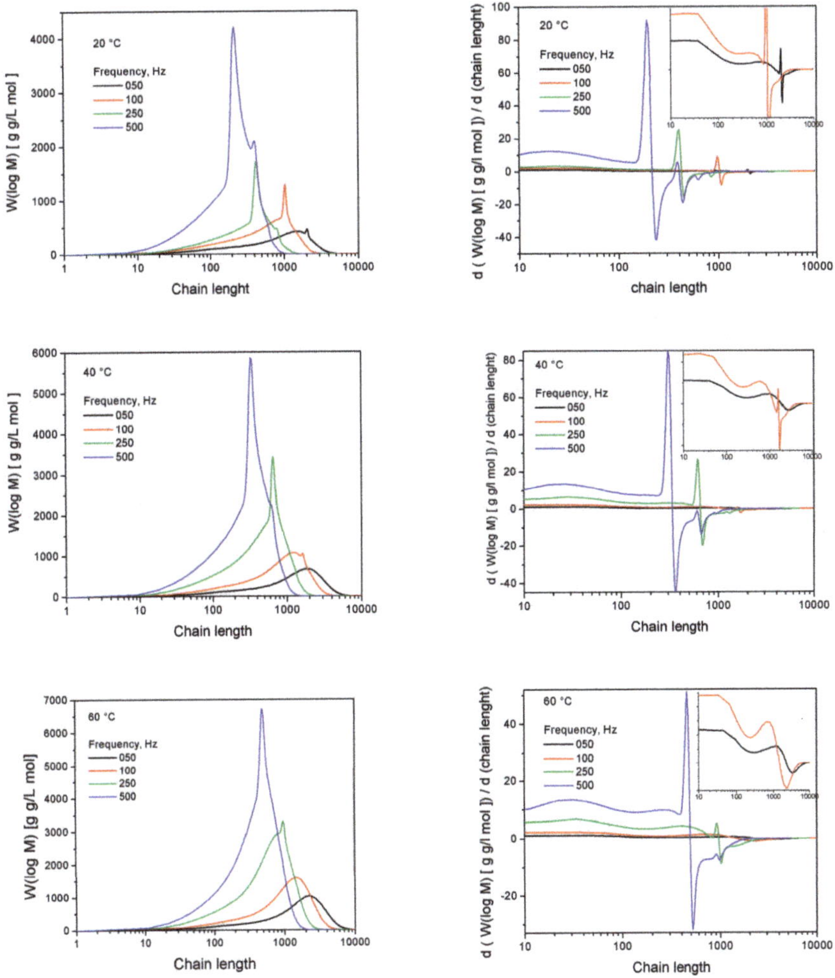

Figure 1. (**left**) Final size exclusion chromatography-chain length distributions (SEC-CLDs), and (**right**) the correspondent derivatives of the polymer chains for different temperatures and laser frequencies. Take into account that the conversion is not the same in each experiment, leading to different area under the curve.

The most relevant information presented in Figure 1 is summarized in Table 1. The chain lengths reported in this table were obtained by reading the maximum of each peak in the derivative of the SEC-CLD (if observable); the first peak is the one at the shortest chain length. In the same table, the k_p was calculated by using the chain length of each peak (k_{p1} corresponds to the peak L_1, and so on) and Equation (1). The concentration of monomer was 7 moL L^{-1}. The last column of Table 1 reports the k_p used as an input in the simulation at the given temperature (calculated with Arrhenius' parameters of Table S1). In all cases, the error between the calculated value of k_p using the data reported in Table 1, along with Equation (1), and the k_p calculated through Arrhenius' parameters was less than 7%. Additionally, the ratio k_{pi}/k_{pi+1} was rather close to 1 in all cases.

Table 1. Chain lengths at which the chain length distribution (CLD) shows critical points, L_i, determined at the maximum of the derivatives of the CLD of Figure 1. k_{pi} calculated with this information and Equation (1) and k_p the input at the given temperature.

T, °C	F, Hz	t_d, s	L_1	L_2	L_3	k_{p1}, L moL^{-1} s^{-1}	k_{p2}, L moL^{-1} s^{-1}	k_{p3}, L moL^{-1} s^{-1}	k_p, L moL^{-1} s^{-1}
20	50	0.020	1968	—	—	14,000	—	—	14,200
20	100	0.010	971	1942	—	13,800	13,800	—	—
20	250	0.004	395	779	1163	14,100	13,900	13,800	—
20	500	0.002	191	382	573	13,600	13,600	13,600	—
40	50	0.020	—	—	—	—	—	—	22,800
40	100	0.010	1542	—	—	22,000	—	—	—
40	250	0.004	612	1223	—	21,800	21,800	—	—
40	500	0.002	303	606	917	21,600	21,600	21,800	—
60	50	0.020	—	—	—	—	—	—	34,400
60	100	0.010	—	—	—	—	—	—	—
60	250	0.004	932	—	—	33,200	—	—	—
60	500	0.002	457	903	—	32,600	32,200	—	—

Figure 2 shows the degree of branching, which was calculated as the cumulative fraction of polymerized acrylic units that have a quaternary carbon for different temperatures and laser frequencies. The quaternary carbons are mainly formed by the MCR propagation (termination by combination also contributes to the formation of quaternary carbons). It is worth mentioning that in order to separate the effect of the patching reaction from the branching level, the quaternary carbons created in the patching reaction were not considered in the degree of branching. In all cases, once the polymerization begins, the degree of branching increases and, after some time, a plateau with positive slope is observed. The degree of branching increases as the temperature becomes higher, which is a reasonable result due to the higher activation energy of the backbiting reaction compared to propagation of the chain-end radicals. By a totally different simulation approach of the PLP of BA at 500 Hz (kinetic Monte Carlo), it has been determined that the degree of branching reaches a value of 0.14% at 32.85 °C and 0.21% at 51.85 °C after 100 pulses [32]. At 51.85 °C and monomer concentration of 6.76 moL L^{-1}, the simulation used in this work reached 0.18% after 1 s, showing a good agreement between both simulation approaches. The differences between both simulation approaches were due to the different kinetic scheme used in the kinetic Monte Carlo simulations (two types of initiator radicals and time dependent initiator concentration). As observed in Figure 2, although slightly lower values of the degree of branching were calculated as the laser frequency increased at constant temperature, there was not a significant decrease of the degree of branching at a fixed temperature. At 20 °C, degree of branching was between 0.10 and 0.08% at 50 and 500 Hz, respectively; at 60 °C, the variation was between 0.22 and 0.21% at 50 and 500 Hz, respectively. From these results, it can be concluded that the degree of branching at a constant temperature does not depend on the laser frequency. In the same figure, for each temperature, the instantaneous probability of backbiting, calculated by $k_{bb}/(k_p[M] + k_{bb})$, is plotted as a solid line, where $[M]$ is the monomer concentration. The ratio of backbiting events to the total propagation events obtained during the simulations at 500 Hz was plotted as a dashed gray line (both values were multiplied by a factor of 100). The instantaneous probability of backbiting under the quasi-steady-state conditions is related to the degree of branching, that is why these calculations were plotted in the same figure. As observed, the degree of branching has approached the instantaneous probability of backbiting at 500 Hz, but within the simulation time they never touched each other; the same behavior was observed with the kinetic Monte Carlo simulation [32]. Regarding these results, the reason for such a difference is that an MCR can terminate with the radicals generated in the laser pulse; the faster the laser pulse, the larger the MCRs that are terminated by this reaction, as discussed below.

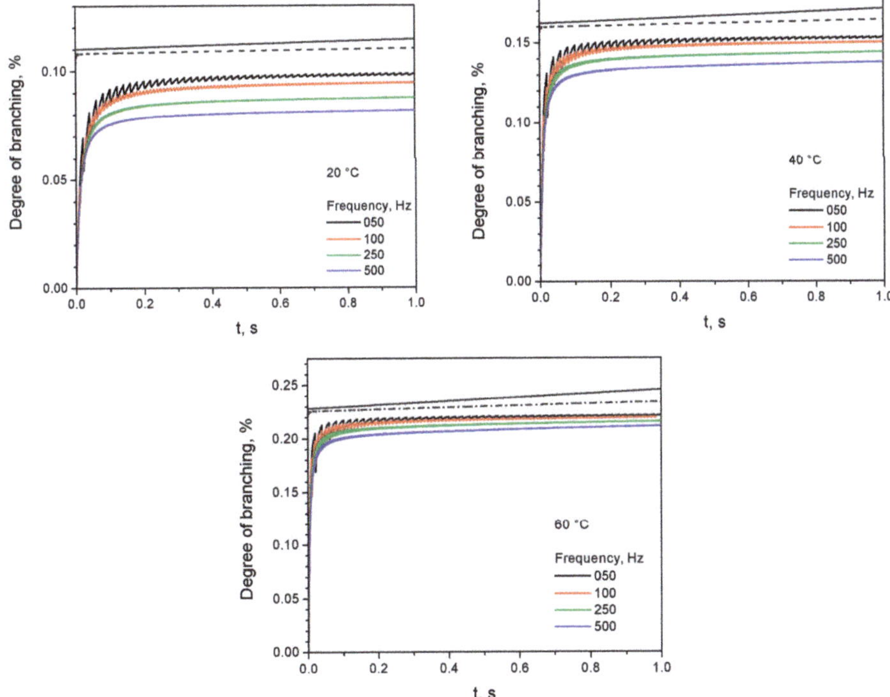

Figure 2. Degree of branching as a function time obtained during the simulation at different temperature and laser frequencies. The solid gray line is the instantaneous probability of backbiting and the dashed line is the ratio of the backbiting events to the total propagation events during the simulation; both calculations multiplied by a factor of 100.

Additionally to MCR's propagation and (cross) termination with another polymeric radical, it might be possible for an MCR to react with the radicals generated during the laser pulse (see Kinetic Scheme S1). It is worth mentioning that in the simulations, initiation rate coefficient was assumed equal to the propagation rate coefficient. Therefore, the difference between the calculated degree of branching in the simulations and the instantaneous probability of backbiting could be explained by the fraction of MCRs that react with the radicals generated in the pulse, giving place to patched MCRs. As the laser frequency increases, the number of radicals at the same interval of time also increases, leading to a larger fraction of patched MCRs. Figure 3 shows the percentage of patched MCRs at different temperatures and laser frequencies. As explained above, it can be seen that increasing laser frequency leads to a larger fraction of patched MCRs. On the other hand, the fraction of patched MCRs decreases by increasing temperature.

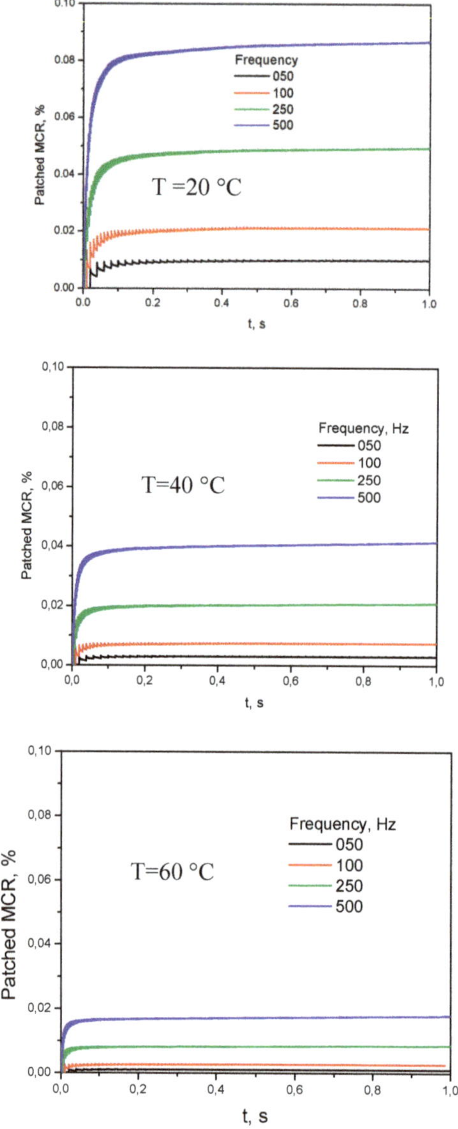

Figure 3. Percentage of mid-chain radicals (MCRs) that are patched by the radicals generated in the laser pulse.

As it was shown in Figure 2, increasing the laser frequency does not significantly change the degree of branching at a constant temperature, and the slight difference can be explained by patching of the MCRs as discussed above. However, the PLP-CLD is recovered by increasing the laser pulse frequency. Nikitin et al. claimed that there is a minimum fraction of chain-end radicals, with respect to total radical concentration, at a constant temperature and time that still allows the observation of the peaks in the CLD [7]. This fraction is proportional to the $\exp(-k_{bb}/f)$, where f is the laser frequency. They also showed that this minimum fraction is independent of polymerization conditions at constant temperature. As a consequence, if the reaction temperature increases, the backbiting rate coefficient will also increase and, therefore, to have a significant fraction of secondary radicals that do not undergo backbiting the laser frequency should increase.

It is worth mentioning here that all the simulations have the same duration (1 s) and monomer concentration. Thus, the number of polymer chains is lower at low frequencies than at high ones. On the other hand, the ratio of backbiting events to propagation is nearly independent of the laser frequency. Therefore, at low laser frequency a larger fraction of chains undergoes backbiting, creating MCRs that could distort the CLD by breaking the time correlation of Equation (1) due to the stability of the MCRs. When the number of polymer chains increases (as in a high frequency experiment with the same duration), nearly the same number of backbiting to propagation events takes place, but affecting a relatively lower fraction of chains; therefore, the PLP-MMD is recovered. This is supported by the results presented in Figure 4, in which the zeroth moment of the chain-end radical species, or the concentration of the chain-end radicals and MCRs, for 20 and 60 °C and different laser frequencies is shown. In this figure, the concentration of each radical is divided by the concentration of radicals generated in each laser pulse (10^{-5} mol L^{-1}). At 20 °C and 50 Hz, immediately after the laser pulse, the concentration of the chain-end radicals starts to decay during the dark period as they mainly undergo termination reaction. Moreover, a fraction of them undergoes backbiting reaction; the concentration of MCRs increases, and then decreases in the last part of the dark period. Towards the end of the dark period, the concentration of MCRs is greater than that of the chain-end radicals. Nevertheless, since some chains did not undergo backbiting during the dark period, they could be terminated in the next laser pulse, leading to a peak in the SEC-CLD. Increasing the laser frequency to 100 Hz leads to a greater concentration of both types of radicals because, in this case, the dark period has half the duration of the previous simulation. Thus, it can be seen that concentration of chain-end radicals is greater, leading to a well-defined peak in the SEC-CLD. It is true that the MCRs are also in greater number but affecting a relatively lower fraction of chains. At 250 Hz, although the concentration of chain-end radicals decreases during the dark period, the next laser pulse hits the reacting mixture when the concentration of chain-end is enough to produce a PLP-CLD of the polymer chains. Again, because there are more radicals, the MCR concentration also increases, but it seems not to affect a significant fraction of chains. With the highest frequency, the concentration of chain-end is even greater, resulting in a higher concentration of MCRs. For this case, such concentration of chain-end radicals leads to a well-structured PLP-CLD. As the laser frequency increases, the fraction of chain-end radicals close to the next laser pulse also increases. At 60 °C, the behavior is rather similar to the previous description; however, as the temperature is higher, a larger fraction of MCRs is present [26]. Nevertheless, as the dark period diminishes and before the next laser pulse hits the sample, a larger fraction of chain-end radicals is present, leading to the recovery of the PLP-CLD.

Additional information that can be extracted from the simulation is the CLD of chain-end radicals, which of course is not obtained by experimental procedures. The results are presented in Figure 5, which show the final SEC-CLD of the chain-end radicals (at 1 s of integration time, before the last laser pulse is applied). It is worth mentioning that it is a cumulative distribution that contains the information of all of the chain-end radicals that were generated at different times during the reaction, and some of the chains might have multiple branching points. For example, at 20 °C, it is observed that increasing the laser frequency leads to the formation of peaks in the SEC-CLD of the chain-end radicals. As the frequency increases, a multimodal CLD is obtained, i.e., some of the chain-end radicals can grow by more than one dark period. These chains would terminate among them or, preferentially, with the radicals generated in the next laser pulse, leading to a well-defined PLP-CLD of the polymer. By increasing the temperature to 40 °C, the multimodal SEC-CLD of the growing chain-end radicals is obtained with frequencies greater than 250 Hz. Because more chain-end radicals undergo backbiting, it is necessary to have enough chains that are not affected by such reaction to recover the PLP-CLD. Increasing the frequency to 500 Hz leads to well-defined peaks in the CLD, because the accumulation of unaffected chain-end radicals is greater. The same behavior is observed at 60 °C, but in this temperature, the multimodality (actually, bimodality) of the CLD is presented only at

500 Hz, in which the accumulation of growing chain-end radicals that are not affected by backbiting is enough to recover the PLP-CLD of the final polymer.

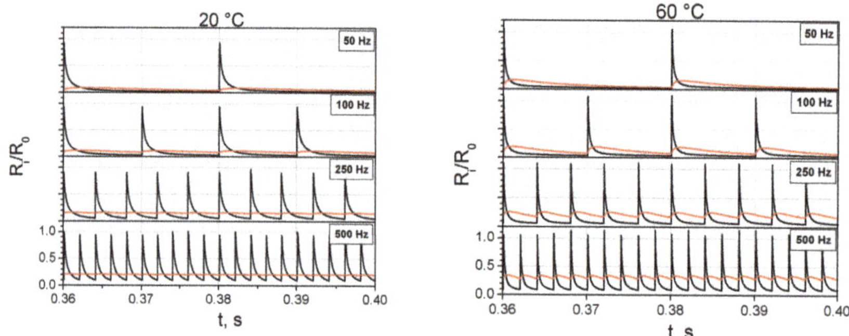

Figure 4. Ratio of the concentration of the radicals i (i: Chain-end radicals in black and MCRs in red), divided by the concentration of the radicals generated in the laser pulse. $[R_0] = 10^{-5}$ moL L^{-1}.

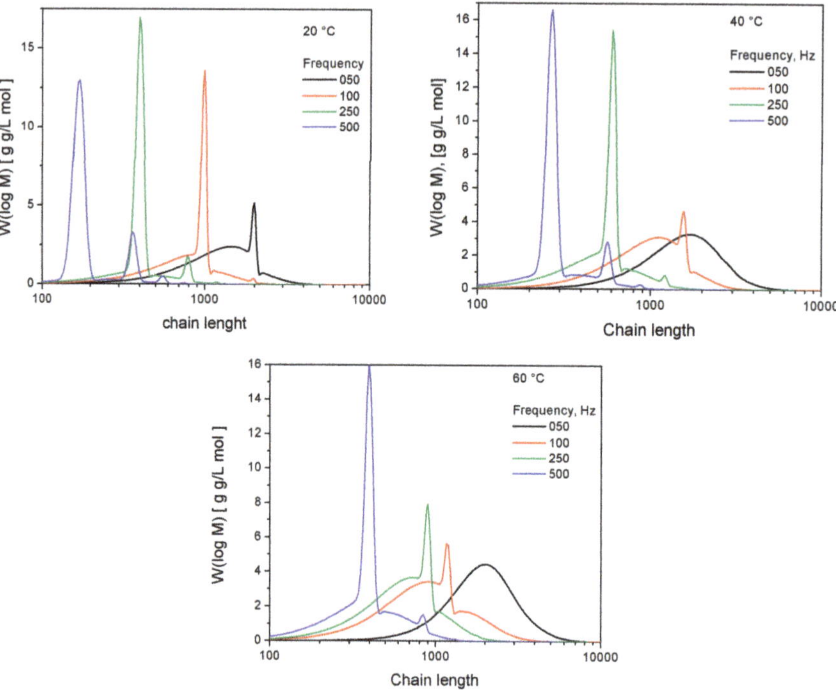

Figure 5. Final SEC-CLD of the chain-end radicals at different temperatures and laser frequencies.

The last part of the results is shown in Figure 6. This figure summarizes the results at 20 °C for laser frequencies of 100 and 500 Hz, but the radical concentration produced in the laser pulse [R$_0$] is changing. In all cases, a PLP-CLD is obtained, meeting the IUPAC criteria for a good k_p determination (more than one peak in the CLD and its derivative, and k_{pi}/k_{pi+1} close to 1). Moreover, the deviation with respect to the k_p used as input at 500 Hz and $[R_0] = 1 \times 10^{-4}$ moL L^{-1} is around 15%; at 100 Hz and the same [R$_0$], the deviation is around 7%. In the other cases, such deviation is smaller, with the smallest being at $[R_0] = 1 \times 10^{-6}$ moL L^{-1} and 500 Hz. Actually, many peaks can be seen at the CLD and its derivative. These results are in agreement with the simulation results of Nikitin et al. [7], in which they demonstrated that at a minimum laser frequency and constant temperature, there is sufficient population of chain-end radicals to produce sharp peak(s) in the CLD, that is indeed independent of $[R_0]$. Also, in the CLD obtained with the largest concentration of radicals, $[R_0] = 1 \times 10^{-4}$ moL L^{-1}, one can see the formation of polymer chains with low molecular mass. The concentration of chains with such low molar mass decreases as $[R_0]$ becomes lower. The reason is that due to the high concentration of growing radicals (chain-end and MCR), the cross termination of these radicals is enhanced during the dark period, still having a fraction of radicals that are terminated at the next laser pulse.

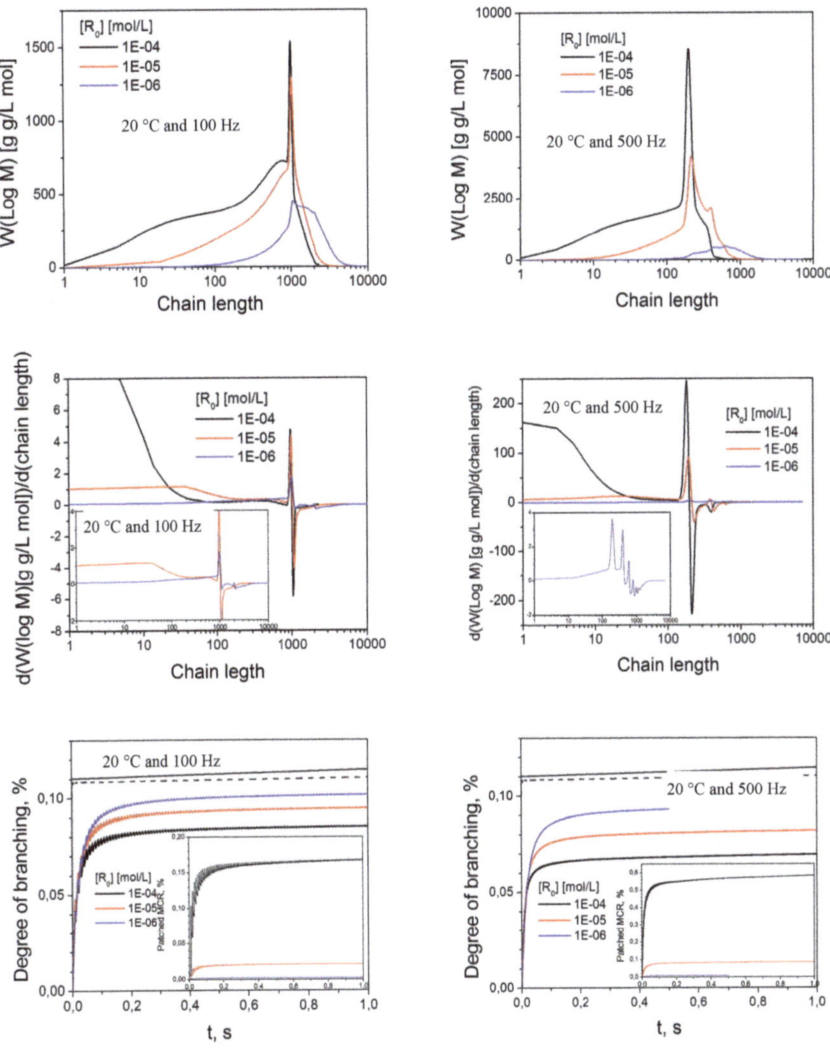

Figure 6. Effect of $[R_0]$ on different parameters. Left side column is at 20 °C and 100 Hz, whereas left side is at 20 °C and 500 Hz. First row, final SEC-CLDs; second row, the correspondent derivatives; third row, the degree of branching (and patched MCR in the inset) at 20 °C, 100 Hz (left side column) and 500 Hz (right side column). At 500 Hz and $[R_0] = 1 \times 10^{-6}$ mol L^{-1}, the simulation was stopped at 0.5 s due to computer memory limitations. Inserts in each graph is a zoom in a desired area.

Regarding the backbiting-to-propagation ratio, the results show that this ratio is not significantly affected. However, the degree of branching decreases as $[R_0]$ increases. Such a decrease is larger at high laser frequency of 500 Hz. In this case, the reduction of the degree of branching is due to the larger fraction of pathed MCR as $[R_0]$ increases. From these results, in order to have reliable k_p determinations, it could be recommended to produce a low concentration of radicals in the laser pulse (by lowering the laser power and/or using a low initiator concentration), but having in mind the necessity of producing enough polymer chains for an adequate SEC measurement.

4. Conclusions

PLP-SEC experiments of n-butyl acrylate were simulated at different temperatures, laser frequencies, and radical concentrations produced at laser pulses. At all temperatures, the PLP-CLD was recovered as the laser frequency increased. It was shown that the recovery of PLP-CLD by increasing laser frequency is a consequence of increasing the fraction of chain-end radicals that does not undergo backbiting, which is in line with the explanation given by Nikitin et al. [7]; therefore, the degree of branching does not significantly decrease as the laser frequency increases at a fixed temperature. According to the simulation results, the ratio of backbiting to propagation reaction was not affected by the increment of the laser frequency; the ratio of quaternary carbons to the total propagation steps was rather close to the instantaneous degree of branching in all conditions. These results did not give support to other explanations given in the literature, which related the recovery of the PLP-MMD only with the decrease in the backbiting rate and, consequently, the decrease of branching density. The difference between the degree of branching with respect to the instantaneous probability of backbiting was explained by the patching of MCRs (termination reaction of the MCRs with the radicals generated in the laser pulse). By increasing laser frequency or the radical concentration at laser pulses, the patched MCRs' percentage increased, which consequently led to a lower degree of branching.

Supplementary Materials: The following are available online at http://www.mdpi.com/2227-9717/7/8/501/s1, Table S1 and Scheme S1: kinetic scheme and rate coefficients used in the simulations of the PLP-SEC of BA.

Author Contributions: Authors contributed equally.

Funding: This research received no external funding.

Acknowledgments: Yuri Reyes acknowledges the financial support of CONACyT (México). Shaghayegh Hamzehlou acknowledges the University of the Basque Country (UPV/EHU) for the "Contratación para la especialización de personal investigador doctor" postdoctoral grant.

Conflicts of Interest: The authors declare no conflict of interest.

References

1. Buback, M.; Gilbert, R.G.; Hutchinson, R.A.; Klumperman, B.; Kuchta, F.; Manders, B.G.; O'Driscoll, K.F.; Russell, G.T.; Schweerg, J. Critically evaluated rate coefficients for free-radical polymerization, 1. Propagation rate coefficient for styrene. *Macromol. Chem. Phys.* **1995**, *196*, 3267–3280. [CrossRef]

2. Beuermann, S.; Buback, M.; Davis, T.P.; Gilbert, R.G.; Hutchinson, R.A.; Olaj, O.F.; Russell, G.T.; Schweer, J.; Van Herk, A.M. Critically evaluated rate coefficients for free-radical polymerization, 2. Propagation rate coefficients for methyl methacrylate. *Macromol. Chem. Phys.* **1997**, *198*, 1545–1560. [CrossRef]

3. Beuermann, S.; Buback, M.; Davis, T.P.; Gilbert, R.G.; Hutchinson, R.A.; Kajiwara, A.; Klumperman, B.; Russell, G.T. Critically evaluated rate coefficients for free-radical polymerization, 3. Propagation rate coefficients for alkyl methacrylates. *Macromol. Chem. Phys.* **2000**, *201*, 1355–1364. [CrossRef]

4. Asua, J.M.; Beuermann, S.; Buback, M.; Castignolles, P.; Charleux, B.; Gilbert, R.G.; Hutchinson, R.A.; Leiza, J.R.; Nikitin, A.N.; Vairon, J.P.; et al. Critically Evaluated Rate Coefficients for Free-Radical Polymerization, 5 Propagation Rate Coefficient for Butyl Acrylate. *Macromol. Chem. Phys.* **2004**, *205*, 2151–2160. [CrossRef]

5. Lacík, I.; Chovancová, A.; Uhelská, L.; Preusser, C.; Hutchinson, R.A.; Buback, M. PLP-SEC Studies into the Propagation Rate Coefficient of Acrylamide Radical Polymerization in Aqueous Solution. *Macromolecules* **2016**, *49*, 3244–3253. [CrossRef]

6. Beuermann, S.; Buback, M. Rate coefficients of free-radical polymerization deduced from pulsed laser experiments. *Prog. Polym. Sci.* **2002**, *27*, 191–254. [CrossRef]

7. Nikitin, A.N.; Hutchinson, R.A.; Buback, M.; Hesse, P. Determination of Intramolecular Chain Transfer and Midchain Radical Propagation Rate Coefficients for Butyl Acrylate by Pulsed Laser Polymerization. *Macromolecules* **2007**, *40*, 8631–8641. [CrossRef]

8. Drawe, P.; Buback, M. The PLP-SEC Method: Perspectives and Limitations. *Macromol. Theory Simul.* **2016**, *25*, 74–84. [CrossRef]

9. Nikitin, A.N.; Lacík, I.; Hutchinson, R.A.; Buback, M.; Russell, G.T. Detection of PLP Structure for Accurate Determination of Propagation Rate Coe ffi cients over an Enhanced Range of PLP-SEC Conditions. *Macromolecules* **2019**, *52*, 55–71. [CrossRef]

10. Lyons, R.A.; Hutovic, J.; Piton, M.C.; Christie, D.I.; Clay, P.A.; Manders, B.G.; Kable, S.H.; Gilbert, R.G. Pulsed-Laser Polymerization Measurements of the Propagation Rate Coefficient for Butyl Acrylate. *Macromolecules* **1996**, *29*, 1918–1927. [CrossRef]

11. Beuermann, S.; Paquet, D.A.; McMinn, J.H.; Hutchinson, R.A. Determination of Free-Radical Propagation Rate Coefficients of Butyl, 2-Ethylhexyl, and Dodecyl Acrylates by Pulsed-Laser Polymerization. *Macromolecules* **1996**, *29*, 4206–4215. [CrossRef]

12. Cuccato, D.; Mavroudakis, E.; Dossi, M.; Moscatelli, D. A Density Functional Theory Study of Secondary Reactions in n-Butyl Acrylate Free Radical Polymerization. *Macromol. Theory Simul.* **2013**, *22*, 127–135. [CrossRef]

13. Plessis, C.; Arzamendi, G.; Alberdi, J.M.; Van Herk, A.M.; Leiza, J.R.; Asua, J.M. Evidence of Branching in Poly(butyl acrylate) Produced in Pulsed-Laser Polymerization Experiments. *Macromol. Rapid Commun.* **2003**, *24*, 173–177. [CrossRef]

14. Nikitin, A.N.; Castignolles, P.; Charleux, B.; Vairon, J.-P.; Vairon, J. Determination of Propagation Rate Coefficient of Acrylates by Pulsed-Laser Polymerization in the Presence of Intramolecular Chain Transfer to Polymer. *Macromol. Rapid Commun.* **2003**, *24*, 778–782. [CrossRef]

15. Van Herk, A.M. Historic Account of the Development in the Understanding of the Propagation Kinetics of Acrylate Radical Polymerizations. *Macromol. Rapid Commun.* **2009**, *30*, 1964–1968. [CrossRef]

16. Marien, Y.W.; van Steenberge, P.H.M.; Kockler, K.B.; Barner-Kowollik, C.; Reyniers, M.F.; D'Hooge, D.R.; Marin, G.B. An alternative method to estimate the bulk backbiting rate coefficient in acrylate radical polymerization. *Polym. Chem.* **2016**, *7*, 6521–6528. [CrossRef]

17. Vir, A.B.; Marien, Y.W.; van Steenberge, P.H.M.; Barner-Kowollik, C.; Reyniers, M.-F.; Marin, G.B.; D'hooge, D. Access to the β-scission rate coefficient in acrylate radical polymerization by careful scanning of pulse laser frequencies at elevated temperature. *React. Chem. Eng.* **2018**, *3*, 807–815. [CrossRef]

18. Barner-Kowollik, C.; Gunzler, F.; Junkers, T. Pushing the Limit: Pulsed Laser Polymerization of n-Butyl Acrylate at 500 Hz. *Macromolecules* **2008**, *41*, 8971–8973. [CrossRef]

19. Dervaux, B.; Junkers, T.; Schneider-Baumann, M.; Prez, F.E.D.; Barner-Kowollik, C. Propagation Rate Coefficients of Isobornyl Acrylate, *Tert*-Butyl Acrylate and 1-Ethoxyethyl Acrylate: A High Frequency PLP-SEC Study. *Polym. Sci. Part A Polym. Chem.* **2009**, *47*, 6641–6654. [CrossRef]

20. Junkers, T.; Schneider-Baumann, M.; Koo, S.S.P.; Castignolles, P.; Barner-Kowollik, C. Determination of Propagation Rate Coefficients for Methyl and 2-Ethylhexyl Acrylate via High Frequency PLP–SEC under Consideration of the Impact of Chain Branching. *Macromolecules* **2010**, *43*, 10427–10434. [CrossRef]

21. Barner-Kowollik, C.; Beuermann, S.; Buback, M.; Castignolles, P.; Charleux, B.; Coote, M.L.; Hutchinson, R.A.; Junkers, T.; Lacík, I.; Russell, G.T.; et al. Critically evaluated rate coefficients in radical polymerization-7. Secondary-radical propagation rate coefficients for methyl acrylate in the bulk. *Polym. Chem.* **2014**, *5*, 204–212. [CrossRef]

22. Ahmad, N.M.; Heatley, F.; Lovell, P.A. Chain Transfer to Polymer in Free-Radical Solution Polymerization of n -Butyl Acrylate Studied by NMR Spectroscopy. *Macromolecules* **1998**, *31*, 2822–2827. [CrossRef]

23. Lovell, P.A.; Shah, T.H.; Heatley, F. Chain transfer to polymer in emulsion polymerization of normal-butyl acrylate studied by C-13 NMR-spectroscopy and GPC. *Polym. Commun.* **1991**, *32*, 98–103.

24. Plessis, C.; Arzamendi, G.; Leiza, J.R.; Schoonbrood, H.; Charmot, D.; Asua, J.M. A Decrease in Effective Acrylate Propagation Rate Constants Caused by Intramolecular Chain Transfer. *Macromolecules* **2000**, *33*, 9786. [CrossRef]

25. Castignolles, P.; Graf, R.; Parkinson, M.; Wilhelm, M.; Gaborieau, M. Detection and quantification of branching in polyacrylates by size-exclusion chromatography (SEC) and melt-state 13C NMR spectroscopy. *Polymer* **2009**, *50*, 2373–2383. [CrossRef]
26. Arzamendi, G.; Plessis, C.; Leiza, J.R.; Asua, J.M. Effect of the Intramolecular Chain Transfer to Polymer on PLP/SEC Experiments of Alkyl Acrylates. *Macromol. Theory Simul.* **2003**, *12*, 315–324. [CrossRef]
27. Kockler, K.B.; Haehnel, A.P.; Junkers, T. Determining Free-Radical Propagation Rate Coeffi cients with High-Frequency Lasers: Current Status and Future Perspectives. *Macromol. Rapid Commun.* **2016**, *37*, 123–134. [CrossRef]
28. Reyes, Y.; Arzamendi, G.; Asua, J.M.; Leiza, J.R. Branching at High Frequency Pulsed Laser Polymerizations of Acrylate Monomers. *Macromolecules* **2011**, *44*, 3674–3679. [CrossRef]
29. Ballard, N.; Rusconi, S.; Akhmatskaya, E.; Sokolovski, D.; De La Cal, J.C.; Asua, J.M. Impact of Competitive Processes on Controlled Radical Polymerization. *Macromolecules* **2014**, *47*, 6580–6590. [CrossRef]
30. Wang, W.; Nikitin, A.N.; Hutchinson, R.A. Consideration of Macromonomer Reactions in n-Butyl Acrylate Free Radical Polymerization. *Macromol. Rapid Commun.* **2009**, *30*, 2022–2027. [CrossRef]
31. Wulkow, M. Computer Aided Modeling of Polymer Reaction Engineering-The Status of Predici, I-Simulation. *Macromol. React. Eng.* **2008**, *2*, 461–494. [CrossRef]
32. Marien, Y.W.; van Steenberge, P.H.M.; Barner-Kowollik, C.; Reyniers, M.-F.; Marin, G.B.; D'hooge, D.R. Kinetic Monte Carlo Modeling Extracts Information on Chain Initiation and Termination from Complete PLP-SEC Traces. *Macromolecules* **2017**, *50*, 1371–1385. [CrossRef]

MDPI

St. Alban-Anlage 66

4052 Basel

Switzerland

Tel. +41 61 683 77 34

Fax +41 61 302 89 18

www.mdpi.com

Processes Editorial Office

E-mail: processes@mdpi.com

www.mdpi.com/journal/processes